A NEOTROPICAL COMPANION

A Neotropical Companion

An Introduction to the
Animals, Plants, and Ecosystems
of the New World Tropics

*Second Edition, Revised
and Expanded*

BY JOHN KRICHER

ILLUSTRATED BY

WILLIAM E. DAVIS, JR.

PRINCETON UNIVERSITY PRESS

Library of Congress Cataloging-in-Publication Data
Kricher, John C.
A neotropical companion : an introduction to the animals,
plants, and ecosystems of the New World tropics /
John Kricher; illustrated by William E. Davis, Jr. —
2nd ed., rev. and expanded.
p.   cm.
Includes bibliographical references and indexes.
ISBN 0-691-04433-3 (cl : alk. paper)
1. Ecology—Latin America.   2. Natural history—
Latin America.   I. Title.
QH106.5.K75   1997
577'.098'0913—dc21   97-9784

This book has been composed in Baskerville
Designed by Jan Lilly

To my parents

Contents

FOREWORD BY MARK J. PLOTKIN, PH.D.   ix

PREFACE   xi

ACKNOWLEDGMENTS   xiv

A PERSONAL NOTE TO THE READER   xvii

CONVERSIONS   xviii

CHAPTER 1. Tropical Climates and Ecosystems   3

CHAPTER 2. Rainforest Structure and Diversity   21

CHAPTER 3. How a Rainforest Functions   44

CHAPTER 4. Evolutionary Patterns in the Tropics   75

CHAPTER 5. Complexities of Coevolution and Ecology of Fruit   126

CHAPTER 6. The Neotropical Pharmacy   144

CHAPTER 7. Living Off the Land in the Tropics   169

CHAPTER 8. Rivers through Rainforest   189

CHAPTER 9. Introduction to the Andes and Tepuis   218

CHAPTER 10. Savannas and Dry Forests   228

CHAPTER 11. Coastal Ecosystems: Mangroves, Seagrass, and Coral Reefs   239

CHAPTER 12. Neotropical Birds   251

CHAPTER 13. A Rainforest Bestiary   295

CHAPTER 14. Deforestation and Conservation of Biodiversity   334

APPENDIX. "AND HEY, LET'S BE CAREFUL OUT THERE"   377

ACRONYMS   389

GLOSSARY   390

NEOTROPICAL FIELD GUIDES AND HANDBOOKS   396

RECOMMENDED BOOK-LENGTH REFERENCES   399

LITERATURE CITED   403

INDEX   435

Foreword

WHEN I first began my work in the Neotropics almost twenty years ago, there was little to read in the way of travel guides. The two books most commonly available were *South America on $15 a Day* and *The South America Handbook*, both focusing mainly on where to eat and sleep. Going to the rainforest armed with only these books was like traveling to Paris with a guide to the fast food restaurants.

Interest in Latin America is at an all-time high. While some people travel there for the food, the music, or the textiles, the single greatest draw is the rainforest, an ecosystem that represents Mother Nature at her most exuberant and spectacular. The single major source of foreign exchange in Costa Rica these days is not coffee, timber, or beef—it is ecotourism.

So up until a few years ago, if you were traveling to see these forests, the best way to prepare yourself was to read the classics of Neotropical natural history: Wallace, Bates, Darwin, and so forth. You could then delve into technical journals to find out what researchers were learning in these most complex of ecosystems. That all changed in 1989 with the publication of the first edition of John Kricher's *A Neotropical Companion*. What Kricher achieved was elegant in design and brilliant in execution: he distilled the best information from all available sources on the natural history of the Neotropics into a highly readable and extremely educational book. How many books are used by kindergarten teachers to write their lesson plans, assigned as required reading in college courses, carried by backpackers from Mexico to Argentina, and devoured by armchair travelers anxious to experience the tropical rainforest at arm's length? Traveling with Kricher's book is a bit like going birdwatching with Ted Parker, investigating ants with E. O. Wilson, collecting plants with Al Gentry, going fishing with Michael Goulding, studying bats with Louise Emmons, measuring trees with Nick Brokaw, and investigating hallucinogenic plants with R. E. Schultes.

The first edition of *A Neotropical Companion* is already considered a classic, and the new edition is even better. The geographical focus of John Kricher's personal research is Central America, and the earlier book reflected that. The new edition features an expanded focus not only on Amazonia but on topics like freshwater ichthyology, ethnobotany, and conservation of biodiversity, which were covered only briefly the first time around. This edition also contains augmented sections on such nonrainforest systems as mountains, savannas, mangroves, and coral reef, which also are of interest to the tropical traveler.

I can think of no other natural history book that is so useful and informative for Central America, South America, and the Caribbean. If you are heading for the Neotropics, don't leave home without it.

Mark J. Plotkin, Ph.D.

Executive Director, The Ethnobotany and
Conservation Team
Author of *The Shaman's Apprentice*

Preface

THIS IS a book about the American tropics, the lands of Central and South America, their great rainforests and other ecosystems, and the creatures that dwell within. This, the second edition of *A Neotropical Companion*, has been substantially revised and greatly expanded to incorporate more of the abundance of information learned about the Neotropics since 1989, when this book was first published. My focus has enlarged to cover Amazonia in far greater detail than in the previous edition. Additions have been made to each chapter, often resulting in major rewriting, requiring the present volume to be printed in a larger format than its predecessor. New chapters have been added on riverine and Andean ecosystems as well as human ecology and deforestation. The evolution chapter has undergone considerable expansion, and thus the topics of coevolution and the ecology of fruit consumption are now combined as a single chapter. Major chapters focus on birds and other Neotropical animals, with an emphasis on vertebrates. As I am by profession an ornithologist, and as many birders and ecology students find birds a compelling reason for a visit to the Neotropics, I have tended to emphasize birds throughout the book. Admittedly, the bird chapter itself is rather long, but I am hoping to provide my readers with a robust introduction to the creatures I know best. The chapters have been rearranged somewhat to accommodate the fact that the book is now used as a text in many courses on tropical ecology. All illustrations were newly prepared for the second edition, photographs are now included, and color has been added.

The New World tropics, or Neotropics, provide remarkable examples of natural history. In recent years, increasing numbers of tourists, students, and researchers have ventured to the tropics in search of exotic birds, mammals, insects, and plants and to see firsthand the amazing biodiversity of the rainforests and the people who live within them. More and more college, high school, and even middle school courses focus on Latin America, and many of these include trips to tropical field stations. Scientific knowledge about the American tropics continues to burgeon. Research performed at many places throughout the Neotropics has provided extraordinary examples of nature's complexity. Many, if not most, of these examples have appeared thus far scattered in the technical literature and have yet to find their way into a general book. This book, continuing in the tradition of its predecessor, attempts to remedy that situation.

The basis for this book developed from an undergraduate course I teach at Wheaton College in Massachusetts as well as from a Chautauqua short course that I taught some years ago. I have relied heavily on my personal experiences throughout the countries of Central and South America. Though this is not meant as a highly technical book, I discuss numerous examples of published research and include a complete literature-cited section so that the interested reader can read firsthand any of the studies mentioned and the student may effectively access the technical literature. By necessity, I have had to be highly selective, and many excellent studies have been omitted. Indeed, since the publication of the first edition, the number of titles in Neotropical ecology has increased seemingly exponentially. I believe, however, that those I chose to include will serve well to introduce you to what it is about the American tropics that makes me and my many colleagues want to keep returning and studying. For the reader with little or no previous knowledge or formal preparation in ecology, I have included a glossary of ecological terms, especially oriented to Neotropical ecology, and considerably expanded from the first edition.

The book begins with an overview of the tropical climate, the importance of seasonality, and a brief survey of the various kinds of ecosystems found in the American tropics. From there I discuss the complex structure, high species richness, and ecological functioning of rainforests. I then focus on evolution in the Neotropics, on why the tropics host such a myriad of species, and how species evolve and interact, including how they sometimes evolve extraordinarily intricate interdependencies. People have had a profound influence on rainforest ecology, and I have substantially expanded my treatment of Neotropical anthropology, which now stands as its own chapter. A chapter titled "The Neotropical Pharmacy" discusses the host of remarkable drugs present in tropical vegetation, the evolutionary influence of these drugs, and the newly revitalized field of ethnobotany. Chapters follow on riverine ecosystems, montane ecosystems, savannas and dry forests, and coastal marine ecosystems. There is a lengthy chapter on the ecology of rainforest birds and one on other animals: the mammals, reptiles, amphibians, and arthropods. These chapters are meant to acquaint you with some of the creatures most commonly encountered in the Neotropics. I close with a chapter, rewritten and much expanded from the previous edition, dealing with the complex conservation issues that will determine the future of the Neotropics.

The tropics are changing rapidly, as rainforests are cleared to make room for agriculture, pasture, and other human activities. Human populations throughout Latin America continue to grow, some at alarming rates, ultimately demanding more from the land. Such changes have significant impact on the ecology of the region, and particularly on various forms of wildlife. Conservationists are saddened that so many species within such groups as mammals, birds, crocodilians, and others are now listed by the Convention on International Trade in Endangered Species of Wild Flora and Fauna (CITES) as threatened or endangered, often throughout their entire range. Some species, such as the magnificent hyacinth macaw (*Anodorhynchus hyacinthinus*), once widespread and abundant, have been dramatically reduced and now occupy a much more restricted range. Other species, such as the glaucous macaw (*A. glaucus*), also once widespread throughout south-central South America, are now feared to be extinct (Collar et al. 1992).

Change is so rapid, in fact, and habitat so much at immediate risk, that researchers from an organization named Conservation International have established what are called Rapid Assessment Programs (RAP), where expert biologists do short-duration but thorough surveys of habitat in order to learn if the area is sufficiently important to be a high priority for preservation. In August 1993, two researchers from CI died when their small airplane crashed into a cloud-enshrouded mountainside in Ecuador as they were completing a RAP survey. Theodore A. Parker III, an ornithologist, and Alwyn H. Gentry, a botanist, had no equals in their respective areas of expertise. Ted Parker and Al Gentry were the best there is at what they did (see Forsyth 1994 for a tribute to Parker, and Hurlbert 1994 for a tribute to Gentry). Their loss is immeasurable, not only to those who knew them, to their colleagues, friends, and families, but to the science of Neotropical ecology. These men paid the ultimate price in their attempt to learn from and preserve Neotropical rainforests. All of us who care about rainforests honor their memories and owe them our deepest respect.

The speed with which rainforests, dry forests, mangroves, and other natural ecosystems are being felled is a matter of deep concern to biologists. These ecosystems are truly magnificent, and there remains so much still to be learned about them. I am one who hopes that this pace can be slowed and that wise decisions will be made with regard to the preservation of essential tracts of rainforest and other Neotropical habitats. I hope that after you have read this book, you not only will understand the American tropics better but will also share my concern for and awareness of the need to preserve the uniqueness that is ecological tropical America.

John Kricher

Novato, California
August 1996

Acknowledgments

ONE researcher in particular is deserving of special mention because he has contributed so much to the fields of tropical ecology and conservation biology and indirectly inspired me to write this book. Without the work of Daniel Janzen, this book would have been very different and far less insightful. It is in part a tribute to Janzen that so many others, myself included, have followed his path to the Neotropics.

I again thank Robert Askins, Nicholas Brokaw, Brian Cassie, William (Ted) Davis, Jr., Stephen Hubbell, and Leslie Johnson, each of whom critically read various parts of the first edition and provided numerous excellent suggestions.

I am most grateful to the following persons, each of whom had a part in helping me as I gathered information for this revision: Peter Alden, Gerhard Beese, Dean Cocking, Victor Emanuel, Richard ffrench, Jürgen Haffer, Edward Harper, John Harwood, Larry Hobbs, Peter Jenson, Robert Meade, Bruce Miller, Carolyn Miller, Charles Munn, Mark Plotkin, Robert Ridgely, James Serach, Scott Shumway, Miles Silman, Robert Stiles, Guy Tudor, James Wetterer, Andrew Whittaker, and Kevin Zimmer.

In addition, I thank the many readers who took the time and trouble to write me with comments, corrections, or suggestions for this revision. I have also benefited from comments published by numerous reviewers of the first edition, and I have adopted many of their suggestions. You all have my deep appreciation. I encourage you to write me about this edition, at Wheaton College, Norton, MA 02766, or jkricher@wheatonma.edu.

James Cronk, Frederick Dodd, Charles Munn, Pepper Trail, Christian Voigt, James Wetterer, and Kevin Zimmer were each kind enough to allow me the use of their transparencies to illustrate this edition, for which I am deeply appreciative. I also thank my good friend William E. "Ted" Davis, Jr., for crafting the splendid pen-and-ink illustrations that grace the new edition. Bob Meade generously sent a wealth of materials on the hydrography and geology of the Orinoco and Amazon. Christopher Neill and Jerry Melillo supplied me with much information on the research they and their colleagues are doing on soil biogeochemistry in Rondonia, Brazil. Robert Askins provided helpful material on ancient Maya farming techniques and modern hydraulic agriculture, as well as insightful discussion on tropical diversity and its possible causes. William Gotwald supplied information on army ants. Burkhard Seubert provided information on Alexander Von Humboldt. I am most grateful for all of this help. The following Wheaton College students assisted in researching various aspects of the second edition: Katherine Banks, Tony Baptista, Sean Fuss,

Jacqueline LaMontagne, Kellie Laurendeau, Lelia Mitchell, Jill Roberge, Jessica Stevens, and Alessandro Vaccaro. I also appreciate the help of Janet Wessel and the cooperation and hospitality of Point Reyes Bird Observatory in permitting me use of their library.

William E. Davis, Jr., thanks James Cronk of the Children's Environmental Trust Fundation International, Fred Dodd, and photographer Michael J. Doolittle for providing photographs that were used as models for many of the illustrations in this book.

This book could not have been written without extensive first hand experience in the tropics. I am grateful to Alice F. Emerson, then president of Wheaton College, for her support and encouragement when in 1977 I proposed the idea of taking students to Belize for a field course in tropical ecology. Since then, my numerous peregrinations in Belize, Guatemala, Panama, Trinidad, Tobago, Mexico, Puerto Rico, Venezuela, Brazil, Peru, Ecuador, and Costa Rica have been made possible by a variety of individuals and granting agencies. I am very grateful to Frederick Dodd of International Zoological Expeditions for his immense help and friendship, especially in the early days of the Belize-Guatemala trips. My thanks also to the Bowman family of the Pelican Beach Resort in Dangriga for their hospitality and friendship in Belize. Grants from Wheaton College, Andrew Mellon Foundation, and The Center for Field Research (Earthwatch) provided funds for some of my trips. Part of the first edition was written during a brief stay at Oxford University, made possible by a grant from GTE/Focus to Wheaton College. Stephanie Gallagher of the Oceanics School facilitated my initial visits to Peru and Ecuador. Society Expeditions provided support for trips to Venezuela and Brazil. I thank the owners and operators of Chan Chich Lodge in Gallon Jug, Belize, for their hospitality toward me and my group. I am very grateful to Children's Environmental Trust Foundation International, and particularly to its president, James Cronk, and national coordinator, Terry Larkin, for the opportunity to revisit the Iquitos area, and for the pleasure of participating in the CET Children's Rainforest Workshops. I thank the owners and operators of Explorama Lodges and Yacumama Lodge for their kind hospitality while in Iquitos. I also acknowledge, with appreciation, the Amazon Center for Environmental Education and Research (ACEER), and its vice president for scientic research, Dr. Stephen L. Timme, for permitting me to visit the unique canopy walkway that is part of the ACEER reserve.

I thank the many participants of my various Chautauqua courses for their enthusiasm and animated discussions. Ann Spearing and William Zeitler were both instrumental in providing me with the opportunity to participate in the Chautauqua program. Among my field companions who joined me as I prepared the first edition, I wish to thank Brian Cassie, Ted Davis, the several dozen Earthwatch volunteers, Wayne Petersen, the Oceanics (Helen, Barbara, Mary Beth, Steve, Andy, Lyman, Chris, Kate, Jon, Enrique, Charo, and Scott), Melinda Welton, and my various Wheaton College classes for providing such memorable and downright joyous times. Special thanks to Linda Kricher, who was a terrific and enthusiastic field companion and an immense help during each of the Wheaton College Belize trips.

Since the first edition, my Neotropical travels have been highlighted by the company of Betsey Davis, Ted Davis, Ann Dewart, Jack Dineen, Pat Eastwood, Ed Harper, Patty O'Neill, Susan Scott, Susan Smith, Martha Steele, Bob Stymeist, Martha Vaughan, and Janet Wessel. All of you made these trips immense fun, as, indeed, they should be.

The book was originally suggested by Mary Kennan, and I am grateful for her encouragement and support. I also thank Judith May, then at Princeton University Press, for her enthusiastic support, and Emily Wilkinson, my editor at Princeton University Press, for unflagging support, encouragement, and editorial skills.

Beyond all others, I thank my wife, Martha Vaughan, for her encouragement, companionship, and tolerance; for her sharp editorial skills and probing questions; for providing me with such a splendid sanctuary in which to write, as well as three enthusiastic feline assistants; for allowing me to share in the joy that is parenthood; and for her unwavering commitment to and obvious love for natural history—and me.

PHOTO CREDITS

Jim Cronk: 43

Frederick J. Dodd: 20, 21, 101, 102, 103, 104, 106, 110, 111, 113, 114, 117, 118, 120, 121, 135, 144, 148, 152

John Kricher: all photos unless otherwise noted

Eduardo Nycander and Charles Munn: 146, 147

Pepper Trail: 136, 137, 138

Christian Voight: 105, 116, 123, 124, 125, 126, 127, 128, 129, 130, 131, 132, 133, 134

James K. Wetterer: 90, 91, 92, 93, 94, 95, 96, 97, 98, 99, 115

Kevin Zimmer: 149

A Personal Note to the Reader

YOU SHOULD really try and visit the Neotropics. Reading this book will, I hope, help you understand some of the complexities of the ecology of this extraordinary region of the world, but there is simply no substitute for "being there." There are many ecotour companies that offer various kinds of tours focused on particular interests, especially birds. I have taken many such tours and have yet to be disappointed—and I have learned so much. In recent years, field stations for ecotourism and research have increased considerably throughout the Neotropics, and more are being built. For information on many of these, see Castner (1990) as well as various ecologically oriented regional travel guides.

A Neotropical Companion can be read in the comfort of your home, in a crowded airliner where the seats are way too narrow and you are cramped and uncomfortable and they still haven't brought the dinner trays, or in the quiet, subdued light in a field station somewhere in the wondrous Neotropical rainforest. I hope you will do all of these. Though the size of the volume has considerably expanded from the first edition, do try and find room for it in your field pack when you travel to the tropics.

Before you go into the field, I strongly advise that you read the appendix, "And Hey, Let's Be Careful Out There." It is a survey of the various potential safety and health hazards that await the visitor to the tropics. Reading it first, well in advance of your trip, should help ensure that you remain safe and healthy and are not eaten by an anaconda or carried off by army ants.

Conversions

1 inch = 2.54 cm

1 mile =1.609 km

1 square mile = 640 acres = 2,590 square km

1 pound = 0.45 kg

1 U.S. ton = 0.9 metric ton = 2,000 lbs

1 kilometer = 0.621 mi

1 meter = 39.37 in = 3.28 ft

1 centimeter = 10 millimeters (mm) = 0.394 in

1 hectare = 10,000 square meters = 2.477 acres

1 kilogram = 2.205 lbs.

A NEOTROPICAL COMPANION

1

Tropical Climates and Ecosystems

NEVER does nature seem more bountiful than in the tropics. Anyone with a passion for natural history must try and visit the tropics and experience Earth's most diverse ecosystems firsthand. This is a book about the New World or *Neotropics*. Alexander von Humboldt, Henry Walter Bates, Charles Darwin, Alfred Russel Wallace, Louis Agassiz, Thomas Belt, Charles Waterton, William Beebe, Frank M. Chapman, and other eminent naturalists have each been profoundly influenced in their beliefs about natural history by visits to the Neotropics. Their spirits of adventure and investigation are no less fervent today. Thousands of tourists annually travel to Neotropical jungles and rainforests in hopes of seeing some of the myriad bird species, colorful butterflies and other diverse insects, noisy monkey troops, and numerous other attractions of these majestic ecosystems. Students and professional researchers by the dozens are patiently and painstakingly unraveling perhaps the most complex Gordian knot in ecology, the multitudes of interactions among plants, animals, and microbes resulting in the vast biodiversity of tropical forests.

There is an urgency about the science of tropical ecology: tropical forests, which occupy approximately 7% of Earth's surface but may harbor as much as 50% of the world's biodiversity (Myers 1988; Wilson 1988), are being cleared at alarming rates (Repetto 1990). Cattle ranches and soybean fields are replacing rainforests. Though tropical rainforests also exist in Africa and Asia, approximately 57% of all rainforests remaining on Earth are in the Neotropics, with 30% in Brazil alone. Many of these are being cut: already only 12% of Brazil's unique Atlantic coastal forest remains (Brown and Brown 1992), and in 1987 alone some 20 million acres of Brazilian rainforest were cut and burned (Miller and Tangley 1991). Other Neotropical areas in danger and judged to require immediate conservation attention include the Colombian Choco, forests of western Ecuador, and the uplands of western Amazonia (Wilson 1992). At the current rate of deforestation, within 177 years all tropical rainforests on Earth could be gone. Right now, less than 5% of the world's tropical forests are protected within national parks or reserves. Though some encouraging data suggest a slowing of Amazonian rainforest clearance (Bonalume 1991), concerns remain about the long-term future of these rainforests as well as other tropical ecosystems, not just in the Neotropics but globally. Obviously, the ecosystems comprising the main subject of this book are potentially endangered. These ecosystems deserve better. Alexander von Humboldt, one of the first of the great naturalists to learn from the tropics, captured the sense of wonder one receives upon seeing rainforest for the first time:

> An enormous wood spread out at our feet that reached down to the ocean; the tree-tops, hung about with lianas, and crowned with great bushes of flowers, spread out like a great carpet, the dark green of which seemed to gleam in contrast to the light. We were all the more impressed by this sight because it was the first time that we had come across a mass of tropical vegetation. . . . But more beautiful still than all the wonders individually is the impression conveyed by the whole of this vigorous, luxuriant and yet light, cheering and mild nature in its entirety. I can tell that I shall be very happy here and that such impressions will often cheer me in the future. (Quoted in Meyer-Abich 1969.)

Most people who have never been to equatorial regions assume them to be continuous rainforest, much as described by Humboldt. Tropical rainforest is, indeed, a principal ecosystem throughout much of the area and is the major focus of this book. Other kinds of ecosystems, however, also characterize the tropics (Beard 1944; Holdridge 1967; Walter 1971). Climate is generally warm and wet but is by no means uniform, and both seasonality and topography have marked effects on the characteristics of various tropical ecosystems. In this chapter I will present an overview of the tropical climate, seasonality, and major ecosystem types occurring in the Neotropics.

The Climate

Definition of the "Tropics"

Should you decide to move to Manaus, Brazil, or perhaps to Iquitos, Peru, both well within the Amazon Basin, you should expect at least 130 days of rain per year, and in some places up to 250 days. Temperature will be consistently warm, often hot (highs of about 31°C [88°F], nighttime low of about 22°C [72°F]), and relative humidity will never be less than 80% (Meggars 1988). Though it can rain on any given day, rainfall, in most places, will be seasonal. That, in a nutshell, is what it's like in the tropics. In the Amazon Basin, the very heart of the Neotropics, climate is permanently hot and humid, with the temperature averaging 27.9°C (82°F) during dry season and 25.8°C (78.5°F) in rainy season. In the tropics, daily temperature fluctuation exceeds average annual seasonal fluctuation (see below) and air humidity is quite high, about 88% in rainy season and 77% in dry season (Junk and Furch 1985).

Geographically, the tropics is an equatorial region, the area between the Tropic of Cancer (23° 27′N) and the Tropic of Capricorn (23° 27′S), an approximately 50-degree band of latitude that, at either extreme, is subtropical rather than tropical. The Tropic of Cancer passes through central Mexico and extreme southern Florida. The Tropic of Capricorn passes through northern

Chile, central Paraguay, and southeastern Brazil, almost directly through the Brazilian city of Sao Paulo. The Neotropics thus include extreme southern North America, all of Central America, and much of South America. You can visit the Neotropics by traveling to southern Mexico, Guatemala, Belize, El Salvador, Honduras, Nicaragua, Costa Rica, Panama, Venezuela, Colombia, Guyana, Surinam, French Guiana, Ecuador, Peru, Brazil, and parts of Bolivia and Paraguay. In the Caribbean Sea, the Greater and Lesser Antilles are within the Neotropics.

The tropics are warm and generally wet because the sun's radiation falls most directly and most constantly upon the equator, thus warming Earth more in the tropics than at other latitudes. As one travels either north or south from the equator, Earth's axial tilt of 23° 27′ results in part of the year being such that the sun's rays fall quite obliquely and for much shorter periods of time, thus the well-known cycles of day length associated with the changing seasons of temperate and polar regions. At the equator, heat builds up and thus the air rises, carrying the warmth. Water is evaporated so water vapor rises as well. The warm, moist air is cooled as it rises, condensing the water, which then falls as precipitation, accounting for the rainy aspect of tropical climates. The normal flow of warm, moisture-laden air is from equatorial to more northern and southern latitudes. As the air cools it not only loses its moisture to precipitation, but also becomes more dense and falls, creating a backward flow toward the equator. At the equator, two major air masses, one from the north and one from the south, along with major ocean currents, form the Intertropical Convergence (ITC), the major climatic heat engine on the planet.

In the Amazon Basin, precipitation ranges between 1,500 (59 in) and 3,000 mm (118 in) annually, averaging around 2,000 mm (79 in) in central Amazonia (Salati and Vose 1984). About half of the total precipitation is brought to the basin by eastern trade winds blowing in from the Atlantic Ocean, while the other half is the result of evapotranspiration from the vast forest that covers the basin (Salati and Vose 1984; Junk and Furch 1985). Up to 75% of the rain falling within a central Amazonian rainforest may come directly from evapotranspiration (Junk and Furch 1985), an obviously tight recycling of water, and a recycling system that clearly demonstrates the importance of intact forest to the cycling of water. This vast precipitation and water-recycling system is essentially in maintaining equilibrium, though large-scale deforestation could significantly upset the balance (Salati and Vose 1984; see also chapter 14).

Tropical areas fall within the trade-wind belts (so named because winds were favorable for sailing ships trading their goods) except near the equator, an area known as the intertropical convergence or doldrums, where winds are usually light, often becalming sailing ships. From the equator to 30°N, the eastern trade winds blow steadily from the northeast, a direction determined because of the constant rotation of Earth from west to east. South of the equator to 30°S, the eastern trades blow from the southeast, again due to the rotational motion of the planet. As Earth, tilted at about 23.5° on its axis, moves in its orbit around the sun, its direct angle to the sun's radiation varies with latitude, causing seasonal change, manifested in the tropics by changing heat

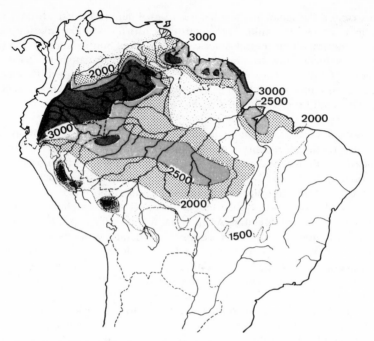

Annual rainfall (mm) in Amazonia. Black areas, near the base of the Andes, represent regions with more than 4,000 mm of rain per year. From Haffer and Fitzpatrick (1985). Reproduced with permission.

patterns of air masses around the intertropical convergence that result in seasonal rainfall. In the Western Hemisphere, from July throughout October, severe wind- and rainstorms called hurricanes can occur in parts of the Neotropics. Similar kinds of storms are referred to as monsoons in the Old World tropics.

Seasonal variations in day length are not nearly as dramatic in the tropics as in the temperate zone. At the equator, a day lasts exactly twelve hours throughout the year. North of the equator, days become a little longer in the northern summer and shorter in winter, but this only means that summer sunset is at 6:15 or 6:20 rather than 6:00 P.M. Temperature fluctuates relatively little in the tropics. Typically, daytime temperature is somewhere around 29°C (85°F), though in many areas it may be 32–37°C (in the 90s), with surprisingly little seasonal fluctuation. In general, there is no more than a 5°C difference between the mean temperatures of the warmest and coldest months. For example, at La Selva Biological Station in Costa Rica, August is the month with the highest mean temperature, 27.1°C (80.5°F), while January has the lowest mean temperature, 24.7°C (77°F) (Sanford et al. 1994). Relative humidity, as noted above, is generally high in the tropics, especially in lowland rainforests where humidities ranging from 90% to 95% at ground level are common. Humidity is less in the rainforest canopy, usually no higher than 70%.

El Niño—Southern Oscillatiion

South American and, indeed, global climates are periodically and some-times dramatically affected by a still poorly understood climatic event called *El Niño* ("The Child"), or the Southern Oscillation. Originally named because it tends to begin around Christmas, El Niño causes sufficient short-term climate change to produce major disruptions to ecosystems, especially marine ecosys-tems (Glynn 1988). An El Niño event involves the unpredictable warming of eastern Pacific Ocean surface waters around the equator.

El Niño occurs periodically, approximately every two to seven years, when a high-pressure weather system that is normally stable over the eastern Pacific Ocean breaks down, destroying the pattern of the westward-blowing trade winds. Trade winds thus weaken severely, sometimes reversing from their nor-mal westward direction. Warm water from the western Pacific flows eastward, enhancing the Equatorial Counter Current and causing an influx of abnor-mally warm water to the western coast of South America. Instead of winds pushing water from the west coast of South America, creating an upwelling of deeper, colder, nutrient-rich water (page 15), the trade winds quit. When that happens, warm waters flow along the normally cold South American coast, global heat patterns vary, and weather systems change, causing floods in some regions and droughts where there should be rainfall, effects that can be any-where from mildly stressful to disastrous to plant and animal populations. For example, some parts of South America experience abnormally heavy down-pours while other areas, particularly in Central America, become drought-stricken. Droughts can also occur in places such as Australia, Indonesia, and southern Africa.

There have been eight major El Niño events since 1945, and at least twenty during this century. In 1982–83 an El Niño considered up to then to be the most powerful of this century caused an estimated $8.65 billion worth of damage worldwide. An even more severe El Niño occurred in 1986–87. A comparable El Niño occurred in the winter of 1994–95. The California coast was inundated by rain, resulting in extensive flooding and mudslides from Los Angeles to the Russian River area north of San Francisco, while New En-gland experienced far less winter precipitation than usual. Satellite data indi-cated that the northern Pacific Ocean was nearly eight inches higher than normal, due to the influx of warm surface waters. The causal factors respon-sible for the periodicity of El Niños are thus far unknown (Canby 1984; Gra-ham and White 1988), but it is clear that the Intertropical Convergence, a complex system of oceanic and air currents, migrates to a lower latitude, rais-ing sea surface temperatures and destroying the normal upwelling pattern along the west coast of South America. The cessation of an El Niño occurs then the ITC returns northward to its normal position (hence the alternate term for El Niño, the Southern Oscillation). Tropical ecosystems, already sen-sitive to seasonal variation (see next section), can be anywhere from moder-ately to severely affected by changes caused by El Niño (Glynn 1988; also see Foster 1982b, below). Indeed, El Niño of 1986–87 has been suggested to have contributed strongly to the apparent extinction of two amphibian species, the golden toad (*Bufo periglenes*) and the harlequin frog (*Atelopus varius*), from Monteverde Cloud Forest Preserve in Costa Rica (Pounds and Crump 1994).

The Importance of Seasonality

Rainfall in tropical latitudes varies seasonally, often dramatically. Because of warm air throughout the year, precipitation is in the form of rain (except atop high mountains such as the Andes, where snow occurs, even at the equator), but the amount of rain varies considerably from month to month and from one location to another. Overall, precipitation is highest in the central Amazon Basin and the eastern Andean slopes and lowlands, and less to the north or to the south, varying from about 6,000 mm (eastern Andean slopes) to 1,500 mm (236–59 in) (extreme north or south) (Junk and Furch 1985). Even within the central Amazon Basin, seasonal rainfall is variable from place to place. For example, Iquitos, Peru, along the Amazon River, receives an average of 2,623 mm (103 in) of rainfall annually, while Manaus, Brazil, also on the Amazon River, receives an average of 1,771 mm (70 in) and experiences a strong dry season. As a more extreme example, Andagoya, in western Colombia, receives 7,089 mm annually (approximately 280 in). (The area that receives the most rainfall on Earth is not in the Neotropics but in the United States! It is Mount Waialeale, Hawaii, averaging 11,981.18 mm [471.7 in] annually.)

Throughout most of the tropics, some rain falls each month, but there is usually a pronounced wet and dry season, and sometimes two wet and dry periods, each of which differs in magnitude. Where the dry season is pronounced, many, often most, trees are deciduous, shedding leaves during that season. Such tropical dry forests are often termed "monsoon forests," since they are in leaf only when the monsoon rains are present. Dry season is defined as less than 10 cm (3.9 in) of rainfall per month, and rainy season features anywhere from 2 to 100 cm (0.8–39 in) (occasionally more) of rainfall per month. A typical tropical rainforest receives a minimum of 150 to 200 cm of rainfall annually (60–80 in).

The rainy season varies in time of onset, duration, and severity from one area to another in the tropics. For example, at Belém, Brazil, virtually on the equator, dry season months are normally August 1 through November, and the wettest months are January through April. In Belize City, Belize, at 17°N, the rainy season begins moderately in early June but in earnest in mid-July and lasts through mid-December and sometimes into January. The dry months are normally mid-February through May. In general, when it is rainy season north of the equator, it is dry season to the south. Because the Amazon River flows in close proximity of the equator, parts of the huge river are experiencing wet season while other parts are in dry season.

The seasonal shift from rainy to dry season has direct effects on plants and animals inhabiting rainforests as well as other tropical ecosystems. One common misconception about the tropics is that seasonality can generally be ignored. Images of year-round sunny skies and soft trade winds are the stuff of myths. The truth is that seasonal shifts are normal and often pronounced, with many ecological patterns reflecting responses to seasonal changes. Some shifts are obvious, but others are subtle and vary considerably depending on the magnitude of the seasonality. During the rainy season, skies are typically cloudy for most of the day and heavy showers are intermittent, often becoming especially torrential during late afternoon and evening. Such cloud cover, blocking sunlight from reaching the forest, can be a strong limiting factor on total photosynthesis; thus plant growth is often greater during dry season, when skies are clear for up to ten hours during the day and showers, though sometimes heavy, are brief.

Seasonal differences are not trivial to organisms. Henry Walter Bates, in *The Naturalist on the River Amazons* (1863), wrote of seasonal patterns as they affect life along the Amazon. At the onset of rainy season, "All of the countless swarms of turtle of various species then leave the main river for the inland pools: sand banks go under water, and the flocks of wading birds then migrate northerly to the upper waters of the tributaries which flow from that direction, or to the Orinoco; which streams during the wet period when the Amazons are enjoying the cloudless skies of their dry season." More recent studies, particularly those carried out by researchers on Barro Colorado Island (BCI) in Panama (Leigh et al. 1982) and La Selva Biological Station in Costa Rica (McDade et al. 1994), have documented the compelling drama of the changing seasons of the tropical forest.

Trees flower more commonly during the dry season (Janzen 1967, 1975) when less frequent and less intense showers permit insect pollinators to be active for longer periods, thus enhancing cross-pollination. Some tree species synchronize their flowering after downpours (Augspurger 1982), which may increase pollination efficiency by concentrating the number of pollinators (Janzen 1975). Dry season pollination also enables more seedlings to survive because they sprout at the onset of rainy season, when there is adequate moisture available to ensure their initial growth. A study of 185 plant species on Barro Colorado Island determined that most seedlings emerged within the first two months of the eight-month rainy season (Garwood 1982). Forty-two percent of the plant species underwent seed dispersal during dry season and germination at the onset of rainy season. Forty percent of the species experienced seed dispersal at the beginning of rainy season, with germination occurring later in rainy season. Approximately 18% of the species produced seeds that were dispersed during one rainy season, were dormant during the next dry season, and germinated at the onset of the second rainy season. The species most sensitive to the onset of rainy season were "pioneer" tree species, lianas, canopy species, and wind- and animal-dispersed species. Understory and shade-tolerant species were less sensitive.

Fruiting patterns, not unexpectedly, are also under strong seasonal influence. In general, most fruiting roughly coincides with peak rainy season, with lowest fruit availability at the onset of dry season (Fleming et al. 1987), though there is much variability among species. Fruiting patterns on Barro Colorado Island are seasonally influenced (Foster 1982a). The timing of fruiting in many species appears to be a compromise between the desirability of seeds germinating at the onset of rainy season and the advantages of flowering early in the rainy season, when insects are most abundant (see below).

Pioneer tree species often germinate at the onset of rainy season, which is when tree falls tend to be most common, opening gaps in the forest where these shade-intolerant species can become established (see chapter 3). At BCI, one large gap per hectare occurs on average every 5.3 years, a frequency sufficient to support a high population of quick-growing pioneer tree species (Brokaw 1982).

Grazing rates on leaves are more than twice as high during the rainy season as during the dry season (Coley 1982, 1983). New leaves are more vulnerable to insect herbivores because they lack protective tissues and chemicals (see chapter 6). Most trees grow their new leaves during early rainy season. Some trees are deciduous during the dry season, dropping their leaves entirely.

As might be expected, arthropods, many of which are highly dependent on plants, also show seasonal changes in abundance. A study conducted among several habitats in southeastern Peru showed that forest floor arthropod biomass was most abundant during the wet season. Virtually all arthropod taxa showed clear seasonal patterns (Pearson and Derr 1986). Similar seasonal effects were noted for Panama (Levings and Windsor 1982) and Costa Rica (Lieberman and Dock 1982), where arthropod abundance peaks at the end of dry season and beginning of rainy season.

Rainforest birds are sensitive to seasonal rhythms. In Costa Rica, most nesting occurs from March through June, the end of the dry season and beginning of the rainy season, with little nesting occurring from October through December (end of rainy season and beginning of dry season), a pattern noted for much of Central America (Levey and Stiles 1994). Seasonal changes in distribution and abundance of nectar-eating, fruit-eating, and understory birds are well documented for Panama (Leck 1972; Karr 1976; Karr et al. 1982) and Costa Rica (Levey and Stiles 1994). Manakins, small birds that feed almost entirely on fruit (page 274), have been found not to breed during seasons of fruit shortage, and, at least at one location near BCI, the manakin population fluctuates with fruit availability (Worthington 1982). On Grenada, the bananaquit (*Coerba flaveola*), a small, nectar-feeding bird (page 265), synchronizes its breeding to coincide with the onset of the wet season (Wunderle 1982). A study conducted on Puerto Rico concluded that birds needed adequate rainfall to breed successfully during their normal season of April–July (Faaborg 1982).

The tamandua (*Tamandua mexicana*), a common forest anteater, shifts its diet from ants in rainy season to termites in dry season (Lubin and Montgomery 1981). Termites are juicier than ants and so afford a higher moisture content to the anteater. Termites (*Nasutitermes sp.*) are also attuned to the seasons, swarming during the onset of rainy season (Lubin 1983). The mass emergence may ensure that each swarming insect has a better chance of reproduction, because it is more likely to encounter another termite quickly. Also, potential termite predators cannot possibly eat all of the swarming masses. Thus some termites survive to initiate new colonies. Many animals, such as monkeys, cats, iguanas, and various lizards, abandon deciduous forests during dry season when leaves have dropped. These creatures move to riverine gallery forests, which remain in leaf.

White-throated (faced) capuchin

On Barro Colorado Island the shortage of fruits at the end of the wet season affects the ecology of two common rainforest rodents. The agouti (*Dasyprocta punctata*), a small, diurnal (daytime-active) rodent, depends on relocating seeds that it has buried to sustain itself through the months of the dry season. Another rodent, the nocturnal paca (*Cuniculis paca*), survives the dry season by browsing more intensively on leaves and living off its stored fat. Both agouti and paca forage for longer periods during dry season, and their populations are indirectly limited by the dry season food shortage. Because they must forage for longer periods and take greater risks to satisfy their hunger, they fall victim to predators more frequently (Smythe et al. 1982).

A most extreme case of seasonal stress was documented at Barro Colorado Island (Foster 1982b). Two fruiting peaks normally occur annually, one in early rainy season and one in mid–rainy season. During 1983, an El Niño year (see above), the second peak failed to occur. Between August 1970 and February 1971 only one-third the normal amount of fruit fell, thus creating a famine. Not all plant species failed to produce a second fruit crop, but enough did to severely affect the animal community. Researchers on BCI noted that normally wary collared peccaries (*Tayassu tajacu*), coatimundis (*Nasua ilarica*), agoutis, tapirs (*Tapirus bairdii*), and kinkajous (*Potos flavus*) made frequent visits to the laboratory area to get food that had been put out for them. Peccaries seemed emaciated, and a kinkajou looked to be starving when it first appeared. Most amazing were the monkeys. To quote Robin Foster, "The spider monkeys, which normally visit the laboratory clearing at least once every day, now launched an all-out assault on food resources inside the buildings, learning for the first time to open doors and make quick forays to the dining room table, where they sought bread and bananas, ignoring the meat, potatoes, and canned fruit cocktail, and brushing aside the startled biologists at their dinner." Foster noted that dead animals were encountered much more frequently than in previous years. "The most abundant carcasses were those of coatis, agoutis, peccaries, howler monkeys, opossums, armadillos, and porcupines; there were only occasional dead two-toed sloths, three-toed sloths, white-faced monkeys, and pacas. At times it was difficult to avoid the stench: neither the turkey vultures nor the black vultures seemed able to keep up with the abundance of carcasses." The reason why the two sloth species, the white-faced monkeys (*Cebus capucinus*), and the pacas were less affected is that they feed on foliage. Fruit, not foliage, was in short supply.

The severe dry season of 1983, due partly to El Niño and partly due to long-term oscillations in climate, also resulted in greatly increased mortality rates among the canopy trees of Barro Colorado (Condit et al. 1995).

Studies cited above contrast strongly with the naive view of the tropics expressed in the Humboldt quotation at the beginning of the chapter. The tropics may appear luxuriant at first glance, but in reality they impose significant seasonal stresses upon the plant and animal inhabitants. Furthermore, the tropics do not host stable, unchanging ecosystems. Tropical ecology, as you will learn, is more than a little dynamic. It's a real jungle out there.

The Importance of Mountains Figures 4, 5, 6, 7, 8, 9

The Andes Mountains began their rise approximately 20 million years ago (Zeil 1979). Orographic uplift has continued unabated to the present day with the Andes chain still one of the most active geological areas on the planet. As recently as a million years ago the northernmost part of the Andes was uplifted, and Charles Darwin (1906), on his voyage of the H.M.S. *Beagle*, bore witness to the awesome power of a major earthquake in Chile.

A bad earthquake at once destroys our oldest associations: the earth, the very emblem of solidarity, has moved beneath our feet like a thin crust over a fluid; one second of time has created in the mind a strange idea of insecurity, which hours of reflection would not have produced.

Darwin's description of his perceptions while experiencing the quake were closer to reality than he probably realized. Geologists now generally agree that Earth's crust consists of huge basaltic plates that continually move, often in opposition to one another, a dynamic pattern termed plate tectonics. Granite continents sit atop these plates. The South American plate, containing the continent of South America, began its split from the African plate about 100 million years ago, creating the south Atlantic Ocean. Since then the South American plate has been moving westward. Eventually it met the eastward-

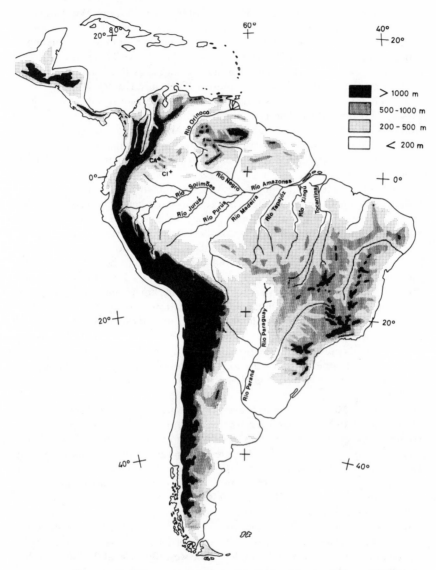

Surface relief of Central and South America, showing lowland nature of Amazonia. The 500 to 1,000 m level is indicated only in regions east of the Andes Mountains. From Haffer (1974). Reproduced with permission.

moving Nazca plate, containing the southeastern Pacific Ocean. When the two plates collided in earnest, the Nazca plate began sliding under the South American plate, creating the thrust that produced the Andes Mountains. This process, called subduction, continues today and is responsible for the geological activity evident in the volcanism and earthquakes that characterize the western part of South and Central America (Dietz and Holden 1972).

The chains of mountains stretching from southernmost Patagonia north through Mexico add to the climatic and, thus, biotic diversity of the Neotropics. Located in western South and Central America, the geologically youthful Andes and Mexican cordilleras host different altitudinal ecosystems and also serve as barriers that isolate populations, thus enhancing the speciation process (see page 109).

The north-south orientation of the Andes results in coastal Peru and Chile having some of the most arid deserts in the western hemisphere, places such as the Atacama Desert south of Lima at Paracas, Peru, and extending southward along the Chilean coast. As I walked across the dry, crusted, reddish soil, I could find no signs of plants or animals, the only time in my years as an ecologist that I've seen a place so devoid of obvious life (other than looking at the moon through a telescope). The Atacama-Sechura Desert extends along the Peruvian and Chilean coast for about 3,000 km (1,865 mi), some of it receiving only about one millimeter of precipitation annually. The driest place on Earth is considered to be Calama, Chile, in the Atacama Desert, where no rain has yet been recorded! Nonetheless, the desert is rather humid due to the fog and clouds produced by the proximity to the cool ocean currents. As with most deserts, temperature fluctuations are often dramatic. The Atacama can drop in temperature from 40°C (104°F) to 0°C (32°F) in little more than an hour.

The Andes Mountains act as a gigantic wall preventing moisture-laden air accumulating in the Amazon Basin from reaching the Peruvian and Chilean coasts. As the clouds are forced up by the tall mountains, the moisture in them condenses to snow or, at mid to low elevations, rain. Rain falls heavily on the eastern slopes of the Andes, creating conditions that support extremely lush montane and lowland rainforest. Snow melt from the Andes is one of the major sources of water for the Amazon Basin.

The Andes, in essence, keep precipitation recycling within the Amazon Basin (Salati and Vose 1984). The Amazon Basin is shaped like an immense horseshoe, with the ancient Guianan Shield bordering to the north, the Brazilian Shield to the south, and the Andes to the west. Because of this topography, all water exits the system to the east, at the huge mouth of the Amazon. This loss is replaced by input from rain and melting snow draining from the high Andes, keeping the Amazon Basin in a state of hydrographic equilibrium (Salati and Vose 1984).

The air that eventually passes over the tall mountains is fundamentally depleted of its moisture; thus, dry deserts occur on the western side. This is called a *rain shadow effect*, and consequently, ecosystems differ dramatically from one side of a mountain to the other, though their elevation may be the same. On a bus ride over a mountain in northern Peru between Jaén and Chiclayo I experienced the rain shadow effect. As our driver engineered his way up the steep mountain slope, we passed through cactus and shrubby desert. Approaching the crest, however, the clear air and blue skies gave way to misty overcast. Tall columnar cactus plants appeared, many heavily laden with bromeliads. At the crest, we were in a miniature cloud forest, bathed in permanent fog, whose ghostlike, stunted trees were adorned with all manner of orchids, bromeliads, and other air plants (epiphytes). As we descended the eastern side, the skies clouded and rain commenced. We left the elfin cloud forest, passing through rich coffee plantations and thick, cloud-enshrouded forests at the same altitudes where desert and dry grassland had occurred on the opposite side of the mountain.

Ironically, the oceanic ecosystem off the coast of southern Peru and Chile is perhaps the richest in the world, the very opposite of a desert. Steady, strong winds blow away surface water of the cold Humboldt current, creating a condition called *upwelling*, the rise to the surface of cold, subantarctic water rich with nutrients and oxygen. These winds are also partly responsible for the terrestrial desert, as they blow from the coast to the sea, and thus no oceanic, moisture-laden air is brought over land. In the sea, vast hordes of tiny plankton are supported by upwelling, and they become food for sardinelike anchovetas (*Engraulis ringens*), which, when they annually numbered well into the high millions, supported a very successful fishing industry until poor fishery management combined with effects of El Niños resulted in an anchoveta crash (Idyll 1973; Canby 1984).

Because of the effects of altitude as it relates to climate, ecosystems change, often dramatically, from the base to the top of a mountain. Working in the western United States in the late 1800s, C. Hart Merriam described what he termed *life zones*, distinct bands of vegetation, each encircling a mountain within a certain range of altitude. Creosote bush and cactus desert or *Lower Sonoran* life zone is replaced by a forest of pinyon pine and juniper or *Upper Sonoran* life zone, which is followed by ponderosa pine *Transition* zone, this giving way to spruce and fir of the *Canadian* and *Hudsonian* life zones. Zonation may appear to be sharp, but in reality one life zone gradually changes into another, often with much overlap. Life zones occur because altitude results in changing climatic conditions that favor different sets of species. It generally gets colder and wetter with altitude.

South American mountains also exhibit zonation patterns, noted in detail by Humboldt in the early nineteenth century. Though Merriam's life zone concept is well known, Humboldt actually preceded Merriam in describing the concept (Morrison 1976). He carefully documented how lowland rainforest gradually changes to montane rainforest, becoming cloud forest at higher altitudes. At its altitudinal extreme, cloud forest may be stunted, becoming a bizarre elfin forest of short, gnarled, epiphyte-laden trees (page 220). Higher still on some mountains is treeless paramo, an alpine shrubland, or puna, an alpine grassland. In general, temperature drops about 1.5°C (4°F) for every 305 m (1,000 ft) rise in elevation along a South American mountainside, an effect that is responsible in large part for the dramatic change in ecosystems. Tropical forest rarely occurs above 1,700 m (approximately 5,000 ft), with subtropical forest between 1,700 m and 2,600 m (5,000–8,500 ft). Above that, climatic conditions are sufficiently severe that only paramo or puna exists.

Zonation patterns are often complex. For example, in southern Peru, near Cuzco, I ascended to about 4,200 m (14,000 ft) and found wet puna, a heathland of orchids, heather, and sphagnum moss intermingled with paramo. Montane ecosystems, their ecology and natural history, are discussed in detail in chapter 9.

Major Neotropical Ecosystems

Hylaea—The Tropical Rainforest   Figures 13, 22, 32

Here no one who has any feeling of the magnificent and the sublime can be disappointed; the sombre shade, scarce illuminated by a single direct ray even of the tropical sun, the enormous size and height of the trees, most of which rise like huge columns a hundred feet or more without throwing out a single branch, the strange buttresses around the base of some, the spiny or furrowed stems of others, the curious and even extraordinary creepers and climbers which wind around them, hanging in long festoons from branch to branch, sometimes curling and twisting on the ground like great serpents, then mounting to the very tops of the trees, thence throwing down roots and fibres which hang waving in the air, or twisting round each other form ropes and cables of every variety of size and often of the most perfect regularity. These, and many other novel features—the parasitic plants growing on the trunks and branches, the wonderful variety of the foliage, the strange fruits and seeds that lie rotting on the ground—taken altogether surpass description, and produce feelings in the beholder of admiration and awe. It is here, too, that the rarest birds, the most lovely insects, and the most interesting mammals and reptiles are to be found. Here lurk the jaguar and the boaconstrictor, and here amid the densest shade the bell-bird tolls his peal.

So wrote Alfred Russel Wallace (1895), who spent four years exploring along the Rio Negro and Amazon and is credited, along with Charles Darwin, for proposing the theory of natural selection (chapter 4). Though rainforest impressed both Wallace and Darwin favorably, it has been depicted in art and literature in ways that range widely, from hauntingly idyllic to the infamous "green hell" image that typified the writings of authors such as Joseph Conrad (Putz and Holbrook 1988). What, exactly, is rainforest?

The Neotropical rainforest was first described by Alexander von Humboldt, who called it *hylaea*, the Greek word for "forest" (Richards 1952). The rainforest is what much of this book is about, so I will merely define it here and save the details for later.

A rainforest, in its purest form, is essentially a nonseasonal forest dominated by broad-leaved evergreen trees, sometimes of great stature, where rainfall is both abundant and constant. Rainforests are lush, with many kinds of vines and epiphytes (air plants) growing on the trees. In general, a rainforest receives at least 200 cm (just under 80 in) of rainfall annually, though it can be much more, with precipitation spread relatively evenly from month to month. Most of the tropics consist, however, of forests where *seasonal* variation in rainfall is both typical and important. Technically, a tropical forest with abundant but seasonal rainfall is called a *moist forest*: an evergreen or partly evergreen (some trees may be deciduous) forest receiving not less than 100 mm (4 in) of precipitation in any month for two out of three years, frost-free, and with an annual temperature of 24°C (75° F) or more (Myers 1980). Since the term *rainforest* is in such widespread and common usage, in this book I will continue to refer to lush, moist, tropical forests, seasonal or not, as rainforests. I've been in many, and, believe me, it rains a lot. Gets pretty muddy too.

The "Jungle"—Disturbed Forest Areas   *Figures 27, 36*

When rainforest is disturbed, such as by hurricane, lightning strike, isolated tree fall, or human activity, the disturbed area is opened, permitting the penetration of large amounts of light. Fast-growing plant species intolerant of shade are temporarily favored, and a tangle of thin-boled trees, shrubs, and vines results. Soon a huge, dense, irregular mass of greenery, or "jungle," covers the gap created by the disturbance. Trees are thin boled and very close together. Palms and bamboos may abound along with various vines, creating thick tangles. To penetrate a jungle requires the skilled use of that most important of all tropical tools, the machete. Jungles are *successional;* they will eventually return to shaded forest as slower-growing species outcompete colonizing species. What has been realized in recent years is that tropical forests are far more subject to natural disturbance than had been previously thought. Disturbance may, in fact, be responsible for many of the ecological patterns evident in tropical forests, including the high diversity of species. I will discuss disturbance patterns and ecological succession in the tropics in detail in chapter 3.

Tropical Riverine and Floodplain Ecosystems   *Figures 155, 156, 160, 162*

Two major river basins profoundly influence the ecology of South America: the Orinoco and the Amazon. These great rivers and their adjacent ecosystems form the subjects of chapter 8.

Forests that border rivers are termed *gallery forests*, and these forests are affected by often dramatic seasonal changes in riverine water level, which occur all along the Amazon, the Orinoco, and their various tributaries. The rainy season typically brings floods. Where rivers drain young mountain systems, such as the Andes, eroded mineral-rich soil from mountain areas is carried long distances, much of it eventually to be deposited along riverbank flood plains. In Amazonia, the term *varzea* is used for floodplain forests that line rivers rich in Andean sediment, and these forests make up only about 2% of the huge Amazon Basin area (Meggars 1988). The sediment-rich rivers tend to be cloudy from the sediment load and are called *whitewater rivers* (though "mocha" would seem a more apt term), as typified by much of the Amazon itself (especially the Solimoes or upper Amazon) as well as some of its major tributaries, such as the Madeira. Some rivers, such as the vast Rio Negro, drain geologically ancient soils that have undergone millions of years of erosion, becoming depleted of minerals. These waters carry almost no sediment, instead being clear but often dark, the so-called *blackwater rivers*. The dark coloration is caused by "humic matter," dissolved organics from vegetation decomposition (page 56). Forests along the floodplain of blackwater rivers are typically called *igapo*. Black- and whitewater rivers represent two opposites on a spectrum. There are also some rivers with low levels of sediment and intermediate concentrations of phenolics, organic compounds from decomposing leaves. These are termed "clearwater" rivers, as typified by the Rio Tapajos, the Rio Xingu, and the Rio Tocantins.

Of course, most forest (about 96–97%!) in Amazonia is found completely off the floodplain, and such forest is referred to as *terra firme*.

Only about 3–4% of the forest area in the Amazon Basin is floodplain. About half of these forests are *varzea* and receive rich sediment from the Andes during the time of flood, with a floodplain extending up to 80.5 km (50 miles) from the river bank. During wet season, the river depth may rise anywhere from 7.6 to 15 m (25–50 ft). Whole islands of vegetation are torn loose from the banks and drift downriver. Quiet pools may harbor groups of giant Victoria waterlilies (*Victoria amazonica*), a remarkable six-foot-wide lily pad with upright edges, the entire plant resembling a gargantuan green coaster.

Rivers and their banks support an exciting diversity of animals, including two species of freshwater dolphins, giant otter, capybara, anaconda, various alligator-like caimans, and many unique bird species. More than 2,400 species of fish, an astounding variety, inhabit the waters of the Amazon and its tributaries.

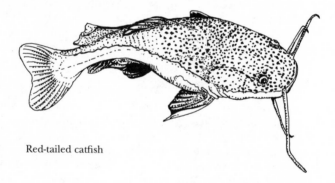

Red-tailed catfish

Savanna and Dry Forest

Figures 10, 11, 12

Part of the Neotropics consists of grasslands scattered with trees and shrubs, an ecosystem called a *savanna*. Savannas may be relatively wet, like the Florida Everglades, or dry and sandy. Neotropical seasonal savannas include the vast Llanos and Gran Sabana of southern Venezuela and the extensive Pantanal of southern Brazil and neighboring Bolivia, as well as much of the Chaco region of Paraguay. A combination of climatic and pronounced seasonal effects, occasional natural burning, and various soil characteristics produce savannas. Human influence also can contribute significantly to their formation. The African plains are an immense area of natural savanna, but savanna is considerably less extensive in the Neotropics, where rainforest dominates. Large expanses of savanna occur also in Central America. They are low-diversity

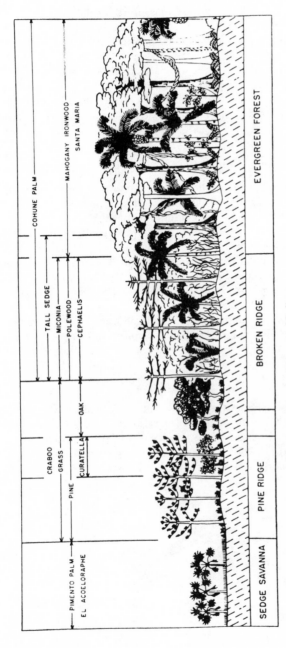

Diagram showing the range of ecosystem types present in parts of Belize.

eocsystems often numerically dominated by one tree species, Caribbean pine (*Pinus caribaea*).

Because they are open areas, savannas afford ideal habitats for seeing wildlife. Though Neotropical savannas lack the large game herds that characterize their African counterparts, there are numerous animal species that depend on savannas.

In addition to savannas, there are habitats of open woodlands, often with many decidous trees. These dry forests typically occur in areas where there is a pronounced dry season, and as such, dry forests often intermingle with savannas. Savanna and dry forest ecology and natural history are treated in chapter 10.

Coastal Ecosystems—Mangal and Seagrass   *Figures 14, 15, 17, 18, 19*

Mangroves are a group of unrelated but highly salt-tolerant tree species forming the dominant vegetation along tropical coastlines, lagoons, deltas, estuaries, and cayes. The ecological community they form is termed *mangal.* Tangled forests of mangroves, some with long prop roots, others with short "air roots" protruding up from the thick sandy mud, are the nesting sites of colonies of magnificent frigatebirds (man-o-war birds) (*Fregata magnificens*), boobies (*Sula spp.*), and brown pelicans (*Pelicanus occidentalis*). Mangroves have an essential role in the ecology of coastal areas and contribute to the health of nearby coral reefs.

Protected by the mangrove cayes, beds of seagrass cover shallow, well-lit coral sand. Like the mangroves, seagrass contributes to the health of the diverse coral reef.

Coral Reef   *Figures 20, 21*

To most visitors, the most exciting of all the coastal ecosystems is the coral reef. Approximately sixty species of coral occur in the Caribbean. Reefs of elkhorn, staghorn, finger, brain, and star corals provide habitats for myriad colorful fish, shrimps, lobsters, sea stars, brittle stars, and sea cucumbers. Mangrove, seagrass, and coral reef ecology is discussed in chapter 11.

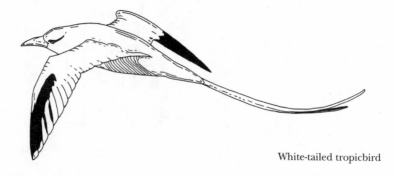

White-tailed tropicbird

2

Rainforest Structure and Diversity

TALL, green, complex-looking, and surprisingly dark inside. . . . Feeling just a little claustrophobic beneath the dense foliage above. . . . Strange bird sounds heard over a cacophony of calling insects. . . . A screeching parrot flock dashes overhead. . . . The tree crowns seem so far away, not easy to see what's in them. . . . Tree trunks propped up by tall, flaring roots. . . . No thick ground or shrub cover, actually rather easy to walk among the widely spaced trees. . . . Except worried about encountering snakes. . . . Lots of palms, some with thorny trunks. . . . Palm fronds russle in the slightest breeze. . . . Occasional openings in the canopy, sunny islands surrounded by a sea of deep shade. . . . Tangled growth in these sunny spots. . . . Lots of sunflecks. . . . Vines draped everywhere, some twisted, looking braided like rope, interconnecting trees. . . . Brilliant, almost neon-colored butterflies. . . . Lizards make thick leathery leaves crackle as they scamper over them. . . . Trails muddy, and mud sticks to boots. . . . Hot, even though shady; oppressively humid.

These are some typical notebook entries that might be made upon initial encounter with tropical rainforest. It doesn't matter whether you're standing in Peruvian, Brazilian, Ecuadorian, Belizean, Costa Rican, or Venezuelan rainforest, it all at first glance looks pretty much the same. It even sounds, smells, and feels generally the same. All over the equatorial regions of the planet where rainforest occurs, the forest tends to have a similar physical structure and appearance. Of course, on closer inspection, numerous differences exist among rainforests both within and among various geographical areas. On a global scale, one does not find orangutans or rattan palms in Venezuela, nor sloths or hummingbirds in Sumatra. And within the Neotropics, rainforests in Costa Rica are different in many significant ways from their counterparts in Brazil. And in Brazil, Amazonian forests show considerable differences from site to site, some sites hosting dense rainforest, some more open forest with palms, some open forest without palms, and some open forest with abundant lianas (Pires and Prance 1985). Rainforests on poor soils differ markedly from those on richer soils, just as rainforests on terra firma are distinct in some important ways from those on floodplains. However, the overall similarities, apparent as first impressions, are striking. Charles Darwin (1839) wrote of his initial impressions of tropical rainforest: "In tropical forests, when quietly walking along the shady pathways, and admiring each successive view, I wished to find language to express my ideas. Epithet after epithet was found too weak to convey to those who have not visited the intertropical regions the sensation of delight which the mind experiences."

Field Trip to a Peruvian Rainforest

Imagine we are standing at the edge of a tropical rainforest near Iquitos, Peru, along the westernmost part of the massive Amazon River in the very heart of equatorial South America. More kinds of plants and animals are to be found here than just about any other place on Earth. It's just after dawn, the hot sun has not yet risen high, and the air is so humid that the dampness makes it seem almost cool. Storm clouds are already gathering, but it's not yet raining. There is a well-marked trail leading us into the forest. We enter. It rained during the night and the trail is muddy and slippery.

Structural Complexity   *Figures 13, 22, 46*

Once inside a rainforest, structural complexity is obvious. How immense it seems, and how dark and enclosing as dense canopy foliage shades the forest interior, especially in the attenuated early morning light. Near a stream beside the forest edge, a pair of blue-and-yellow macaws (*Ara araraura*), their brilliant plumage muted, perch high on a moriche palm frond. With a pale sky overhead and shade inside the forest, highly colorful birds like these large macaw parrots often look subdued. Even at midday, when the sun is high overhead, only scattered flecks of sunlight dot the interior forest floor. Shade prevents a dense undergrowth from forming, and we certainly do not need our machete to move about. Plants we've seen only as potted houseplants grow here "in the wild." There's a clump of *Dieffenbachia* directly ahead on the forest floor. Large arum vines, philodendrons like *Monstera*, with its huge, sometimes deeply lobed leaves, are climbing up tree trunks. The biggest trees tend to be widely spaced, many with large, flaring buttressed roots, some with long, extended prop roots. All the trees are broad-leaved. There seem to be no equivalents of the needle-leaved trees of the temperate zone, the pines, spruces, and hemlocks. Palms abound, especially in the understory, and many have whorls of sharp spines around their trunks. Tree boles are straight and most rise a considerable height before spreading into crowns, which, themselves, are hard to discern clearly because so much other vegetation grows among them. Clumps of cacti, occasional orchids, many kinds of ferns, and an abundance of pineapple-like plants called bromeliads adorn the widely spreading branches. It's frustrating to try to see the delicate flowers of the orchids so high above us, but binoculars help. Vines, some nearly as thick as tree trunks back home, hang haphazardly, seemingly everywhere. Rounded, basketball-sized termite nests are easy to spot on the trees, and the dried tunnels made by their inhabitants vaguely suggest brown ski trails running along the tree trunks.

North American broadleaf forests are often neatly layered. There is a nearly uniform canopy, the height to which the tallest trees, such as the oaks and maples, grow, a subcanopy of understory trees such as sassafras and flowering dogwood, a shrub layer of viburnums or mountain laurel, and a herbaceous layer of ferns and wildflowers.

The tropical rainforest is not neatly layered (Richards 1952), and up to five poorly defined strata can be present (Klinge et al. 1975). The forest structure (called physiognomy) is complex (Hartshorn 1983a). Some trees, called emergents, erupt above the canopy to tower over the rest of the forest. Trees are of varying heights, including many palms, in both understory and canopy. Most trees are monotonously green, but a few may be bursting with colorful blossoms, while others may be essentially leafless, revealing the many epiphytes that have attached themselves to their main branches. Shrubs and other herbaceous plants share the heavily shaded forest floor with numerous seedling and sapling trees, ferns, and palms. It is difficult to perceive a simple pattern in the overall structure of a rainforest. Complexity is the rule.

Typical Tropical Trees

Figures 25, 34

A mild irony of nature in the tropics is that, though there are more different tree species than anywhere else (see below and chapter 3), many are sufficiently similar in appearance so that one can meaningfully describe a "typical tropical tree." Broad-leaved trees inside a rainforest tend, on first inspection, to look much alike, though an experienced observer can accurately identify many, if not most, at least to the level of family (and often genus). What follows is a general description of tropical tree characteristics, a description that will apply not only in the Neotropics but also to rainforests in tropical Africa and Southeast Asia.

STATURE

Figures 33, 34, 35

Tropical rainforests have a reputation for having huge trees. Old engravings depict trees of stunning size with up to a dozen people holding hands around the circumference of the trunk. No pun intended (well, truth be told, it is intended), but such accounts generally represent "tall tales." Tropical trees can, indeed, be both wide and lofty, but bear in mind that many look taller than they really are because their boles are slender (just as a thin person gives the appearance of being taller than a stocky person of equal height), and branches tend not to radiate from the trunk until canopy level, thus enhancing the tall look of the bole. The tallest tropical trees are found in lowland rainforests, and these range in height between 25 and 45 meters (roughly 80–150 ft), the majority around 25–30 m. Tropical trees occasionally exceed heights of 45 m (150 ft), and some emergents do top 61 m (200 ft) and may occasionally approach 90 m (300 ft), though such heights are uncommon. I have been in quite a few temperate-zone forests with equally tall or taller trees. In the United States, Sierra Nevada giant sequoia groves, coastal California redwood groves, and Pacific Northwest old-growth forests of sitka spruce, common Douglas fir, western red cedar, and western hemlock all routinely exceed the height of the trees comprising the majority of tropical forests. So do the temperate bluegum eucalyptus forests in southeastern Australia. Neither the

Profile diagram of primary mixed forest, Moraballi Creek, Guyana. The diagram represents a strip of forest 41 m (135 ft) long and 7.6 m (25 ft) wide. Only trees over 4.6 m (15 ft) high are shown. From Richards (1952). Reproduced with permission.

tallest, the broadest, nor the oldest trees on Earth occur in rainforest: the tallest is a California redwood, at 112 m (367 ft); the broadest is a Montezuma cypress in subtropical Mexico, with a circumference of almost 49 m (160 ft); and the oldest is a bristlecone pine in the White Mountains of eastern California, about 4,600 years old.

BUTTRESSES AND PROP ROOTS   *Figures 57, 58, 59, 60*

A buttress is a root flaring out from the trunk to form a flangelike base. Many, if not most, rainforest trees have buttressed roots, giving tropical rainforests a distinctive look in comparison with temperate forests (though old-growth temperate rainforest trees such as are found in the Pacific Northwest are sometimes weakly buttressed). Several buttresses radiate from a given tree, surrounding and seeming to support the bole, often making cozy retreats for snakes. Buttress shape is sometimes helpful in identifying specific trees. Buttresses can be large, often radiating from the bole six or more feet from the ground.

The function of buttressing has been a topic of active discussion among tropical botanists. Because buttressing is particularly common among trees of stream and river banks as well as among trees lacking a deep taproot, many believe that buttressing acts principally to support the tree (Richards 1952; Longman and Jenik 1974). I was once told of a team of botanists in Costa Rica who were discussing several esoteric theories for the existence of buttresses when their local guide offered the comment that buttresses hold up the tree. When the guide's opinion was dismissed lightly, he produced his machete and adeptly cut away each of the buttresses from a nearby small tree. He then casually pushed the tree over. Whether it is true or false that buttresses function principally for support, they may indeed serve other functions related to root growth patterns (page 54). Some trees lack buttresses but have stilt or prop roots that radiate from the tree's base, remaining above ground. Stilt roots are particularly common in areas such as floodplains and mangrove forests (page 240) that become periodically innundated with water. Some tropical trees, including the huge Brazilnut tree (*Bertholletia excelsa*), lack both buttresses and prop roots and have instead either horizontal surface roots or deeper underground roots. In a few cases, large taproots occur.

TRUNKS AND CROWNS   *Figures 25, 26, 31*

As we look around the Peruvian forest, we notice that many trees have tall, slender boles. The bark may be smooth or rough, light colored or dark, almost white in some cases, almost ebony in others. Bark is often splotchy, with pale and dark patches. There is much variability. Tropical tree bark may be thin, but on some trees it can be thick (and the wood inside may be very hard— remember that wood-eating termites abound in the tropics). Bark is not usually a good means of identifying the tree, as many different species may have similar-appearing bark. Some trees, however, such as the chicle tree (*Manilkara zapota*) of Central America (the original source of the latex base from which chewing gum is manufactured), have distinct bark. Chicle bark is black and vertically ridged into narrow strips, the inner bark red, with white resin. The color and taste of the underlying cambium layer is sometimes a good key to identifying the tree species (Richards 1952; Gentry 1993).

Many canopy trees have a spreading, flattened crown (Richards 1952). Main branches radiate out from one or a few points, somewhat resembling the spokes of an umbrella. Each of these main radiating branches contributes to the overall symmetrical crown, an architectural pattern called *sympodial* construction. Of course, the effect of crowding by neighboring trees can significantly modify crown shape. Single trees left standing after adjacent trees have been felled often have oddly shaped crowns, a result of earlier competition for light with neighboring trees. Many trees that grow both in the canopy and in the shaded understory have foliage that is *monolayered*, where a single, dense blanket of leaves covers the tree. Trees in the understory are often lollipop-shaped and monolayered. Because they have not yet reached the canopy, their crowns are composed of lateral branches from a single main trunk, a growth pattern termed *monopodial*. Lower branches will eventually drop off through self-shading as the tree grows and becomes a sympodial canopy tree. Trees growing in forest gaps where sunlight is much more abundant (see below and chapter 3) are *multilayered*, with many layers of leaves to intercept light (Horn 1971; Hartshorn 1980, 1983a). The architecture of tropical trees is discussed further in Halle et al. (1978).

Many tropical trees, not just in the Neotropics but globally, exhibit a unique characteristic termed *cauliflory*, meaning the flowers and subsequent fruits abruptly grow from the wooded trunk, rather than from the canopy branches. Cauliflory generally does not occur outside of the tropics. Cocoa (*Theobroma cacao*), from which chocolate is produced, is a cauliflorous understory tree (page 185). Some trees may be cauliflorous due to the large, heavy fruits that are produced, the weight of which could not be supported on outer branches (though it is equally arguable that the opposite may be the case—the fruits may have grown large and heavy *because* they were growing from the trunk, not the outer branches). The presence of cauliflorous flowers may facilitate pollination by large animals such as bats, or, equally likely, cauliflorous fruiting may facilitate dispersal of seeds from fruit consumption by large, terrestrial animals that could not reach canopy fruits. A similar phenomenon, ramiflory, is the bearing of flowers on older branches or occasionally underground.

LEAVES

Figure 37

Leaves of many tropical tree species are surprisingly similar in shape, making species identification difficult (but see below, *Identifying Neotropical Plants*). The distinctive lobing patterns of many North American maples and oaks are missing from most tropical trees. Instead, leaves are characteristically oval and unlobed, and they often possess sharply pointed ends, called *drip tips*, which help facilitate rapid runoff of rainwater (page 49). Leaves of most species have smooth margins rather than "teeth," though serrated leaves are found in some species. Both lowland and montane tropical forest trees produce heavy, thick, leathery, waxy leaves that can remain on the tree for well over a year. Many tropical species produce palmate leaves, where the leaflets radiate like spokes from a center, forming a shape similar to that of a parasol. Some leaves, particularly those on plants that are found in disturbed areas such as gaps, are very large, well in excess of temperate zone species. Though many trees have simple leaves, compound leaves are by no means uncommon, particularly due to the abundance of legumes, a highly species-rich plant family (page 70). Tropical leaves, with some exceptions, tend to show little obvious insect damage (see chapter 6).

FLOWERS

Many tropical trees have colorful, fragrant blossoms, often large in size. Typical examples include such species as coral tree (*Erythrina* spp.), pink poui (*Tabebuia pentaphylla*), cannonball tree (*Couroupita guanianensis*), frangipani (*Plumeria* spp.), and morning glory tree (*Ipomoena arborescens*). Many striking trees that are abundantly represented in the Neotropics are actually imported from other tropical regions. For instance, the gorgeous and widespread flamboyant tree (*Delonix regia*), the national tree of Puerto Rico, is actually native to Madagascar. The bottle brush tree (*Callistemom lanceolatus*) is from Australia, and the Norfolk Island pine (*Araucaria excelsa*) is from, well, you guessed it. (In case you didn't, it's from Norfolk Island in the southern Pacific Ocean.)

Red, orange, and yellow are associated with bird-pollinated plants such as *Heliconia*, while lavender flowers such as *Jacaranda* are more commonly insect-pollinated. Some trees, such as silk-cotton or kapok (*Ceiba pentandra*), flower mostly at night, producing conspicuous white flowers that, depending on species, attract bats or moths. Fragrant flowers are mostly pollinated by moths, bees, beetles, or other insects. Bat-pollinated flowers smell musty, kind of like the bats themselves (page 129). Because of the high incidence of animal pollination, especially by large animals such as birds, bats, and large lepidopterans, flowers tend not only to be large but also to be nectar-rich and borne on long stalks or branches away from leaves, or else on the trunk (cauliflory, above). Many flowers are tubular or brushlike in shape, though some, particularly those pollinated by small insects, are shaped as flattened bowls or plates. Though animal pollination is fairly general, wind pollination occurs in some species of canopy trees.

FRUITS AND SEEDS

Many tropical trees produce small to medium-sized fruits, but some produce large, conspicuous fruits and the seeds contained within are large as well. Indeed, another distinctive characteristic of tropical forests is the abundance of trees that make large fruits. Many palms, the coconut (*Cocos nucifera*) for example, produce large, hard fruits in which the seeds are encased. The monkey pot tree (*Lecythis costaricensis*) produces thick, 20-cm (8-inch)-diameter "cannon ball" fruits, each containing up to fifty elongated seeds. The seeds are reported to contain toxic quantities of the element selenium (Kerdal-Vargas in Hartshorn 1983b), perhaps serving to protect the tree from seed predators (see below). The milk tree (*Brosimum utile*) forms succulent, sweet-tasting, edible fruits, each with a single large seed inside. This tree, named for its white sap (which is drinkable), may have been planted extensively at places like Tikal by Mayan Indians (Flannery 1982 and page 183). The famous Brazil nut comes from the forest giant *Bertholletia excelsa*. The nuts are contained in large, woody, rounded pods that break open upon dropping to the forest floor. Many tree species in the huge legume family package seeds in long, flattened pods, and the seeds tend to contain toxic amino acids (page 147). Among the legumes, the stinking toe tree (*Hymenaea courbaril*) produces 12.7-cm (5-in) oval pods with five large seeds inside. The pods drop whole to the forest floor and often fall prey to agoutis and other forest mammals as well as various weevils.

Large fruits with large seeds are a major food source for the large animals of the forest. Among the mammals, monkeys, bats, various rodents, peccaries, and tapirs are common consumers of fruits and seeds, sometimes dispersing the seeds, sometimes destroying them. Agoutis, which are rodents, skillfully use their sharp incisors to gnaw away the tough, protective seed coat on the Brazil nut, thus enabling the animal to eat the seed contained within. Some extinct mammals, such as the giant ground sloths and bovinelike gomphotheres, may have been important in dispersing large seeds of various tropical plants. Birds such as tinamous, guans, curassows, doves and pigeons, trogons, toucans, and parrots are also attracted to large fruits and the seeds within them. Along flooded forests, some fish species are important fruit consumers and seed dispersers (page 204). Insects especially are frequent predators of small seeds.

Some trees have wind-dispersed seeds and thus the fruits are usually not consumed by animals. The huge silk cotton or kapok tree is so-named because its seeds are dispersed by parachute-like, silky fibers that give the tree one of its common names. Mahogany trees (*Swietenia macrophylla* and *S. humilis*), famous for their superb wood, develop 15-cm (6-in) oval, woody fruits, each containing about forty seeds. The seeds are wind-dispersed and would be vulnerable to predation were it not for the fact that they have an extremely pungent, irritating taste.

Palms

Figures 16, 29

Palms, which occur worldwide, are among the most distinctive Neotropical plants, frequenting interior rainforest, disturbed areas, and grassy savannas. They are particularly abundant components of swamp and riverine forest. There are 1,500 species of palms in the world and 550 in the Americas (Henderson et al. 1995). Alfred Russel Wallace (1853) made a detailed study of South American palms and published an important book on the subject. All palms are members of the family Palmae, and all are monocots, sharing characteristics of such plants as grasses, arums, lilies, and orchids. The most obvious monocot feature of palms is the parallel veins evident in the large leaves, which themselves are referred to as palm fronds. Palms are widely used by indigenous peoples of Amazonia for diverse purposes: thatch for houses, wood to support dwellings, ropes, strings, weavings, hunting bows, fishing line, hooks, utensils, musical instruments, and various kinds of food and drink. Indeed, many palm species have multiple uses and are thus among the most important plant species for humans.

Palms are often abundant in the forest understory and are frequently armed with sharp spines along the trunks and leaves. Be especially careful not to grab a palm sapling as the spines can create a wound and introduce bacteria.

Identifying Neotropical Plants

Figure 41

Palms are fairly easy to identify, at least to the level of genus, but what about all those other trees and shrubs in the rainforest? The bad news is that for the vast majority of students of Neotropical biology, it will not be possible to identify accurately most plants (including palms) to the level of species. There are just too many look-alike species, and the ranges of many species are not precisely known; thus species identification must be left to taxonomic experts. Also, there are essentially no field guides to Neotropical plants, at least not at the level of species. Lotschert and Beese (1981) is a useful but very general guide to many of the most widespread and conspicuous tropical plants, and Henderson et al. (1995) is a complete guide to palms of the Americas. Gentry (1993) is currently the most useful guide to Neotropical woody plants, but, though 895 pages in length and weighing in at about three pounds (softcover), it includes only the countries of Colombia, Ecuador, and Peru and deals with identifications only at the level of family and genera (a smaller-format version is now available). Croat (1978) is a large volume (943 pages) on the flora of Barro Colorado Island in Panama. Hopefully, as the Neotropics become better known, more guides to plants will be published for various regions.

The good news is that it is indeed possible to identify many, if not most,

Neotropical plants to the level of family, and many of those to the level of genus (Gentry 1993). Using combinations of characteristics such as leaf shape (palmate, pinnate, bipinnate), compound versus simple leaves, opposite versus alternate leaves, presence or absence of tendrils, presence or absence of spines, smooth or serrate leaf edges, fruit and/or flower characteristics, and even, in many cases, odor and taste, you can, with a guide such as Gentry's, master the flora, no mean feat since Gentry describes 182 flowering plant families.

Climbers, Lianas, Stranglers, and Epiphytes

As we continue our perambulations through the Peruvian rainforest we cannot help but notice the plethora of vines and epiphytes. Trees are so laden with these hitchhikers that it is often a challenge to discern the actual crown from the myriad ancillary plants. With binoculars and practice, however, we can begin to make some sense of what is growing where and on what. In this lowland Peruvian forest epiphytes are abundant, but there is much variability from one forest site to another. Generally epiphytes and vines are most abundant where humidity is highest, declining in frequency in forests that experience a strong dry season.

VINES   *Figures 22, 23, 50, 51, 52, 58*

Vines are a conspicuous and important component of most tropical rainforests (though vine density is often quite variable from site to site), and they come in various forms. Vines are a distinct and important structural feature of rainforests, in a sense literally tying the forest together. They account for much of the biomass in some rainforests, they compete with trees for light, water, and nutrients, and many are essential foods for various animals. In the Neotropics, 133 plant families include at least some climbing species. Some, called *lianas*, entwine elaborately as they dangle from tree crowns. Others, the bole climbers, attach tightly to the tree trunk and ascend. Still others, the stranglers, encircle a tree and may eventually choke it. All told, there are nearly 600 species of climbers in the Neotropics (Gentry 1991). Tropical vines occur abundantly in disturbed sunlit areas as well as in forest interiors, at varying densities and on virtually all soil types. Humans make extensive use of vines for foods, medicines, hallucinogens, poisons, and construction materials (Phillips 1991). For a comprehensive account of vine biology, see Putz and Mooney (1991).

A liana usually gets its start when a forest opening called a gap is created (page 33), permitting abundant light penetration. Lianas typically begin life as shrubs rooted in the ground but eventually become vines, with woody stems as thick or thicker than the trunks of many temperate zone trees. Tendrils from the branches entwine neighboring trees, climbing upward, reaching the tree crown as both tree and liana grow. Lianas spread in the crown, and a single liana may eventually loop through several tree crowns. Lianas seem to drape limply, winding through tree crowns or hanging as loose ropes parallel to the main bole. Their stems remain rooted in the ground and are oddly shaped, often being flattened, lobed, coiled like a rope, or spiraled in a helixlike shape.

The thinnest have remarkable springiness and often will support a person's weight, at least for a short time. Some liana stems are hollow, containing potable water, attainable through the use of a machete.

Liana is a growth form, not a family of plants, and thus lianas are represented among many different plant families (Leguminosae, Sapindaceae, Cucurbitaceae, Vitaceae, Smilacaceae, and Polygonaceae, to name several). Lianas, like tropical trees, can be very difficult to identify, but some lianas can be identified to the level of genera by noting their distinctive cross-sectional shapes (Gentry 1993).

In Panama, a single hectare (10,000 square meters, or about 2.5 acres) hosted 1,597 climbing lianas, distributed among 43% of the canopy trees (Putz 1984). In the understory, 22% of the upright plants were lianas, and lianas were particularly common in forest gaps. A heavy liana burden reduced the survival rate of trees, making them more likely to be toppled by winds. Fallen lianas merely grew back onto other trees.

Other vines, such as the well-known ornamental arum *Monstera deliciosa* or various philodendrons, are bole climbers. They begin life on the ground. Their seed germinates and sends out a tendril toward shade cast by a nearby tree. The tendril soon grows up the tree trunk, attaching by aerial roots, and the vine thus moves from the forest floor to become anchored on a tree. There it continues to grow ever upward, often encircling the bole as it proliferates. In humid tropical forests it is quite common to see boles totally enshrouded by the wide, thick leaves of climbers. As it grows the plant ceases to be rooted in the ground and becomes a climbing epiphyte (technically referred to as a hemi-epiphyte), its entire root system invested on the tree bark.

The most aggressive vines are the stranglers (*Ficus* spp.). There are approximately 150 species of *Ficus* (figs) in the Neotropics, and an additional 600 or so in the Old World tropics. In the Neotropics, most species are strangers, begining as a seed dropped by a bird or monkey in the tree crown among the epiphytes. Tendrils grow toward the tree bole and downward around the bole, anastomosing or fusing together like a crude mesh. The strangler eventually touches ground and sends out its own root system. The host tree often dies and decomposes, leaving the strangler standing alone. The mortality of the host tree may be caused by constriction from the vine or the shading effect of the vine. It is a common sight in Neotropical forests to see a mature strangler, its host tree having died and decomposed. The strangler's trunk is now a dense fusion of what were once separate vines, now making a single, strong, woody labyrinth that successfully supports a wide canopy, itself now laden with vines.

Vines of many kinds frequent disturbed areas where light is abundant. Members of the family Passifloraceae, some 400 species of passion-flowers (page 155), most of which are native to the Neotropics, are among the most conspicuous vines in the tangles that characterize open areas and roadsides.

EPIPHYTES   *Figures 38, 39, 55, 56*

As the prefix *epi* implies, epiphytes (air plants) live *on* other plants. They are not internally parasitic, but they do claim space on a branch where they set out roots, trap soil and dust particles, and photosynthesize as canopy residents. Rainforests, both in the temperate zone (such as the Olympic rainforests of

Washington and Oregon) and in the tropics, abound with epiphytes of many different kinds. Cloud forests also host an abundance of air plants. In a lowland tropical rainforest nearly one quarter of the plant species are likely to be epiphytes (Richards 1952; Klinge et al. 1975), though the representation of epiphytes varies. As forests become drier, epiphytes decline radically in both abundance and diversity.

Many different kinds of plants grow epiphytically. In Central and South America alone, there are estimated to be 15,500 epiphyte species (Perry 1984). Looking at a single tropical tree can reveal an amazing diversity. Lichens, liverworts, and mosses, many of them tiny (see below), grow abundantly on trunk and branches, and often leaves. Cacti, ferns, and colorful orchids line the branches. Also abundant and conspicuous on both trunk and branch alike are the bromeliads, with their sharply pointed, daggerlike leaves. The density of epiphytes on a single branch is often high. I witnessed this under somewhat alarming circumstances when, following a heavy downpour in Belize, a tree limb fell from onto my (fortunately for me) unoccupied tent. Though the tent was ruined, I at least (sort of) enjoyed seeing the many delicate ferns and orchids growing among the dense mosses and lichens that completely covered the upper surface of the branch that could have killed me.

Epiphytes attach firmly to a branch and survive by trapping soil particles blown to the canopy and using the captured soil as a source of nutrients such as phosphorus, calcium, and potassium. As epiphytes develop root systems they accumulate organic matter, and thus a soil-organic litter base, termed an epiphyte mat, builds up on the tree branch. Many epiphytes have root systems containing fungi called mycorrhizae. These fungi greatly aid in the uptake of scarce minerals (see below). Mycorrhizae are also of major importance to many trees, especially in areas with poor soil (page 50). Epiphytes efficiently take up water and thrive in areas of heavy cloud cover and mist.

Though epiphytes do not directly harm the trees on which they reside, they may indirectly affect them through competition for water and minerals. Epiphytes get first crack at the water dripping down through the canopy. However, some temperate and tropical canopy trees develop aerial roots that grow into the soil mat accumulated by the epiphytes, tapping into that source of nutrients and water. Because of the epiphyte presence, the host tree benefits by obtaining nutrients from its own canopy (Nadkarni 1981). Perry (1978) suggests that monkeys traveling regular routes through the canopy may aid in keeping branches from being overburdened by epiphytes.

Bromeliads are abundant epiphytes in virtually all Neotropical moist forests. Leaves of many species are arranged in an overlapping rosette to form a cistern that holds water and detrital material. Some species have a dense covering of hairlike trichomes on the leaves that help to absorb water and minerals rapidly. The approximately 2,000 New World bromeliad species are members of the pineapple family, Bromeliaceae, and, like orchids (below), not all grow as epiphytes. There are many areas where terrestrial bromeliads make up a significant portion of the ground vegetation. Epiphytic bromeliads provide a source of moisture for many canopy dwellers. Tree frogs, mosquitos, flatworms, snails, salamanders, and even crabs complete their life cycles in the tiny aquatic habitats provided by the cuplike interiors of bromeliads (Zahl 1975;

Wilson 1991). One classic study found 250 animal species occurring in brome-liads (Picado 1913, cited in Utley and Burt-Utley 1983). Some species of small, colorful birds called euphonias (page 264) use bromeliads as nest sites. Bro-meliad flowers grow on a central spike and are usually bright red, attracting many kinds of hummingbirds (page 260).

Orchids are a global family (Orchidaceaé) abundantly represented among Neotropical epiphytes (Dressler 1981). There are estimated to be approxi-mately 25,000–35,000 species worldwide (World Conservation Monitoring Centre 1992), a huge plant family indeed. In Costa Rica, approximately 88% of the orchid species are epiphytes, while the rest are terrestrial (Walterm 1983). Many orchids grow as vines, and many have bulbous stems (called pseu-dobulbs) that store water. Indeed, the origin of the name "orchid" is the Greek word meaning "testicle," a reference to the appearance of the bulbs (Plotkin 1993). Some have succulent leaves filled with spongy tissue and covered by a waxy cuticle to reduce evaporative water loss. All orchids depend on mycor-rhizae during some phase of their life cycles. These fungi grow partly within the orchid root and facilitate uptake of water and minerals. The fungi survive by ingesting some of the orchid photosynthate; thus, the association between orchid and fungus is mutualistic: both benefit. A close look at some orchids will reveal two types of roots: those growing on the substrate and those that form a basket, up and away from the plant. Basket roots aid in trapping leaf litter and other organic material that, when decomposed, can be used as a mineral source by the plant (Walterm 1983). Orchid flowers are among the most beautiful in the plant world. Some, like the familiar *Cattleya*, are large, while others are delicate and tiny. (Binoculars help the would-be orchid ob-server in the rainforest.) Cross-pollination is accomplished by insects, some quite specific for certain orchid species. Bees are primary pollinators of Neo-tropical orchids. These include long-distance fliers, like the euglossine bees that cross-pollinate orchids separated by substantial distances (Dressler 1968). Some orchid blossoms apparently mimic insects, facilitating visitation by in-sects intending (mistakenly) to copulate with the blossom (Darwin 1862). Aside from their value as ornamentals, one orchid genus is of particular impor-tance to humans. There are 90 orchid species in the genus *Vanilla*, of which two are of economic importance, their use dating back to the Aztecs (Plotkin 1993). Dressler (1993) provides a field guide to orchids of Costa Rica and Panama.

In many tropical moist forests, even the epiphytes can have epiphytes. Trop-ical leaves often are colonized by tiny lichens, mosses, and liverworts, which grow only after the leaf has been tenanted by a diverse community of microbes: bacteria, fungi, algae, and various yeasts, as well as microbial animals such as slime molds, amoebas, and ciliates. This tiny community that lives upon leaves is termed the *epiphyllus* community (Jacobs 1988), and its existence adds yet another dimension to the vast species richness of tropical moist forests. Epiphylls also grow liberally on moist wood, including the spines on trunks of many understory palms and other tree species. This is a good reason to use disinfectant promptly if you are scratched by tropical thorns, as they may have innoculated you with bacteria that could result in an infection (see appendix).

The Understory and Forest Gaps *Figures 24, 27, 36*

Much of the understory of a tropical forest will be so deprived of light that plant growth is limited. Low light intensity is a chronic feature of rainforest interior and is an important potential limiting factor for plant growth. This is why it is fairly easy to traverse a closed-canopy rainforest. Many of the seedlings and shoots that surround you are those of trees that may or may not eventually attain full canopy status, and a small, unpretentious sapling could be well over twenty years old.

Certain families of shrubs frequently dominate rainforest understory. These include members of the family Melasomataceae (e.g., *Miconia*), the Rubiaceae (e.g., *Psychotria*), and the Piperaceae (e.g., *Piper*). In addition there are often forest interior Heliconias (page 69) and terrestrial bromeliads. Many ferns and fern allies, including the ancient genus *Selaginella*, can carpet much of the forest herb layer.

The understory is frequently far from uniform. The deep shade is interrupted by areas of greater light intensity and denser plant growth. The careful observer inevitably notices the presence of many forest gaps of varying sizes, openings created by fallen trees or parts thereof (like the tree branch that fell on my tent in Belize). Gaps permit greater amounts of light to reach the forest interior, providing enhanced growing conditions for many species. Though understory plants and juvenile trees are adapted to grow very slowly (Bawa and McDade 1994), many are also adapted to respond with quickened growth in the presence of a newly created gap. Recent research at La Selva has revealed a surprisingly high disturbance frequency caused by treefalls and branchfalls, where estimates are that the average square meter of forest floor lies within a gap every hundred years or so (Bawa and McDade 1994). As described by Deborah Clark (1994),

> The primary forest at La Selva is a scene of constant change. Trees and large branches are falling to the ground, opening up new gaps and smashing smaller plants in the process. Smaller branches, bromeliads, and other epiphytes, 6-m-long palm fronds, smaller leaves, and fruits fall constantly as well. The lifetime risk of suffering physical damage is, therefore, high for plants at La Selva.

Gap dynamics has become an important consideration in the study of plant demographics in the rainforest (see chapter 3).

High Species Richness *Figures 32, 116*

Looking around inside the Peruvian rainforest, we cannot help but wonder just how many things we are looking at and, for that matter, how many are looking back at us. Both animal and plant life are abundant and diverse. The terms *species richness* and *biodiversity* refer to how many different species of any given taxon inhabit a specified area; thus we speak of such things as the species richness of flowering plants in Amazonia, or ferns in Costa Rican montane forests, or birds in Belize, or mammals in Rio Negro igapo forest, or beetles in the canopy of a single ceiba tree, or whatever. High species richness among many different taxons is one of the most distinctive features of tropical forests worldwide and Neotropical lowland forests in particular. In a temperate zone forest it is often possible to count the number of tree species on the fingers of both hands (though a toe or two may be needed). Even in the most diverse North American forests, those of the lush southeastern Appalachian coves, only about 30 species of trees occur in a hectare (10,000 square meters, or about 2.5 acres). In the tropics, however, anywhere from 40 to 100 or more species of trees can occur per hectare. Indeed, one site in the Peruvian Amazon has been found to contain approximately 300 tree species per hectare (Gentry 1988). Brazil alone has been estimated to contain around 55,000 flowering plant species (World Conservation Monitoring Centre 1992). Altogether, about 85,000 species of flowering plants are estimated to occur in the Neotropics (Gentry 1982). This is roughly double the richness of tropical and subtropical Africa, about 1.7 times that of tropical and subtropical Asia, and 5 times that of North America.

British naturalist Alfred Russel Wallace (1895) commented upon the difficulty of finding two of the same species of tree nearby each other. He stated of tropical trees:

> If the traveller notices a particular species and wishes to find more like it, he may often turn his eyes in vain in every direction. Trees of varied forms, dimensions and colour are around him, but he rarely sees any one of them repeated. Time after time he goes towards a tree which looks like the one he seeks, but a closer examination proves it to be distinct.

As Wallace implies, though richness is high, the number of individuals within a single species often tends to be low, which is another way of saying that rarity is usual among many species in the lowland tropics. Though some plant species are abundant and widespread (for example, kapok tree), the majority are not, existing in small numbers over extensive areas. The concept of identifying a forest type by its dominant species, which works well in the temperate zone (i.e., eastern white pine forest, redwood forest), is much less useful in the tropics, though not always. On the island of Trinidad one can visit a *Mora* forest where the canopy consists almost exclusively of but a single species, *Mora excelsa*, a tree that can reach the height of 46 m (150 ft). The understory is also dominated by *Mora* saplings, but examples of such low-diversity forests are extremely rare in the Neotropics. At La Selva Biological Station in Costa Rica, one leguminous tree, *Pentaclethra macroloba*, is disproportionately abundant compared with all other species (Hartshorn and Hammel 1994); nonetheless, many other species are present. Among animal taxa, high species richness and rarity also tend to correlate, especially at lowest latitudes (page 86).

Within the Neotropics, species richness, though generally high, shows clear variability. Knight (1975), working on Barro Colorado Island in Panama, found an average of 57 tree species per 1,000 square meters (10,764 sq ft) in mature forest and 58 species in young forest. Knight found that in the older forest, when he counted 500 trees randomly, he encountered an average of 151 species. In the younger forest, he encountered an average of 115 species in a survey of 500 trees. Hubbell and Foster (1986b) have established a 50-

hectare (500,000 sq m) permanent study plot in old-growth forest at BCI. They surveyed approximately 238,000 woody plants with stem diameter of 1 cm (2.5 in) diameter breast height (dbh) or more and found 303 species. They classified 58 species as shrubs, 60 as understory treelets, 71 as midstory trees, and 114 as canopy and emergent trees. Gentry (1988), working in upper Amazonia and Choco, found between 155 and 283 species of trees greater than 10 cm (25.4 in) dbh in a single hectare. When he included lianas of greater than 10 cm dbh, he found that the total increased to between 165 and 300 species. Prance et al. (1976) found 179 species greater than 15 cm (38.1 in) dbh in a 1-ha plot near Manaus, Brazil, on a terra firme forest characterized by poor soil and a very strong dry season.

If all vascular flora are taken together (trees, shrubs, herbs, epiphytes, lianas, but excluding introduced weedy species), the inventory for BCI is 1,320 species from 118 families (Foster and Hubbell 1990; Gentry 1990b). By comparison, the total number of vascular plant species documented at La Selva Biological Station in nearby Costa Rica is 1,668 species from 121 families (Hammel 1990; Gentry 1990b). Let's compare these totals with those from Amazonian rainforests. A floodplain forest on rich soils at Cocha Cashu Biological Station along the Rio Manu, a whitewater tributary of the vast Rio Madeira in southeastern Peru, was found to contain 1,856 species (in 751 genera and 130 families) of higher plants (Foster 1990a). At Reserva Ducke, a forest reserve on poor soils near Manaus in central Amazonia, 825 species of vascular plants from 88 families were inventoried (Prance 1990a; Gentry 1990b).

When the two Central American sites described above (BCI and La Selva) are compared with the two Amazonian sites (Cocha Cashu and Reserva Ducke), there are several important differences. Tree species richness is far greater in Amazonia (Gentry 1986a, 1988, 1990a), but the richness of epiphytes, herbs, and shrubs is greater in Central America. At La Selva, 23% of all vascular plant species are epiphytes, the highest percentage recorded among the closely studied sites (Hartshorn and Hammel 1994). The most species-rich of any of the four sites is Cocha Cashu, located on fertile varzea soils in western Amazonia. A total of 29 plant families that are present at BCI, La Selva, and Cocha Cashu are absent from Reserva Ducke, presumably because of the poor soil conditions at that site. However, the similarities among these four geographically separated forest sites are perhaps more compelling than the differences. The dozen well best-represented plant families are essentially the same for each of the sites. Legumes (Leguminosae), for instance, are the most species-rich family at BCI, Cocha Cashu, and Ducke, and the fifth richest family at La Selva. Of the 153 vascular plant families represented in at least one of the four sites, 66 (43%) are represented at all four sites, a high overlap (Gentry 1990a).

Plants are not the only diverse groups. Insects, birds, amphibians, and most other major groups also exhibit high species richness. A guide to birds of Colombia lists 1,695 migrant and resident species occurring in that country (Hilty and Brown 1986). At Cocha Cashu Biological Station in Amazonian Peru, in an area of approximately 50 square km (19 sq mi), the total species list of birds is approximately 550 (Robinson and Terborgh 1990). At La Selva Biological Station in Costa Rica, an area of approximately 1,500 ha (3,705 acres), 410 species of birds have been found (Blake et al. 1990). In Amazonia, at the Explorer's Inn Reserve in southern Peru, about 575 bird species have been identified within an area of approximately 5,500 ha (13,585 acres) (Foster et al. 1994). By comparison, barely 700 bird species occur in all of North America. More species of birds exist in the Neotropics than in the temperate zone largely because of the unique characteristics of the rainforest (Tramer 1974, and page 95). Bird species richness drops dramatically as soon as you leave the rainforest.

At one site in the Ecuadorian Amazon, the species richness of frogs is 81, which is exactly how many species occur in all of the United States. Indeed, the researcher collected 56 different species on a single night of sampling and reports that it is routine to find 40 or more species in areas of rainforest as small as two square kilometers (Duellman 1992).

Assassin bug

Insect species richness can seem staggering. For the small Central American country of Costa Rica, Philip DeVries (1987) describes nearly 550 butterfly species. At La Selva alone, 204 butterfly species have been identified, and 136 species have been documented for BCI (DeVries 1994). At Explorer's Inn Reserve, 1,234 butterfly species have been identified from an area about 2.0 square km within the reserve (Foster et al. 1994). Edward O. Wilson (1987) reported collecting 40 genera and 135 species of ants from four forest types at Tambopata Reserve in the Peruvian Amazon. Wilson noted that 43 species of ants were found in one tree, a total approximately equal to all ant species occurring in the British Isles! Terry Erwin studied the insect species richness of Neotropical rainforest canopies (page 41) and found 163 beetle species occurring exclusively in but one Panamanian tree species, *Leuhea seemannii.* Erwin then multiplied this figure by the number of different tree species present in the global tropics and concluded that the potential species richness of beetles alone was over 8 million! Since beetles are estimated to represent approximately 40% of all tropical terrestrial arthropod species (including spiders, crustaceans, centipedes, millipedes, and insects), Erwin suggested that the total arthropod species richness of the tropical canopy might be as high as 20 million, and that figure climbs to 30 million when you add in the ground and understory arthropods (Erwin 1982, 1983, 1988; Wilson 1992). Such a species richness seems staggering given that only a total of 1.4 million species of plants, animals, and microbes have as yet been named and described, and it is by no means clear that Erwin's assumptions in making his calculations are accurate. It is nonetheless obvious that many, if not most, aspects of insect species richness remain poorly known, in much need of additional research. New species are virtually guaranteed from every collecting trip.

Species richness and biodiversity patterns of the Neotropics are discussed further in chapters 4 and 14.

A Rainforest Walk: Sights and Sounds of Animals   *Figures 93, 112, 146*

The rainforest, unlike the African savanna, does not provide easy views of its abundant animal life. Erwin (1988) noted that most beetles and their six-legged and eight-legged colleagues are in the canopy, far from where you are standing on the forest floor. You really have to work at it to see rainforest animals well. Many are highly cryptic, a result of evolution in a predator-rich environment (page 79). Even the most gaudy birds may appear remarkably dull in the dense forest shade. To make matters worse, some tropical birds, such as trogons and motmots, tend to sit very still for long periods and can easily be missed even when close by. Monkeys noisily scamper through the canopy, but tree crowns are so dense that we can only catch a glimpse of the often hyperkinetic simians. Iguanas remain still, suggesting reptilian gargoyles stretched out on tree limbs high above the forest floor. The animals are there, but finding them is a different matter.

In searching for rainforest animals you should try to adhere to the following guidelines: First, dress in dark clothing. You don't need to wear military-type camouflage, but dark clothing is definitely preferable to light. A bright white T-shirt that says "Save the Rainforest" in scarlet Day-Glo letters is fine back at the field station, but it will give away your presence in the forest. Second, move slowly and quietly, keeping your body motions minimal. Take a few steps along a trail and then stop and look around, beginning with the understory and working your eyes up to the canopy. Third, look for movement and listen for sounds. Leaf movement suggests a bird or other animal in motion. Listen for the soft crackle of leaves on the forest floor. Secretive birds such as tinamous and wood-quail as well as mammals such as agoutis and coatis are often best located by hearing them as they walk.

Sounds reveal some of the forest dwellers: there is often a dawn chorus of howler monkeys, the various troops proclaiming their territorial rights to one another, their tentative low grunts soon becoming loud, protracted roars, their combined voices creating one of the most exciting, memorable sounds of Neotropical forests. Cicadas provide a much different kind of background din, their monotonous stridulations reminding one of the oscillating (and irritating) high-low pitch of a French ambulance siren: "HHEeeee-ooooh, heeeee-ooooh, eeeee-ooooh." Parrots, hidden in the thick foliage of a fruiting fig tree, reveal themselves by an occasional harsh squeek, sounding like a door hinge in desperate need of oil. Scarlet macaws, flying serenely overhead with deep, dignified wing beats, so close to us that they fill our binocular field, suddenly emit a gutteral, high-decibel squawk, about as musical as screeching brakes. Macaws are a feast for the eyes but an assault on the ears. Peccaries, Neotropical relatives of wild pigs, grunt back and forth to one another in low tones as they root for dinner. A white-tailed trogon (*Trogon viridis*) calls softly, "cow, cow, cow, cow." Much louder, a sharp, ringing, highly demonstrative "PEA-HE-HAH," sounding vaguely like the crack of a whip, is the mating call of a common though drab understory bird, the screaming piha (*Lipaugus vociferans*).

We walk along the muddy forest trail, careful to listen and look. At several places we can't help but notice lines of leafcutter ants, their well-worn trails crossing ours. Leafcutters are abundant throughout the Neotropics and occur nowhere else. We notice that the ants come in various sizes, the largest bearing big leaf fragments, neatly clipped in a circular pattern. The leaves won't be eaten by the ants but will, instead, be taken to a vast underground colony, where they will be used to cultivate a fungus species that the ants farm. It is the fungus that is the real food of the ants (page 133). Rain begins, soft at first, soon heavy. We are surprised at how little of it seems to wet us. The rainforest canopy is, indeed, a fine umbrella. Soon the downpour ceases, though we are at first fooled by the steady dripping from the canopy, making it seem like it is still raining. A loud sound, not too distant, indicates that a big branch, or perhaps a tree, has fallen, a common event in rainforests.

A small blackish-brown animal resembling a cross between a tiny deer and an oversized, tailless mouse tentatively prances across the trail, pausing just long enough for us to get a binocular view of it. It's an agouti (*Dasyprocta fuliginosa*), a common fruit-eating rodent unknown outside of the Neotropics. We come to a stream and walk along it a short distance. Overhead, thin lianas hang limply downward, though one seems abnormally short and stiff. Binoculars reveal that it's not a vine at all, but a long, thin tail, belonging to an iguana (*Iguana iguana*). Before we have finished looking at the arboreal lizard, a bright green and rufous bird zooms purposefully by, an Amazon kingfisher (*Chloroceryle amazona*). Following in rapid, bouncing flight behind it is the large, brilliantly colored blue morpho butterfly (*Morpho didius*), a huge lepidopteran, dazzling electric blue in flight as its shimmering inner wing surfaces are illuminated among the sun flecks.

After rejoining the trail we begin to notice the quiet. Rainforests often seem all too serene, especially toward midday. Even bird song and insect cacophony cease, the seemingly tireless screaming piha being perhaps the one exception. Things don't really become active again until dusk.

Were we here as darkness fell we might catch sight of a great tinamou (*Tinamus major*), a chunky, ground-dwelling bird with a seemingly undersized, dove-like head, which greets vespers with a melancholy, whistled song known to move emotionally more than one Neotropical explorer. We might encounter a family group of South American coatis (*Nasua nasua*) resembling sleek, pointy-nosed raccoons. We might hear the odd treetop vocalizations of romping kinkajous (*Potos flavus*), arboreal members of the raccoon family. We might even encounter an ocelot (*Felis pardalis*) hunting stealthily in the cover

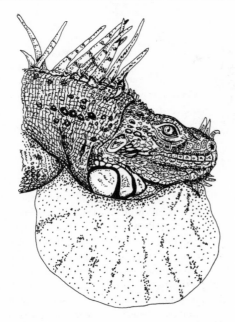

Common iguana

of darkness. And, of course, there is always the possibility of glimpsing a jaguar (*Panthera onca*). We probably won't, but we can always hope. There are cat tracks along the streambed, too small for a jaguar, but quite possibly left by an ocelot. Finally, were we here at night, there would be many species of bats flying about the canopy and understory. In the Neotropics, bats own the night skies. But none of these can we find, at least not easily, during the day.

The silence is suddenly broken by birdcalls. Incredibly, birds seem everywhere, when minutes ago there was none to be found. Soon we locate the reason for the flock. The trail is being crossed by several columns of a large troop of *Eciton*, army ants. Being careful not to step where the ants are, we don't want to miss the opportunity of seeing the ant-following bird flock. Antbirds, unique to the Neotropics, join with many other bird species to feed on the numerous arthropods, the insects, spiders, and their kindred, flushed by the marauding ant horde (page 326). A medium-sized brown bird with a black throat and face and bright, rusty head, a rufous-capped antthrush (*Folmicarius colma*), walks methodically beside the ants. From the lower branches we briefly glimpse a black-spotted bare-eye (*Phlegopsis nigromanulata*), an ebony bird with rusty markings on its wings and bright orangy bare skin surrounding each eye. The frenetic bare-eye skillfully snatches a katydid, launched from its hiding place by the oncoming ants. Birds are everywhere, or so it seems. Perched on an understory branch, three nunbirds (*Monasa nigrifrons*) are loudly calling, their whole bodies shaking as they sing in chorus, emphasizing their bright orangy-red bills adorning an otherwise black bird. Another ant-bird appears, this one utterly outrageous looking, deep rusty orange with gray wings and a tufted headdress of upright, white, shaggy feathers above its bill, nicely framed by a thick, white, feathery beard below it. This, the white-plumed

White-plumed antbird

antbird (*Pithys albifrons*), often the most abundant of the ant followers throughout much of Amazonia, is a constant follower of army ants. Indeed, we can hear two others responding to the loud calls of the bird we are watching. On a tree trunk we find a large, woodpecker-like bird, deeply rufous brown, a barred woodcreeper (*Dendrocolaptes certhia*). On a horizontal limb of a nearby small tree a large, rufous motmot (*Baryphthengus ruficapillus*) sits upright, swinging its long, pendulum-like tail methodically from side to side. Nearby a trogon flips off a branch in pursuit of a dragonfly. The trogon, seen only momentarily, has a yellow breast. Several species have yellow breasts, and we don't get a good enough look to identify it. That will happen more than once. Before we leave the ants, we've seen at least a dozen bird species, and possibly more are around.

The trail has brought us out to a clearing, a large forest gap (page 57), where it seems suddenly much hotter, especially with the accompanying high humidity. We encounter a dense clump of thin, spindly trees with huge, umbrella-like, lobed leaves. These distinctive trees, whose slender trunks are reminiscent of bamboo, seem to occur wherever an opening exists, and they are certainly common along roadsides. They are cecropias (*Cecropia* spp.), among the most abundant tree species on disturbed sites. We'll look at these in more detail later (page 71), but for now we pay little attention since, sitting idly in the midst of a large cecropia, is a serene-looking three-toed sloth (*Bradypus variegatus*). Sloths have such a slow metabolism that they barely move, and this one is no exception. Slowly it raises its left forearm, a parody of slow-motion photography. Like the Tin Man in the *Wizard of Oz* before he was oiled, the sloth's muscles seem to begrudge it the ability to move.

The sloth's cecropia is flowering, the slender, pendulous blossoms hanging down under the huge leaves. Soon a mixed-species flock of tanagers, honeycreepers, and euphonias fills its branches, gleaning both insects and nectar from the tree. Unlike the antbird flock, this group is brilliantly colored: metallic violets, greens, and reds.

Beneath the cecropia grows a clump of heliconias, with huge, elongate, paddlelike leaves quite similar to those of banana plants. A long-tailed hermit hummingbird (*Phaethornis superciliosus*) plunges its elongate, sickle-shaped bill into the small flowers that are highlighted by cuplike, bright orange bracts that surround them. A mild commotion along the forest edge is caused by a small troop of saddleback tamarins (*Saguinus fuscicollis*), miniature monkeys that frequent edges and areas of dense growth. The gnomelike simians seem to slide up and down the tree branches. They are active and wary, not easy to see well.

The sky begins to cloud up again. The high humidity has taken its toll, and we are feeling a bit tired. One more small trail leads back into the rainforest. Should we do just a little bit more exploring? It's going to rain again soon, that's obvious, but still, we take the trail. As we approach a large, buttressed tree alongside the trail, we hear an odd sound ahead of us, like dry leaves buzzing. It's better not to go on until we locate the sound. Soon we find the source of the buzzing, and, in spite of the heat, it inspires a few chills. Coiled alongside the trail, in the protection of a large buttress, is a 1.5-m-long (approx. 5 ft) *Bothrops atrox*, a pitviper similar to the well-known snake called fer-de-lance. It has seen us and is vibrating its tail rattlesnake-style in the leaves. Highly venomous, this animal is to be avoided, as its bite can be lethal. It's a beautiful and exciting animal, however, and its soft browns and black diamond-shaped pattern impressively camouflage it against the shady brown background of the litter. However, its large and distinctive triangular head and slitted, catlike eyes warn us of its potential for harm. We look at the serpent from a respectful distance, admire it, know we have been lucky to see it, and carefully retreat, leaving it very much alone and undisturbed.

The rain begins again in earnest, feeling cool, helping offset the high humidity. We put our binoculars in tightly sealed plastic bags and begin walking back to the field station, ever so alert, having already seen one pitviper. But when you really think about it, seeing a pitviper safely is cause for celebration. It's exciting to see a poisonous snake. And it's quite safe to walk in rainforest if you know how to keep alert for possible danger (now might be a good time to read the appendix). We did. It continues pouring rain. Obviously time for a beer. And, as the beer is consumed, our newly refreshed minds wander to the rainforest canopy itself. We had a great walk, but we were never really close to that vast layer of green, with all its varied inhabitants. What would it be like up there?

The Rainforest Canopy, Up Close and Personal

Figures 42, 43, 44, 45, 47, 48

Even with binoculars, seeing the tropical rainforest canopy from the forest floor is a real challenge. Imagine that you wish to take a close look at something, perhaps a bird, an insect, a flower. Or, from a scientific perspective, imagine that you wish to take data on it. Then imagine that you must do so from a hundred or more feet away. It's kind of like looking at a beetle walking across home plate when you are on second base or even in the outfield! That, of course, is exactly the situation when you are on the forest floor attempting to study something in the upper canopy. But, at least in a few places, it is now possible to access the rainforest canopy directly. And, to borrow a phrase from astronaut John Glenn when he lifted off on his *Friendship 7* orbital mission in 1962, "Oh, that view is tremendous."

There is a tower not far from Manaus, Brazil, located within primary rainforest, a tower sufficiently tall that it just exceeds the canopy. From the gentle sway atop this structure one can enjoy miles of vista, seeing vast tracts of forest, while at the same time surrounded by the crowns of dozens of canopy trees. It is from such a vantage point that one might actually see a harpy eagle (*Harpia harpyia*) soar overhead, or catch a glimpse of the rare crimson fruitcrow (*Haematodevus militaris*). Colorful birds, such as the pompadour cotinga (*Xipholena punicea*), sit upright, perched on emergent branches from emergent trees. From the ground, you'd never know they were there. Mixed foraging flocks of canopy birds are now at eye level. Colorful butterflies, many of them strict canopy dwellers rarely or never seen in the understory, are easy to observe. Monkeys, squirrels, and other canopy-dwelling creatures can be seen from above, as you actually look down on them. But a tower, for all its advantages, is limited. It occupies a very restricted area. An even better way to see and study the canopy would be to walk within it, kind of like what howler monkeys do. And there is a place where that is possible.

The Amazon Center for Environmental Education and Research (ACEER) is located in one of the most species-rich areas in upper Amazonia, along the Napo River about 161 km (100 mi) east of Iquitos, Peru. The feature that makes ACEER unique among field stations is that the site includes a superbly engineered canopy walkway of over 0.4 km (0.25 mi) in length, an elaborate arboreal pathway interconnected with fourteen emergent trees, permitting one literally to walk through the rainforest canopy. Each of the trees used in the walkway is fitted with strong wooden platforms allowing several people to stand and look out at the canopy. The narrow spans between the platformed trees are built rather like suspension bridges, supported by strong metal cable and meshed at the sides to provide total security and safety (see figures). The spans vibrate a bit, especially when more than one person is walking across. One of the spans is nearly the length of a football field, affording a breathtaking, if shaky, look at the rainforest below. The first platform is about 17 m (55 ft) above the forest floor, but the spans eventually take you to a platform that is fully 36 m (118 ft) above the ground. From that privileged position, you gaze upon a panorama of unbroken rainforest for many, many miles. And yes, that view is tremendous.

From within the canopy you get an immediate, almost overwhelming impression of the richness of the rainforest. Trees are anything but uniform in height—and there are so many species, you wonder if, in the quarter mile of walkway, you pass two that are the same or if every tree you pass is different from every other. You notice the many different leaf sizes and shapes and see that some leaves are damaged by leafcutter ants, the insects having patiently walked 30 m (100 ft) up the tree bole to collect food for their subterranean fungus gardens. Now you can really look at the fine details of epiphytic plants such as orchids and bromeliads. You can see down into the cisternlike bromeliads and learn what kinds of tiny animals inhabit these microhabitats high above the forest floor. You note the uneven terrain below and realize that the canopy is by no means continuous, but is punctuated by frequent gaps, openings of various sizes. A male collared trogon (*Trogon collaris*) is perched 6 m (20 ft) below the walkway. How odd it is actually to look down on such a creature. A male spangled cotinga (*Cotinga cayana*) sits in display at eye level, a stunning turquoise bird whose plumage seems to shimmer with iridescence in the full sunlight.

A tree near one of the platforms is in heavy fruit, hundreds of small, orange, berrylike fruits peppering the branches. Fruit trees normally attract a crowd, and this one is no exception. Colorful tanagers of six different species fly in to feast on the fruits, at most 3 m (10 ft) away from us. Equally gaudy aracaris and toucanets join the tanagers. Two sedate, long-haired saki monkeys (*Pithecia monachus*), apparently a female and an adolescent, stop at the fruiting tree. The long, bushy tails of the monkeys hang limply below the branch on which they sit, as these simians do not have prehensile tails, like howlers, spiders, and woolly monkeys. The simians soon realize they are not alone. The female sees us and rubs her chin on the branch. She stands fully erect and emits a short, demonstrative hoot to warn us to come no closer. She needn't worry. We are not about to leave the security of the walkway. And we marvel at how monkeys have adapted the requisite skills to move effortlessly through such a tenuous three-dimensional world as rainforest canopy. A frenetic Amazon dwarf squirrel (*Microsciurus flaviventer*), a chipmunk-sized evolutionary relative of the northern acorn collectors, scurries with nonchalance on the underside of a branch over 30 m (100 ft) from the ground below. A thought reoccurs, and has occurred many times: from the ground, we'd never know this little animal was up here.

The canopy walkway affords a unique and broad window into the life above the forest understory. It is exciting to visit it, to be on it at dawn, when the forest below is still clothed in mist, or to watch the sun set over what seems like an endless vista of rainforest. But it also affords an opportunity for the kind of research that needs to be done to ascertain accurately an understanding of the rhythms of life in this essential habitat. We'll soon know more about rainforests because of the ACEER canopy walkway and others like it that are being or have been constructed in various other rainforest localities (Wilson 1991; Moffett 1993).

Tree frog

3

How a Rainforest Functions

T HE REMARKABLE structural complexity of tropical rainforest provides the infrastructure for one of the most intricate ecological machines on Earth. In the course of any given year, the world's diverse rainforests capture more sunlight per unit area than any other natural ecological system. A small but highly significant fraction of that solar radiation is incorporated into complex molecules, ultimately providing energy and structure that support the rainforest community. Tropical soils, much of them delicate and mineral-poor, are nonetheless efficiently tapped for nutrients by root systems aided, in most cases, by symbiotic fungi. Dead plant and animal tissue quickly decays and is recycled to the living components of the ecosystem. The torrential downpours that characterize the rainy season could erode already mineral-poor soil, but forest vegetation has adapted to deluges and their effects. There is much to be learned from a study of plant ecology. As Alfred Russel Wallace (1895) put it,

> To the student of nature the vegetation of the tropics will ever be of surpassing interest, whether for the variety of forms and structures which it presents, for the boundless energy with which the life of plants is therein manifested, or for the help which it gives us in our search after the laws which have determined the production of such infinitely varied organisms.

Productivity

Figure 1

Ecologists use the term *productivity* to describe the amount of solar radiation, sunlight, converted by plants into complex molecules such as sugars. The biochemical process by which this energy transformation is accomplished is, of course, *photosynthesis*. Plants capture red and blue wavelengths of sunlight and use the energy to split water molecules into their component atoms, hydrogen and oxygen. To do this, plants utilize the green pigment chlorophyll. The reason plants look green is that chlorophyll reflects light at green wavelengths, while absorbing light in the blue and red portions of the spectrum. The essence of photosynthesis is that energy-enriched hydrogen from water combines with the simple, low-energy compound carbon dioxide (CO_2, an atmospheric gas) to form high-energy sugars and related compounds. This process provides the basis upon which virtually all life on Earth ultimately depends. Oxygen from water is given off as a byproduct. Photosynthesis, occurring over the past three billion years, has been responsible for changing Earth's atmosphere from one of virtually no free oxygen to its present 21% oxygen.

Of all natural, terrestrial ecosystems on Earth, none accomplishes more photosynthesis than tropical rainforests. A hectare (10,000 m^2) of rainforest is more than twice as productive as a hectare of northern coniferous forest, half again as productive as a temperate forest, and between four and five times as productive as savanna and grassland (Whittaker 1975).

Ecologists distinguish between gross primary productivity (GPP) and net primary productivity (NPP). The former refers to the total amount of photosynthesis accomplished, while the latter refers to the amount of carbon fixed in excess of the respiratory needs of the plant; in other words, the amount of carbon (as plant tissue) added to the plant, for growth and reproduction. By way of example, if you watch a field of corn grow from seed to harvest, you are seeing net primary productivity. You don't actually know how much energy the corn has used to maintain itself during its growing season. Such respiratory energy has been radiated back to the atmosphere as heat energy. And if you were to fly over the cornfield and photograph it with an infrared camera, you would see from the deep red image that lots of heat is continually coming from the corn. This is the energy of respiration. Normally, net primary productivity is much easier to calculate than gross primary productivity, since NPP can be measured as easily as weighing biomass over a period of time.

Tropical rainforests exhibit high net productivities, essentially the highest of any terrestrial ecosystem. Estimates from Brazilian grasslands and rainforest suggest that rainforests are about three times more productive than grasslands (Smil 1979). In addition, rainforests have rates of respiration that exceed those of other ecosystems, presumably due to temperature stress (Kormondy 1996). Rainforests expend as much as 50–60% of their gross primary productivity in maintenance. What this means, of course, is that gross primary productivity, the total rate of photosynthesis (net productivity plus energy used for respiration), is vastly higher in rainforests than in virtually any other ecosystem on the planet.

Using a highly complex, mechanistically based computer simulation called the Terrestrial Ecosystem Model (TEM), a team of researchers has estimated the range in NPP among various major ecosystem types in South America (Raich et al. 1991). Unsurprisingly, of the total NPP of the continent, more than half of it occurs in tropical and subtropical broadleaf evergreen forest. Mean annual NPP estimates for tropical evergreen forest ranged from 900 to 1,510 grams per meter squared per year, with an overall average of 1,170 g/m^2/yr. The most productive forests were clearly those within the Amazon Basin, particularly those close to the river or its major tributaries. Compared to these figures, South American shrublands had a NPP estimate of 95 g/m^2/yr and savannas averaged 930 g/m^2/yr. Obviously, broadleaf tropical forests are far more productive than either savanna or shrublands. NPP varied seasonally, correlating with moisture availability, and strongly influenced by seasonal differences in cloudiness in tropical evergreen forests (Raich et al. 1991). Cloud cover, which intercepts significant amounts of solar radiation, is a major factor in reducing rates of productivity.

Considering the total global area covered by rainforests, these ecosystems are estimated to produce 49.4 billion tons of dry organic matter annually, compared with 14.9 billion tons for temperate forests (Whittaker 1975).[1] In the course of one year, a square meter of rainforest captures about 28,140 kilocalories of sunlight. Of this total, the plants convert a minimum of 8,400 kilocalories (about 35%) into new growth and reproduction, using the remainder for metabolic energy.

It is worth noting that as rainforests are cut and replaced by anthropogenic (human created and controlled) ecosystems (chapter 14), much more NPP is directed specifically toward humans (in the form of agriculture or pasturage) and some is lost altogether (fields and pastures are less productive than forests), making less energy available for supporting overall global biodiversity. One research team has estimated that almost 40% of the world's NPP has been either co-opted by humans or lost due to human activities of habitat conversion (Vitousek et al. 1986). It is estimated that tropical forests store 46% of the world's living terrestrial carbon and 11% of the world's soil carbon (Brown and Lugo 1982). No other ecosystem in the world stores so much carbon in the form of living biomass.

Ecologists express leaf density as a figure called *leaf area index* (LAI), the leaf area above a square meter of forest floor. In a mature temperate forest such as Hubbard Brook in New Hampshire, LAI is nearly 6, meaning that the equivalent of 6 square meters of leaves cover one square meter of forest floor. For tropical rainforest at Barro Colorado Island in Panama, the figure is about 8 (Leigh 1975). Typically, LAI in the humid tropics ranges from about 5.1 (a forest on poor soil, Amazon Caatinga, at San Carlos, Venezuela) to a high of 10.6–22.4 (a lush forest on rich soil at Darien, Panama) (Jordan 1985a). In forests with extreme high LAI, it is probable that the intensity of shading is so great that many, if not most, understory leaves do not approach optimum NPP because they are severely light limited.

Tropical leaves also have greater biomass than temperate zone leaves. In the tropics, one hectare of dried leaves weighs approximately one ton, about twice that of temperate zone leaves (Leigh 1975). Litterfall was measured at over 9,000 kg/ha/yr for tropical broadleaf forest compared with just over 4,000 for a warm, temperate broadleaf forest, and 3,100 for a cold, temperate needleleaf forest (Vogt et al. 1986). Because tropical forests vary in productivity, so must leaf litter amounts. Leaf litter production on rich tropical soils can exceed twice that on nutrient poor soils (Jordan 1985a).

The high productivity of broadleaf tropical rainforests is facilitated by a growing season much longer than in the temperate zone. Growth in the tropics is not interrupted by a cold winter. Temperature hardly varies, water is usually available, and, because the year is frost-free, there is no time at which all plants must become dormant, as they do in much of the temperate zone in winter. The dry season does, however, slow growth (sometimes dramatically), and where it is severe most trees are deciduous, dropping leaves at the onset of dry season and growing new leaves with the onset of rainy season.

---

[1] This figure is now two decades old. The increasing loss of rainforest means it is undoubtedly smaller today.

Given the prolonged growing season typical of the tropics, it may be tempting to conclude that productivity *per unit time* is no greater in the tropics than in the temperate zone. In other words, the tropics are more productive because there is more time to produce. But does a gram of plant tissue in the tropics take exactly (or nearly) as long to produce as a gram in the temperate zone? The answer is poorly known, but some data suggest that at least some tropical trees seem to grow much faster than ecologically similar species in the temperate zone. A study by Kobe (1995) documented that *Cecropia* can increase its radius by as much as 15 times in a year of growth. When compared with species such as red oak (*Quercus rubra*) and red maple (*Acer rubrum*), and when corrected for length of growing season, tropical species studied grew by an order of magnitude more than those from the temperate zone, an indication that per tree productivity is considerably enhanced in the tropics.

Productivity depends upon adequate light, moisture, and carbon dioxide, plus sufficient amounts of diverse minerals from soil. In the first three of these essentials, tropical rainforest seemingly fares well, though low light intensity certainly limits the growth of plants below the canopy. In the fourth category, sufficient minerals, however, rainforests are often (but not always) deprived. In many areas within the American tropics, soils are old and mineral poor, factors that could limit productivity. However, rainforest trees have adapted well to nutrient-poor soils.

Nutrient Cycling and the Soil Community   *Figure 40*

Because Earth has no significant input of matter from space (a year's worth of meteorites adds up to very little), atoms present in dead tissue must be reacquired, recycled back to living tissue. Decomposition and subsequent recycling is the process by which materials move between the living and nonliving components of an ecosystem. Recycling occurs as a byproduct of decomposition, and decomposition occurs as a means by which organisms acquire energy. Consider that in a rainforest a unit of energy fixed during the net productivity can move in one of two major directions: either it can be consumed as part of living tissue, as when a caterpillar chews a leaf, or it can remain as part of the leaf until the leaf eventually drops from the tree, at which time the energy becomes available to the soil community. This latter direction moves energy directly to what is termed the *decomposer food web*. A glance at a lush, green rainforest plus a dash of pure logic is enough to show that the vast majority of the energy fixed during photosynthesis eventually enters the decomposer food web. If it were otherwise, trees, shrubs, and other green plants would show far more leaf damage than they do. Instead, most energy remains as potential energy in leaf, bark, stem, and root tissue, only to be eventually released by a host of soil community organisms as they unpretentiously make their livings below your muddy boots on and in the forest floor.

Fungi and bacteria are the principal actors in this ongoing and essential drama of decomposition that is one of nature's most fundamental processes. It is they who convert dead organic tissue back into simple inorganic compounds that are then reavailable to the root systems of plants. Of course there is a supporting cast: slime molds, actinomycetes, algae, and hordes of animals

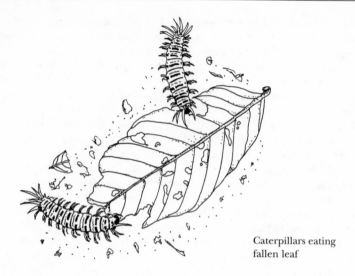

Caterpillars eating
fallen leaf

ranging from vultures to numerous arthropods, earthworms, and other invertebrates as well as many kinds of protozoans all have varying roles in the complex, hierarchical process of converting a dead leaf, a dead agouti, or peccary
feces back to those most basic chemical elements.

Knowledge of the structure and function of microbial decomposer communities in tropical rainforests is still rather rudimentary. It is well understood
that fungi are immensely abundant in the tropics. An individual fungal strand
is called a hypha, and a network of hyphae is called a mycelium. In some
tropical forests mycelial mesh seems to cover parts of the forest floor. The
creatures that make up the tropical soil community may rival the biodiversity
found in the leafy canopy. But there are relatively few detailed studies that
make estimates of such parameters as fungal biomass or pathways of energy
movement among the constituent flora and fauna of the decomposer community. More are needed. For an exemplary study, see Lodge (1996).

Organisms facilitate a process called *humification*, in which complex soil organic matter is maintained at the interface between the tree roots and soil.
Humus is important in forming colloids that cement soil particles, in helping
aerate the soil, in possessing a negative net electrical charge, an important aid
in retaining critical minerals in the soil (by electrostatic attraction), and in
overall conservation of nutrients (Lavelle et al. 1993) The soil itself represents
a temporary repository for essential minerals such as nitrogen, calcium, magnesium, phosphorus, and potassium. Each of these and other minerals is necessary for biochemical reactions in organisms, and a shortage of any one of
them can significantly limit productivity. For example, phosphorus and nitrogen are important in the structure of nucleic acids (DNA and RNA) as well as
proteins and other necessary molecules. Magnesium is an essential part of the
chlorophyll molecule, without which photosynthesis could not occur.

Consider how an atom is cycled. Suppose a dead leaf falls to the ground.
Inside the leaf are billions of atoms, but we will select, for example, just one,
an atom of calcium. This calcium atom may initially pass through a millipede

or other invertebrate, only to be returned to the litter through elimination of waste or the death of the creature itself. Or the atom may be taken up directly by a fungus. This same atom eventually will pass through several dozen fungal and bacterial species, each of which gains a modicum of energy by ingesting, digesting, and thus decomposing the deceased leaf (or millipede). Within days the calcium atom becomes part of the inorganic components of the soil. Almost immediately, other types of fungus (called mycorrhizae, see below), usually growing from within a tree root, take up the calcium and pass it along to the tree, where the calcium atom may well end up in another leaf. The cycle is complete and will now go around again.

Nutrient cycling is often termed *biogeochemical cycling* to describe the process of chemicals moving continuously between the bios (living) and the geos (nonliving) parts of an ecosystem. The movement of minerals in an ecosystem is strongly influenced by temperature and rainfall, the major features of climate. In the tropics, both high temperature and abundant rainfall exert profound effects on the patterns of biogeochemical cycling (Golley et al. 1969, 1975; Golley 1983).

Heat stimulates evaporation. As plants warm they evaporate water, cooling the plants and, thus, returning a great deal of water to the atmosphere in this heat-related pumping process called *transpiration.* Water from rainfall is taken up by plants and transpired, returned to the atmosphere, under the stress of tropical heat. Nowhere is this continuous process of transpiration more obvious than along the wider stretches of the Amazon River. At midday, skies immediately above the big river tend to be clear and blue, but should you look over distant forest on either of the river banks, you will likely see big, puffy, white clouds, formed by the condensing moisture transpired by the forest; you are literally watching the forest breathe. Indeed, approximately 50% of the precipitation falling on the Amazon Basin is directly recycled via transpiration from the myriad vegetation (Salati and Vose 1984).

Since minerals are always taken up through roots via water, the uptake of water is essential to the uptake of minerals as well. But evaporation can be a mixed blessing. Plants can lose too much water when subjected to constant high temperature. Many tropical plants retard evaporative water loss both by closing their stomata (openings on the leaves for gas exchange) and by producing waxy leaves.

Leaching

Water can wash essential minerals and other chemicals from leaves, a process called *leaching.* Leaching can be especially severe in areas subject to frequent heavy downpours. The protective waxy coating of tropical leaves contains lipid-soluble (but water-insoluble) secondary compounds such as terpenoids that act to retard water loss and discourage both herbivores and fungi (Hubbell et al. 1983, 1984). Drip tips probably reduce leaching by speeding water runoff. Such adaptations enable a typical tropical leaf to retain both its essential nutrients and adequate moisture.

Rainfall also leaches minerals from the soil, washing them downward into the deeper soil layers. Clay particles and humus have negative electrostatic charges that attract minerals with positive charges such as calcium and potassium. Because water adds hydrogen atoms to the soil, which are also positively charged, these abundant atoms can exchange with those of elements such as calcium or potassium, which then wash to a deeper part of the soil or may wash out of the soil entirely. Rainfall strongly influences soil acidity because the accumulation of hydrogen atoms, on either humus or clay, lowers the pH, thus raising the acidity of the soil. In the tropics, the combination of high temperatures and heavy rainfall can often result in much leaching and strongly acidic soils. Typical Amazon soils are frequently mineral-poor, high in clay, acidic, and low in available phosphorus (Jordan 1982, 1985b), and the nutrient-poor nature of the soil is a major limiting factor to plant productivity (Uhl et al. 1990). One estimate suggests that nearly 75% of the soils in the Amazon Basin are acidic and generally infertile (Nicholaides et al. 1985). Much water movement occurs among the atmosphere, the soil, and the organisms. Tropical plants are adapted to be very stingy about giving up minerals. Consequently, one of the major differences between tropical and temperate forests is that in tropical forests most of the rapidly cycling minerals are in the living plants, the biomass. Most of the calcium, magnesium, and potassium in an Amazon rainforest is located, not in the soil, but in the living plant tissue (Richards 1973; Jordan 1982; Salati and Vose 1984). For example, in a study performed near San Carlos de Rio Negro in Venezuela, the distribution of calcium was as follows: 3.3% in leaves; 62.2% in wood; 14.0% in roots; 3.1% in litter and humus; and only 17.4% in soil (Herrera 1985). Another study concluded that 66% to 80% of potassium, sodium, calcium, and magnesium is in aerial parts of plants, not in soil (Salati and Vose 1984). However, this same study concluded that most nitrogen and phosphorus (somewhere around 70%) is in soil, roots, and litter. It is not surprising that most tropical soils are considered generally nutrient-poor. In the temperate zone, minerals are more equally distributed between the vegetation and soil bank.

Mycorrhizae

Throughout the tropics as well as most of the temperate zone, there is an intimate, mutualistic association between tree roots and a diverse group of fungi collectively termed *mycorrhizae*. Many of these fungi grow directly into tree roots, using some of the plant's photosynthate as food. In this regard, the fungi would seem to be parasitic, much like the athlete's-foot fungi that many tropical visitors come to experience between their toes. But though the fungi take food from the tree, they are essential to the tree's welfare as they facilitate the uptake of minerals from the forest litter. Trees dependent on mycorrhizae typically have poorly developed root hairs; the fungal strands substitute for the missing root hairs (St. John 1985). Most of the mycorrhizal fungi within rainforests are grouped together as vesicular-arbuscular mycorrhizae (VAM), meaning that they grow within tree roots. Some mycorrhizae, particularly those found in poor soils (such as white sandy soils) or in disturbed areas, grow outside of the tree roots and are referred to as ectomycorrhizae. The extensive surface area of the fungal mycelium is efficient in the uptake of both minerals and water, as experiments have demonstrated (Janos 1980, 1983). VAM are particularly important in aiding in the uptake of phosphorus, which tends to be of limited availability in rainforest soils (Vitousek 1984). They may also have a role in direct decomposition and cycling, moving minerals from dead leaves into living trees without first releasing them to the soil (Janos 1983; St. John 1985), and they may affect competitive interactions among plants, thus influencing the biodiversity of a given forest (Janos 1983). Mycorrhizae are also essential to certain epiphytes such as orchids. In early successional ecosystems, waterlogged areas, and high elevation regions, mycorrhizae may be less essential (Parker 1994), though it has been suggested that ectomycorrhizal fungi, which dominate at least in successional areas, may provide their host plants with a competitive advantage over VAM host plants (Lodge 1996).

VAM spores may be widely distributed by certain rodent species such as spiny rats (*Proechimys* spp.) and rice rats (*Oryzomys* spp.). A study performed in rainforest at Cocha Cashu, in Manu National Park in Peru, demonstrated that that VAM spores are well represented in the feces of spiny and rice rats (Janos et al. 1995). Though most mycorrhizae spread by direct infection from root to root, the authors suggest that long-distance dispersal of VAM may be significantly facilitated by mammalian spore transport. Given that VAM are essential in the uptake of minerals by the majority of rainforest tree species, the health and species richness of rainforest may depend, at least in part, on the wanderings of some unpretentious little rodents.

Rapid Recycling

There is often surprisingly little accumulation of dead leaves and wood on rainforest floor, making for a generally thin litter layer. Unlike the northern coniferous forests, for example, which are endowed with a thick, spongy carpet of soft, fallen needles, or the broadleaf temperate forests where layer after layer of fallen oak and maple leaves have accumulated, a rainforest floor is, by comparison, often sparsely covered by fallen leaves. This becomes particularly interesting when you keep in mind that more and heavier leaves occur in rainforest. The solution to this seeming paradox is that decomposition and recycling of fallen parts occur with much greater speed in rainforests than in temperate forests. Just as productivity can be relatively continuous, uninterrupted by the frozen soils of a northern winter, so can biogeochemical cycling continue unabated throughout the year. Studies indicate that in tropical wet forests, particularly those on richer soils, litter is decomposed totally in less than one year, and minerals efficiently conserved (Jordan 1985a). Forests on poorer soils show reduced rates of decomposition (Lavelle et al. 1993). Rainforests also cycle minerals very "tightly." The resident time of an atom in the nonliving component of the ecosystem is very brief (Jordan and Herrera 1981; Jordan 1982, 1985a, 1985b). One study estimated that approximately 80% of the total leaf matter in an Amazon rainforest is annually returned to the soil (Klinge et al. 1975). Leaf litter does accumulate in tropical dry forests, especially during dry season (Hubbell, pers. com. 1987).

Rainforest Soil Types and Mineral Cycling   *Figures 28, 76*

One should not be quick to generalize about tropical soils or about patterns of nutrient cycling in the tropics (Vitousek and Sanford 1986). In some regions, such as the eastern and central Amazon Basin, soils are very old and mineral-poor (oligotrophic), while in other regions, such as volcanic areas of Costa Rica or much of the Andes, soils are young and mineral-rich (eutrophic) (Jordan and Herrera 1981). Soil characteristics vary regionally because soil is the product of several factors: climate, vegetation, topographic position, parent material, and soil age (Jenny 1941, cited in Sollins et al. 1994). Because these factors vary substantially throughout Central and South America, so do soil types. Even within a relatively limited region there can be high variability among soil types. For instance, a single day's ride in southern Belize will take you from orange-red soil to clayey gray-brown soil. The gray-brown soil is largely from limestone, common throughout much of Belize.

In general, much of the soil throughout the humid tropics, often reddish to yellowish in color, falls into one of three classifications: ultisols, oxisols, or alfisols. Ultisols are generally well-weathered, meaning that minerals have been washed from (leached) the upper parts of the soils. Oxisols, also called ferralsols or latasols, are deeply weathered, old, acidic, and found on well-drained soils of humid regions; typically, these soils occur on old geologic formations such as the ancient Guianan Shield. Oxisols are common throughout the global tropics and are typically heavily leached of minerals as well as quite acidic (Lucas et al. 1993). Alfisols are common in the subhumid and semiarid tropics and are closer to a neutral pH (though still acidic), with less overall leaching than typical oxisols. It is estimated that ultisols, oxisols, and alfisols, taken together, comprise about 71% of the land surface in the humid tropics worldwide (Lal 1990). This is generally similar to estimates made by Vitousek and Sanford (1986), which suggested that 63% of moist tropical forests are atop soils of moderate to very low fertility. In the Amazon Basin, about 75% of the area is classified as having oxisols and ultisols (Nicholaides et al. 1985).

Not all tropical soils are old or heavily weathered or infertile. Vitousek and Sanford (1986) estimated that 15% of moist tropical forests are situated on soils of at least moderate fertility. Inceptisols and entisols are young soils of recent origin, rich in minerals near the surface, with higher pH (still acid, but closer to neutral). Soils generated from deposits during the flood cycle (alluvial soils) or from recent volcanic activity typify these categories (Lal 1990; Sollins et al. 1994).

Soil types are not absolute; in most areas soil types grade into one another along a continuum. For example, at La Selva Biological Station in Costa Rica (a 1,500-ha nature reserve of premontane rainforest), it is estimated that approximately one-third of the soils are fertile inceptisols (some of recent volcanic origin) and some Entisols of alluvial origin, while the remainder of the soils are older, more acidic, and less fertile Ultisols (Parker 1994).

Semiarid and arid regions in the tropics, because of climatic differences, have somewhat different soil types from those of humid and semihumid regions. Some of these soils are dark, heavily textured, and calcareous, sometimes subject to salt accumulation (Lal 1990). Because of frequent occurrences of burning, and sometimes animal grazing, litter is thin and poorly developed on savanna soils, and the decomposer ecosystem is more limited. Termites, however, can be particularly abundant in arid, grassy areas (Lavelle et al. 1993).

The general pattern throughout much of the humid tropics is that heat and heavy moisture input cause the formation of oxides of iron and aluminum (which are not taken up by plants), giving the soil its reddish color. Clay content is usually high, evident as you slip and slide your way over a wet trail. Mountain roads become more dangerous and often impassable during rainy season because wet clay makes them slippery. In the Amazon Basin, sediments eroded from highland areas during the late Tertiary period were deposited in the western end of the basin, forming a flat surface about 250 m (820 ft) above sea level. Much of this surface, called the Amazon Planalto, is made up of kaolinitic clay, a substance devoid of most essential minerals but rich in silicon, aluminum, hydrogen, and oxygen (Jordan 1985b). In the eastern part of Amazonia, soils are quite sandy, though remaining acidic and nutrient poor.

One extreme situation, called *laterization*, results from the combined effects of intensive erosion and heat acting on soil. If vegetation cover is totally removed and the bare soil is exposed to extreme downpours and heat, it can bake into a bricklike substance, ruining it for future productivity. Tropical peoples around the world have long used laterization to make bricks used in buildings as impressive and as venerable as some of the ancient temples in Cambodia. Though laterization has been widely reported as demonstrating the extreme delicacy of tropical soils and thus the futility of farming such soils, such a generalization is untrue. Laterization only occurs with repeated wetting and drying of the soil in the absence of any vegetative cover. The loss of roots (which utilize aeration channels in the soil) plus repeated wetting and drying act to break up soil aggregates of bound clay particles. Only when these aggregates are broken up and the soil thus subject to compaction, does laterization ensue (Jordan 1985b). In Amazonia, only about 4% of the soils even run the risk of laterization (Nicholaides et al. 1985).

Nonetheless, even without the extreme of laterization, attempts to farm the tropics using intensive agriculture have often failed because of quick loss of soil fertility. This need not be the case. Much of the soil composition in Amazonia is, in fact, surprisingly similar to that found in the southeastern United States, where successful agriculture is routinely practiced (Nicholaides et al. 1985). Soil infertility is generally common throughout the Amazon Basin (Irion 1978; Nicholaides et al. 1985; Uhl et al. 1990), though most soils will support some form of limited agriculture. Where Amazonian soils are most fertile, they will support continuous cultivation by small-scale family units (subsistence agriculture), with crops such as maize, bananas, sweet potatoes, as well as small herds of cattle (Irion 1978). Various approaches have been shown to be successful in achieving continuous farming of low-fertility Amazonian soils (Nicholaides et al. 1985; Dale et al. 1994). Amazonian agriculture will be discussed further in chapters 7 and 14.

In parts of the Amazon Basin, white and sandy soils predominate, most of which are derived from the Brazilian and Guianan Shields, both ancient, eroded mountain ranges. Because these soils have eroded for hundreds of millions of years, they have lost their fertility and are extremely poor in mineral content. The paradox is that lush broadleaf rainforests grow on these infertile soils. I stress *on* and not in the soil because recycling occurs on the soil surface.

Mineral Cycling on Oligotrophic Soil

Oligotrophic refers to "nutrient deprived." Poor soil forests can be located on terra firme or on igapo floodplain (see below). Forests on oligotrophic soils are less lush and of smaller stature than forests on rich soils. Henry Walter Bates (1862) commented about forest on poor-soil igapo (which he spelled "Ygapo") floodplain, comparing it with the forest on the rich-soil delta: "The low-lying areas of forest or Ygapos, which alternate everywhere with the more elevated districts, did not furnish the same luxuriant vegetation as they do in the Delta region of the Amazons."

In forests with oligotrophic soil, up to 26% of the roots can be on the surface rather than buried within the soil (Jordan 1985), and *root mats* as thick as several centimeters sometimes develop (Lavelle et al. 1993). This obvious mat (you can actually trip over it) of superficial roots, intimately associated with the litter ecosystem, is much reduced or entirely absent from forests on rich, eutrophic soil. Surface roots from the trees are quite obvious as they radiate from the many boles and virtually cover the forest floor. A thin humus layer of decomposing material also covers the forest floor, and thus the root mat of surface roots, with the aid of mycorrhizal fungi, directly adsorbs available minerals (Lavelle et al. 1993). Carl F. Jordan and colleagues have made extensive studies of Amazon forest nutrient conservation (Jordan and Kline 1972; Jordan 1982, 1985a). Using radioactive calcium and phosphorus to trace mineral uptake by vegetation, they found that 99.9% of all calcium and phosphorus was adsorbed (attached) to the root mat by mycorrhizae plus root tissue. The root mat, which grows very quickly, literally grabs and holds the minerals. For example, in one study from Venezuela, the decomposition of fallen trees does not result in any substantial increase in nutrient concentration of leachate water, suggesting strongly that nutrients leached from fallen vegetation move essentially immediately back into living vegetation (Uhl et al. 1990).

Phosphorus is seemingly problematic in Amazonian soils because it complexes with iron and aluminum and, due to the high acidity, is held in stable compounds that make it unavailable for uptake by plants (Jordon 1985b). Indeed, it may be the nutrient most difficult for plants to procure (Vitousek 1984). However, vesicular-arbuscular mycorrhizae apparently greatly enhance the uptake of phosphorus (St. John 1985).

Microorganisms living within the root mat are essential in aiding the uptake of available minerals. The forest floor functions like a living sponge, preventing minerals from being washed from the system (Jordan and Kline 1972; Jordan and Herrera 1981; Jordan 1982).

Such a unique rapid recycling system may be one reason for the presence of buttresses. The buttress allows the root to spread widely at the surface, where it can reclaim minerals, without significantly reducing the anchorage of the tree. This is probably the tightest recycling system in nature. If the thin layer of forest humus with its mycorrhizal fungi is destroyed, this recycling system is stopped, and the fertility is lost. Removal of forest from white sandy soils can result in the regrowth of savanna rather than rainforest because of the destruction of the tight nutrient cycling system.

Though a dense surface root mat seems to be an obvious adaptation to the need for rapid and efficient recycling on highly oligotrophic soils, the generalization is not universal. A research team working on Maraca Island, a lowland evergreen rainforest site in Roraima, Brazil, found that in spite of low nutrient concentration in the dry, sandy soil, the vegetation did not exhibit the types of adaptations described above. There was no surface root mat, nor was the root biomass unusually high, but the leaves were nonetheless relatively rich in nutrient content. Trees grew rapidly and litterfall was high. The rate of leaf decomposition was also quite high, indicating a rapid recycling mechanism, but what intrigued and baffled the researchers was that this island forest showed none of the presumed adaptations of rainforests elsewhere on highly oligotrophic soils, yet it seemed to be functioning efficiently and without nutrient limitation (Thompson et al. 1992; Scott et al. 1992). The work is an example of the need for caution in generalizing about rainforest ecology and adaptations.

Other Nutrient-Retention Adaptations

Some tropical plants have root systems that grow vertically upward, from the soil onto the stems of neighboring trees. These *apogeotropic roots* grow as fast as 5.6 cm (2.5 in) in seventy-two hours (Sanford 1987). The advantage of growing on the stems of other trees may be that the roots can quickly and directly absorb nutrients leached from the trees, as precipitation flows down the stem. This unique system, thus far described only for some plants growing on poor-quality Amazon soils, results in recycling *without* the minerals ever entering the soil!

A somewhat similar process, called *arrested litter*, has been documented at La Selva in Costa Rica (Parker 1994). Both epiphytes and understory plants, especially the wide crowns of certain palms (nicknamed "wastebasket plants"), catch litter as it falls from the canopy. The litter subsequently decomposes above ground, enriching the mineral content of stemflow and thus fertilizing the soil in the immediate vicinity of the wastebasket plant.

Jordan and his colleagues also learned that canopy leaves play a direct role in taking up nutrients. Calling them nutrient scavengers, Jordan pointed out that algae and lichens that cover the leaves adsorb nutrients from rainfall, trapping the nutrients on the leaf surface. When the leaf dies and decomposes, these nutrients are taken up by the root mat and returned to the canopy trees (Jordan et al. 1979).

Nitrogen Fixation

Some plants, particularly those in the huge legume family, which is abundantly represented in both biomass and biodiversity throughout the global tropics, can take up gaseous nitrogen directly from the atmosphere and convert it to nitrate, a chemical form in which it can be used by the plant. This process is termed *nitrogen fixation*. In legumes, nitrogen fixation occurs in mutualistic association with bacteria called *Rhizobium* that colonize nodules in the plants' root systems. It is not yet clear exactly how much nitrogen fixation occurs in tropical forests and other tropical ecosystems, but indications are that it is far from inconsequential (Parker 1994). For example, one study estimated that, on average, there is around 20 kg (40 lbs) of nitrogen fixed per hectare per year throughout the Amazon Basin, which was considered a conservative estimate. An estimate of the total annual nitrogen input into the Amazon Basin concluded that nitrogen fixation accounts for over three times the nitrogen input that comes from precipitation (Salati and Vose 1984).

Certain epiphytic lichens convert gaseous nitrogen into usable form for plants in a manner similar to that of leguminous plants, which have nitrogen-fixing bacteria in their roots (Forman 1975). Between 1.5 and 8 kg of nitrogen per hectare (1–7 lbs per acre) was supplied annually by canopy lichens. These nitrogen-fixing lichens provide an important and direct way for nitrogen, vital to most biochemical processes, to enter the rainforest vegetation. Other studies have suggested that leaf-surface microbes and liverworts may facilitate uptake of gaseous nitrogen (Bentley and Carpenter 1980, 1984). Nitrogen fixation also occurs in termites because of the activities of microbes in termite guts (Prestwich et al. 1980; Prestwich and Bentley 1981). Because of the abundance of termites in the tropics, they may add a substantial amount of nitrogen to the soil.

Blackwaters and Whitewaters   *Figures 157, 162*

White, sandy soils are usually drained by blackwater rivers, best seen at areas such as the Rio Negro near Manaus, Brazil, or Canaima Falls in southeastern Venezuela. The tealike, dark, clear water is colored by tannins, phenolics, and related compounds, collectively called humic matter. Blackwaters are not confined to the Neotropics but occur in many places, including North America, especially such habitats as boreal peatlands and coniferous forests drained by mineral-poor, sandy soils (Meyer 1990). Part of the humic matter in blackwaters consists of *defense compounds* leached from fallen leaves. I'll discuss defense compounds in detail in chapter 6, but for now I want to point out that leaves are costly to grow on such poor soils because it is not easy to find raw materials to replace a fallen or injured leaf. Therefore, leaves on plants growing on white, sandy soils tend to concentrate defense compounds that help discourage herbivory. Leaf production can be less than half that in forests on richer soils, and leaf decomposition time can be in excess of two years (Jordan 1985a). When the old leaf finally does drop, the rainfall and microbial activity eventually leach out the tannins and phenols, making the water dark, a kind of "tropical tannin-rich tea." This water is also very clear because there is little unbound sediment to drain into streams and rivers. Gallery forests (igapo forests) bordering blackwater rivers are subject to seasonal flooding and their ecology is intimately tied to the flooding, cycle (chapter 8). Ecological relationships among species inhabiting blackwater forests are different in many ways from those of species in forests situated on richer soils (Janzen 1974).

In contrast to white, sandy soils, soils in places such as Puerto Rico, much of Costa Rica, and much of the Andes Mountains are not mineral-poor but mineral-rich. These eutrophic soils are much younger, mostly volcanic in origin, some up to 60 million years old, some much more recent. Though exposed to high rainfall and temperatures, they can be farmed efficiently and will maintain their fertility if basic soil conservation practices are applied. Because so much sediment leaches by runoff from the land into the river, waters that drain rich soils are typically cloudy and are called *whitewaters*. Please do not let this terminology confuse you. Whitewater rivers do not drain white, sandy soils; blackwaters do. Whitewaters drain nutrient- and sediment-rich Andean soils, and the term "white" refers to the cloudy appearance of the water, loaded as it is with sediment. Some have suggested that "mochawater" would be a more accurate descriptive term.

A dramatic example of the difference between blackwater and whitewater rivers occurs at the confluence of the Amazon River and the Rio Negro near Manaus, Brazil. The clear, dark Rio Negro, a major tributary draining some of the white, sandy soils of the ancient Guianan Shield, meets the muddy, whitewater Amazon, rich in nutrient load, draining mostly from the youthful though distant Andes. The result, locally called the "wedding of the waters," is a swirling maelstrom of soupy brown Amazonian water awkwardly mixing with clear blackwater from the Negro, a process that continues downriver for anywhere from 15 to 25 km (9–15 mi), until the mixing is complete. The most remarkable feature is that both soil types support impressive rainforest, igapo in the blackwater areas, varzea in the whitewater areas. See also chapter 8.

Rainforest Gaps and Tree Demographics

Forest Gaps   *Figures 24, 27, 30, 36*

Rainforest trees are not immortal, and each and every one of them will eventually die. Some remain in place, becoming dead, decaying snags, and others fall immediately to the ground, some bringing their root systems to the surface as they fall, some snapping off along the trunk and thus leaving their roots in the ground. A tree may blow down by windthrow or topple when weakened by termites, epiphyte load, or old age. Large branches can break off and drop. Indeed, one of the more common sounds heard when walking through rainforest is the sudden sound of a falling tree or large branch. When a rainforest tree or significant part of it falls, it creates a canopy opening, a *forest gap*. In gaps, light is increased, causing microclimatic conditions to differ from those inside the shaded, cooler, closed canopy. Air and soil temperatures as well as humidity fluctuate more widely in gaps than in forest understory.

Gaps can be of almost any size, and even the ecologists who study gaps have not yet agreed on a uniform range of gap sizes, particularly what defines a minimum area gap (Clark and Clark 1992). The general pattern is that most wet forests are characterized by many small gaps and few large gaps, where a large gap is defined as having an area in excess of 300 or 400 m² (3,200–4,300 sq ft) (Denslow and Hartshorn 1994). Large gaps, few in number but with much greater total area, can nonetheless comprise a large percentage of total gap space within a forest. An emergent tree, should it fall, can take several other trees with it, creating quite a large gap. Lianas, connecting several trees, increase the probability of multiple tree falls. When one tree goes, its liana connection to others can result in additional trees falling, or large branches being pulled down (Putz 1984). Treefalls are often correlated with seasonality. On Barro Colorado Island, tree falls peak around the middle of rainy season, when soils as well as the trees themselves are very wet and strong, gusty winds blow (Brokaw 1982). At La Selva, most gaps occur in June–July and November–January, the wettest months (Brandani et al. 1988, cited in Denslow and Hartshorn 1994). Landslides along steep slopes can open an entire swath of forest. In the Stann Creek Valley in central Belize, hurricanes have periodically leveled hundreds of acres of forest, a giant gap indeed. Gaps occur normally in all rainforests. In Amazonia, for instance, it has been estimated that 4–6% of any forest will be made up of recently formed gaps (Uhl 1988).

Hubbell and Foster (1986a, 1986b) have censused over 600 gaps in the BCI forest in Panama. They learned not only that large gaps are less common than small gaps, but also that gap size and frequency change significantly as one moves vertically, from forest floor to canopy. They assert that a typical gap is shaped roughly like an inverted cone, a pattern resulting in expansion of gap area as one moves higher in the canopy, and adding yet another component of structural complexity to an already complex forest. Since both horizontal and vertical heterogeneity of a forest are significantly increased by gaps, gaps become a potentially important consideration in explaining high biodiversity (page 95).

Simply because they admit light, gaps create opportunities for rapid growth and reproduction. Many plant species utilize gaps to spurt their growth, and at least a few are dependent upon gaps (Brokaw 1985; Hubbell and Foster 1986a; Murray 1988; Clark and Clark 1992). Of 105 canopy tree species studied as saplings at La Selva, about 75% are estimated to depend at least in part on gaps to complete their life cycle (Hartshorn 1978).

Gaps create a diverse array of microclimates, affecting light, moisture, and wind conditions (Brokaw 1985). Measurements made at La Selva in Costa Rica indicate that gaps of 275–335 sq m (3,000–3,600 sq ft) experience 8.6–23.3% full sunlight, compared with interior forest understory, which receives only 0.4–2.4% full sunlight (Denslow and Hartshorn 1994). Thus a large gap can offer plants up to fifty times as much solar radiation as interior forest. Further, it is "high-quality" sunlight, with wavelengths appropriate for photosynthesis. By contrast, the shaded forest understory is generally limited not only in total light intensity but in wavelengths from 400 to 700 nannometers, the red and blue wavelengths most utilized in photosynthesis (Fetcher et al. 1994). Most high-quality solar radiation (61–77%) within a shady rainforest understory is received in the form of short-duration sun flecks (Chazdon and Fetcher 1984). The total amount and quality of solar radiation is probably the single largest limiting factor to plant growth inside tropical forests, thus the importance of gaps. This restriction may be evident in the fact that many understory herbs have leaves that are unusually colored: blue iridescence, velvety surface sheen, variegation, and red or purple undersides (Fetcher et al. 1994). The suggestion has been made that abaxial anthocyanin, the pigment responsible for the red underside of some leaves, is physiologically adaptive in aiding the plant in absorption of scarce light (Lee et al. 1979), but this has yet to be demonstrated.

Julie Denslow (1980) suggested that rainforest trees fall roughly into three categories, depending upon how they respond to gaps and gap size: (1) large-gap specialists whose seeds require high temperatures of gaps to germinate and whose seedlings are not shade tolerant, (2) small-gap specialists whose seeds germinate in shade but whose seedlings require gaps to grow to mature size, and (3) understory specialists that seem not to require gaps at all.

Since Denslow's study, other researchers have attempted to classify rainforest tree species on the basis of their degrees of dependence on gaps. It is clear that there exist pioneer species that require gaps (see below). But the picture has become more complicated since Denslow first suggested her three-category schema. Many, if not most, shade-tolerant tree species show no gap association but rather demonstrate high levels of growth plasticity, meaning that they can survive and even slowly grow under the deeply shaded conditions of the forest interior, but still grow much more rapidly in gaps (Clark and Clark 1992; Denslow and Hartshorn 1994). The leguminous tree *Pentaclethra macroloba*, common at La Selva, is typical of many trees in that it is highly tolerant of deep shade but will nonetheless grow rapidly in high light conditions provided by gaps (Fetcher et al. 1994). Only species that are completely shade-intolerant require gaps for growth and reproduction. For many years it has been known that sapling trees of some species are capable of remaining in the understory, small but healthy, continuing their upward growth when adequate light becomes available (Richards 1952). Understory specialists do not necessarily require gaps but utilize them when the opportunity is presented.

Once in a gap, many tree and shrub species show higher reproductive outputs, and thus larger fruit crops create more competition among plants for frugivores to disperse their seeds (Denslow and Hartshorn 1994, and see page 92).

The ecology of gap-dependent pioneer species is generally well understood. Brokaw (1982, 1985) studied regeneration of trees in thirty varying-sized forest gaps on Barro Colorado Island. Pioneer species produced an abundance of small seeds, usually dispersed by birds or bats, and capable of long dormancy periods. In another study, Brokaw (1987) focused only on three pioneer species and learned that the three make up a continuum of what he called regeneration behavior. One species, *Trema micrantha*, both colonized and grew very rapidly, growing up to 7 m (22.7 ft) per year. This species only colonized during the first year of the gap. After that, it could not successfully invade, presumably due to competition with other individuals. The second species, *Cecropia insignis*, invaded mostly during the first year of the gap, but a few managed to survive and enter large gaps during the second and third year. This species grew more slowly (4.9 m [16 ft] per year) than *Trema*. The third colonist was *Miconia argentea*, which grew the most slowly of the three (2.5 m [8.2 ft] per year) but was still successfully invading the gap up to seven years following gap formation. Brokaw's study reveals how the three species utilize different growth patterns to reproduce successfully within gaps. Such subtle distinctions may help explain the apparent coexistence of so many different species within rainforest ecosystems.

Forest Demographics

How long do rainforest trees survive? How long does it take for a canopy giant to grow from seedling to adult? Does most growth occur in rainy or dry season? How do short-term climatic fluctuations influence forest dynamics? What forces determine the probable survivorship of any given tree? These and many other questions comprise the study of rainforest demographics. To answer these questions it is necessary to initiate long-term, detailed studies of specific tracts of forest, monitoring the fate of literally each tree. Studies of this nature have been ongoing at La Selva in Costa Rica (Clark and Clark 1992; McDade et al. 1994) and Barro Colorado Island in Panama (Hubbell and Foster 1990, 1992; Condit et al. 1992). What follows is a summary of these exhaustive, continuing efforts.

La Selva: The Life Histories of Trees

The total inventory of vascular plants known to occur at La Selva now totals 1,458 species, the vast majority of them present in low numbers, if not outright rare. Suppose you happened to be one of these plants, say a *Dipteryx panamensis*, a common emergent tree that favors alluvial soils. If longevity is your goal, you'd be far better off being a Great Basin bristlecone pine (*Pinus longaeva*), atop the cold, windswept White Mountains of the western Great Basin Desert. Were you a bristlecone pine, you could anticipate living more than 4,000 years. Tropical trees show no comparable longevity. Ecologists discuss forest turnover, which, though subject to slightly differing definitions among researchers, generally means the average time that a given tree (defined within a certain size range) will survive in a particular spot. So, if you randomly select any place on a rainforest floor and imagine you are now a *D. panamensis*, which is at least 10 cm (about 4 in) in diameter, how long before you are somehow destroyed or die? The answer, for La Selva, is known: The rate of local disturbance is sufficiently high that the entire forest is estimated to turn over approximately every 118 plus or minus 27 years (Hartshorn 1978), and 6% of the primary forest is in young gaps at any one time (Clark 1994). One study, from 1970 to 1982, indicated an annual mortality rate of 2.03% for trees and lianas greater than 10 cm diameter (Lieberman et al., cited in Clark 1994). Overall, adult survivorship of more than 100–200 years seems rare for subcanopy and canopy trees at La Selva (Clark 1994). Estimates from other forests are similar. At Cocha Cashu in Peru, a forest on rich soils, mortality rate of adult trees (>10 cm diameter) was 1.58% per year, implying an average life of 63.3 years (Gentry and Terborgh 1990). At San Carlos de Rio Negro in Amazonian Venezuela, mean annual mortality rates for trees greater than 10 cm diameter breast height (dbh) was 1.2% (Uhl et al. 1988a). Most trees died in such a way as to create small gaps (large gaps were much rarer), and approximately 4–6% of the forest area was in gap phase at any given time. At Manaus, Brazil, mortality was 1.13% for adult trees, with a turnover time of from 82 to 89 years (Rankin-De-Merona et al. 1990). Keep in mind these turnover rates are for adult trees with a minimum size of more than 10 cm diameter. A tree often lives many years before attaining such a diameter, so the total age, from seedling to death, can be considerably longer. In the Manaus study it was learned that, in general, the larger a tree grew to be in diameter, the longer its probably life span from that point onward. In other words, for trees as large as 55 cm dbh, turnover time increases to 204 years.

It is even tougher, however, to be a seedling or sapling than an adult tree. Any recently germinated seedling stands a fairly high chance of being smashed by a falling branch, or a single fruit, or whole tree, or perhaps buried beneath a fallen palm frond or some other leaf. Or, it could be the next meal for a herbivore. For *Dipteryx panamensis*, seedlings ranging in age from 7 to 59 months experienced a 16% mortality rate from litter fall alone (Clark and Clark 1987). Of course many seeds never germinate because they are destroyed by a wide diversity of seed predators as well as fungal pathogens. Mortality rates are consistently highest in juvenile plants, sometimes very high indeed, declining steadily as the plants age (Denslow and Hartshorn 1994). For example, in a study of six tree species, highest mortality rates, from 3% to 19% per year, occurred in the smallest saplings (Clark and Clark 1992). Mortality rates were much lower for intermediate to large juvenile sizes. In all, it requires probably more than 150 years for growth from small sapling stage to canopy (Clark and Clark 1992), which, when considering the estimated mortality rates of adult trees, indicates that fully adult trees do not persist all that long.

For most of a tree's life cycle, light is a strongly limiting factor. Growth rates of young trees in shaded interior forest are very much less than in more lighted, open areas. Trees such as *Dipteryx panamensis* show extremely slow growth in low light conditions but are capable of growing taller and wider very quickly in a gap. For this reason, growth rates tend to fluctuate several times during the typical life cycle of a tree. Gaps open, close, and can reopen, so that any given tree might experience several periods of rapid growth (when in gaps) alternating with periods of extreme slow growth (under fully closed canopy). As would be expected, most tropical forest trees and shrubs show high levels of shade tolerance, with an accompanying high degree of growth plasticity: the ability to survive very low light levels of the forest understory and grow rapidly in gaps (Denslow and Hartshorn 1994).

The existence of emergent trees has long been recognized as a characteristic of rainforests. Of what possible benefit is it to a tree to invest additional energy to grow above the majority of other trees in the canopy? Added light availability is certainly a possibility. But in a La Selva study of five emergent tree species, these species showed significantly lower adult mortality rates than nonemergent trees (Clark and Clark 1992). Perhaps emergents are more protected from being damaged by other falling trees, given that their crowns rise above the rest.

Barro Colorado Island: The Dynamics of Drought

Beginning in 1980, a 50-ha (123.5-acre) permanent plot was established at BCI. All free-standing woody plants that were at least 1 cm dbh were identified to species, measured, and mapped. Censuses were done in 1982, 1985, and 1990. Over the three censuses, 310 species were recorded in the plot, with data on 306,620 individual stems (Condit et al. 1992). In the brief timeframe of this study, weather was an unexpectedly strong factor. An unusually protracted dry season coincident with a strong El Niño (see chapter 1) brought a severe drought to BCI in 1983.

Mortality rates were strongly elevated in the years immediately following the drought. From 1982 to 1985, trees with diameters in excess of 8 cm experienced a mortality rate of 3.04% per year, nearly three times higher than measured before the major drought (Clark 1994). Compared with mortality during the interval 1985–1990, annualized forestwide mortality from 1982 to 1985 was elevated 10.5% in shrubs, 18.6% in understory trees, 19.3% in subcanopy trees, and 31.8% in canopy trees. For trees with dbh greater than 16 cm, mortality was elevated fully 50% (Condit et al. 1992). The increased death rate among vegetation species was attributed to the drought, and approximately two-thirds of the species in the plot experienced elevated mortality from 1982 to 1985.

Those plants surviving the drought tended strongly to show elevated growth rates. For example, growth of trees of 16–32 cm was more than 60% faster in 1982–1985 than in 1985–1990. While this result might be surprising at first, it is really to be expected. The death of so many trees permitted much more light into the forest (the gap effect) and reduced root competition for water and nutrients among plants. Though total gap area on the plot initially increased after the drought, it had returned to its predrought level by 1991, an indication of how rapidly the surviving plants responded to the influx of light.

Many species' populations experienced changes in abundance during the period of the study, 40% of them changing by more than 10% in the first three years of the census (Hubbell and Foster 1992). Ten species were lost from the plot and nine species migrated into the plot from 1982 to 1990. Nonetheless, there was remarkable constancy in the number of species and number of individuals within the plot at any given time: 1982 = 301 sp., 4,032 ind.; 1985 = 303 sp., 4,021 ind.; 1990 = 300 sp., 4,107 ind. What happened is that the drought killed many trees but created opportunities for additional growth such that the deceased plants were very quickly replaced. The speed of the replacement process was a surprise to the researchers (Condit et al. 1992).

The analysis of the BCI data suggests two major and important conclusions: first, that the forest is highly responsive to short-term fluctuations caused by climate and that the forest as a whole remains intact, though many species undergo population changes; and second, that the forest may be undergoing a longer-term change in species composition. This latter conclusion is based on the fact that there has been a decline of approximately 14% in annual precipitation over the past seven decades, dropping from 2.7 m (8.9 ft) total in 1925 to 2.4 m (7.9 ft) in 1995. The researchers hypothesize future local extinctions of 20–30 species, each of which requires a high level of moisture. Another reason for suggesting a long-term change in species composition is that after the 1983 drought, rare species declined more than common species, suggesting, of course, that not only might the community be changing, but plant species richness might be in decline (Condit et al. 1992).

The BCI study has demonstrated the dynamics of change as they relate to both a climatic drying period and short-term acute drought. The researchers summarize:

> Are there stabilizing forces in tropical forest communities that might buffer them against perturbations caused by climate change or other human activities? The Barro Colorado Island forest suffered a severe drought, yet the overall structure of the forest bounced back. There is a regulating force at work here: remove a tree and a tree grows back. But this force only preserves the forest as a forest, not the diversity of tree species it contains. (Condit et al. 1992)

The BCI study has also added valuable insight into forces that affect biodiversity, and thus we shall revisit this study in the following chapter.

Disturbance and Ecological Succession

As you traverse the Neotropics you will undoubtedly notice much open, brushy habitat as well as areas in which plants grow densely but not yet to forest height. In many places, plant cover is so dense as to be impenetrable without a well-sharpened machete. Living blankets of vines envelope thorny brush as tall, spindly trees and feathery palms push aggressively upward above the tangled mass. Clumps of huge-leaved plants, named heliconias for their sun-loving habit, compete for solar radiation against scores of legumes and other fast-growing plants. This sunny, tangled assemblage of competing plants is the habitat we can correctly call "jungle." Jungles are representative of disturbed rainforest, and, to the trained eye, evidence of varying degrees of forest disturbance is seemingly everywhere.

During the eighteenth and nineteenth centuries, eastern North American forests were felled so that the land could be converted to agriculture and pasture. Approximately 85% of the original New England forest was cut and in use for homestead, agriculture, or pasture at any given time during the early to mid-1800s. Following the abandonment of agricultural land as large-scale farming moved to the midwestern states, the open lands were recolonized in a natural way by various tree and shrub species and so forests gradually renewed their claim on the New England landscape. Henry David Thoreau was one of the first authors to comment about this process of vegetation replacement dynamics, now called ecological succession.

Disturbed land, whether tropical or not, gradually returns to its original or near original state (prior to disturbance) through a somewhat haphazard but nonetheless roughly predictable succession of various species over time. An herbaceous field left undisturbed eventually becomes woody forest through a process of species replacement. Succession is complex and affected by many factors, including chance. What is fundamentally involved is that each species is adapted somewhat differently to such factors as light, temperature fluctuation, and growth rate, and thus species with effective dispersal or long seed lives that grow quickly in high light tend to invade first, followed by slower-growing but more competitive species. In the tropics, because of greater richness of species, variable levels of soil fertility, and differing levels of usage, successional patterns demonstrate complex and differing patterns from site to site (Ewel 1980; Buschbacher et al. 1988).

Succession does not have to be initiated by human activity, as nature regularly disturbs ecological systems. Many species are adapted to exploit disturbed areas; some species, in fact, cannot grow unless they colonize a disturbed site. And disturbed sites are anything but uncommon. Heavy rainfalls, hurricanes, fires, lightning strikes, and high winds destroy individual canopy trees and create forest gaps, sometimes leveling whole forest tracts. Isolated branches, often densely laden with epiphytes, can break off and crash down through the canopy. Natural disturbances within a forest open areas to sunlight, and a whole series of plant species are provided a fortuitous opportunity to grow much more rapidly.

Many native peoples in tropical America have skillfully exploited the tendency of the land to return to its original state following disturbance and have adapted their agricultural practices to follow nature's pattern (chapter 7).

The Jungle—Early Succession in the Neotropics

Figures 36, 65, 66, 67, 68

The dictionary definition of jungle is "land overgrown with tangled vegetation, especially in the tropics" (*Oxford American Dictionary*, 1980). This definition, though descriptively accurate, does not say what a jungle is ecologically. Jungles represent large areas where rainforest has been opened because of some disturbance event that has initiated an ecological succession. Bare land is quickly colonized by herbaceous vegetation. Seeds dormant in the soil now germinate. Within Amazonia, a typical square meter of soil is estimated to contain 500–1,000 seeds (Uhl 1988). In addition to the soil seed bank, new seeds are brought in by wind and animals. Soon vines, shrubs, and quick-growing palms and trees are all competing for a place in the sun. The effect of this intensive, ongoing competition for light and soil nutrients is the "tangled vegetation" of the definition above.

Just as in rainforests, high species richness is true of successional areas, and species composition is highly variable from site to site (Bazzaz and Pickett 1980). It is nonetheless possible, however, to provide a basic description and point out some of the most conspicuous and common plants seen during tropical succession (Ewel 1983). Though successions on rich soils usually result in the eventual redevelopment of forest, on poor soil, repeated elimination of rainforest and depletion of soil fertility can sometimes result in conversion of the ecosystem to savanna rather than forest.

Gap openings provide conditions conducive to the growth of seedlings and saplings, and large gaps are colonized by shade-intolerant species. However, gaps, especially small ones, do not follow the outline of ecological succession presented below, which is based on vegetation development beginning with generally bare soil. Most forest gaps have resident seedling and sapling shrubs and trees as well as other understory plants (see below).

Herbaceous weeds, grasses, and sedges of many species are first to colonize bare soil. Soon these are joined by shrubs, vines, and woody vegetation, whose seeds may have been present all along but require longer to germinate and grow. Large epiphytes are almost entirely absent from early successional areas. Plant biomass usually increases rapidly as plants compete against one another. In one Panamanian study, biomass increased from 15.3 to 57.6 dry tons per hectare from year 2 to year 6 (Bazzaz and Pickett 1980). This rapid growth reduces soil erosion as vegetation blankets and secures the soil. Studies in Veracruz, Mexico, have shown that young (10 months and 7 years old) successional areas take up nutrients as efficiently as mature rainforest (Williams-Linera 1983). Young successional fields actually have more nutrients per unit biomass than closed canopy forests. In Amazonia, succession on abandoned pastures does not result in significant depletion of soil nutrient stocks, though there are major differences between nutrient stocks in mature forest and those in disturbed areas. Successional sites have higher nutrient concentrations in their biomass than is the case in mature forests, and there are more extractable soil nutrients on successional sites; thus successional sites have a lower proportion of total site nutrients stored in biomass than does mature forest (Buschbacher et al. 1988). Because of the density of competing plants, the leaf area index may reach that of a closed canopy forest within 6–10 years, although the vegetation is still relatively low growing, and the species composition at that time is not at all similar to what it will be as the site returns to forest. With such a high LAI, high competition among plants for access to light would be expected. By the time the site is about 15 years from the onset of succession, the ground conditions can be similar to those in closed canopy forest, though, again, the species composition is not the same. In only 11 months from burning, study plots underwent a succession such that vegetation attained a height of 5 m (16.4 ft) and consisted of dense mixture of vines, shrubs, large herbs, and small trees (Ewel et al. 1982).

Some major and some subtle physiological changes occur in plants that live in early successional, high-light environments (Fetcher et al. 1994). Photosynthesis rates in early successional species are much higher in full sun than in partial or full shade; these plants are obviously adapted to grow quickly. Some early successional plants that can grow in both shade and sun develop significantly thicker leaves in full sun. Some studies indicate that stomatal densities increase when a species is grown in full sun versus partial or full shade. Increased stomata permit increased rates of gas exchange, necessary when photosynthesis rate is elevated. In addition, successional species tend to allocate considerable energy to root production, an aid in rapid uptake of soil nutrients, which are then used to the utmost efficiency (Uhl et al. 1990).

During early succession, many plant species called *colonizers* tend to be small in stature, grow fast, and produce many-seeded fruits. In later succession, most plants tend to be larger, grow more slowly, and have fewer seeds per fruit. These plants, often called *equilibrium species*, are adapted to persist in the closed canopy (Opler et al. 1980). While this overview is generally illuminating, these two broad categories are probably insufficient to describe the true complexity of succession. Because of physiological plasticity in various light regimens, distinctions between successional categories blur (Fetcher et al. 1994).

Succession to an equilibrium forest requires many years and in some places, because of disturbance frequency, may really never be attained. Dennis Knight (1975), in his study on Barro Colorado Island in Panama, found that plant species diversity of successional areas increased rapidly during the first 15 years of succession. Diversity continued to increase, though less rapidly, until 65 years. Following that, diversity still increased, though quite slowly. Knight concluded that, after 130 years of succession, the forest was still changing. He was correct, though he underestimated the actual age of the site. Hubbell and Foster (1986a, 1986b, 1986c) note that forest at BCI is actually between 500 and 600 years old, and they agree with Knight that it is not yet in equilibrium. They conclude that though initial succession is rapid, factors such as chance, climatic changes, periodic drought, and biological uncertainty from interactions among competing tree species act to prevent establishment of a stable equilibrium (Hubbell and Foster 1986c, 1990; Condit et al. 1992). This means that BCI remains in a dynamic state, continuing to change. Such a condition is probably the norm for tropical rainforests. The Rio Manu floodplain forest in Amazonian Peru shows perhaps a very long term successional pattern. Pioneer tree species such as *Cecropia* (see below) dominate the early succession, to be followed by large, statuesque *Ficus* and *Cedrelia*. These are eventually replaced by slow-growing emergent trees such as *Brosimum* and *Ceiba*, most of which have been present essentially since the succession began. The overall pattern of succession at Manu may require as much as 600 years (Foster 1990b).

Regeneration Pathways in Amazonia

By now it should be apparent that disturbances that initiate ecological successions range from small scale to large scale in a continuum-like pattern. In addition, disturbance effects and subsequent regeneration patterns are strongly influenced by duration of the disturbance as well as disturbance frequency (Uhl et al. 1990). Extensive studies in Amazonia conducted by Christopher Uhl and colleagues (Uhl and Jordan 1984; Uhl et al. 1988a; Uhl et al. 1990) have demonstrated differences between regeneration patterns in small-scale and large-scale disturbances. Small scale is defined as a disturbance area of 0.01–10 ha (0.025–25 acres), areas typical of most tree fall gaps. Large scale is 1–100,000 km² (38,310 sq mi), with causal factors being principally floods and fires. These scales are roughly analagous to human disturbances caused by slash-and-burn agriculture (small scale, see chapter 7) and conversion of forest to pasture (large scale, see chapter 14).

Following disturbance, recovery and regeneration can occur from the following possible regeneration pathways: (1) from seedlings and saplings already present in the forest understory (termed the "advance regeneration" pathway); (2) from vegetative sprouting from stem bases and/or roots (which remain after trees are disturbed); (3) from recolonization by germination of seeds already present in soil (called the "seed bank"); (4) from the arrival of new seeds brought by wind or animal dispersal (Uhl et al. 1990).

In cases of small-scale disturbance, the advance regeneration pathway dominates throughout the Neotropics. In Amazonia, there are usually between 10 and 20 seedlings and small saplings (<2 m tall) in every square meter of forest floor, most of which can persist for very long periods in the darkened understory. These account for over 95% of all trees more than 1 m tall four years after gap formation (Uhl et al. 1988a; Uhl et al. 1990). The second pathway, sprouting, is also common in many tree species in small gaps. Large-scale gaps can result in the death of understory trees, destroyed by immersion in flood or by fire. Regeneration in large gaps is from a combination of vegetative sprouting plus germination of seeds in the soil, plus import of seeds by dispersal mechanisms.

Critical to regeneration is the presence of viable seeds in the soil seed bed plus the added distribution of seeds into disturbed sites (carried either by wind or by animal dispersers). Further, once the seeds are so located, they must germinate and the seedlings must survive. Research at Cocha Cashu Biological Station has indicated some of the ecological complexities that accompany regeneration from seed (Silman 1996). This study has significant implications for understanding tree species richness in Amazonia and will be discussed further in the next chapter.

Fire in Amazonia

While you are standing in a forest experiencing 100% relative humidity, watching in wonder the intense deluge of the pouring rain, the thought of the rainforest catching fire and burning seems at best a fanciful notion. Well, stick around long enough and you may change your mind. Evidence has accumulated suggesting that for the past few thousand years, the most important natural, large-scale disturbance factor throughout Amazonia has been fire (Uhl 1988; Uhl et al. 1990). There is an abundance of charcoal residue in central and eastern Amazonia soils, and studies from the Venezuelan Amazon along the upper Rio Negro employing radiocarbon dating of the sediments indicate that during the past 6,000 years there have been several major fires, occurring perhaps during periods of extended dryness (Absy 1985; Sanford et al. 1985; Uhl et al. 1988b; Uhl et al. 1990). The reality of large-scale Amazonian fires, even if infrequent, adds yet another disturbance dimension to the dynamics of rainforests, a dimension that may help us to explain how the high tree biodiversity of the region came to be and is maintained.

Resilient Pastures—A Lesson from Amazonia

Most students of Neotropical ecology are aware of the fact that large forested areas of Amazonia have been cut and converted to pasture (see chapter 14). What happens when cattle pastures are abandoned? Does the natural vegetation recover and reestablish a forest? Studies from Para, Brazil, in eastern Amazonia indicate that successional patterns do normally result in the reestablishment of forest (Buschbacher et al. 1988; Uhl et al. 1988c).

Each of the sites in the Amazonian study had been cut and burned and then used for cattle pasture. Sites ranged in age (from abandonment) from two to eight years and, depending upon the site, had received either light, medium, or heavy use for up to thirteen years. Vegetation composition, structure, and biomass accumulation were carefully documented. In areas subject to light use, succession was quite rapid, with a biomass accumulation of about 10 tons per hectare annually, or 80 tons after eight years. Tree species richness was high, with many forest species invading the sites. Moderately grazed pastures also recovered rapidly when abandoned, but biomass accumulation was only about half what it was on lightly grazed sites, and tree species richness was lower as well. Heavily grazed sites remained essentially in grasses and herbaceous species, with few trees invading and a biomass accumulation of only about 0.6 ton annually per hectare. The conclusion drawn from this study was that most Amazonian lands subjected to light or moderate grazing can recover to forest. Only in areas subject to intensive grazing for long periods, areas that were estimated to represent less than 10% of all pastureland in northern Para, was there a probability that forest recovery might not occur. Nonetheless, the authors caution that even the recovered successional sites contained neither the exact physiognomy nor exact species composition of the original undisturbed sites. Moreover, heavy, continued disturbance clearly affected the successional pattern negatively (Uhl et al. 1988c). The subject of pasture regeneration will be revisited in chapter 14.

A Resilient Rainforest—A Lesson from Tikal   *Figure 61*

Tikal, a great city of the classic period of Mayan civilization, provides an example of how rainforest can return after people abandon an area that has been largely deforested and used for agriculture and urbanization. Located on the flat Petén region of western Guatemala, Tikal was founded around 600 B.C. and flourished from about A.D. 200 until it was mysteriously abandoned around the year 900. Anthropologists are still far from agreement over the odd total abandonment of the classic city and the subsequent deterioration of Mayan society (well in advance of the Spanish conquest). At its peak, Tikal served as a major trade center. Maize (corn), beans, squash, chile peppers, tomatoes, pumpkins, gourds, papaya, and avocado were brought from small, widely scattered farms to be sold in the busy markets of the city.

An estimated population of 50,000 lived in Tikal, which spread over an area of 123 km^2 (47 sq mi), protected by earthworks and moats. As is the case in cities today, Tikal was surrounded by densely populated suburbs. Further, the society practiced sophisticated intensive agriculture (LaFay 1975; Flannery 1982; Hammond 1982, and see page 182). The majestic, pyramid-like temples, excavated relatively recently in this century, now serve as silent memorials where tourists come to see what remains and to reflect on the past. This long-deceased civilization had developed a calendar equally accurate as today's, a complex writing system that still has not been entirely deciphered, and a mathematical sophistication that included the concept of zero. The sight of the Great Plaza and Temple I, the Temple of the Giant Jaguar, enshrouded in the cool, early morning tropical mist, romantically transports the mind's eye back to the brief time when Tikal was the Paris, the London, the New York City of Mesoamerica.

Today Tikal is isolated, surrounded, enclosed really, by lush rainforest. The city itself had to be rediscovered and excavated, so much had the rainforest enveloped it. This metropolis was literally under rainforest, and much of it still is, its once crowded plazas, thoroughfares, and temples overgrown by epiphyte-laden milk trees *(Brosimum alicastrum)*, figs, palms, mahoganys, and chicle trees, to name but a few. From atop the sacred temples, one can watch spider and howler monkeys cavort in the treetops. Agoutis and coatis shuffle through the picnic grounds, amusing tourists while searching for food scraps. Toucans and parrots pull fruits from trees growing along what was once the central avenues leading to and from the city. Birdwatchers search the tall comb of Temple II, trying to spot nesting orange-breasted falcons *(Falco deiroleucus)*. My point is that this once great metropolis of 50,000 Mayans, covering many square miles, was abandoned and subsequently reclaimed by rainforest. Tikal was one of the largest forest gaps in the history of the American tropics, but, given hundreds of years, the gap closed.

Though many areas of rainforest (i.e., those on white, sandy soils) are fragile, Tikal demonstrates that, at least on more fertile soils, the process of ecological succession can restore rainforest, even after profound alteration. All of Tikal is second-growth forest, and in some respects it is certainly different from what it probably was before Mayans converted it to city and farmland. Nonetheless, it is now diverse, impressive forest, with a biodiversity that seems generally reflective of the region.

Recent studies suggest that Tikal is not an isolated case of rainforest regeneration. The Darien of southern Panama, a remote region that is today rich, diverse rainforest, was subject to extensive human disturbance. A study of the pollen and sediment profiles from the region reveal that much of the landscape was historically planted with corn and subject to frequent fires, probably set by humans. Only after the Spanish conquest was the region abandoned, allowing succession to occur. Thus the lush and seemingly pristine rainforest that defines the Darien today is only about 350 years old (Bush and Colinvaux 1994). It's a successional forest, still regrowing.

The occurrence of disturbed areas and gaps of various sizes has probably always been true of rainforests, and many species have adapted to this fact. The high biodiversity of rainforest trees as well as other taxa may be caused in large part by frequent and irregular disturbances of varying magnitudes that make it possible for a range of differently adapted species to coexist within the heterogeneous conditions created by the disturbance regime. Much more on this topic will be discussed in the next chapter.

Some Examples of Widespread Successional Plants

Heliconia *Figures 27, 49*

Among the most conspicuous tenants of successional areas are the heliconias (*Heliconia* spp., family Helicaniaceae), recognized by their huge, elongate, paddle-shaped leaves (bananas are closely related) and their distinctive, colorful red, orange, or yellow bracts surrounding the inconspicuous flowers (in some species, bracts are reminiscent of lobster claws, hence the common name "lobster-clawed" heliconia). Though some heliconias grow well in shade, most grow best where light is abundant, in open fields, along roadsides, forest edges, and stream banks. They grow quickly, clumps spreading by underground rhizomes. Named for Mt. Helicon in ancient Greek mythology (the home of the muses), these plants are all Neotropical in origin, with approximately 150 species distributed throughout Central and South America (Lotschert and Beese 1983).

Colorful, conspicuous bracts surrounding the smaller flowers attract hummingbird pollinators, especially a group called the hermits (page 262), most of which have long, downcurved bills permitting them to dip deeply into the twenty yellow-greenish flowers within the bracts (Stiles 1975). When several species of heliconia occur together, they tend to flower at different times, a probable evolutionary response to competition for pollinators such as the hermits (Stiles 1975, 1977). Sweet, somewhat sticky nectar oozes from the tiny flowers into the cuplike bracts where it is sometimes diluted with rainwater.

Heliconias produce green fruits that ripen and become blue-black in approximately three months. Each fruit contains three large, hard seeds. Birds are attracted to heliconia fruits and are important in the plant's seed dispersal. At La Selva in Costa Rica, Stiles (1983) reports that twenty-eight species of birds have been observed taking the fruits of one heliconia species. The birds digest the pulp but regurgitate the seed whole. Heliconia seeds have a six- to seven-month dormancy period prior to germination, which assures that the seeds will germinate at the onset of rainy season.

Piper

Figure 66

Piper (*Piper* spp., family Piperaceae) is common in successional areas as well as forest understory, with about 500 species occurring in the American tropics (Fleming 1983). Most are shrubs, but some species grow as herbs, and some are small trees. The Spanish name, *Candela* or *Candellillos*, means "candle" and refers to the plant's distinctive flowers, which are densely packed on an erect stalk. When immature, the flower stalk droops, but it becomes stiffened and stands fully upright when the flowers are ripe for pollination. Piper flowers are apparently pollinated by many species of bees, beetles, and fruit flies; their pollination seems inspecific. On the other hand, seed dispersal is highly specific. Small fruits form on the spike and are eaten, and the seeds subsequently dispersed by one group of bats in the genus *Carollia*, called "piperphiles." Several species of Piper may occur on a given site, but evidence suggests that they do not all flower at the same time; thus, like heliconias, competition among them for pollinators is reduced as well as the probability of accidental hybridization (Fleming 1985a, 1985b).

Mimosas and Other Legumes

Along roadsides and in wet pastures and fields throughout the Neotropics grow mimosas, spreading, spindly shrubs and trees. Mimosas are legumes (family Leguminosae), perhaps the most diverse family of tropical plants. Virtually all terrestrial habitats in the tropics are abundantly populated by legumes, including not only mimosas but acacias (*Acacia* spp.), beans, peas, and trees such as *Samanea saman, Calliandra surinamensis,* and *Caesalpinia brasiliensis* (which gave Brazil its name). The colorful, flamboyant tree *Delonix regia,* a native of Madagascar, has been widely introduced in the Neotropics. Amazonian rainforests typically contain more legume species than any other plant family (Klinge et al. 1975). Legumes have compound leaves and produce seeds contained in dry pods. Many legumes have spines and some, like the sensitive plant *Mimosa pudica,* have leaves that quickly lose their turgor pressure and close when touched.

Mimosa pigra, an abundant species, has round pink flowers and is unusual because it flowers early in the rainy season. The flowers, which are pollinated by bees, become flattened pods 8–15 cm (3–6 in) m length that are covered by stiff hairs. Stems and leaf stalks (petioles) are spiny and are not browsed by horses or cattle. Experiments with captive native mammals such as peccaries, deer, and tapir show that these creatures refuse to eat Mimosa stems on the basis of odor alone (Janzen 1983b). Given its apparent unpalatability, it is easy to see why *Mimosa pigra* thrives in open areas. Janzen (1983b) reports that seeds are spread by road construction equipment, accounting for the abundance of this species along roadsides.

Cecropias   *Figures 30, 53, 54*

As a group, cecropias (*Cecropia* spp., family Moraceae) are one of the most conspicuous genera of trees in the Neotropics. Cecropias occur abundantly in areas of large light gaps or secondary growth. Pioneer colonizing species, cecropias are well adapted to grow quickly when light becomes abundant. Studies in Surinam have revealed that seeds remain viable in the soil for at least two years, ready to germinate when a gap is created. Cecropia seeds are anything but rare. An average of 73 per square meter were present on one study site in Surinam (Holthuijzen and Boerboom 1982). Because there are so many viable seeds present, cecropias sometimes completely cover a newly abandoned field or open area. They line roadsides and are abundant along forest edges and stream banks.

Cecropias are easy to recognize. They are thin-boled, spindly trees with bamboolike rings surrounding a gray trunk. Their leaves are large, deeply lobed, and palmate, somewhat resembling a parasol. They look a bit like gigantic horse chestnut leaves. Leaves are whitish underneath and frequently insect damaged. Dried, shriveled cecropia leaves that have dropped from the trees are a common roadside feature in the tropics. Some cecropias have stilted roots, but the trees do not form buttresses.

Cecropias are effective colonizers. In addition to having many seeds lying in wait in the soil, once germinated, cecropias grow quickly, up to 2.4 m (8 ft) in a year. Nick Brokaw recorded 4.9 m (16 ft) of height growth in one year for a single cecropia. They are generally short-lived, old ones surviving about thirty years, although Hubbell (pers. comm. 1987) reports that once established in the canopy *Cecropia insignis* can persist nearly as long as most shade-tolerant species. Cecropias are moderate in size, rarely exceeding 18.3 m (60 ft) in height, though Hubbell (pers. comm. 1987) has measured emergent cecropias 40 m (131 ft) tall. They are intolerant of shade, their success hinging on their ability to grow quickly above the myriad vines and herbs competing with them for space. To this end, cecropias, like many pioneer tree species, have a very simple branching pattern (Bazzaz and Pickett 1980) and leaves that hang loosely downward. Vines attempting to grow over a developing cecropia can easily be blown off by wind, though I have seen many small cecropias that were vine-covered (see below). Cecropias have hollow stems that are easy to sever with a machete. I've watched Mayan boys effortlessly chop down 5-m-tall cecropias. Hollow stems may be an adaptation for rapid growth in response to competition for light, as they permit the tree to devote energy to growing tall rather than to the production of wood.

Cecropias have separate male and female trees and are well adapted for mass reproductive efforts. A single female tree can produce over 900,000 seeds every time it fruits, and it can fruit often! Flowers hang in fingerlike catkins, with each flower base holding four long, whitish catkins. Research in Mexico (Estrada et al. 1984) showed that forty-eight animal species, including leafcutting ants, iguanas, birds, and mammals, made direct use of *Cecropia obtusifolia*. A total of thirty-three bird species from ten families, including some North American migrants, feed on cecropia flowers or fruit (page 137). I have frequently stopped by a blooming cecropia to enjoy the bird show. Such trees function for tropical birds as fast-food restaurants. Mammals from bats to monkeys eat the fruit, and sloths gorge (in slow motion) on the leaves. One North American migrant bird, the worm-eating warbler (*Helmitheros vermivorus*), specializes in searching for arthropod prey in dried leaf clusters, often those of cecropias (Greenberg 1987b).

Estrada and his coresearchers aptly described cecropias:

> Apparently, *Cecropica obtusifolia* has traded long life, heavy investment in a few seeds, and the resulting high quality of seed dispersal, for a short life, high fecundity, a large investment in the production of thousands of seeds, extended seed dormancy, and the ability to attract a very diverse dispersal coterie that maximizes the number of seeds capable of colonizing a very specific habitat. *Cecropia* seed's ability to "wait" for the right conditions is probably an adaptation to the rare and random occurrence, in the forest, of gaps of suitable large size and light conditions sufficient to trigger germination and facilitate rapid growth.

Cecropias have obviously profited from human activities, as cutting the forest provides exactly the conditions they require.

One final note on cecropias: Beware of the ants, especially if you cut the tree down. Biting ants (*Azteca* spp.) live inside the stem. These ants feed on nectar produced at the leaf axils of the cecropia, on structures called *extrafloral nectaries*. I will describe these on page 131, but for now note that these ants sometimes protect the cecropia in a unique way. Many cecropias are free of vines or epiphytes once they've reached fair size, which is good for them since such hangers-on could significantly shade the cecropia. Janzen (1969a) observed that *Azteca* ants clip vines attempting to entwine cecropias. The plant rewards the ants by providing both room and board, a probable case of evolutionary mutualism (page 127). However, some cecropias hosting abundant ants are, indeed, vine covered, and the ants seem to patrol only the stem and leaf nodes, not the main leaf surfaces (Andrade and Carauta 1982).

The Kapok, Silk Cotton, or Ceiba Tree

One of the commonest, most widespread, and most majestic Neotropical trees is the ceiba or kapok tree (*Ceiba pentandra*, family Bombacaceae), the sacred tree of the Mayan peoples. Mayans believe that souls ascend to heaven by rising up a mythical ceiba whose branches are heaven itself. Ceibas are sometimes left standing when surrounding forest is felled. Throughout much of Central America, the look of today's tropics is a cattle pasture watched over by a lone ceiba.

The ceiba is a superb-looking tree. From its buttressed roots rises a smooth gray trunk often ascending 50 m (164 ft) before spreading into a wide flattened crown. Leaves are compound, with five to eight leaflets dangling like fingers from a long stalk. The major branches radiate horizontally from the trunk and are usually covered with epiphytes. Many lianas typically hang from the tree.

Ceibas probably originated in the American tropics but dispersed naturally to West Africa (Baker 1983). They have been planted in Southeast Asia as well, so today they are distributed throughout the world's tropics.

Ceibas require high light intensity to grow and are most common along forest edges, river banks, and disturbed areas. Like most successional trees, they exhibit rapid growth, up to 3 m (10 ft) annually. During the dry season they are deciduous, dropping their leaves. When leafless, masses of epiphytes and vines stand out dramatically, silhouetted against the sky.

Leaf drop precedes flowering, and thus the flowers are well exposed to bats, their pollinators. The five-petaled flowers are white or pink, opening during early evening. Their high visibility and sour odor probably help attract the flying mammals. Cross-pollination is facilitated since only a few flowers open each night, thus requiring two to three weeks for the entire tree to complete its flowering. Flowers close in the morning, but many insects, hummingbirds, and mammals seeking nectar visit the remnant flowers (Toledo 1977). A single ceiba may flower only every five to ten years, but each tree is capable of producing 500–4,000 fruits, each with approximately 200 or more seeds. A single tree can therefore produce about 800,000 seeds during one year of flowering (Baker 1983). Seeds are contained in oval fruits, which open on the tree. Each seed is surrounded by silky, cottonlike fibers called kapok (hence the name "kapok tree" and also "silk-cotton tree"). These fibers aid in wind-dispersing the seeds. Kapok fibers are commercially valuable as stuffing for mattresses, upholstery, and life preservers (Baker 1983). Since the tree lacks leaves when it flowers, wind can more efficiently blow the seeds away from the parent. Seeds can remain dormant for a substantial period, germinating when exposed to high light. Large gaps are ideal for ceiba, and the tree is considered successional, though it may persist indefinitely once established in the canopy.

Ceiba leaves are extensively parasitized and grazed by insects. Leaf drop may serve not only to advertise the flowers and aid in wind-dispersing the seeds but also to help periodically rid the tree of its insect burden.

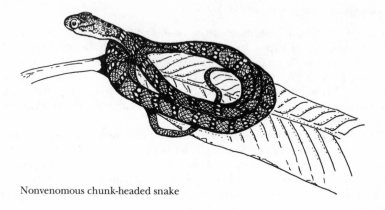

Nonvenomous chunk-headed snake

4

Evolutionary Patterns in
the Tropics

BIOLOGICAL evolution is the process responsible for the way organisms look, function, and act. The arctic tundra, boreal and temperate forests, grasslands, and deserts all are composed of plants, animals, and microbes that represent varying solutions to problems imposed upon living systems by different environments throughout the eons. Environments change and life forms change too, or become extinct. Indeed, the life forms themselves are often influential in causing environmental change, as when oxygen was added to the atmosphere by photosynthetic organisms billions of years ago. The tropical rainforest has long been assumed to be a "laboratory of evolution" because of its extraordinary diversity of species and the complex relationships among its members.

How Evolution Works

A Primer on Natural Selection

On November 24, 1859, the Britisher Charles Robert Darwin, then in his fifty-first year, published perhaps the most important book ever written on the subject of biology. This work, which bore the cumbersome and typically Victorian title *On the Origin of Species by Means of Natural Selection or the Preservation of Favoured Races in the Struggle for Life*, has since become known simply as *Origin of Species*, or the *Origin*.

In *Origin Of Species*, Darwin argued that biological evolution has occurred throughout the history of life on Earth and that the forms of life existent today all share common ancestors with those that have lived in the past. Thus, species are not immutable but rather are changeable over time. They can and do evolve from previous ancestors, ultimately going all the way back to the first appearance of life on the planet. Darwin envisaged evolution as a dense, highly branching bush with a common base, where each branch represents an evolving lineage. The process of how the bush developed through long stretches of time he called "descent with modification."

Darwin further argued that it is the physical and biological environment of each species that imposes the conditions under which it must adapt to survive. Calling his theory *natural selection*, and making an analogy with artificial selection used in the process of plant and animal domestication, Darwin stated that all organisms tend to reproduce more offspring than can survive within the limits imposed by their environments and are therefore engaged among themselves in a "struggle for existence." Those individuals best able to survive the particular conditions of the environment will tend to leave more offspring than those individuals less suited to a particular set of conditions. Any superior variant will, therefore, be naturally selected and survive. Any genetic characteristic that enhances either an individual's survival or ability to reproduce would thus tend to be passed disproportionately to the next generation. Those less adapted tend to perish. This theory, termed by Darwin the "survival of the fittest"(a phrase he borrowed from the philosopher Herbert Spencer), could explain how changing environments select for characteristics that ultimately alter a species, in some cases to the degree that it becomes a different species. Should an environment change, adaptation will tend to reflect the parameters of the changed environment. And, of course, there is the possibility that a lineage may fail to adapt. Extinction is extremely common over geologic time.

Darwin's theory of natural selection was the first evolutionary theory to win wide acceptance in the scientific community. It is a seemingly simple but nonetheless elegant theory. Darwin's tenacious defender, Thomas Henry Huxley, upon hearing it for the first time, declared, "How stupid of me not to have thought of it." Not really. Natural selection is not all that obvious; it is a statistical truth that follows only when one recognizes how to formulate the logic. Indeed, it was not until the mid-twentieth century that natural selection was unambiguously demonstrated to be true in a field situation. Natural selection continues today as the major explanation for the processes of adaptation and, to a lesser degree, speciation.

Modern evolutionary theory defines natural selection as "differential reproduction among genotypes." Genes, the long, coiled molecules of DNA that contain hereditary information, were quite unknown to Darwin, although he knew that physical traits could be inherited. As long as a population contains genetic variability among its members, natural selection can act, since there are genetic variants from which to "choose." The process, of course, is blind, so the word "choose" is metaphorical. No intention of direction is implied. Natural selection, as pointed out above, is a general statistical truth: individuals with genes that confer reproductive advantage will tend to leave the most progeny, and thus those genes will proportionally increase relative to others in the population's gene pool. This is little different from saying that a coin weighted to heads will, when flipped, come up tails more often than heads.

Genetic variability originates through the random process of mutation, a sudden and unpredictable change in a gene. Mutation is nondirectional: environments do not cause or produce useful mutants in response to need. Selection can act only on whatever genetic variants are present. If some are adapted more so than others, they will be passed in higher proportion to the next generation. Using the raw material of genetic variation, natural selection, like an invisible hand, shapes organisms to fit their environments. Selection, unlike mutation, is not a random process (any more than a weighted coin flip is considered random in its outcome), because only certain members of a population are best suited for a given environment and thus have a nonrandom chance of survivorship and reproduction. In recent years both geneticists and molecular biologists have established beyond doubt that large amounts of genetic variability exist in most populations. Thus there is ample raw material for natural selection to act upon.

Darwin was not alone in his discovery of natural selection. Another Britisher, some years younger than Darwin, independently developed exactly the same theory, even to the point of using nearly identical terminology. In 1858 Alfred Russel Wallace sent an essay to Darwin that he had composed during a bout with fever while on the island of Ternate in what is today Indonesia (Wallace 1895). Imagine Darwin's chagrin when, on June 18, he opened and read Wallace's essay and seemed to be reading his own words, words about which he had been thinking for more than two decades but had not yet published. Darwin had been at work on his theory virtually from his return in October 1836 from the five-year global voyage of the HMS *Beagle*. He had written private, unpublished essays on natural selection in 1842 and 1844. The latter essay, a longer and more detailed version of the first, formed the backbone of *The Origin*. Darwin instructed his wife Emma to see to its publication in the event of his death. Two of Darwin's closest friends and scientific colleagues, the geologist Charles Lyell and the botanist Joseph Hooker, both eminent scientists, knew of Darwin's work and of his priority in claiming the discovery of natural selection. At the joint urging of Lyell and Hooker, Darwin and Wallace had their essays read together before the prestigious Linnaean Society in London on July 1, 1858. Wallace graciously accepted Darwin's priority and insisted that Darwin be awarded the lion's share of the credit for discovering natural selection. Darwin's book, published a little over a year later, assured such would be the case.

Both Darwin and Wallace were strongly influenced by their studies in the American tropics. Both made voyages into rainforests to study natural history, and each man was astonished at the immensity, diversity, and complexity that he witnessed. Wallace spent four years exploring the Amazon and voyaging upriver along the Rio Negro. Darwin explored not only Amazonia but much of the rest of South America as well (Darwin 1906). He rode with gauchos over the Patagonian pampas, scaled the Andes, and became as deeply intrigued by the geology of the region as by the biology. He unearthed fossils of animals long extinct but bearing uncanny resemblances to living forms such as sloths and armadillos. Darwin also visited the Galapagos Archipelago, 1,127 km (700 mi) west of Ecuador on the equator. These barren volcanic islands are named for their curious collection of giant tortoises (*galapagos* means "tortoise" in Spanish). The different variants of tortoises, finches, and mockingbirds that Darwin noticed as he visited several of the islands stimulated his thinking in ways that, once back in England, soon crystallized into the theory of evolution by natural selection.

Adaptation

An adaptation is any anatomical, physiological, or behavioral characteristic that can be shown to enhance either the survival or reproduction of an organism. Adaptations with a genetic basis are thought to result from the action of natural selection. Any organism may be viewed as a cluster of various adaptations. For instance, opossums, Neotropical porcupines, kinkajous, as well as many monkeys of the American tropics possess a prehensile tail. Such a structure functions effectively as a fifth limb, lending security and mobility to the animal as it moves through the treetops. It is easy to see intuitively that the prehensile tail is an adaptive structure. Tailless monkeys or opossums would face a smaller lifetime reproductive success because of the added risk of falling. Monkeys also have excellent stereoscopic vision, because their eyes have widely overlapping fields of vision. This enables them to judge distances precisely, as when jumping from branch to branch.

Not all adaptations are obvious. In parts of Africa, many humans carry a gene that, when present in double dose (one inherited from the mother, one from the father), produces defective hemoglobin molecules, resulting in misshapen red blood cells called "sickle cells." These victims of sickle cell anemia usually die before reaching reproductive age. However, other individuals in the population who possess only one dose of the sickle cell gene and one dose of the normal gene for beta hemoglobin are not seriously disabled and are more resistent to malaria than others in the population, including those individuals with two normal genes for hemoglobin. Thus the sickle cell gene is adaptive (since the protozoan that causes malaria is part of this environment), but only when in a single dose. When in double dose it is lethal.

Because they can seem so obvious, adaptations are often inferred, and such inference may be true, but rigorous testing is the only way in which adaptation can actually be demonstrated. For example, orb-weaving spiders throughout the Neotropics and temperate zone make large webs in which there are some areas of obvious, thickened, zigzag strands called stabilimenta. Spiders must use considerable energy to synthesize the silk for the web, especially the dense stabilimenta. Why should spiders invest in making stabilimenta? Are these conspicuous zigzag strands adaptive? The spiders that make stabilimenta are those whose webs remain intact throughout the daylight hours (many spiders make new webs each evening and take them down at dawn). Biologists have hypothesized that stabilimenta, which make webs easily visible to humans, have the same effect on birds (which are also visually oriented), thus allowing a flying bird to avoid an orb-weaver spider's web, saving the spider from having to remake the web, which would be significantly damaged should a flying bird strike it. It would thus be to the spider's advantage to invest additional energy in making stabilimenta rather than risk having to start from scratch and make a whole new web. Therefore, stabilimenta could represent an adaptation to energy saving in an environment where birds pose a risk to the security of the web. As such, the hypothesis sounds plausible, but, without testing, it is just a story, a mere educated guess. How could it be tested?

First, an observer could simply watch spider webs and note bird behavior. This was done, both in Panama and in Florida, and birds were observed to take short-range evasive action when approaching webs with stabilimenta (Eisner and Nowicki 1983). Second, using webs that do not have stabilimenta, researchers altered some webs, adding artificial stabilimenta, keeping other webs without stabilimenta as controls. The webs without stabilimenta did not remain intact during the day, while those with stabilimenta generally did (Eisner and Nowicki 1983). Further direct observations implicated birds as the

Figure 1. Rainforest overview, Arima Valley, Trinidad. Lowland rainforests, with their vast biomass and uninterrupted growing season, accomplish more photosynthesis (gross productivity) per unit area than any other ecosystem on Earth.

Figure 2. Geologically ancient table mountains called tepuis characterize parts of northeastern South America. Note the brownish color of the waterfalls, evidence of phenolics present in decomposing leaves washed into the waters. From Canaima National Park, Venezuela.

Figure 3. A tepui in southeastern Venezuela. The flattened table mountain shape is typical of these ancient, highly eroded mountains. Unique species of plants and animals can be found isolated on the tops of tepuis.

Figure 4. The central Andes Mountains as seen from the air between Lima, Peru, and Guayaquil, Ecuador. The sharp relief is indicative of the relatively recent origin. Snowcover is due to high elevation.

Figure 5. Cloud forest covers the east slopes of the Andes from mid-elevation until becoming lowland forest. This ecosystem is characterized by nearly perpetual mist, an abundance of epiphytes, and high species richness, including such avian taxa as hummingbirds and tanagers. From southern Peru, near Cuzco.

Figure 6. Interior of cloud forest dominated by tree ferns.

Figure 7. High-elevation paramo (moist shrubland) ecosystems have low biomass and low species richness and are dominated by such widely spaced shrubs as *Espeletia*, a member of the composite family, shown here. From Venezuela, near Merida.

Figure 8. Espeletias dominate the misty paramo ecosystem in high elevations throughout much of the Andes Mountains.

Figure 9. Puna is high-elevation grassland. More arid than paramo, it is also low in biomass and species richness and is typically composed of various bunch grass species. Puna characterizes much of the Andean Antiplano. From Peru, near Cajamarca.

Figure 10. Savannas, ecosystems of widely spaced, low-statured trees interspersed among grasses, sedges, and various other species, characterize much of Central and South America. This photo, from southern Belize, is of a savanna dominated by *Pinus caribea*, an area subject to frequent natural fires, and where the soil is generally poor.

Figure 11. Deciduous trees that flower at the onset of dry season are common in savanna ecosystems. Note the abundant termite mounds among the low vegetation. From Mato Grosso, Brazil, near Pantanal.

Figure 12. *Cerrado,* such as this example from Venezuela, south of the Orinoco River, is the term for ecosystems of dense, short-statured trees that typically are found in environments with a pronounced dry season or on relatively poor soils.

Figure 13. Lowland tropical rainforests are the most species diverse and most structurally complex of the world's terrestrial ecosystems. From Mato Grosso, Brazil, near Alta Floresta.

Figure 14. Mangroves of various species line coastal areas and make up whole islands (cayes) throughout the Neotropics. Most of the mangroves visible are *Rhizophora mangle*, red mangrove. This island, Man-O-War Caye in Belize, has a nesting colony of magnificent frigatebirds. Visible are the red throat pouches of the males perched in the mangroves.

Figure 15. Man-O-War Caye one year after the previous photo, showing extensive hurricane damage. The magnificent frigatebirds continue to nest, however. Mangrove ecosystems adapt to periodic severe disturbances. This caye recolonized immediately.

Figure 16. Coconut palms.

Figure 17. Inside a black mangrove swamp.

Figure 18. Colonizing clump of red mangrove.

Figure 19. Old hurricane damage, Cocoa Plum Caye, Belize.

Figure 20. Coral reefs rival rainforests in diversity and productivity. As marine ecosystems, however, they are organized quite differently from rainforests. Elkhorn coral, which dominates this photo from the Belizean barrier reef, has commensal algae that makes the coral colony act functionally as an autotroph.

Figure 21. Queen angelfish, one of dozens of species of colorful and conspicuous coral reef–inhabiting fish found throughout the Caribbean.

Figure 22. The interior of a rainforest is structurally complex, often abundant with vines and epiphytes and lacking clearly demarcated tree strata. Such complexity and high biomass reflect high productivity and species richness. From southern Ecuador, near Tena.

Figure 23. Rainforests often abound in vines of various sorts, as shown in this photo from southern Belize. Note how both vines and epiphytes seem to envelope the tree crown.

Figure 24. Rainforest gaps like this one from Guatemala near Tikal National Park occur due to tree falls or other natural disturbances, allowing light to flood the forest floor and thus promote growth among certain plant species. Gap phase dynamics describes the process whereby natural gaps form a mosaic of differing-aged disturbance points within a forest.

Figure 25. "Typical tropical trees" like this one are slender and tall, with main branches radiating from the bole at canopy level. From Panama, near Pipeline Road.

Figure 26. Each major branch of this tree (whose leaves are just opening) hosts a diverse community of epiphytes ranging from mosses and liverworts to bromeliads, cacti, and orchids.

Figure 27. Large forest gaps such as are seen along roadcuts provide habitat for such species as heliconias, which have huge leaves similar to those of bananas. Note the density of plant cover, a true "jungle" effect. From Panama, near Colon.

Figure 28. This roadcut along the Hummingbird Highway in Belize demonstrates the reddish color that typifies many, but by no means all, Neotropical soils. Reddish color is mainly due to oxides of iron and aluminum.

Figure 29. Moriche palm (*Mauritia* sp.) is abundant in such wet areas as moist savannas and along rivers throughout the Neotropics. Aboriginal people utilize the products of this palm for many needs that range from food to shelter. From the lower Amazon.

Figure 30. Cecropia (*Cecropia* spp.) abounds throughout the Neotropics and is an important large gap species that thrives in highly disturbed areas abundant with sunlight. Large palmate leaves and spindly trunks make this genus easy to identify. From southern Belize.

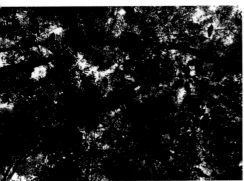

Figure 31. Early-morning light strikes an emergent tree along a stream in Amazonian Peru.

Figure 32. This mature rainforest in Mato Grosso, Brazil, has among the highest plant species richness of any ecosystem.

Figure 33. Inside a closed, high-canopy rainforest. Only flecks of sunlight illuminate the forest floor. From Mato Grosso, Brazil, near Alta Floresta.

Figure 34. An emergent mahogany tree from Amazonian Peru.

Figure 35. The author standing before the trunk of the same mahogany tree as shown in figure 34.

Figure 36. Dense vegetation growth in an old forest gap in the Peruvian rainforest.

Figure 37. Tropical leaves, though exposed to herbivores throughout the year, are not significantly damaged, presumably due to the presence of defense compounds.

Figure 38. Rainforests typically contain a diverse array of such vines and epiphytes as is shown here in southern Belize.

Figure 39. Some rainforest trees periodically shed all their leaves, revealing the complex community of epiphytes that line their branches.

Figure 40. As rainforest productivity is very high, so is decomposition rate. A thin layer of fallen leaves and other organic litter often characterize forest floors.

Figure 41. Typical examples of Neotropical leaf shapes.

major threat to webs, though other large animals, ranging from butterflies to deer, could also be forewarned by the presence of stabilimenta. This work demonstrated the adaptiveness of stabilimenta.

Not all traits are adaptive. Organisms represent the combined effects of thousands of genes working in concert. The anatomy, physiology, and behavior of an organism represent various compromises imposed by the interactive effects of the genes that formed it. Is it adaptive that humans generally lack body hair? Maybe, but no generally acceptable adaptive explanation for hairlessness has been forthcoming. Our hairlessness could be a mere byproduct of our development, with no adaptive function now or ever in the past. A given trait may, however, have been adaptive in the past but not any longer. The human appendix probably once functioned as a cecum, an additional area for fermentation of tough plant fiber.

An adaptation must be viewed in the context of the environment. Obviously, the sickle cell gene is nonadaptive in human populations from environments free of malaria. And a beautifully adapted creature may appear awkward, a "mistake of nature," if seen out of its proper environment. Charles Waterton, a Britisher who explored the Amazon during the early nineteenth century, noted that the three-toed sloth (*Bradypus variegatus*) appears very ill-suited to survive when seen struggling over the ground. Sloths, however, do not normally wander about on the ground. They are totally arboreal, skillfully moving upside down from branch to branch. In his well-known account, *Wanderings in South America*, Waterton (1825) wrote of the sloth, "This singular animal is destined by nature to be produced, to live and to die in the trees; and to do justice to him, naturalists must examine him in this his upper element." Waterton then went on to describe in detail how well adapted the sloth is for its arboreal life.

Cryptic Coloration—The Art
of Disguise
Figures 96, 97, 98, 99, 100, 101, 102, 103

Cryptic coloration, or crypsis, is nature's camouflage. Thousands of species of tropical insects, spiders, birds, mammals, and reptiles exhibit cryptic coloration to various degrees. Any of these creatures, if removed from its environment, would appear obvious. For instance, the common boa constrictor (*Boa constrictor*), if placed on a white table, would be seen as boldly patterned and complexly colored in browns and golds, with stripes, diamonds, and other markings. Place the snake on the rainforest floor, among the decomposing leaves and sunflecks, and watch it melt into the background. A seven-foot serpent can seemingly disappear before your eyes. Place your hand on a tree trunk and part of the bark can suddenly erupt in flight as a blue morpho butterfly (*Morpho* spp.), resting on the tree trunk, flies off at the disturbance. Once in flight, the deep blue wings make the large insect easy to see. At rest, however, the blue is covered by wings virtually indistinguishable from bark. Many insect species have wings patterned to look remarkably like leaves. Certain katydids mimic living leaves, and many lepidopterans (and some katydids) mimic dead leaves, even to the point of including lighter areas simulating insect damage and leaf skeletal damage. An example is the butterfly species

Zaretis itys, in the family Charaxinae, a species that feeds part of the time on the forest floor, on rotting fruits. The underside of the female butterfly bears an uncanny resemblance to a skeletonized leaf (DeVries 1987).

Cryptic coloration functions widely to protect animals from detection by predators. In the tropics, what appears at first to be a green leaf may be a katydid, a twig may be a walkingstick or a mantis, a thorn may be a treehopper, dried leaves may be a butterfly (and a bunch of dead leaves may even be a coiled fer-de-lance), a green vine may be a vine snake, bark may be a butterfly or moth, and a tree stump may be a bird. The common potoo (*Nyctibius griseus*), a large nocturnal bird, spends its days sitting utterly still in plain sight. The bird so much resembles the end of a thick branch that it is easily overlooked. It even postures its body in such a manner as to resemble more closely a branch.

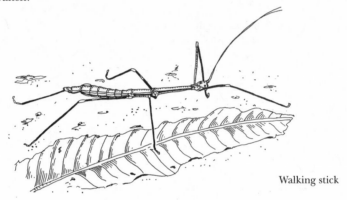

Walking stick

Like many birds in the temperate zone, only the males of many tropical bird species are brightly colorful while females are cryptically colored. The little manakins, discussed in chapter 12, are excellent examples of this form of sexual dimorphism. Most male manakins are very colorful and easy to spot even in the rainforest interior, but females are universally olive or dull yellow and are much more easily overlooked. It is only the female that builds and attends the nest. The female's cryptic coloration is undoubtedly adaptive in keeping the nest more secure from discovery by potential predators.

Tropical cats also demonstrate cryptic coloration. The spotting and/or banding patterns of the ocelot, margay, and jaguar, so obvious when the animals are observed in zoos, help disguise them as they stealthily move through the rainforest interior. The coat patterns, when seen in the shaded, sunflecked forest interior, tend to break up the animal's outline, rendering it less visible. Although cats are predators, cryptic coloration is no less an advantage to them, as it aids them in moving undetected toward their prey. The three-toed sloth, the animal that so impressed Charles Waterton, is rendered far less visible in the treetops by virtue of the fact that algae growing in grooves in the sloth's hair give a distinct greenish tinge to the animal. In many Neotropical forests the three-toed sloth is abundant, but easily overlooked because it is so effectively camouflaged.

The first unambiguous demonstration of natural selection in action involved cryptic coloration of moths in Great Britain. The well-known example of industrial melanism, where over time moths with light-colored wings (mostly of the species *Biston bitularia*) were replaced by dark or melanistic individuals, was caused by soot from industrial pollution changing the background pattern of the tree trunks where the moths spent the day. Darkened trunks rendered light-winged moths more visible to birds, their principal predators. Dark-winged moths, once quite visible because of their melanistic pattern, were now much less visible and thus were more fit in the polluted woodlands. Melanistic moths soon predominated in polluted areas. When pollution was reduced, light-winged moths increased again (Kettlewell 1973; Bishop and Cook 1975).

Warning Coloration—Don't Tread on Me!   *Figure 105*

Although many animals survive by camouflage, some seem to send exactly the opposite message. Certain tropical butterflies, caterpillars, snakes, and frogs are brilliantly colored and stand out dramatically in any environment.

Among the most strikingly colored of any animals are the small frogs of the family Dendrobatidae, the so-called poison-dart frogs. Within this group there are about 30 species in the genus *Dendrobates* and 5 in the genus *Phyllobates* that contain alkaloids in their skin secretions. All of these diminutive frogs of Central and northern South American rainforests are characterized by bold, striped patterns of orange, red, or yellow that glow almost neonlike against a dark background. The Choco Indian tribes of western Colombia use toxic alkaloid compounds called batrachotoxins extracted from the frogs' skins as potent arrow poison (Myers and Daly 1983; Maxson and Myers 1985). One species, aptly named *Phyllobates terribilis*, is reputed to be potentially lethal to the touch and may be the most toxic of all living creatures (Moffett 1995a). The fact that the skin is extremely toxic and the fact that the frogs are very colorful are probably not unrelated (magnificent photographs of these frogs appear in Moffett 1995a). These frogs represent a case of *aposematic*, or warning, coloration. Their bold patterning serves as an easy-to-remember warning to potential predators to leave this animal alone. Thus far, almost 300 noxious or toxic alkaloids have been isolated from various species of amphibians (Daly et al. 1993), demonstrating that protection by poison is a general characteristic

Poison-dart frog

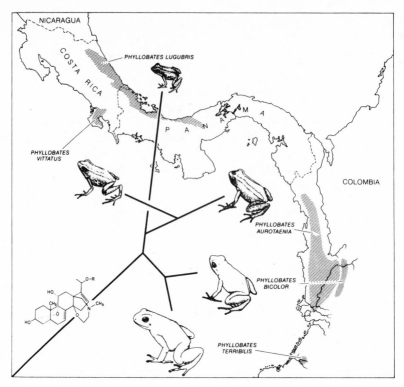

Ranges of species of poison-dart frog (*Phyllobates*). From Maxson and Myers (1985). Reproduced with permission.

of much of this taxon. Batrachotoxins are the most toxic of the various alkaloid compounds. Recently (and remarkably), batrachotoxins have been found in the feathers of three brightly colored bird species of New Guinea rainforests, the pitohuis (*Pitohui dichrous, P. kirkocephalus, P. ferrugineus*) (Dumbacher et al. 1992). The pitohuis represent the first known example of toxic bird feathers, and their bright colors suggest the possibility of warning coloration. The batrachotoxins in pitohuis were, of course, evolved independently from those in frogs.

How could the poison-dart frogs have evolved such toxicity? Evidence exists that the precursors of the batrachotoxin can be obtained in the insect diets of the frogs, particularly among ants. Frogs kept in captivity show reduced levels of toxicity, suggesting that diet is, indeed, a major factor in synthesizing the skin toxins (Crump 1983; Daly et al. 1993).

Coral snakes represent yet another example of boldly patterned and colorful animals that are also extremely toxic. Coral snakes tend to be unaggressive unless threatened. Even though deadly, a coral snake could still suffer extensive harm if attacked. However, its well-defined red, black, and yellow pattern is easy to recognize, presumably to remember, and to avoid. Both the turquoise-browed motmot (*Eumomota superciliosa*) and the great kiskadee fly-

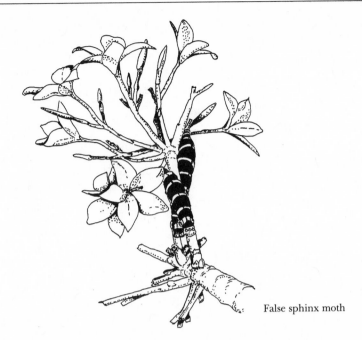

False sphinx moth

catcher (*Pitangus sulfuratus*) have been shown instinctively to avoid coral snakelike patterns (Smith 1975, 1977). Certain nonpoisonous snakes, such as some kingsnakes, closely resemble coral snakes and are therefore thought to be coral snake mimics (Greene and McDiarmid 1981). Even caterpillars of the false sphinx moth (*Pseudosphinx tetrio*) have patterning reminiscent of coral snakes (Janzen 1980a).

Many butterflies are among the most obvious and colorful tropical animals. However, some species store toxic compounds ingested from plants that in turn make them toxic to birds and other predators. Thus the colorful patterning serves to warn potential predators that these creatures are unsafe to eat. Nontoxic species often mimic toxic species (see chapter 6). Some colorful caterpillars are covered with toxic hairs. Warning coloration is a visual manifestation of evolutionary chemical warfare in the tropics. Even if an animal is able to acquire or synthesize a noxious chemical, doing so confers little protection unless potential predators can learn and remember to avoid the poisonous species. Thus most visible and boldly patterned species seem to be sending an easily memorized signal that they are dangerous. The patterns and brilliant colors warn predators to stay away.

Tropical Selection Pressures

Factors in the environment of an organism that influence the probability of its survival or reproductive success are called *selection pressures*. Selection pressures may come from climate or physical disturbance, in the form of seasonal stresses, fire, or hurricane, or from the soil in the form of nutrient shortage or imbalance. These are abiotic selection pressures, since they are generated by the physical environment, not by organisms. Selection pressures may also come from competing individuals of the same species or other species, or from predators, parasites, disease organisms, mutualistic partners, or food species. These are biotic selection pressures, since they are imposed by other organisms.

Usually, combinations of selection pressures are at work. A population of howler monkeys may be ravaged by yellow fever, making some of the animals easier for predators such as jaguars or harpy eagles to capture. Or they may be ousted from a section of forest by another, more powerful howler troop. Or some of them may be affected by toxins in the leaves they consume. Selection pressures are rarely constant and many vary seasonally. The dry season often imposes significant stresses, especially in savanna areas, through both water shortage and increased likelihood of fire.

To survive, an organism must be reasonably well adapted to all of the selection pressures it is likely to encounter during its life cycle. Those individuals within the species best able to cope with the constellation of selection pressures imposed by the environment will tend to leave the most offspring, thus preserving the genetically caused traits that proved most adaptive. It is this simple statistical truth that is the essence of Darwin's and Wallace's theory of natural selection.

Abiotic Selection Pressures

There is a real reason why people like me who live in Boston like to go to places like Belize in January. Generally speaking, the climate of a tropical region is more conducive to living things than terrestrial climates outside of the tropics. Although there is usually a pronounced dry season (chapter 1), tropical areas experience no period of frost or snowfall, when water is in solid form and is unavailable as a medium for cellular growth and development. Precipitation falls as rain twelve months a year. Temperature is relatively high and nearly constant. Leaves are almost always available to herbivores and, though some individual trees may drop all leaves, there is no season in which photosynthesis shuts down entirely. Flowers and fruits are, likewise, a relatively constant resource in a tropical rainforest. Life tends to prosper under such conditions. This is not to say, however, that the tropical climate imposes no abiotic selection pressures, far from it. Hurricanes may blow, rainfall may leach nutrients, and heat stress can occur. Even large-scale fires may burn tropical forests. Nutrient-poor tropical soils may impose significant selection pressures. As discussed in chapter 3, tropical plants have evolved various adaptations helping them adapt. Yet overall the physical climate is far more equitable and gentle in equatorial regions than in temperate, polar, and desert climates. Physiologically, it is probably easier to live in the tropics than out of the tropics.

Biotic Selection Pressures

The tropics abound with species that exert biotic selection pressures on each other. For instance, probably the least favorable place for a seed to fall is beneath its parent plant. Such proximity to an already established individual could doom the seed to be discovered and eaten by some seed predator attracted to the fruiting parent or, should it sprout, to be outcompeted by its own parent (Janzen 1971a).

Many, if not most, tropical species are probably subject to significant levels of competition from other species. The large number of species in almost every taxa in the tropics (see below) suggests strongly that the effects exerted by the presence of other species with similar ecological needs impose a potentially substantial selection pressure. Interspecific competition may be an important influence in shaping the patterns of species diversity in the tropics. Because many different species can each exert a small but not insignificant competitive selection pressure on any given species, the term *diffuse competition* has been suggested to describe the cumulative effect of competitive selection pressures imposed by other species (MacArthur 1972). Diffuse competition, a function of biodiversity, is undoubtedly more common in the tropics than elsewhere.

Predators represent yet another significant source of biological selection pressures in the tropics. Parasites and pathogens are considered here as they, too, represent a form of predation. Indeed, one of the most sobering and worrisome possibilities of the continued global deforestation of tropical areas is the release of potentially lethal viruses such as Marburg and Ebola that have been harbored within rainforest species (Garrett 1994).

If a tropical plant or animal manages to survive in competition with others of its own species as well as other competing species, it could still fall to a predator (or herbivore in the case of plants) or pathogen. For instance, birds suffer from high levels of nest predation in the tropics, probably higher than in the temperate zone (Lill 1974). Many tropical species are secretive nesters, the females making few daily trips to feed nestlings and nestlings being less vocal than their temperate zone counterparts. If one begins to add up potential nest predators, the list soon becomes staggering. Toucans, forest falcons, snakes, monkeys, kinkajous, cats, even army ants may attack bird nests. A study of the nesting behavior of the turquoise-browed motmot in the Mexican Yucatan revealed the importance of predation. Motmot nests were located in the walls of Mayan ruins. Approximately 36% of the nests studied were terminated by predators (Scott and Martin 1983).

It has been suggested that one reason tropical bird species lay fewer eggs per clutch than temperate species is that the attention created by attending a larger brood would increase the risk of attracting predators to the point that

Turquoise-browed motmot

most broods would be discovered and devoured (Ricklefs 1969a, 1969b, 1970). Smaller clutches are more adaptive because they can be raised more secretly. In other words, because of predators, the probability of raising two chicks is substantially higher than raising three, even if food is abundant. Food, however, may not in actuality be abundant for tropical nesting birds, and food shortage has been suggested as an alternate explanation to account for low clutch size in the tropics (Lack 1966).

Frog sex represents another of many possible examples of the importance of predation as a tropical selection pressure. Male frogs must call at night to attract females. Lucky males then get to mate. Unlucky males, however, attract bats that are specialized to capture frogs by homing in on their amorous calls. In Panama, the fringe-lipped bat, *Trachops cirrhosus*, responds to a wide variety of frog calls. The more frequently or longer a frog calls, the more likely it is to attract a predatory bat. Bats also differentiated between poisonous and nonpoisonous frogs, strongly preferring to attack nonpoisonous animals (Tuttle and Ryan 1981). Male frogs risk much for the sake of the next amphibious generation.

Predatory bats have also been shown to influence strongly the calling behavior of katydids (Belwood and Morris 1987). Bats home in on the mating calls of male katydids (which call at night, when the bats are active). Interestingly, some katydid species have evolved antibat adaptations, such as shorter songs with complex, species-specific tremulations that generate vibrations that bats apparently cannot hear (though, of course, they are audible to female katydids).

Even something as simple as preparing for a trip to the tropics reflects in a subtle way the importance of biotic versus abiotic selection pressures. If you were going to the arctic, you would spend time and dollars on thermal clothing, snowshoes or heavy boots, and various other paraphernalia designed to protect against the potentially harsh elements. For a human visitor to "adapt" to the arctic requires profound respect for the climate (though mosquitos can be an awesome nuisance in the arctic summer). Preparing for a trip to the tropics, in contrast, requires a trip to the doctor for the various shots and pills that protect against the pathogens and parasites that await the human tropical visitor (see appendix). My point is that the presence of large numbers of diverse organisms means that most important selection pressures in the tropics tend to be biotic. When we speak of interesting adaptations of tropical species, we most often speak of adaptations involving interactions among species. Most organisms, physiologically, can live easily in the warm and wet tropical climate. In the tropics, evolution has been most influenced by interactions among the myriad plants, animals, and microbes that occupy it.

Species Richness and Diversity Gradients

It is not known exactly, or even approximately, how many species inhabit Earth (May 1988, 1992; Wilson 1992). But what is known, at least for terrestrial and freshwater ecosystems, is that, with few exceptions, the greatest number of species for most major taxa—the flowering plants, ferns, mammals, birds, reptiles, amphibians, freshwater fish, insects, spiders, and snails—is in the tropics.

Further, some tropical moist forests seem to contain so many species in relation to any other kind of ecosystem or any other region that they exhibit what ecologists have come to call "hyperdiversity." Why do tropical ecosystems, and rainforests in particular, contain so many species?

Charles Darwin was impressed by high tropical species diversity, which he defined as the number of species found in a given area. Darwin realized that diversity declines latitudinally as one travels north or south from the equator, a point he noted in chapter 3 of *Origin of Species*. The reduction in diversity with increasing latitude is now termed a *latitudinal diversity gradient* (Connell and Orias 1964; MacArthur 1965; Pianka 1966). Dobzhansky, in his important 1950 paper titled "Evolution in the Tropics," noted that only 56 species of breeding birds occurred in Greenland, while New York had 195 breeding species. However, Guatemala had 469, Panama 1,100, and Colombia 1,395! Breeding bird diversity increased almost twenty-five times from Greenland in the arctic to Colombia on the equator. For snakes, Dobzhansky noted that 22 species occurred in all of Canada, whereas 210 were found in Brazilian forests and savannas.

Dobzhansky suggested that since the animals and plants are all products of evolution, any differences between tropical and temperate organisms must be the outcome of differences in evolutionary patterns. What selection pressures and other causes have brought about the greater richness and variety of the tropical faunas and floras, compared with those of temperate and polar lands? How does life in tropical environments influence the evolutionary potentialities of the inhabitants?

Dobzhansky believed that part of the answer to high tropical diversity rested in the benign nature of the tropical climate. He argued that polar and even temperate climates impose such significant physical stresses that fewer organisms have been able to adapt over evolutionary time. The tropics, in contrast, offer a more equitable climate, permitting speciation to exceed extinction. Specialization among organisms has resulted in high diversity as more and more species have "packed" into the tropics over time. Dobzhansky's hypothesis was speculative, but it stimulated thinking about latitudinal diversity gradients.

Tropical moist forests have the highest species diversities of any terrestrial ecosystem. Tropical savannas and grasslands are also generally more diverse than their temperate counterparts, but not nearly to the degree seen when comparing temperate with tropical forests.

Among the various taxonomic groups experiencing maximum diversity in Neotropical forests, mammals represent a somewhat unusual example because most of the increase in mammalian diversity is due to bats. Although monkeys, sloths, anteaters, and various marsupials all contribute to the enhanced diversity of the Neotropics, bats add by far the most species, representing about 39% of all Neotropical mammal species (Emmons 1990). Bats will be discussed later in this chapter.

Components of Species Richness and Diversity

Ecologists recognize a distinction between the terms "species diversity" and "species richness." *Diversity* (in the strict sense) normally refers to a consideration of both the number of species present in a given ecosystem *and* the population size of each species relative to the others. The term *richness* refers to simply the number of species present. For example, suppose we compare two hypothetical communities, each of which has three species. In Community 1, the abundances of each of the three species are: Species A = 102, Species B = 97, Species C = 101. The total number of organisms in Community 1 is thus 300, and each species has about the same abundance. Should you be studying this community, you would have essentially the same chance of encountering Species A as you would Species B or Species C, and it would not take very much sampling before you would record each of the three species. Now consider Community 2, where the abundances of each of the three species are: Species A = 3, Species B = 295, Species C = 2. Both communities, 1 and 2, have identical species richness and identical numbers of organisms, but the diversities, as affected by population sizes, make these two communities "feel" very different. In Community 2, you would quickly conclude that only one species, Species B, comprised the community, though with diligent sampling you would eventually encounter Species A and Species C, both of which are rare in Community 2.

Obviously the communities with the highest species diversities are those in which (1) there are many species and (2) the population sizes of each of the species are relatively equal. There are few natural communities where not only are there large numbers of species but population sizes are relatively equal, though lowland tropical forests in Amazonia may approach this condition, at least for such taxa as trees and birds. For example, in a 1-hectare plot within an upper Amazonian forest, Gentry (1988) found that 63% of the species were represented by single individuals and only 15% of the species were represented by more than two individuals. In a 97-hectare plot in the Peruvian Amazon, an astonishing total of 245 resident birds were found, plus 74 additional species that occasionally visited from other habitats or appeared only as migrants (Terborgh et al. 1990). Most of these species had small populations, and about 10% were considered uniformly rare throughout the study area. Up to 160 species were found at single locations in some portions of the study area.

Using both the number of species present and the relative abundance of each species, ecologists calculate a "diversity index," which, in a single number, allows ecologists to compare diversities among various ecological communities. Several different formulas are in common usage for calculating diversity indexes. Lowland tropical forests exhibit the highest species diversity indexes, much higher for such taxa as flowering plants, birds, mammals, butterflies, for instance, than those calculated for temperate regions. Additional information on the calculation and use of various diversity indexes is found in Magurran (1988).

Diversity and richness can be viewed at different spatial scales, all of which reach their peaks in tropical forests (Whittaker 1975; Boulière 1983). First, there is *within habitat* diversity, also called alpha diversity. This is the total number of species (usually within a given taxon such as birds, for example) in a given area of forest. Tropical forests are cluttered with species. No ecosystems are known that have higher alpha diversities. Second, there is *between habitat* diversity, or beta diversity, the change in species composition from one habitat to another, similar habitat. Neotropical forests rank very high in this diversity component also. This means that one can find many different species of hummingbirds, tanagers, antbirds, and so on as one moves from one forest site to another, even when the sites are in relatively close proximity. Finally, there is *regional* or gamma diversity, the total number of species found in all habitats over a large geographic region. Again, the tropics, and perhaps more than any other tropical area the Neotropics, exceed all other regions in gamma diversity. The vast Amazon Basin, the Atlantic Coastal Rainforest of Brazil, and the rainforest sites west of the Andes Mountains in Ecuador and Colombia and into Central America together make a vast region.

Several explanations have been posited for how tropical forests have come to be so diverse (Connell and Orias 1964; MacArthur 1965; Pianka 1966). No simple or single explanation has yet emerged. Causal factors for species diversity are difficult to test in a way that limits the test to only one variable, excluding all other possibilities. Below I review the major hypotheses for high tropical forest diversity.

The Stability-Time Hypothesis

It has long been argued that the tropics have been around for such a long time that large numbers of species have had ample time to evolve. This has been coupled with the argument that the tropics are climatically stable and large in area, thus permitting high speciation with low extinction, resulting in a steady accumulation of species. This idea is difficult to test since it is basically historical.

It is clear that the tropics contain far more species of tree ferns, orchids, bromeliads, bats, birds, freshwater fish, reptiles, and amphibians than elsewhere. In the case of arthropods, fish, reptiles, and amphibians, the warm, wet tropical climate probably does contribute to their high species richness, as they are all ectothermic, unable physiologically to regulate their body temperatures. Amphibians, for example, require moisture to keep their skins wet. The tropics are thus climatically far better suited for their needs than are temperate or polar regions. Nonetheless, the present species richness of reptiles and amphibians in the tropics may be the result of recent speciation, not the slow accumulation of species due to a long-stable climate. Even if speciation patterns among reptiles and amphibians have an ancient history (see Heyer and Maxson 1982), the causal factors of speciation may have nothing to do with a stable climate (see below). The same could be true for other taxonomic groups.

Speciation is not necessarily dependent on vast periods of time. There are five species of kingfishers that inhabit Neotropical rivers and streams (page 215), an assemblage that represents the total species richness of kingfishers in the Neotropics. Kingfishers are birds that dive into water and capture fish with their beaks. The five species of Neotropical kingfishers differ fundamentally in body and bill size. The fossil record, as well as biogeographic studies, show that kingfishers evolved in the Old World and found their way to the Neotropics relatively recently, probably during the Pleistocene. Nonetheless, the five species of Amazonian kingfishers span the same size range and within habitat

Ringed kingfisher

species richness as that found anywhere in the Old World (Remsen 1990). This means that in the two million or so years since kingfishers invaded the New World tropics, they have evolved an alpha diversity equal to that found anywhere else kingfishers occur. This pattern, if correct, suggests that major speciation events can occur in geologically short time periods, and thus that species richness is not directly dependent on long, stable climatic conditions.

The major premise of the stability-time hypothesis is in all likelihood invalid. Indeed, there is now almost universal agreement among climatologists, geologists, and biogeographers that tropical regions have not, in fact, been climatically stable. A diverse assemblage of data from climatology, geology, and biology suggests otherwise (Haffer 1993). True, the Ice Age undoubtedly affected polar and north temperate regions more dramatically than it did the equatorial region. It is a mistake, however, to conclude that the tropics were unaffected by the dramatic movement of the northern glaciers, or even by climatic events that occurred in the Tertiary or Cretaceous, long before the Ice Age. Indeed, as I shall discuss later in this chapter, ongoing disturbances, most recently those caused by the Ice Age, may have provided a great stimulus to speciation in the tropics.

The argument that high tropical diversity is caused by the region being old and climatically stable is not seen as a sufficient explanation.

The Interspecific Competition Hypothesis

One of the most difficult measures to make in ecological research is the degree to which competition occurs between two or more species. One must be able to demonstrate that two (or more) species are seeking the same limited resource and then measure the degree to which each species negatively affects the other(s) contesting for the resource. It is essential to identify the resource being contested and demonstrate that it is a *limited* resource. If it is not limited, there will be enough for each species and thus no competition. (So if you were to observe two different species each using the resource, you could not automatically conclude that they are in competition, any more than two people ordering burgers at a fast-food restaurant are in competition.) It is also essential to show how the competition affects each species. Does one outcompete the other? Do they somehow divide the resource? Does one expand its population in the absence of the other, competing species?

One prevalent hypothesis to explain high species diversity in the tropics argues that high levels of competition among species have, over time, resulted in increased specialization. Each species evolves into a specialist focusing on a specific resource that it and it alone is best at procuring. This trend toward specialization due to interspecific competition leads to the "packing" of greater and greater numbers of species into tropical ecosystems while at the same time reducing the intensity of competition among species as each specializes to its exclusive pool of resources. Ecologists describe the total constellation of resources required by a given species to be that species' *ecological niche.* The interspecific competition hypothesis argues that niches are narrower in the tropics than in the temperate zone because competition has compressed them.

Note that there is a strong historical assumption to this hypothesis. Though competition may have exerted a major influence in the past, now that specialization and niche compression have occurred, competition may be quite minimal or even nonexistent. It should be obvious that this is a very difficult hypothesis to test. For example, both the ocelot (*Felis pardalis*) and margay (*F. wiedii*) occur throughout most Neotropical lowland forests. These two small cats are similar in body size, the ocelot being a bit larger. Both are essentially nocturnal, and it is not unreasonable to assume that they feed on many of the same prey items: rodents, birds, snakes, lizards, other small mammals. The ocelot and margay differ somewhat in their foraging behavior as the ocelot is almost entirely terrestrial while the margay routinely climbs trees in the course of its hunting behavior. Thus the ocelot and margay have foraging niches that do not precisely overlap. Did competition between these two small felines cause the divergence of foraging niche, selecting for the smaller cat to become more arboreal? There is no way to know. It cannot be known if resources were limited, and even if they were, it cannot be known if that limitation was, indeed, an active selection pressure. In other words, it cannot be known if margays that climbed reproduced better than those that did not *because* they climbed, and even if they did, that could have been merely because of the

Ocelot

availability of arboreal resources presenting an opportunity for procuring more calories and protein, not because ocelots were also foraging on the forest floor. I am not suggesting that ocelots and margays have never competed, but I am suggesting that making the assumption that they did, based on current foraging niches, is exactly that, an assumption. Nothing more.

As the above example was meant to illustrate, evidence for the interspecific competition hypothesis is mostly circumstantial. Direct demonstrations of interspecific competition are generally lacking in the tropics. Certain patterns suggest, however, that competition among species has been an influential component of tropical evolution. Varying bill shapes and gradations in body sizes within many bird groups (see page 104 and chapter 12) suggest that competition may have influenced the evolutionary history of these groups. But different body sizes and bill characteristics also reflect specialization for capturing differing food items. The very act of food capture per se could and probably does select for such specializations, though it is quite likely that the presence of similar species with similar ecological needs could act as an additional strong selection pressure in producing divergence among species. Insectivorous bats, for example, feed on different-sized prey items and also forage at different heights in the rainforest, a pattern possibly reflective of avoidance competition among the bat species.

The fact that clusters of similar species develop differences in foraging areas is an indirect indication that competition may have been at work. In a study of four antbird species of the genus *Myrmotherula*, it was found that each species foraged at a preferred height above the ground and that foraging heights had relatively little overlap (Terborgh, in MacArthur 1972). This suggests that each species is being restrained by the others from expanding its foraging height niche. As another example, each of three flatbilled flycatcher species found in Costa Rica forages at a different height in the forest, and certain flatbill species replace others in specific habitats, one species occurring in forest, another very similar species only in successional areas (Sherry 1984). Andean bird communities also show possible evidence for interspecific competition. In a study of two Andean mountain peaks in Peru, one of which was quite isolated, colonization by birds was less frequent on the isolated peak, and thus this peak had a much reduced bird species richness, missing an estimated 80–82% of the bird species that would have occupied the isolated peak had it been part of the main body of the Andes. However, of the species that did occur, 71% had expanded their altitudinal range (compared with the other, more diverse mountain peak) presumably because of the absence of similar species that would have been competitors (Terborgh and Weske 1975). The researchers concluded that the combined effects of direct and diffuse competition account for approximately two-thirds of the distributional limits of Andean mountain-dwelling birds.

Interspecific competition for pollinators and seed dispersers may have aided in producing staggered flowering patterns among plants. In Trinidad's Arima Valley, eighteen species of the shrub *Miconia* have flowering times that are staggered in such a way that only a few species are flowering in any given month, a possible evolutionary result of competition among *Miconia* for access to birds that eat the fruit and thus disperse the seeds (Snow 1966). Any *Miconia* species that flowered when most others didn't would be able to attract more birds to disperse its seeds, thus it would have a selective advantage compared with others of its own as well as other species. Over time, the staggered flowering pattern emerged. A similar pattern among plants that are bat pollinated has been observed in Costa Rica (Heithaus et al. 1975). Of twenty-five commonly visited plant species, an average of only 35.3% flowered in any given month. Also in Costa Rica, flowering peaks of hummingbird-pollinated plants are staggered and they shift somewhat from one year to another, suggesting that seasonal and annual variations in rainfall are also important in influencing flowering peak sequence (Stiles 1977).

There is no doubt that ecological specialization, particularly for food acquisition, is a widespread phenomenon in the tropics. Examples of such specialization could fill this book and several other volumes. But the degree to which direct or diffuse competition with other cohabiting species was and continues to be *the cause* of the specialization is unclear. Specialization could be largely the result of straightforward adaptation to food resources, quite apart from any influence of competition. It is probably true, however, that in such hyperdiverse communities as lowland tropical forests, competition among species (especially diffuse competition, where a given species may have slight negative influence on many other species) has figured substantially in and continues to contribute to the evolution of diversity patterns in the tropics. That's my educated *guess*.

The Predation Hypothesis

One result of intensive competition can be the extinction of one or more of the competing species. What can prevent such extinction? One factor is predators. Suppose four caterpillar species are competing for the same plant. One species begins to win, and the others are being driven toward extinction. What was a four-species system is about to become a one-species system. But suppose birds prey on the caterpillars. Which of the four species are the birds likely to take? The most obvious and abundant species would seem the likely choice. The result of predation by the birds would be to reduce the growing population of the "winning" species, allowing the other, "losing" species to regain some control of the resources and increase in population. This scenario describes the predation hypothesis of diversity.

The argument of the predation hypothesis is that predators prevent prey species from competing within their ranks to the point where extinction occurs. Predators constantly switch their attentions to the most abundant prey, thus the rarer the species, the safer it is from predators. This idea is a form of *frequency-dependent selection* because the intensity of selection (in the form of predation) depends directly on the abundance of the prey. The result of predator pressure is to preserve diversity by preventing extinction by competition.

Note that the predation hypothesis is basically the opposite of the interspecific competition hypothesis. The competition hypothesis says that competition among species promotes diversity by leading to specialization, narrower niches, and tighter species packing. The predation hypothesis says that predators reduce interspecific competition, thus permitting coexistence among competing species. The predation hypothesis does not predict extreme specialization. Indeed, specialization would be less likely to occur because predators keep competition levels low. Likewise, the predator hypothesis predicts wide niche overlap among similar species.

There is little direct evidence supporting the predation hypothesis, but there is some. It has been shown that birds and other predators develop *search images* of prey and thus do indeed focus on a specific prey type when it is abundant (Cain and Sheppard 1954). Predators often switch their attentions according to the relative abundances of their prey species. Tropical forests contain impressive predator richness, circumstantial evidence for the possible importance of predator effects. In a study conducted in a tropical moist forest in Puerto Rico, lizards in the genus *Anolis* were shown to have strong effects on the arthropod community of the rainforest canopy. When lizards were experimentally excluded, large arthropods such as orb spiders, cockroaches, beetles, and katydids all increased significantly (Dial and Roughgarden 1995). This is not sufficient proof, however, that predation is a major process influencing species diversity. But it is suggestive.

Terborgh (1992) has suggested an intriguing example of how predators, specifically cats, might be inadvertently structuring rainforest communities. Citing Emmons' (1987) work, Terborgh points out that jaguar, puma, and ocelot all forage nonselectively, taking whatever they encounter and can catch. In other words, the populations of prey species correlate directly with the frequency upon which each prey species is taken by a cat. If agoutis represent 40% of prey species, agouti remains show up in cat scat 40% of the time. Terborgh goes on to argue that since prey species differ in their fecundities (rates of reproduction), predation by nonselective cats could significantly reduce certain low-fecundity prey populations. In other words, peccaries can produce more offspring annually than pacas, so if cats do not ever discriminate between peccaries and pacas, pacas must decline more than peccaries, since they cannot replenish their losses as quickly. Because pacas and peccaries both eat many of the same things, an ecologist might be tempted to conclude that paca reduction was due to losing in competition for food with peccaries, never guessing that cat predation was the real reason. Terborgh's argument demonstrates that predation can result in a loss of diversity, as well as act to maintain it, if predation is, in fact, nonselective rather than frequency dependent. Terborgh points out that on Barro Colorado Island, where there are no jaguars or puma, species such as paca and agouti have dramatically increased in population density. Since both agoutis and pacas are considered major seed predators, the loss of large cats can eventually result in significant changes in plant species richness because of seed predation by the large and abundant rodents.

Parasites also act as predators of a sort and could have a strong influence on structuring rainforest communities. Epizootics of yellow fever, for example, can significantly reduce certain monkey populations. One would expect that parasites would also show increased species richness in equatorial compared with temperate latitudes, but such may not be the case. An analysis of terrestrial and aquatic parasite communities in vertebrates by Poulin (1995) has shown no correlation between richness and latitude. Indeed, only the size of the host correlated with the richness of its parasite community. However,

Poulin noted that there are precious few studies of tropical parasite communities, vertebrate or otherwise, and thus any conclusions about the role of parasites in structuring rainforest communities must await the acquisition of much more data.

Both the interspecific competition and predation hypotheses could operate simultaneously in the tropics. Predators may compete among themselves and specialize on various prey groups that are prevented from competing by predation! Anyone who finds such an example, please publish it.

The Productivity-Resources Hypothesis

One frequent suggestion to explain high diversity in the tropics is that high productivity permits more species to be accommodated (Wilson 1992). The idea here is that the high tropical productivity and biomass translate into more space for more species. Daniel Janzen (1976), in a paper titled "Why Are There So Many Species of Insects?" concluded by saying, "I think that there are so many species of insects because the world contains a very large amount of harvestable productivity that is arranged in a sufficiently heterogeneous manner that it can be partitioned among a large number of populations of small organisms." Janzen was not restricting his speculation to the tropics, but his remark fits the tropics particularly well (and see May 1988). There is indeed a tremendous potential harvestable productivity, and there are lots of spaces for small animals in the three-dimensionally complex rain and cloud forests. Support for this view comes from studies such as one done on birds of the Peruvian Amazon, where it was demonstrated that the total number of breeding birds was equivalent to that in many temperate forests but their combined biomass was about five times as great in the tropical forest (Terborgh et al. 1990). But what is it about tropical forests that permits such a greater avian biomass?

Bird species diversity often correlates with foliage height complexity in the temperate zone (MacArthur and MacArthur 1961). The more layers of trees there are, the more bird species that can be accommodated. In the tropics, however, bird species diversity does not correlate well with foliage height diversity (Lovejoy 1974). This indicates that the tropical forest must be more spatially complex than temperate forests, offering resources not measured by simple structural analysis. Tropical forests do indeed offer for birds certain resources found in no other ecosystem, at least not to the same degree of abundance or constancy (Karr 1975). Among these additional resources are vines, epiphytes, and dry leaf clusters (page 292), all of which add both space and potential food for birds, large insects, and other arthropods, and the year-round availability of nectar and fruit. Neither nectar specialists such as hummingbirds nor fruit-eating specialists such as cotingas, guans, curassows, or parrots can exist nearly as successfully outside of the tropics, since they are so dependent on constant availability of nectar and/or fruit. Army ants are a unique resource in that their activity provides the basis for the existence of "professional antbirds," species that exclusively follow army ant swarms, feeding on the arthropods driven out by the ants (page 284). Forest gaps also represent a resource of sorts. Many species of plants and animals may be essentially "gap-specialists." Even bats are a resource for predators. The abundance

Black-and-white owl

of Neotropical bats that inhabit rainforest has permitted the evolution of specialization in the diet of the black-and-white owl (*Ciccaba nigrolineata*). This species, which ranges from southern Mexico to northwestern Peru, is a bat specialist, feeding almost entirely on bats. As a final example, monkeys are a resource of sorts for another predatory bird, the double-toothed kite (*Harpagus bidentatus*). This bird routinely follows monkey troops, feeding on the various animals disturbed by the movement of the troop through the canopy (Ridgely and Gwynne 1989).

Some data exist that support the productivity-resources hypothesis. In a study comparing the woodpecker communities of Maryland, Minnesota, and Guatemala, Robert A. Askins (1983) recorded seven woodpecker species in Guatemala compared with only four in each of the temperate study areas. Although there were more species of tropical woodpeckers, these were in reality no more specialized in their foraging behavior than their temperate relatives (thus, their niches were not narrower in the tropics). They did utilize a wider range of resources, however. Some of the tropical species fed on termites and ants, probing into the excavations made by these insects, and the tropical species fed much more heavily on fruit than the temperate woodpeckers. Two species in particular, the black-cheeked (*Melanerpes pucherani*) and the golden-olive (*Piculus rubiginosus*), utilized resources not available to temperate species. The black-cheeked frequently fed on fruit, and the golden-olives probed moss and bromeliads.

Bamboo stands provide another example of additional tropical resources that enhance species richness. In tall bamboo stands, where plants reach heights of up to 15 m (50 ft), as many as twenty-one bird species are specialized in some way to feed only in bamboo. Nine species specialized on eating bamboo seeds, and twelve were insect foragers. An additional sixteen species of insect foragers were found mostly in bamboo but also in other habitats (Remsen and Parker 1985).

Ephemeral Amazonian river islands provide yet another important resource to which birds have specialized. Fifteen bird species have been shown to be restricted to island habitats such as sandbar scrub and young *Cecropia* forest (Rosenberg 1985).

Mammalian diversity also correlates with productivity and habitat characteristics. The density and number of species of Amazonian mammals (excluding bats) correlate positively with soil fertility and undergrowth density. Large mammalian species tend to range widely and maintain relatively constant densities over large areas, but small species vary dramatically in numbers and diversity from one study site to another (Emmons 1984).

There seems little doubt, when examining the diversity of birds or other animal taxa, that the lushness and largeness of tropical forest is at least in part responsible for the high diversity. But in a sense this begs the question, because it does not explain why the plants are so diverse. In the Amazon Basin, there are approximately 1,700 species of trees! Why are there so many kinds of trees? I will take up this question later in the chapter.

No convincing and all-encompassing theory of tropical species diversity has as yet been put forth. One major difficulty with both the interspecific competition and predation hypotheses is that it is not clear why tropical regions should be more influenced by these processes than temperate ecosystems. Some have argued that disturbance is less of a force in the tropics, thus permitting more intensive and continuous ecological interactions. There are many, however, who doubt that disturbance is less important in the tropics (see below). It may, in fact, be just the opposite. There are likely to be elements of truth in each of the diversity hypotheses. Perhaps that is the real key to tropical diversity. Several ecological forces are working simultaneously in promoting and preserving species diversity.

Mild seasonality and consequent predictability of resources are widely believed to contribute to the high degree of specialization exhibited by tropical species. In the tropics, species can afford to evolve as specialists because resources vary less dramatically with the seasons (i.e., something is always flowering, something is always fruiting). In the temperate zone, species must be more adapted as "generalists," since they are confronted with selecting different resources in different seasons. The fact that the tropics are frost-free could be a major factor responsible for the higher diversity. In a sense, being frost-free could be one of the resources of most importance.

Productivity, as noted above, may be the second resource of most importance. The extraordinarily high levels of carbon fixation that occur in tropical ecosystems (page 44) provide a food base that substantially exceeds that in most other ecosystems. Evolutionary forces have partitioned that food base among many species in ways that are not fully clear, but the fact remains that without so much available organic matter, it is difficult to see how so many life forms could be sustained.

Wilson (1992) offers what he calls the ESA theory for high tropical biodiversity, where E stands for energy, S for stability (in the sense of equitable climate), and A for area. The high productivity of equatorial regions, plus the warm and wet nature of the environment, plus the large area available for species to inhabit (and for speciation to occur) are primary categories of factors that together interact to produce very high biodiversity. Quite possibly so.

Adaptive Radiation Patterns

The abundance of species in the tropics is far from random. Many obvious and important patterns emerge when studying tropical diversity. One such pattern is *adaptive radiation*, which occurs when one type of organism gives rise to many different species, each adapted to exploit a different set of environmental resources. Darwin observed adaptive radiation of a group of small finches on the Galapagos Islands, a group so important to his theory of natural selection that they now bear the common name of Darwin's finches (Lack 1947). These small, chunky, black and brown birds vary in bill size and shape, though otherwise they appear similar. In his account of the voyage of the *Beagle*, Darwin wrote of these birds, "Seeing this gradation and diversity of structure in one small, intimately related group of birds, one might really fancy that from an original paucity of birds in this archipelago, one species had been taken and modified for different ends." Two excellent examples of adaptive radiation are provided by Neotropical bats and tyrant flycatchers.

Bats   Figures 122–34

Bats, in the order Chiroptera, are the only mammals anatomically capable of true flight. Bats first appear in the fossil record as far back as the Eocene, approximately 60 million years ago (Carroll 1988). Their closest relatives are insectivores, the moles and shrews. The most important adaptation of bats is the modification of the forelimb (arm) as a wing through elongation and enclosure of the arm and finger bones within a membrane of tough but flexible skin. Like birds, bats have large hearts, light body weight, and high metabolism. While most birds are diurnal, all bats are nocturnal. In terms of species richness, there are approximately 950 bat species in the world (Emmons 1990), but birds are almost ten times as species rich.

Bats in the American tropics are all members of the suborder Microchiroptera, the "insectivorous" bats. Microchiropterans capture prey on the wing and avoid obstacles by using echolocation, a highly sophisticated sensory adaptation. These animals emit loud, high-pitched vocalizations (mostly inaudible to humans) that bounce off objects of approximately the same size as the wavelength of the emitted sound, thus providing the bat with an effective sonar system for locating small nearby objects, such as flying insects. Most microchiropterans have highly prominent pinnae or external ears that aid in receiving the echolocation signals. The bat's brain, of course, is capable of processing and integrating the sonar inputs while maintaining full flight.

The adaptive radiation of microchiropteran bats in the American tropics is impressive. These animals were originally adapted to feed on insects captured in the air using echolocation. Indeed, many Neotropical species still feed in this evolutionarily traditional insectivorous way. However, there are also fruit-eating, nectar-eating, pollen-eating, fish-eating, frog-eating, bird-eating, lizard-eating, mouse-eating, bat-eating, and even blood-lapping bats. From insectivorous ancestors, Neotropical bats have greatly diversified, not only in number of species but also into dramatically different feeding niches, taking evolutionary advantage of the tremendous diversity of rainforest resources.

The highest bat species richness in the world is in the Neotropics, where there are 75 bat genera, of which 48 are in the family Phyllostomidae (Wilson 1989). The remaining genera are spread among eight other families. The small country of Belize, with a total area about equal to the state of Massachusetts, has 84 bat species (B. Miller, pers. comm. 1996) compared with 40 species in the entire United States. Costa Rican bats provide a well-documented example of bat foraging diversity, with 103 species, of which 43 are insectivorous, 25 are frugivorous, 11 nectarivorous, 2 carnivorous, 1 piscivorous (fish-feeder), 3 sanguivorous (blood feeders), and 18 species that feed on some combination from the above list (Janzen and Wilson 1983). Some bats, such as the spear-nosed bats (*Phyllostomus* spp.), are foraging generalists, feeding on a combination of fruits, large insects, pollen, and nectar (Emmons 1990).

Data on bat community structure is generally difficult to obtain. Bats are nocturnal, reclusive during daylight hours, and their vocalizations are inaudible to the human ear. Much research is based on netting bats and identifying them in the hand. However, two researchers in Gallon Jug, Belize, Bruce and Carolyn Miller, have successfully employed a computerized sound-detection system that can be taken into the field to record bats on the wing (Possehl 1996). The system records vocalizations of free-flying bats and displays a sonographic pattern on the computer screen that can be stored in memory and used to identify a given species of bat wherever it is heard. Once all species' vocalizations have been recorded, analyzed, and identified to species (which requires capturing the animals at least once to verify their vocalizations, species by species), the Millers can take their microphone and portable computer anywhere in the rainforest, record the free-flying bats, and know exactly which species are present. This technique has already shown that some species, previously believed to be rare, are actually quite numerous (B. Miller, pers. comm. 1996).

Among the many insectivorous bats, some capture prey by aerial foraging and some by foliage gleaning. Aerial foragers catch insects on the wing, while foliage gleaners pick their prey off leaves, branches, and even the ground. Some bats specialize on tiny insects, some focus on certain insect types, such as beetles, while others forage on insects attracted to water.

The false vampire bat, *Vampyrum spectrum*, captures sleeping birds as well as rodents and other bats, probably locating some of its prey by olfaction. Like many bats, false vampires have a keenly developed sense of smell. This is one of the largest of the Neotropical bats, with a wingspread of approximately 76 cm (30 in). It has prominent ears, a long snout, a large "leaf" nose, and long, sharp canine teeth. Its generally ferocious appearance misled people into believing it to be a vampire, which it is not.

Flattened, leaflike noses are common not only among carnivores but also among fruit-eating bats (subfamilies Carolliinae and Stenodermatinae) and nectarivorous bats (Glossophaginae). The prominent, flattened nose may be an adaptation aiding the bat with echolocation. Many leaf-nosed bats (all in the family Phyllostomidae) emit sounds from their nostrils that enable them to echolocate, and the fleshy fold on the nose may act as a kind of megaphone to concentrate the sound (Emmons 1990).

Fruit bats tend to have large eyes and sensitive noses as they locate their fruit by both sight and smell. Fruit bats are essential seed dispersers, just as nectarivorous bats are essential cross-pollinators.

The greater fishing bulldog bat, *Noctilio leporinus*, is an excellent example of the diverse nature of bat adaptive radiation. This highly successful species, which ranges extensively from Mexico to northern Argentina and Uruguay, uses its sonar to locate fish just beneath the water's surface (Brandon 1983). It gets the name "bulldog bat" from its flattened but puffy face, with small eyes, short, pointed ears, and prominent cheek pouches. A large bat, with wingspread of 61 cm (24 in), it has long toes with prominent, sharp claws. Using its sonar, the bat detects small fish and crustaceans breaking the surface of calm rivers and pools. It swoops down, gaffing the fish with its large, well-clawed feet. It then transfers the piscine to its mouth, where it stuffs it in its large cheek pouches as it grinds it up bones and all. I once spent a night in a motel room with a *Noctilio* that a colleague was bringing back for study in the United States. The bat fed on fish from a pail in the bathroom. It consistently devoured fish headfirst and hung contentedly upside down (feet attached to a picture frame) as it noisily munched away, fish tail protruding from its stuffed mouth. The fact that the bat reeked of fish and strong musk, plus the constant sound of fish bones being ground up, made for fitful sleeping that evening.

The most infamous Neotropical bat is the common vampire bat, *Desmodus rotundus*. This extraordinary animal feeds entirely on the blood of mammals such as tapirs and peccaries. In recent years vampires have prospered in some areas because of increased numbers of cattle and swine, both of which afford an easily accessible source of blood (D. C. Turner 1975). Vampires fly from their roosting caves at night to locate prey. The bat finds its victim both by olfaction and vision. Remarkably agile (for a bat), the vampire scurries over the ground on its hind legs and thumbs, climbing onto the sleeping animal. Using specialized incisors, the bat slices into the superficial skin layers and initiates bleeding. The cut is so sharp that the prey animal rarely awakens. The bat's saliva contains an anticoagulant, so the blood flows freely while the bat feeds. The bat's digestive tract is modified to deal with blood (as well as being capable of significant distension as the creature fills itself), which is extraordinarily high in protein. Another vampire species, *Diaemus youngii*, feeds only on birds.

Common vampire bats routinely regurgitate blood to those members of the roost who failed to feed (Wilkinson 1990). Both males and females engage in this behavior (DeNault and McFarlane 1995). Given the high metabolism of bats, a single bat cannot survive for more than seventy-two hours without food. Considering that the food resource is frequently hard to find, but also considering that when it is found (a sleeping tapir, peccary, or cow) the bat can often

Common vampire bat

fill itself with more blood than it requires, it is to each bat's personal advantage to share food with others, as it helps ensure that it will be fed should it fail to find food. Vampires typically share blood with close relatives, such as siblings (Wilkinson 1990). Doing so obviously enhances the probability that these individuals, with which the donating bat shares a very high percentage of its genes, will survive. Since a bat, like any other sexually reproducing creature, shares 50% of its genes by inheritance with each sibling (providing the mother and father are the same for each), if a bat feeds three siblings, it has aided 1.5 sets of its genes, a process called kin selection. Natural selection ought to select for kin loyalty in colonial species, especially when food is such a clumped, unpredictable resource. It should be no surprise, therefore, that vampires share blood with their blood relatives, so to speak (sorry).

But vampires also share with individuals not closely related to them (Wilkinson 1990). In turn, these nonrelated animals also share with nonrelatives when they have fed. Such mutually advantageous sharing is termed "reciprocal altruism." But note that reciprocal altruism is no more altruistic than trading a hamburger for a candy bar. Both parties in the transaction benefit, but at a cost to both. In the case of common vampire bats, food acquisition on a daily basis is problematic, and thus these bats have evolved a social system of sharing behavior in which unsuccessful foragers can survive and they, in turn, can feed others when they succeed at finding prey.

Vampires sometimes attack people. I have seen Mayan children in Belize who have been bitten about the face and fingers by vampires. Fortunately for these people, none of the bats was a carrier of rabies, though in some places, particularly where there are large cattle herds (thus artificially inflating vampire populations), vampire bats do transmit this serious viral disease. Unfortunately, because of fear of vampires, many other bat species are wantonly killed by local peoples.

In evolutionary terms, it is intriguing to speculate as to how such an unusual habit as blood lapping could have evolved. One idea is that ancestors of today's vampires were attracted to feed on ectoparasites of large mammals and gradually acquired a taste for mammalian blood (Timson 1993). Another suggestion is that the blood-feeding habit may have developed from an attraction to insects that in turn were attracted to wounds on large mammals (Fenton 1992).

Bats exhibit adaptive radiation not only in their feeding behaviors but also in their choice of roosting sites. The roost is the place in which a bat spends about half its life, and roosts are the sites of most social interactions, mating, rearing young, and food digestion, as well being refuges from adverse weather conditions and predators (Kunz 1982). Bats may be colonial or solitary roosters, and roost sites include caves, crevices, or hollow trees. Some bats roost in foliage, often modifying it to suit their needs. The small bat *Thyroptera tricolor* roosts in furled *Heliconia* and banana leaves, attaching to the slick leaf with adhesive disks on the legs and wrist joints of the wings. *Ectophylla alba*, a small, beautiful, all-white bat, goes one step further. It is one of several species actually to construct a tent, in this case out of a *Heliconia* leaf. The white bat forces the huge leaf to droop by carefully chewing veins that are perpendicular to the midrib. The leaf is only partially chewed, and the result is a protective, thick tent where a half dozen or so of these diminutive bats can cuddle in safety (Kunz 1982).

Bat courtship and mating is also diverse. Some bats are monogamous, and some are highly promiscuous. The male Puerto Rican bat *Artibeus jamaicensis* establishes a "harem" of females (Kunz et al. 1983). Bat clusters in caves consist of two to fourteen pregnant, lactating females and their offspring plus a single adult male. Older and heavier (larger) males typically have larger harems than younger animals. Nonreproductive females and haremless males form their own separate groups. One reason for harem formation may be that females select caves or hollow trees in which to raise young, and these areas are small enough to be effectively defended by a single dominant male. Males therefore compete against each other for access to females. A successful male obviously enjoys a tremendous reproductive advantage.

Bat adaptations will be discussed further in the section on coevolution, page 129.

Tyrant Flycatchers

Figure 143

One of the most diverse families of birds in the world is the tyrant flycatchers (family Tyrannidae), and collectively they provide another excellent example of the evolutionary process of adaptive radiation. There are 90 genera and an amazing 393 species of Tyrannidae (Parker et al. 1996), all of which are confined to the New World and most of which are in Central and South America. The taxonomy of tyrannids is an active topic of research. Traylor and Fitzpatrick (1982), in a comprehensive review of tyrant flycatcher diversity, have estimated that one of every ten bird species in South America is a tyrant flycatcher. Tyrant flycatchers make up 20% of the perching bird species in Colombia and 26% in Argentina. They abundantly occupy all habitats: rainforests, cloud forests, savannas, puna, and paramo. Though the name *flycatcher* is meaningful in most cases, as most are, indeed, insectivorous, the methods of capture and the types of insects captured vary tremendously among species. Some have diverged entirely from capturing insects, becoming fruit eaters.

The immense success of the tyrannids may be in part attributable to their versatility of feeding methods. The basic feeding behavior is called sally-glean-

Social flycatcher

ing, meaning that the bird sits on a perch and flies out to capture an insect, either in the air, on the ground, or on a leaf surface. Because flycatchers prefer exposed perches, they are well suited to invade any open habitat, such as forest edges, riverbanks, and savannas. From the fundamental sally-gleaning technique has evolved many specializations. There are hawkers, ground feeders, runners, hoverers, water's edge specialists, perch-gleaners, and fruit eaters (see also Fitzpatrick 1980a, 1980b, 1985).

Body size as well as bill size and shape vary widely among tyrannids, much more so than in the temperate zone. Among insectivorous birds in general, and tyrannids in particular, large-billed species are much more evident in the Neotropics, probably due to greater availability of large arthropod prey (Schoener 1971). Interspecific competition within the group may also have provided selection pressures resulting in divergence of body size, bill, and feeding characteristics.

Some tyrant flycatchers have uniquely shaped bills. The northern bentbill (*Oncostoma cinereigulare*) has a short but distinctly down-curved bill. The northern royal flycatcher (*Onychorhynchus mexicanus*) has a long and flattened bill, and the tiny spadebills (genus *Platyrinchus*) have, as the name implies, extremely wide, flattened bills.

Size and bill variation are strikingly evident in a group of species that all have yellow bellies and striped heads. Collectively they look much alike, except for body size and bill shape. In Panama alone, seven species occur. The great kiskadee (*Pitangus sulphuratus*) and boat-billed (*Megarhynchus pitangua*) are similar in size, but the latter has a wide, flattened bill. The lesser kiskadee

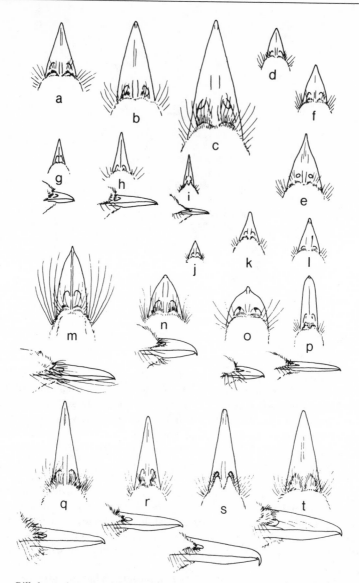

Bill shapes in some of the tyrannid flycatchers. From Traylor and Fitzpatrick (1982). Reproduced with permission.

(*Pitangus lictor*) looks like a small version of the greater and differs from the very similarly sized rusty-margined flycatcher (*Myiozetetes cayanensis*) by having a longer bill. The white-ringed (*Conopias parva*) and social (*Myiozetetes similis*) flycatchers differ only in minor facial characteristics, and the gray-capped (*Myiozetetes grandadensis*) has a gray cap that helps distinguish it from the others (Ridgely and Gwynne 1989).

Stub-tailed spadebill

Tyrannids abound in rainforests and disturbed areas. In my Belize study areas, I find twenty-three species, ranging from the 10-inch great kiskadee to the tiny 9-cm (3.5-in) stub-tailed spadebill (*Platyrinchus cancrominus*). Traylor and Fitzpatrick (1982), citing a survey done by Terborgh, note that at a single site in Manu National Park in the Peruvian Amazon, sixty-five (23%) of the 281 species of perching birds are tyrannid flycatchers! In a comprehensive study of the structure of a bird community in the Peruvian Amazon, there were thirty-four tyrant flycatcher species. Of the forty-three families of birds that were found to occur in the forest, only the antbirds (Formicariidae) had a greater richness, with forty-four species (Terborgh et al. 1990). Tyrannids are most abundant in the canopy and along the forest edge, but there are species that occur virtually anywhere from the ground up.

Some flycatchers switch their diets regularly, and others are basically opportunistic, switching according to the relative abundances of various insect prey (see predation hypothesis, above). One seasonal and dramatic switch is seen with the eastern kingbird, *Tyrannus tyrannus.* One of thirty-two tyrannids that migrate to breed in North America, the eastern kingbird feeds on insects on its summer breeding grounds. When on its wintering grounds in southwest Amazonia, however, it forms large flocks and feeds mostly on fruit, the flocks

Royal flycatcher

wandering nomadically in search of fruiting trees (Fitzpatrick 1980b). Another flycatcher, which resides year-round in the Neotropics, has evolved a fruit-eating diet. The nondescript ochre-bellied flycatcher (*Pipromorpha oleaginea) is* the most abundant forest flycatcher in Trinidad (and it occurs in great numbers elsewhere in the Neotropics). Its inordinate abundance compared with other tyrannids may be due to its diet shift from arthropods to fruit, which is a bountiful and easily "captured" resource (Snow and Snow 1979).

Overall, the ability of varied tyrannids to find food in virtually all habitats has likely been a major factor promoting speciation within the group. Species have specialized on certain types of arthropods and other food, captured in distinct ways.

Speciation in the Tropics

The impressive adaptive radiations apparent in bats, tyrant flycatchers, and other groups in the tropics indicate that the process of adaptation is closely paralleled by the process of speciation. Over time one species may evolve into a different species, or into several different species. *Speciation,* the splitting of one species into two or more new species, where a single gene pool becomes divided into several distinct gene pools, is the process that has produced the vast species richness of the Neotropics.

How Speciation Occurs

A biological species is traditionally defined as an actually or potentially interbreeding population, reproductively isolated from other such populations (Mayr 1963). The key to identifying a species therefore rests on the fact of reproductive isolation. Only members of the same species can mate and produce fertile offspring. A horse and a donkey are capable of mating, but the union results in a mule, a robust but infertile animal. Therefore, the gene pools of the horse and donkey remain isolated from one another, and the horse and donkey are considered to be separate species. Other species concepts, not based on the establishment of reproductive isolation, have been proposed in recent years because it is often difficult or impossible to know with certainty if separated populations are capable of interbreeding (Zink and McKitrick 1994). There are also cases where two seemingly distinct species do hybridize (to varying degrees depending upon the example) where their ranges come together, posing a problem in assigning species status. Nonetheless, the biological species concept, focused as it is on the presumption of reproductive isolation, is still the most widely used species definition, and it is the one that will be used here.

The speciation process involves the establishment of reproductive isolation between two populations. In order for this to happen, the flow of genes between these populations must be prevented for a sufficient number of generations to permit genetic divergence adequate to establish reproductive isolation. Once two populations have genetically diverged, they will be unable to produce fertile offspring and thus will be distinct species.

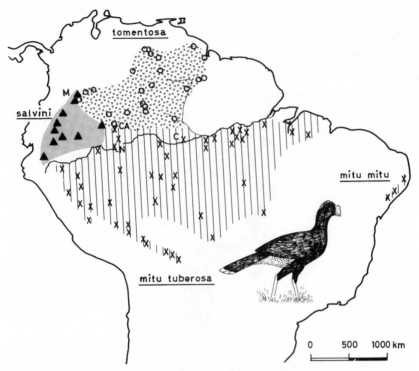

Distribution of the razor-billed curassow and allies (*Mitu mitu* superspecies, Carcidae).
M. mitu (male illustrated): *crosses and vertical hatching*. M. *salvini*: *closed triangles and
shading*. M. *tomentosa*: *open circles and dots*. Symbols denote locality records. C = Codajás,
Brazil; N = Puerto Nariño; CA = Caquetá region; M = Macarena region of Colombia.
From Haffer (1985). Reproduced with permission.

One important means by which populations can become fragmented and
gene flow thus interrupted is *geographic isolation,* or *vicariance*. Mountains, rivers, deserts, and savannas all represent possible barriers between rainforest
sites. Other factors, most notably climate changes, can fragment species'
ranges, creating vicariance. Should a mountain be formed by uplifting of
Earth's crust, as has happened extensively in the Andes Mountains, what was
once a contiguous area of forest will be fragmented by the mountain range.
Individuals on one side of a mountain are prevented from mating with those
on the other side because they simply cannot cross the mountain. Once populations are separated by geographic factors, there is the strong possibility of
genetic divergence over time among the subpopulations. For instance, a population of tapirs on the west side of a mountain range may be subjected to
different selection pressures from the population on the east side of the range.
The result of this selection pressure difference is that each population will be
selected for different characteristics, and thus for different genes. A probable
example of this sort of speciation can indeed be seen with the tapirs (family

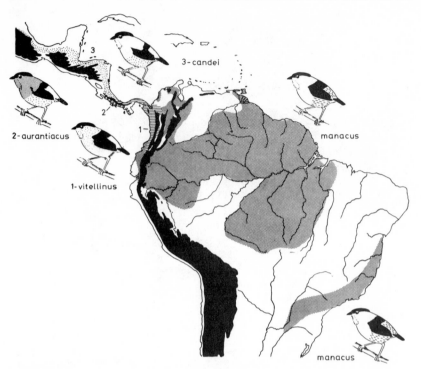

Distribution of the bearded manakin, *Manacus manacus* superspecies. The forms east of the Andes are still conspecific; three strongly differentiated taxa inhabit differing ranges west of the Andes. Color key: *unmarked* = white; *solid* = black; *dashed* = gray; *lightly stippled* = yellow; *densely stippled* = orange; *hatched* = green. Sketches are of adult males. From Haffer (1974). Reproduced with permission.

Tapiridae, page 306). Baird's tapir (*Tapirus bairdii*) is found only in lowland forest on the west side of the Andes, extending into Central America as far north as tropical Mexico. The similar Brazilian tapir (*T. terrestris*) occurs only on the east side of the Andes, occupying the entire Amazon Basin. The Andes Mountains act as the major factor in geographically isolating these two species and may have provided the main factor in splitting an ancestral species into separate species. Finally there is a third tapir species, the mountain tapir (*T. pinchaque*), which, as the common name suggests, inhabits montane forests at middle to high elevations of the central and eastern cordilleras of the Andes, in Colombia and Ecuador (Eisenberg 1989). The mountain tapir is isolated by both range and elevation from the other two species.

Small, isolated populations are particularly subject to chance factors affecting their gene pool, a process called genetic drift. For example, an isolated population of, say, twenty individuals, where only one of the twenty contains gene "A," could lose gene "A" very easily if the individual carrying it either fails to mate or does mate but does not pass on that particular gene. Genetic drift is quite different from natural selection in that changes in gene frequencies are caused entirely by chance. With natural selection, changes in gene frequencies are the result of differential reproduction of traits favorable in a particular array of environmental selection pressures, the very opposite of chance.

Geographic isolation facilitates speciation because it allows for the accumulation of genetic differences between populations by preventing gene flow among them. Both natural selection and genetic drift can act simultaneously to differentiate vicariant populations genetically. Speciation is promoted when populations fragment or when a small subset of a population "breaks off" from the parent population and becomes the founder of a new population in a different location. This was undoubtedly the case for Darwin's finches. A small flock (or perhaps a single pregnant female) flew (probably "helped" by a storm) from mainland South America to a landfall on one of the Galapagos Islands, becoming the founding population of the Darwin's finches.

The Effects of Topography

Central and western South America are both geologically active areas. The Andes mountain chain, which has been uplifting for essentially all of the Tertiary and Quaternary periods (approximately 65 million years), is responsible for creating a diversified complex of habitats as well as providing numerous climatic and physical barriers that greatly enhance geographic isolation among populations. The massive Amazon River and its numerous wide tributaries have also served to isolate tracts of forest and savanna, and, given the sedentary nature of many animal populations in Amazonia, rivers have probably served as important forces of geographic isolation. The width of the Amazon and its tributaries is sufficient to isolate populations of birds whose individual members are reluctant to cross such a wide expanse of water (Haffer 1985; Haffer and Fitzpatrick 1985). For example, blue-crowned manakin (*Pipra coronata*) populations from opposite banks of both the Napo and Amazon rivers have distinct genetic differences (Capparella 1985). The bearded manakins (the name refers to the feathering on the throat of the males) provide a good example of the effect of geographical separation in promoting speciation. This group, comprising four species, is referred to collectively as a *superspecies* because each of the four looks very similar, but each is allopatric (geographically separated) from the others. Another example of a superspecies complex is the curassows, where four similar species occupy separate ranges. These and other examples are detailed in Haffer (1974). The Amazon River largely isolates two similar species of antbirds. The dusky antbird (*Cercomacra tyrannina*) occurs north of the Amazon, while the similar blackish antbird (*C. nigrescens*) occurs south of the river. Both species occur together only along a section of the northern bank of the river, but here the blackish inhabits wet varzea forests, and the dusky favors second-growth vegetation of the terra firme (Haffer 1985). Because of complex topography, especially in northern and western South America, geographic isolation has provided an ideal template for speciation. It is no accident that the Neotropics has a bird species richness of approximately 3,300 (Haffer 1985), highest in the world.

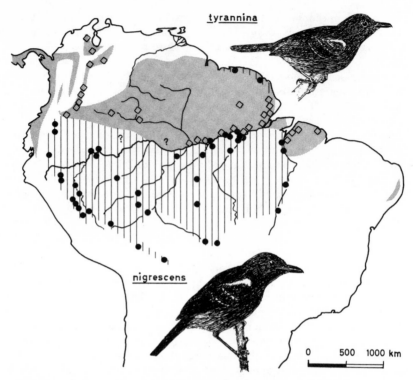

Distribution of the *Cercomacra tyrannia* superspecies (typical antbirds, Thamnophilidae). *Cercomacra tyrannina* (*above*): *open squares and shading.* Individual records for this widespread and common species are shown for selected regions only. *C nigriscens* (*below*): *solid circles and hatching.* The birds illustrated are adult males. Symbols denote locality records. From Haffer (1985). Reproduced with permission.

Diverse habitats ranging from lowland moist forests to alpine puna and paramo, plus tracts of habitat topographically isolated from similar tracts, provide ideal situations for promoting genetic divergence and speciation.

A perusal of any of the current guides to birds for areas in South America reveals almost extreme similarity among species within certain groups, evidence that speciation has probably been relatively recent. This is one reason why identification to species level is often so difficult. The antshrikes and antbirds, which are discussed more in chapter 12, illustrate that much of the species diversity of the tropics involves many species that look remarkably alike. Antbirds and antshrikes of the genera *Thamnophilus, Percnostola, Cercomacra,* and *Thamnomanes* are examples. All are basically little gray and brown birds of the rainforest understory. Males of all species are sparrow-sized, slate-gray birds (the intensity of the gray varies with species) with varying amounts of white spots on the wings and tail. Females are chestnut brown with or sometimes without white spots, depending on species. Such close morphological similarity among groups of species suggests recent speciation (see below). Additional patterns of similarity among bird species groups are seen in the woodcreepers, hermit hummingbirds (see chapter 6), and many of the tyrant fly-catcher genera, to give but a few examples.

A pattern termed "leapfrog" geographic variation is seen among a group of Andean birds (Remsen 1984). This involves a discontinuity in the appearance of three subspecies of the mountain tanager, *Hemispingus superciliaris*. A *sub-species* is a population not yet reproductively isolated from other populations within the species but different enough genetically that it does look distinct. The word *race* is essentially equivalent to subspecies. Subspecies formation is often, but not always, a precursor to speciation, and it is often quite difficult to know whether or not different subspecies are interbreeding or are, in fact, reproductively isolated (in which case they should be classified as separate species). The northern *H. superciliaris* subspecies in Colombia and Ecuador is very yellowish, a pattern shared with the subspecies in southern Peru and Bolivia. However, the subspecies of central Peru, which occurs *between* the yellowish northern and southern subspecies, is dull gray. The northern, central, and southern Andean subspecies are all closely related. Indeed, they are still considered members of the same species. Why should the central population diverge in color from the other two? Perhaps the complex topography of the Andes has largely isolated the three subspecies and thus reduced gene flow among them. Occasional random mutations, such as one for plumage color, could occur and not necessarily be selected for or against. In short, subspecies could form and diverge by chance, producing the odd "leapfrog" pattern of a central population looking distinct from two bordering populations, both of which look essentially alike. Some other bird species with three or more distinct subspecies also show the leapfrog pattern. Many species of Neotropical birds exhibit complex patterns of subspeciation (Haffer and Fitzpatrick 1985).

A Brief Evolutionary History of South America

The Breakup of Gondwanaland

South America's origin as a single continent began slowly, approximately 120 million years ago. At that time, when dinosaurs were (to put it mildly) conspicuous components of terrestrial ecosystems, it would be accurate to describe much of the world's terrestrial ecosystems as "Jurassic Parks." It was, after all, the Jurassic Period of the Mesozoic Era. Immense, elephantine apatosaurs and bizarre-looking stegosaurs were stalked by swift, predatory allosaurs. Some small, active dinosaurs were evolving feathers, derived from their reptilian scales, which probably functioned to allow them to maintain a warm body temperature. These creatures soon evolved flight, and birds came into the world. What would eventually become South America was still united with what would become Africa, Madagascar, Australia, New Zealand, Antarctica, and India, all fused together into a gigantic continental mass known as Gondwanaland. A walk through a Gondwanaland forest would be far different from a walk in today's Amazonian rainforest. Instead of flowering trees and shrubs pollinated by insects and hummingbirds, there would be gymnosperms, such as tall redwoods, and cypresses, all pollinated by wind. Cycads would abound in the understory. Instead of jaguars and sloths, there would be dinosaurs.

During the Jurassic Period the breakup of Gondwanaland began in earnest, and it continued throughout the Cretaceous Period, the close of the Mesozoic Era, which ended approximately 65 million years ago. Continents continued to drift apart during the succeeding Cenozoic Era as, indeed, they continue at present time. Biogeographers have long noted the striking similarities that exist among some of the plants and animals of the now widely separated southern continents, particularly such among groups as freshwater fish, crayfish, treefrogs, and certain turtles, lizards, birds, and trees (Webb 1978). For example, the southern beech tree, *Nothofagus*, can be found today in southern Patagonia (Chile and Argentina), south Australia, Tasmania, and New Zealand. Fossil pollen grains of *Nothofagus* occur abundantly in Antarctic sediments. This curious distribution, especially for a kind of plant that has limited dispersal powers (Hill 1992), provides very strong biogeographic evidence for the fact of continental drift, a process more properly termed plate tectonics (page 12). Other plant groups, among them the Heliconiaceae and Musaceae, are shared between the Old and New World tropics, another probable historical accident of the drifting continents (Raven and Axelrod 1975). Among birds, genetic studies of the large flightless birds (collectively termed ratites because they lack a large bony keel on the sternum) suggest strongly that the South American rheas are very close genetic cousins of the African ostrich, and only a bit more distant cousins of the Australasian cassowaries and emu, as well as the kiwis of New Zealand (Sibley and Monroe 1990). These species each evolved from a common ancestor, but their evolutionary histories were then determined by their subsequent geographic isolation as their continents separated, a gigantic vicariance indeed. The fact that marsupial mammals are found today only in South America and Australia (the Virginia opossum is a recent immigrant to North America) is yet another biogeographic outcome of continental drift following the breakup of Gondwanaland. The plants and animals of South America have experienced a dynamic evolutionary history, tied to the equally dynamic geology of the continent (Raven and Axelrod 1975).

The Unique (and Mostly Extinct) Creatures of South America

Once the separation of South America from Africa was complete, South America became an isolated continent. It remained isolated for most of the Cenozoic Era (Marshall 1988), an ample time for the evolution of many unique groups of creatures. Among the mammals, these groups include some that still remain familiar today: the armadillos, tree sloths, anteaters, and opossums. But they also include others now extinct: the giant ground sloths, predatory borhyaenoids, immense armored glyptodonts, and groups with such unfamiliar names as litopterns, notoungulates, condylarths, astrapotheres, pyrotheres, gomphotheres, and pantodonts. These odd-looking, facinating mammals made South American ecosystems unique, evolving in what evolutionist George Gaylord Simpson described in somewhat poetic terms as "splendid isolation" (Simpson 1980). Joining these groups during the mid-Cenozoic were such others as monkeys (which first appear in the fossil record 26 million years ago) and cavimorph rodents such as the capybaras and porcupines (which first appear 34 million years ago). These latter groups may have dispersed accidently over water, carried on miniature rafts of vegetation broken lose from rivers in flood. They came perhaps from Africa (the two continents were then much closer than now), but more likely from North America (Marshall 1988).

It is beyond the scope of this volume to detail each of the extinct groups, but some brief comments should serve to give you a sense of them. Many were relatively large, camel- and horse-sized, and most were savanna dwellers (which suggests how extensive savannas were at that time). The litoptern *Thoatherium* bore a remarkable resemblance to a horse. The notoungulate *Toxodon* was a thickly built, large-skulled creature somewhat similar to a combination rhinoceros and hippopotamus. The litoptern *Macrauchenia* had a singularly long neck vaguely resembling that of a giraffe and a face that featured a short, elephantine proboscis. The remainder of the beast looked like a camel, which, indeed, it was once suspected to be—by none other than Charles Darwin. The giant ground sloth *Megatherium* looked like a huge, bulky version of today's tree sloths. And the glyptodont *Doedicurus* resembled an utterly gigantic armadillo with a thick, armored tail terminating in a medieval-like, spiked mace. For much more on these odd creatures, see Simpson (1980).

Mammals were not the only group to evolve in unique ways on the isolated South American continent. Following the extinction of the dinosaurs, the prevalent large carnivores of South America were, of all things, birds. From approximately 62 million years ago until just 2.5 million years ago, a group of large, flightless birds were the dominant carnivores of savanna ecosystems. This group bears the name Phorusrhacoids, or simply "terror birds" (Marshall 1994). Phorusrhacoids are extinct, but some of their genes live on in the form of long-legged, carnivorous birds called seriemas, which inhabit the grasslands and cerrado from northern Argentina to southern Brazil (page 231). Phorusrhacoids were flightless but could run perhaps as fast as 60 km (37 mi) per hour. The largest of the 25 phorusrhacoid species was about 3 m (9.8 ft) tall; thus, were it still around today, it could look down on the tallest basketball player (just before eating him). For most of the Cenozoic, phorusrhacoids prospered. The fossil record suggests that by 5 million years ago they had largely replaced the borhyaenoids, the only group of large, carnivorous mammals on the continent. The demise of the terror birds was in all likelihood attributable to the rapid invasion of South America by such predators as sabertoothed cats, jaguars, and wild dogs, an invasion that started with the exposure of a land bridge over the Isthmus of Panama beginning about 2.5 million years ago (Marshall 1994). The Panamanian land bridge is responsible for what has become known as the Great American Interchange (Marshall 1988).

The Great American Interchange

The Panamanian land bridge formed because of a combination uplift of the northern Andes and a global drop in sea level, perhaps as much as 50 m (164 ft), as a result of the increasing size of the polar ice caps (Marshall 1994). Thus it was just prior to the onset of the Ice Age that the continents of North and South America were no longer separated by water. The Panamanian land bridge profoundly altered the ecology of South America, much more so than it did that of North America. Consider that the faunas of North and South America had evolved independently of one another for at least 40 million years, but their mingling, once the land bridge developed, was completed within a mere 2 million years (Webb 1978).

Once the land bridge formed, some South American animals moved northward as early as 2.5 million years ago, literally walking to North America. These included two armadillo species, a glyptodont, two species of ground sloths, a porcupine, a large capybara, and one phorusrhacoid bird. These animals were followed by other invaders, which included at least one toxodon species as well as *Didelphis*, the familiar opossum, which invaded approximately 1.9 million years ago (Marshall 1988). The collective impact of the South American invaders was modest at best. Of their host, only armadillos and opossums remain, both of which are thriving. The ground sloths were probably killed as the human population spread southward. The others just drop out of the fossil record.

But many North American animals walked across the land bridge to South America, and there the impact was profoundly greater. The list of invaders alone is extensive and sobering: skunks, peccaries, horses, dogs, saber-toothed cats, other cats, tapirs, camels, deer, rabbits, tree squirrels, bears, and an odd group of elephant-like creatures called gomphotheres. Add to this list the field mice, or cricetid rodents, whose travel route to South America is still debated, but who have since radiated into 54 living genera, and you begin to see why the effect of North American animals on South American ecosystems was so great (Marshall 1988).

Climate, which varied dramatically between cold and warm periods during the Pleistocene, also affected the faunal interchange. Typically most of the exchange occurred during times of expansion of savanna habitat (when conditions became cooler and drier), forming an Andean corridor that permitted savanna-dwelling creatures to disperse (Webb 1978). When savannas expanded, forests contracted (see below). Unfortunately, there is a far better fossil record of savanna species than forest species (Marshall 1988), so much less is known about how the interchange affected rainforest ecology.

One outcome of the exchange is incontrovertible. The look of the South American fauna is now much more like that of the North American. It is thus tempting to conclude that the faunal interchange provides a clear example of competitive replacement. Such a conclusion would be premature. The picture is more complicated. For example, many hoofed mammals invaded from North America. However, their ecological counterparts in South America, the litopterns and notoungulates, were declining in numbers and diversity long before the camels and horses arrived. It is not considered likely that the North American ungulates were the primary cause of the extinction of the South American groups (Marshall 1988). Among predatory mammals, the borhyaenoids were essentially outcompeted by the terror birds, not by mammalian invaders from the north. As mentioned above, however, the phorusracoids may have subsequently lost out to invading placental mammals. But there is really only one relatively clear-cut example where it appears that direct competition with a North American species may be responsible for the extinction of a South American species. This is the case of the marsupial saber-toothed cat,

Thylacosmilus. This animal bore a striking anatomical similarity to the placental saber-toothed cat, *Smilodon*, a case of what evolutionists call "convergent evolution." *Smilodon* crossed the land bridge into South America, and the extinction of *Thylacosmilus* coincides closely with the arrival of *Smilodon*. Finally, much of the extinction of large mammals on both continents is probably explainable by the proliferation of humans during the Pleistocene (Marshall 1988), not by competition among the animal groups themselves.

What is indisputable about the interchange is the degree to which the South American fauna was permanently altered. As Marshall (1988) so aptly summarizes (about mammals), "the Great American Interchange resulted in a major restructuring. Nearly half of the families and genera now on the South American continent belong to groups that emigrated from North America during the last 3 million years." It is thus to the subsequent events of the last three million years that we now turn our attention.

The Ice Age and Tropical Refuges

Speciation rates in the American tropics may have increased substantially during the so-called Ice Age, a period more properly known as the Pleistocene, which began about 2.5 million years ago and ended a mere 10,000 years ago, with the onset of the present interglacial period, the Holocene. Many people naively believe that the equatorial tropics were climatically stable and constant throughout the Pleistocene, undisturbed by the giant glaciers bearing down upon northern temperate areas. Such was not the case, however, a point recognized long ago by Thomas Belt, who, in 1874, discussed possible effects of northern glaciation on the tropics. More recently, much evidence from geomorphology (the historical development of present landforms), paleobotany (the study of past patterns of plant distribution), and biogeography indicates that dramatic changes took place in Amazonia during the Pleistocene (Haffer 1969, 1974, 1985, 1993; Simpson and Haffer 1978; Prance 1985a; Colinvaux 1989a, 1989b).

Evidence suggests that during glacial advances in northern latitudes, the Neotropics became markedly cooler and perhaps much drier. For example, during part of the Pleistocene, temperature in the Ecuadorian foothills, east of the Andes, was 4–6°C (5–9°F) cooler than at present (Colinvaux 1989a, 1989b). The ecological result of the climatic cooling was to alter and shift the distribution of ecosystems. Ecosystems such as grasslands, savannas, and cerrado areas enlarged, presumably at the expense of more lush moist forests. Large continuous tracts of lowland rainforest were therefore fragmented into varying-sized "forest islands" surrounded by "seas" of savanna or dry woodland. Because of the repeated shrinking and fragmenting of forests, forest organisms periodically became geographically isolated from populations in other forest areas, promoting speciation (Nores 1992). The Amazon Basin became a climatically dynamic "archipelago" of varying-sized rainforest islands.

Areas where rainfall remained high are believed to have persisted as rainforest even during the driest, coolest periods, forming what have been termed *refugia,* and it is in these scattered rainforest refuges that rainforest species continued to thrive. Studies based on the current distribution of certain kinds

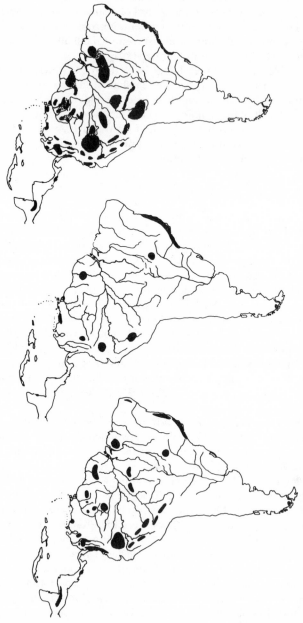

Distribution of presumed forest refugia in the Neotropics during dry climatic phases of the Pleistocene. *Left*, reconstruction based on the distribution pattern of Neotropical birds (Haffer 1967a, 1969). *Center*, based on the population structure of Amazonian lizards (Anolis chrysolepis species group, Vanzolini et al. 1970; Vanzolini 1970). *Right*, based on an analysis of Heliconius butterflies (Brown et al. 1974). From Haffer (1974). Reproduced with permission.

of birds (Haffer 1969, 1974, 1985, 1991; Haffer and Fitzpatrick 1985; Simpson and Haffer 1978) postulate that at least nine major and numerous smaller forest island refuges were present in Amazonia during the Pleistocene. Because the refuges were geographically separated by wide expanses of grassland savanna or other nonrainforest ecosystems, populations were sufficiently isolated (a situation called *allopatric*) from those in other refuges that conditions for speciation (and subspeciation) were ideal. Different refuge regions, isolated from one another geographically (vicariance), may have been subject to differing selection pressures, promoting rapid genetic divergence among isolated populations. Many groups subsequently went through periods of rapid speciation because there were repeated episodes of rainforest shrinkage and expansion. During interglacial periods, forests expanded and secondary contact was established between newly speciated populations, explaining why so many extremely similar species can be found today in Amazonia. For example, in Colombia there are six species of toucanlike aracaris in the genus *Pteroglossus*, all of which look similar and four of which are highly *sympatric* (their ranges significantly overlap). These largely frugivorous birds are yellow underneath and dark above. Depending upon species, there are combinations of dark and red bands on the breast and belly, and the long, banana-like bill has varying amounts of yellow and black. Evolutionists regard *Pteroglossus* as representing several superspecies complexes where various recently evolved, similar species exclude one another geographically (Haffer, pers. comm.). As mentioned previously, there are many examples of superspecies complexes among Neotropical birds (Haffer 1974, 1985; Haffer and Fitzpatrick 1985), which accounts for why Neotropical bird identification is so challenging.

The importance of the refugia hypothesis is that it attributes a significant part of the high species richness of Neotropical rainforests to the effects of *instability* rather than the older concept of the stable undisturbed tropics. Speciation was stimulated by geographic isolation brought about by the scattered refuges. For a review of the broader applications of this process, see Haffer (1990).

It might be asked why speciation exceeded extinction. Why did the existence of forest refuges result in an acceleration of speciation? If, indeed, speciation is both recent and due to the refuge effect in Amazonia, the other side of the coin is that most populations must have suffered some contraction from the loss of area due to rainforest shrinkage. Should not many species have gone extinct? Maybe not.

Interestingly, populations today can suffer major reductions without necessarily becoming extinct (though the potential risk for extinction obviously does increase). For example, although Atlantic coastal forests in Brazil have been reduced by at least 88% of their former area, only six species of resident animals have been documented to have thus far become extinct (page 352). Small populations can persist and rebuild if suitable habitat again becomes available. Evolutionary biologists have estimated that a population of but a few hundred individuals may be sufficient to maintain evolutionary fitness and survive (Soule 1980; Lande and Barrowclough 1987). Anyone who has studied islands realizes that colonist species normally arrive in very low numbers, but those few can effectively be the founders of a new population, often with significant evolutionary results (Mayr 1963). Shrinkage of Amazonian rainforest, analogous to isolation of small populations on islands, could indeed have resulted in bursts of speciation with relatively little extinction, helping to create the hyperdiversity that characterizes the region today.

The refugia model has been examined in detail (see Prance 1982a, 1985a), and evidence supporting it is based mostly on the inference that present distribution and diversity patterns can be used to ascertain where most speciation occurred (and thus where the refuges were). Areas rich in endemic species are thought to be indicative of the historical location of refuges. For example, Prance (1982b) has closely examined woody plant diversity and concluded that 26 probable forest refuges existed for these plants. Kinzey (1982) concluded that present primate distribution fits predictions of the refuge model, and Haffer (1974), Haffer and Fitzpatrick (1985), and Pearson (1977, 1982) have presented evidence from present bird distribution in support of the model. In a series of papers by Keith Brown, cited by Prance (1985a), heliconid butterflies have been shown to have 44 centers of endemism throughout Amazonia and surrounding areas. Note that centers of endemism, the presumed Pleistocene refuges, do not always precisely coincide among taxa, which, to some degree, lessens support for the refuge model. But supporters of the refuge model argue that different taxa have different dispersal powers and different generation times and thus would be expected to differ somewhat with regard to the degree of regional endemism (Prance 1985a).

The refuge model has been subject to strong criticism. Colinvaux (1989a) summarizes the arguments against the Pleistocene-based refuge model, pointing out that at least one study has suggested that some refuge locations coincide with areas in which sampling of plants for herbaria specimens has been historically most intense. This, of course, would suggest that the refuges are artifactual derivatives of uneven sampling effort. However, even the authors of that study note that their analysis (based, as it was, entirely on plant data) in no way weakens the arguments for endemism centers (and refugia) for animal groups such as birds, lizards, and butterflies (Nelson et al. 1990).

This aside, there are other objections that have been raised. One analysis of Amazonian sedimentary patterns, for example, suggests that there was no substantive climatic change or reduction in forest cover in Amazonia during the Pleistocene (Irion 1989). However, what is suggested by sediment patterns is not a stable Amazonia, but strong oscillations throughout the Pleistocene in the distribution of land, water, floodplain forests, and *terra firme*. Sea-level changes are thought to have led to the alternation of huge Amazonian lakes with strong valley cutting, geological events that would presumably have a strong impact on flora and fauna (Irion 1976). This view sees the Amazonian rainforest as subject to vicariance, though not necessarily reduced to scattered refuges.

Not all present species distributions support the refuge model. A complex of frog species of the genus *Leptodactylus* exhibits high species richness that dates back farther than the Pleistocene. Most of the frogs apparently speciated in the mid-Tertiary Period, before the Pleistocene began and refuges were presumably created (Maxson and Heyer 1982; Heyer and Maxson 1982).

Figure 42. The 42-m (105 ft)-tall tower near Manaus, Brazil, that provides access to the rainforest canopy.

Figure 43. A forest canopy walkway, such as this example at the Amazonian Center for Environmental Education and Research (ACEER), near Iquitos, Peru, affords researchers the opportunity to study the treetop ecology.

Figure 44. Looking down toward the rainforest floor from the canopy walkway in Amazonian Peru.

Figure 45. Amazonian rainforest near Iquitos, Peru, at ACEER.

Figure 46. Looking up into the rainforest canopy.

Figure 47. Canopy walkway at ACEER near Iquitos, Peru.

Figure 48. The author's wife, Martha Vaughan, demonstrating the ease and safety of viewing the rainforest canopy from the unique ACEER canopy walkway.

Figure 49. Lobster claw heliconia attracts pollinating hummingbirds and insects with colorful and conspicuous bracts. Parts of the small flowers can be seen protruding from the bracts.

Figure 50. Many broad-leaved philodendrons (*Monstera* spp.) typify the kinds of vines that grow as climbers on the boles of Neotropical trees.

Figure 51. Lianas are vines that often have a twisted rope-like shape such as is shown here. Note also the small climber with its green leaves flattened on the tree trunk.

Figure 52. Strangler figs anastomose, surrounding the trunk, and may eventually overtake the host tree.

Figure 53. A solitary cecropia abundantly adorned with pendulous flowers will attract many kinds of insects and nectarivorous birds. The palm growing below the tree is a cohune palm (*Orbigyna cohune*), common in much of Central America.

Figure 54. Even though cecropia flowers dangle below the large, palmate leaves, they are nonetheless conspicuous and easy for pollinators to reach. Likewise are the tree's fruits.

Figure 55. Large brome-
liads, recognized by their
clusters of spiky, pine-
apple-like leaves, are
common sights through-
out rainforests. The
cistern-like shape collects
water and supports many
kinds of tiny animals.

Figure 56. Bromeliads
have a conspicuous flow-
er cluster atop a central
spike, making it easy for
such would-be polli-
nators as hummingbirds
to visit the flowers.

Figure 57. Many species
of rainforest trees have
strongly buttressed
roots, a characteristic
that develops even when
the tree is still relatively
small, such as in this
example.

Figure 58. Strongly
buttressed roots and
twisted, draping vines are
both strong indicators of
a tropical rainforest.

Figure 59. Prop or stilt
roots, a probable
adaptation to periodic
flooding.

Figure 60. Stilt roots
are common to many
species of trees that
inhabit floodplains,
such as this one in
Amazonian Peru.

Figure 61. Temple II, Tikal.

Figure 62. Blue Creek Village, Belize.

Figure 63. Slash-and-burn or *milpa* agriculture in southern Belize. Note small plot size and surrounding successional vegetation.

Figure 64. Slash-and-burn banana plantation in Trinidad. Note charred stumps of trees felled when land was cleared.

Figure 65. Early successional abandoned plot in southern Belize, with numerous fast-growing vines and small trees that will soon overtop banana plants.

Figure 66. *Piper* is a common successional species throughout the Neotropics. Various insects pollinate the erect flowers, and bats eventually disperse the seeds.

Figure 67. Succession to closed forest in southern Brazil shows the abundance of thin boled trees and high light intensity at ground level.

Figure 68. Cohune palms, cecropias, and gumbo limbo trees are among the various species that comprise this young second-growth forest in southern Belize.

Figure 69. Thin stemmed and spindly, the roots of these small cassava or manioc trees (*Manihot esculenta*) provide an essential carbohydrate throughout the Neotropics.

Figure 70. Cassava (manioc, yuca) root. Note the thickness of the root.

Figure 71. This young banana grove in Venezuela is fenced in by living trees connected by wire fencing. Beans, peppers, squash, and other crops are planted as well.

Figure 72. This shade-grown coffee (*Coffea arabica*) plantation in Gallon Jug, Belize, is an example of how economic and ecological interests can be coordinated. Many animal species can be found because the canopy remains intact.

Figure 73. Ripening coffee beans from a shade-grown coffee plantation near San Blas, Mexico.

Figure 74. Cocoa (*Theobroma cacao*) is an understory tree with cauliflorous flowers and fruits. Chocolate is eventually extracted from these as yet unripened fruits.

Figure 75. Chocolate "on the vine": ripened cacao, a cauliflorous fruit.

Figure 76. Many tropical soils, such as this soil from the cloud forests of northern Panama, are rich in nutrients and serve agriculture well. Note, however, the potential for erosion.

Figure 77. Billowing clouds of smoke signal yet another tract of forest that has been felled and burned to make space available for agriculture or pasture. From south of the Orinoco River in Venezuela.

Figure 78. Preparing freshly cleared tract of successional forest in southern Belize to plant upland rice.

The following four photographs are from the state of Mato Grosso in southern Brazil, September 1994.

Figure 79. Newly made roads provide access for forest clearance.

Figure 80. The "burning season" refers to the density of smoke that accumulates from extensive burning of forests during the dry season. This photograph was taken at midday and shows the density of acrid smoke at ground level. Note extensiveness of forest clearance.

Figure 81. Loaded lumber trucks, common sights along the Transamazonian highway system, transport rainforest giants to the numerous lumber mills.

Figure 82. This roadside lumber mill will soon convert these stacked logs into anything from toothpicks to paper pulp.

Figure 83. Crowded houses near Caracas, Venezuela. Throughout Latin America urban populations are often dense with many citizens economically disadvantaged. Population growth and poverty add to the economic pressure to exploit rainforests and other tropical ecosystems.

Figure 84. Pelts of jaguar, ocelot, mountain lion, and red howler monkey in a storage shed in southeastern Venezuela. The pelts will be sold to either North Americans or Europeans.

Figure 85. Indian child with ocelot kitten. The cat was orphaned when its mother (a pelt in the previous photo) was killed.

Figure 86. Chan Chich Lodge in Gallon Jug, Belize, is an example of an ideal ecotourist facility where all wildlife enjoys total protection.

Figure 87. Kekchi-Mayan children from Blue Creek Village in Belize listen as the author describes the manakin he is holding.

Haffer (1993), however, enlarges the time frame of the refugia model to encompass the Tertiary as well as the Quaternary, basing his logic on the continuous nature of Milankovitch cycles, large-scale astronomical oscillations attributable to slight but important cyclic changes Earth's orbit (precession cycles), the tilt of the equator (obliquity cycles), and the shape of Earth's solar orbit (eccentricity cycles). The effect of Milankovitch cycles has been that climate has continually changed (to varying degrees) throughout the Cenozoic, not merely during the Pleistocene (though change was most extreme during the Pleistocene). Haffer, in essence, now views forest and nonforest refugia as having probably existed throughout the Cenozoic, and even before (J. Haffer, pers. comm. 1996).

Biogeographical studies of Amazonian birds have suggested that events prior to the Pleistocene do indeed account for some of the speciation patterns observed (Cracraft 1985; Cracraft and Prum 1988). For example, parrots in the genus *Pionopsitta* and toucans in the genus *Selenidera* show essentially identical broad-scale taxonomic distributions. Cracraft and Prum (1988) trace the pattern of vicariance from the original uplift of the northern Andes and subsequent fragmentation of populations by major rivers, such as the Amazon itself, events that collectively occurred millions of years prior to the onset of the Pleistocene. Of course, if many avian speciation patterns are mostly attributable to pre-Pleistocene events, then most Amazonian bird species are considerably more ancient than a strictly Pleistocene-based refuge model would allow. A strict interpretation of the Pleistocene refuge model would suggest that most Amazonian speciation events are no older than about two million years. Studies of DNA divergence times among related bird species could be helpful in assessing the refuge model. But again, Haffer (1993) has recently expanded the refugia model to encompass a much longer expanse of time than the Pleistocene, stating, "The origin of high species diversity in the tropics probably is linked to the fairly long-term disturbance cycles caused by Milankovitch processes throughout the history of the earth."

The argument over the reality or falsity of refugia remains current. Colinvaux et al. (1996), in a study examining 40,000 years of pollen history from Lake Pata, in northern Amazonia, strongly assert that "The data suggest that the western Amazon frontier was not fragmented into refugia in glacial times." Instead, the authors argue that rainforest, never interrupted by savanna, formed continuous cover during that time period. Haffer (pers. comm.), however, interprets the study differently. He argues that the area around Lake Pata is historically part of the Imerí Refugium, a lowland area that others have assumed remained densely forested throughout the dry climatic phases. In other words, this area was not assumed by proponents of the refugia hypothesis ever to have been interrupted by savanna.

Though some students of South American biogeography remain unconvinced of the existence of periodic rainforest refuges, virtually none doubts that the climatic history of the region has been subject to significant levels of disturbance both during and preceding the Pleistocene. Pollen profiles from lakes or bogs in Andean regions have shown that grasses and other types of savanna and woodland species once grew where today there is moist forest (Absy 1985; Colinvaux 1989b). Some plants now characteristic of montane forests grew at much lower elevations during these colder episodes (Colinvaux 1989b). It is thus probable that lowland rainforest may indeed have contracted, an effect that certainly could have promoted periods of active speciation. Unfortunately, there are few pollen profiles from within Amazonia itself (Connor 1986; Colinvaux 1989a), and these are insufficient to rigorously test the refuge model. Because present distribution patterns are open to various interpretations, evidence for the refuge model for Neotropical diversity remains circumstantial. But some combination of circumstances generated a great many species over a wide range of taxa throughout Amazonia and surrounding areas, and those circumstances probably had much to do with periodic environmental disturbance.

Are the Neotropics Now in Equilibrium?

Forest shrinkage and expansion before and during the Pleistocene may have helped shape present-day patterns of Neotropical species richness, but to what degree are the tropics now stable? I reiterate, yet again, that many people have a belief that the tropics, particularly lowland rainforests, are representative of stable, mature ecosystems. But are they? Are Neotropical ecosystems today in a state of equilibrium, as saturated with species as they can be? This view envisions the tropical rainforest to be maximally packed with specialized species, the vast majority of which have narrow ecological niches, essentially a restatement of Dobzhansky's hypothesis (see above). In recent years, however, evidence has been accumulating to suggest that rainforest ecosystems are not in equilibrium, though they are indeed packed with species. It is time now to take up again the question of why there are so many tree species in Neotropical rainforests.

Surveys in Amazonia demonstrate clearly that tree species richness is extremely high, being generally highest in the upper Amazon, in the vicinity of the Andes Mountains (Gentry 1988). Gentry surveyed a series of 1-hectare (2.5-acre) forest plots in Peru and along the Venezuela-Brazil border. His most species-rich site was Yanomamo, Peru, where he recorded 580 individual trees of 283 species, all on a single hectare! On this site, 63% of the tree species were represented by but a single individual. Gentry speculated that upper Amazonian forests may contain the most diverse floral and faunal assemblages on Earth. Other work throughout the Neotropics confirms the overall high species richness of trees, even on poor soils (Gentry 1990b).

Why are tropical tree species so widely dispersed? Janzen (1970) hypothesized that a "seed shadow effect" occurs where seeds near a parent tree are subject to high rates of predation compared with seeds that somehow manage to be dispersed well away from the parent tree. Thus the surviving seedlings are only those located far from the parent, producing a wide dispersion pattern for each species. Evidence for this hypothesis as a major determinant of high species richness remains unconvincing.

It is also difficult to explain high tree species richness with the argument that each tree species has its own unique ecological niche (but see Ricklefs 1977 and Grubb 1977b). For groups such as trees, it is unlikely that resources can be divided finely enough to permit equilibrium coexistence of dozens of species on a single site. After all, trees in a rainforest compete only for light, moisture, and basically the same array of minerals. Nonetheless, it has been suggested that extremely small distinctions among sites, microsite variation, could be sufficiently diverse to provide for a large number of relatively distinct tree niches, each of which could be occupied by one or a few species (Grubb 1977b). But the view that is becoming most embraced by researchers is that the packing of numerous species into a site is probably not due to narrow, non-overlapping niches, even allowing for possibly high light, moisture, and soil heterogeneity among sites. Rather, trees are most likely to have widely overlapping niches (thus any number of tree species could occupy a given site), and species are often in competition. But trees grow slowly (chapter 3), and competition therefore takes a long time.

Suppose that intermittent, moderate-scale disturbances such as hurricanes, lightning strikes, landslips, or fires, occurring fairly frequently, have prevented competition among species from proceeding sufficiently far to result in extinction of "loser" species. For instance, if five tree species are in competition in an area of forest, given sufficient time, one species would be expected to predominate and outcompete the others. However, a disturbance, such as a hurricane, would perhaps annihilate all or at least most adult trees, thus removing the competitive edge of the "winning" species. The seeds and seedlings of all five species survive to recolonize. Without the hurricane, a single species might eventually prevail, and the site would be in equilibrium with but one species. However, with the hurricane, the "game of competition" is restarted, and five species continue to occur on the site, not just one. This hypothesis is somewhat similar to the predator hypothesis (see above) except that natural disturbance is the "predator" in this case, and the "predator" is nonselective.

Disturbance *intensity* and *frequency* are critical variables in considering disturbance per se as a force for maintaining high species richness (Connell and Slatyer 1977). Obviously, if an area is disturbed harshly and frequently the severe physical conditions that will always prevail would act to preclude high species richness. Not much can persist on a site subject to an annual catastrophic hurricane! Perhaps less obviously, if an area is never disturbed, richness will eventually decline due to interspecific competition, as suggested above. Disturbance frequency and intensity must be neither too severe nor too benign in order for disturbance to result in high species richness. This model, termed the "intermediate disturbance hypothesis," was first argued by Connell (1978) to account for high species richness in both rainforests and coral reefs. It has been well described (though not for tropical forests) by Reice (1994) and Pickett and White (1985). It postulates that intermediate-level disturbance is locally patchy but regionally continuous, and that the overall disturbance regime is sufficient to maintain high species richness by preventing extinction of competing species. The model envisions the tropics as a mosaic of differing aged disturbance patches, some quite local, some more extensive (Condit et al. 1992; Hubbell and Foster 1986b, 1986c, 1990). Bottom line: essentially every place is in some state of recovery from disturbance—there is no equilibrium, nor is there often climax forest.

Mounting evidence suggests that the frequency of local disturbance in the Neotropics is sufficiently high to maintain the tropics largely in a nonequilibrium state (Connell 1978; Boucher 1990; Clark and Clark 1992; Condit et al. 1992, 1995).

Studies of coral reefs as well as tropical moist and dry forests provide evidence in support of Connell's disturbance-diversity hypothesis (Stoddart 1969; Connell 1978; Boucher 1990; Tanner et al. 1994). Areas of coral reef in Belize that have not been struck by hurricanes have lower diversities of coral and other species than areas that have experienced disturbances. Disturbance frequency on the Australian Great Barrier Reef is much higher than the time required for coral communities to reach equilibrium (Tanner et al. 1994). Coral species compete for space and sunlight (because some contain photosynthesizing algae), and, without disturbance, some coral species outcompete and exclude others. Disturbance, however, opens up the area and provides sites for many species to colonize. A major hurricane created highly patchy distributions on Jamaican coral reefs, supporting the notion that more mature reefs show high levels of species heterogeneity due to a past history of disturbance (Woodley et al. 1981).

In recent years, several major hurricanes have moved across the Caribbean, each causing widespread destruction, and ecologists have been ready for them. The destruction and disturbance caused by these hurricanes have spawned numerous ongoing studies that have contributed significantly to our understanding of just how important hurricanes are to the ecology of coral reef, coastal, and all manner of terrestrial ecosystems. For excellent summaries of Caribbean hurricane effects, see the special issue of *Biotropica* (vol. 23, no. 4, part A, December 1991) and the volume edited by Reagan and Waide (1996).

Some (not many) rainforest sites have been observed to have a low species diversity of canopy trees, in stark contrast to the usual high diversity observed. These low-diversity sites may be representative of the final outcome of long-term interspecific competition in the absence of disturbance (Connell 1978). They are perhaps rare examples of forests in equilibrium. On some of these low-diversity sites, the seedling trees are of the same species as the trees of the canopy, indicating that equilibrium, a self-replacing ecosystem, has been reached. But these forests are exceptional.

Hubbell (1979) described the pattern of tree species distribution in a Costa Rican tropical dry forest and concluded that periodic disturbances were strongly affecting the sixty-one tree species present in the study area. Trees were randomly distributed and clumped, and rare species had low reproductive success. Hubbell (1979, 1980) rejected the seed shadow hypothesis for both dry and moist forests and instead argued that periodic community disturbance was the factor most responsible for the distribution pattern.

Hubbell and Foster (1986c) suggest that combined effects of diffuse interspecific competition, periodic disturbance, and historical factors (high rates of speciation) have resulted in the high species richness patterns evident throughout the Neotropics. This conclusion asserts that most tree species are ecological generalists, not specialists. True, many species are primarily light-dependent colonizers, many may require high moisture, many may be shade tolerant, but these are general differences among groups, not sufficiently precise to finely separate species on a niche gradient. Diffuse competition may be strong among species but does not result in the evolution of narrow niches. Instead, the average tropical tree is a generalist of sorts, occupying a broad adaptive zone.

It is important to realize, however, that the intermediate disturbance model does not predict that periodic physical disturbance will be the *only* causal factor in determining species composition of rainforests. It would indeed seem unrealistic to assume that in such species-rich ecosystems as rainforests, biological interactions wouldn't be important as well. A model of the nonequilibruim hypothesis, based entirely on the assumption that there are no significant biological determinants of tree community structure (and thus tree distribution is essentially random, essentially determined by chance factors), was tested for floodplain forests in the Peruvian Amazon and failed. The results of the study showed that biological interactions, particularly based upon population density of the various species, must have significant effects on the directionality of floodplain succession patterns and subsequent forest community structure (Terborgh et al. 1996).

In a comprehensive study conducted at Cocha Cashu Biological Station (in Manu National Park, Peru), Miles Silman (1996) demonstrated the importance of biological factors. An evolutionary trade-off exists among *the seeds* of various tropical tree species between competitive and colonization abilities. Silman studied colonization patterns and survivorship among seeds that ranged in size over six orders of magnitude. He found that once a seed germinated and grew to the stage of seedling, its probability of survival was about the same, regardless of species. In essence, such a result supports the belief that tree survivorship is random, determined essentially by luck. But biological factors had strong influence prior to the seedling stage of development. Silman censused seeds collected in seed traps and found 196 species from 117 genera and 46 families, a total of 20,694 seeds. Of this total, 75% of the species were only present at one or two sites (out of 40), showing that dispersal is generally quite limited. In experimental plots using eight species of trees representing a wide range in seed size, Silman learned that large-seeded species showed significantly higher survivorship than small-seeded species. In other words, tree species are not equivalent in survival potential *when they are seeds.* But large seeds were dispersed much less and were thus significantly poorer colonizers than small seeds. Small-seeded species, on the other hand, though the most numerous colonizers, were poorest in subsequent survival. Thus there was a trade-off between seed size and colonizing ability. Silman learned that but two factors, seed size and germination strategy (sensitivity to light levels and herbivory by mammals), accounted for 71% of the variation in survivorship among the plants.

Silman's study supports the contention that strong biological factors (seed size, dispersal ability, response to light, resistance to herbivory by mammals) have significant influence in determining diversity patterns. But, as Silman stated, "environmental filtering in the experimental community occurred during the seed and germinating seedling stages, with seedlings having nearly equivalent survivorship." In other words, if one was to study trees any time during their life cycles from seedling to adult, survivorship would appear to be influenced essentially by chance, with little biological interaction. Biological factors play their significant roles very early in a tree's life cycle. As Silman concluded, "Tropical trees are not ecological equivalents. While adults may have no perceptible interspecific differences in their response to the environment, the early life-stages differ dramatically."

If, as evidence now suggests, long-term periodic intermediate disturbance results in diffuse coevolution of generalist tree species, tropical forests may accumulate tree species simply because forests of generalists may not be particularly resistant to invasion, and because there may be few forces besides chance that can lead to the systematic elimination of generalist tree species, once they become established as seedlings (Hubbell and Foster 1986c).

Studies comparing temperate and tropical forests show that whether in the tropics or temperate zone, disturbances such as landslides, fire, vulcanism, and hurricanes, where relatively common (as the intermediate disturbance model suggests), correlate with forest communities that tend to show a predominance of random distributions among tree species. In forests rarely subjected to catastrophic disturbance, trees tend to be less random, and the process of gap creation and colonization is of major importance. In a forest dominated by "gap-phase dynamics," there is a shifting mosaic of patches within the forest as gaps are created, colonized, and closed (Armesto et al. 1986; Haffer 1991).

Animals would be expected to show similar patterns. A study of birds in Panama found them to be highly sensitive to both moisture gradients and gradients of vegetation structure. Differences in microclimate from site to site were of major influence in determining which bird species occurred. Since microclimates change regularly, so does the bird community at any given site, evidence for nonequilibrium in Panamanian rainforest bird communities. Bird species respond individually in predictable ways to microclimate and vegetation change; thus, the composition of the avian community fluctuates with time due to changes in vegetation and microclimate (Karr and Freemark 1983). Detailed studies in both montane and lowland forests in Costa Rica demonstrate considerable annual variation in many bird species' populations (Loiselle and Blake 1992). Studies of bird communities in Amazonia demonstrate that many species are adapted to the reality of a mosaic of habitats, the nature of which depends on disturbance history (Haffer 1991; Andrade and Rubio-Torgler 1994). In one study performed in the Colombian Amazon, intensive netting was used to sample bird communities, and differences in species richness and number of individuals captured between successional areas and primary forest were both low, indicating that many forest species were adapted to utilize successional habitats (Andrade and Rubio-Torgler 1994). I have noted a similar pattern in work I and a colleague did in Belize (Kricher and Davis 1992).

The Neotropics are dynamic. Past climate shifts likely have caused rainforest shrinkage and stimulated episodes of speciation. Today, climatic cycles coupled with relatively frequent storms, windthrow, and other disturbances create gaps, reinitiate succession, and preserve high diversity (Condit et al. 1994; and see chapter 3).

To return to the question posed at the beginning of this section, are Neotropical rainforests in equilibrium? No, they are not. With disturbance and nonequilibrium affecting an already complex picture of tropical diversity discussed earlier in the chapter, it becomes apparent that many factors must combine to produce and maintain these most diverse of ecosystems. Factors affecting the diversity of one taxon may not be those affecting another, at least not to the same degree. There is no single, simplistic answer to the question, why are there so many species in the tropics? But there are answers.

Razor-billed curassow

5

Complexities of Coevolution and
Ecology of Fruit

The Evolution of Coevolution

SPECIES can affect each other in intricate ways. Herbivores are dependent
on plants as food, but plants are often spiny and noxious, or toxic, an
apparent defense against herbivores (see chapter 6). On the other hand,
herbivores aid plants as they help recycle nutrients by producing organic waste
products, serve as pollinators (which amounts to being a surrogate sex organ),
and disperse seeds. The relationship between herbivores and plants is therefore anything but simple. By the same token, predators are adapted to capture
prey, but prey are adapted to avoid capture. The ecological "cat-and-mouse
game" between predators and prey has given rise to elegant adaptations, including both cryptic and warning coloration. Parasites and hosts also engage
in an evolutionary war of adaptation. Many parasites seem to evolve increasing
degrees of host specificity, and hosts tend to evolve greater tolerance to parasites to whom they have had protracted evolutionary exposure. The "ecological web" woven by the myriad interactions among species within an ecosystem
has an important historical aspect. As mentioned earlier, biotic selection pressures, the influences of predators on prey, of herbivores on plants, parasites on
hosts, and so on, are influential in determining the evolutionary directions
taken by natural selection in the tropics. When one species has a trait that acts
as a selection pressure on another, and the second species in turn possesses a
trait that acts as a counterselection pressure back upon the first, the evolutionary fates of both species can become permanently interlocked. This situation
is called *coevolution* (Ehrlich and Raven 1964; Janzen 1980b; Futuyma and Slatkin 1983).

Coevolutionary intereactions can be expressed as parasitic or predatory relationships, in which one species benefits at the expense of another, or as
mutualistic relationships, in which both species benefit. Some mutualisms may
be faculative, others obligatory. Note that not all, or even most, species interactions are necessarily coevolutionary (Janzen 1980b). Indeed, assumptions
about coevolution continue to be challenged, as there are alternative explanations for what may appear to be coevolutionary relationships (Bernays and
Graham 1988). Only when species evolve clearly in relation to mutual selection pressures is the relationship properly defined as coevolutionary. For example, sloths commonly harbor tiny moths that reside within the recesses of

the sloth's fur. These diminutive lepidopterans are well attuned to the sloth's behavior. They lay their eggs on sloth droppings, where the larvae eventually feed upon hatching. These moths, while clearly adapted to life on a sloth, do not represent coevolution because there is no evidence that they have exerted any form of selection pressure on the sloth that has resulted in an adaptive response. These moths are *commensal*: they benefit from the sloth as a resource but have no significant impact (at least none that has been demonstrated), positively or negatively, on the sluggish mammal. In addition to little moths, some sloths may have an abundance of algae living within their often moist fur, and the algae confer a degree of crypsis on the sloth, making it more difficult to see in the canopy. The sloth supplies a suitable habitat for the algae, and the algae aid the sloth's potential survival by adding an element of crypsis. Is this coevolution? Probably not. There is no evidence of any adaptive accommodation between the two participants. Their association is a kind of commensal mutualism, where both benefit in an essentially accidental way, without exerting any known selection pressures on each other (though one could argue that if sloths lacking algae are more susceptible to predation, the presence of algae is an *indirect* selection pressure that affects sloth survival probability).

Pollination

One of the most obvious examples of coevolution is pollination of plants by various animals. Many flowering plants, particularly in the tropics, are dependent upon insects, birds, or bats for survival. Animals seek out flowers as a food source, ingesting nectar and, in some cases, pollen. By their travels from plant to plant, animals also disperse pollen, making cross-pollination efficient and insuring reproduction of the plants. This relationship is *mutualistic* because both plant and pollinating animal benefit. Charles Darwin, in *Origin of Species*, wrote about the coevolved relationship between bees and the clover they pollinate: "The tubes of the corollas of the common red and incarnate clovers (*Trifolium pratense* and *incarnatum*) do not on a hasty glance appear to differ in length; yet the hive-bee can easily suck the nectar out of the incarnate clover, but not out of the common red clover, which is visited by humble-bees alone." Darwin discussed pollination and coevolution further in a monograph, *On the various contrivances by which British and foreign orchids are fertilised by insects, and on the good effects of intercrossing*, published in 1862.

In the tropics, animal pollination is widespread, particularly in rainforests. In contrast to temperate forests where wind pollination is common, rainforests are sufficiently dense that wind pollination would tend to be ineffective, except perhaps for emergent trees. It is not surprising that grasses, sedges, pines, and other open-area savanna species are the only tropical plant groups dominated by wind pollination. Vertebrates and insects are both major pollinators in the Neotropics. Among insects, pollination is accomplished by bees, flies, beetles, butterflies, and moths (Prance 1985b). Among vertebrates, hummingbirds and bats are specialized to feed on nectar of many flower species. Hummingbird-pollinated flowers tend to have rather long tubes and are colored red, orange, purple, or yellow. Bat-pollinated flowers are often white (easy to locate in the dark) and may have a musky odor, an attractant to the bats. Many flowers are pollinated by a variety of vertebrates and insects. In Trinidad, for example, I watched seven hummingbird species feed on vervain, which was also being visited by various bees and assorted flies.

Pollinators are most advantageous to plants when they tend to fly long distances. Such behavior helps ensure effective cross-pollination between widely separated plants (Janzen 1975). Euglossine bees are long-distance fliers, and males pollinate certain orchids that are widely separated. Compounds in the orchid flower that are absorbed by the male bees contribute to the longevity of the insects. Male euglossine bees can live up to six months, a long life for a bee (Janzen 1971b). Female euglossine bees, as well as carpenter and bumble bees, are important long-distance pollinators of many rainforest trees (Prance 1985b).

Some important commercial Neotropical trees, such as Brazil nuts and cocoa, are insect pollinated. Various species in the family of Brazil nut trees (Lecythidaceae) are bee pollinated, and cocoa (*Theobroma cacao*) is pollinated by a small midge (Prance 1985b).

Pollination coevolution can become sophisticated. The huge Victoria or royal waterlily (*Victoria amazonica*) is pollinated by beetles. The large, conspicuous white flowers open in synchrony, emit a strong odor, and are often warmer than ambient temperature. These characteristics combine to attract certain beetles (*Cyclocephala* spp.), which enter the flower only to become trapped inside as night falls, when the large petals tightly close. The imprisoned beetles, presumably comfortable in the warm, food-rich interior of the flower, feed on nectar-rich structures throughout the night, getting thoroughly sticky, and also becoming covered with pollen. The next day the flowers open, having changed in petal color from white to red, as well as lost their scent and cooled in temperature, thus being no longer an attractant to beetles. The formerly incarcerated pollen-bearing beetles leave the flower and fly off to seek out another white flower from another Victoria waterlily, where they will inadvertently deposit pollen as they feed (Prance 1985b).

An elaborate coevolution has occurred between figs and fig wasps (Janzen 1979; Wiebes 1979). Various species of figs (*Ficus* spp.) all produce a bulbous green flower that looks like a fruit rather than a flower. This kind of odd flower is called a synconium and is really a cluster of enclosed flowers within a gourdlike covering. Externally, the fig flower reveals nothing of its unusual inner structure. Internally, it is a dense carpet of tiny flowers, some male, some female, and some sterile (called gall flowers). Though male and female flowers exist side by side, males cannot pollinate females in the same synconium because female flowers mature earlier than males. Since the synconium is utterly enclosed, it would appear impossible for fig pollination to occur without the assistance of an animal pollinator, and a mighty skilled one at that. Enter (literally!) the diminutive fig wasps (family Agaonidae).

Sterile gall flowers contain fig wasp eggs laid by females of the previous generation who died after depositing their eggs. Males hatch first and burrow into gall flowers to inseminate the still unhatched females. Each male commonly inseminates several females. (That's right, the females are impregnated before they are born!) Females hatch, already pregnant with the next wasp generation. Newly hatched females wander over the stamens of male flowers, which reached maturity precisely when the females hatched. Laden with pollen, either the female wasps make an exit tunnel from the flower or, in some species, the flower opens enough to permit females to exit. Females have but a day or so to locate another flowering fig and tunnel into a flower. The female is often physically damaged from burrowing into the flower, but once inside she has only to locate a gall flower in which to deposit her eggs. After that she dies. In her search for a gall flower, the female passes over fertile female flowers and deposits the pollen she brought from the flower of her birth.

But the story is more complex, with a touch of the bizarre. Parasitic fig wasps also inhabit synconiums. These insects use the fig as a food source but do not pollinate it. Parasitic males roam about in the pitch black interior of the synconium using their huge jaws to dismember other wasps as they all search for females with which to mate. William Hamilton (1979), who spent a great deal of time looking into fig synconiums, wrote, "a situation that can only be likened in human terms to a darkened room full of jostling people among whom, or else lurking in cupboards and recesses which open on all sides, are a dozen or so maniacal homicides armed with knives."

It is apparent that everything about the reproductive biologies of figs and fig wasps is interrelated. Such an intricate level of interaction indicates a long evolutionary association between plants and insects, a true coevolutionary mutualism. Note also that functionally there really is no such thing as a fig tree without fig wasps. It would be a sterile plant. Note also that there is functionally really no such thing as a fig wasp without a fig tree. These species cannot exist in isolation from one another and thus have evolved into, affectively, a single entity. This is what coevolution can accomplish.

Chiropterophily
Figures 123, 126, 133, 134

Pollination of flowers by bats is common in the tropics, with over 500 plant species wholly or partly dependent on bats as pollinators (Heithus et al. 1974). Plants adapted to host bats are termed *chiropterophilous*, meaning "bat-loving" (bats are in the mammalian order Chiroptera). This coevolution has occurred at both the physiological and anatomical levels in both bats and plants (Howell 1976).

As noted above, plants with bats as primary pollinators tend to have large white flowers, often with a musky, "batlike" odor. These flowers, of course, open at night when bats are active. Flowers are often shaped like a deep vase, though some are flat and brushy, loading the bat's face with pollen as it laps up nectar. Many bat flowers are cauliflorous, growing directly from the tree trunks. Flowers may hang from long, whiplike branches (flagelliflory) or hang downward as streamers (penduliflory), a condition common in many vines. Cauliflory, flagelliflory, and penuliflory all have in common the fact that the flowers are positioned in such a manner that they are easily accessible to bats, which are generally large and must hover in the open as they feed. In Guanacaste Province, Costa Rica, two bat species feeding on flowers of *Bauhinia pauletia* exhibited markedly differing feeding behaviors. *Phyllostornus discolor* visited high flowers, grasping the branch beneath the flower and pulling it down. *Glossophaga soricina* visited both high and low flowers and hovered as it fed. Both bat species aided in cross-pollinating *Bauhinia* (Heithaus et al. 1974).

Nectar-feeding bats tend to have large eyes and relatively good vision. They visually locate white flowers that show up well in the dark. The sonar of nectar-feeders is often quite reduced, but the olfactory sense is well developed. Sight and scent, not sound, are how the animals find their sugary dinners. They also have long muzzles and weak teeth, both advantageous in probing deeply into flowers. Finally, they have long tongues covered with fleshy bristles that can extend well into the flower, and their neck hairs project forward, acting as a "pollen scoop." When nectar-feeding bats feed, they pick up a great deal of pollen.

The pollen from bat plants tends to be significantly higher in protein than in non-bat-pollinated plants, and bats ingest pollen as well as sugary nectar. Pollen contains the amino acids proline and tyrosine, useless to the plant but important to the bats. Proline is necessary to make connective tissue such as is used in wing and tail membranes, and tyrosine is essential to milk production. Bats are not the only animals to eat and derive important protein from pollen. Heliconius butterflies (chapter 6) collect pollen on a small brush near their mouth parts, ingesting it after dissolving it with nectar.

Once ingested, nectar helps dissolve the tough pollen coat, but bats aid this process because their stomachs secrete extraordinarily large amounts of hydrochloric acid. These bats also drink their own urine, which helps dissolve the pollen coat, liberating the essential protein. Nectar-feeding bats and bat-pollinated plants, like the figs and fig wasps, are evolutionarily locked together in a mutualistic relationship in which each party is essential to the other.

Ants and Ant-Plants   *Figure 89*

Some tropical plant species possess nectar-secreting glands as well as other structures that collectively act to attract ants. Indeed, a diversity of plant species ranging through nineteen families (and including ferns, epiphytes, vines, and trees) have been classified as "ant-plants" or *myrmecophytes* because of their ant-attractant properties (Benson 1985). Various ant-plants also occur in the Old World tropics, especially Southeast Asia. Ant-plants usually have some kind of shelter for ants (ant domatia) as well as some form of nutrition for them.

Domatia range from mere hollow stems to more sophisticated shelters such as specialized pouches or thorns. Food glands, termed *extrafloral nectaries*, are found on leafblades, leaf petioles, stems, or other locations on the plant. These glands manufacture various energy-rich sugary compounds as well as certain amino acids. In addition, some plants have bead bodies, which are modified hairs rich in oil (Benson 1985).

The odd sugar- and oil-producing bodies initially puzzled botanists, who could identify no obvious function for them. It was soon observed, however, that many of the plants with extrafloral nectaries are liberally populated by various ant species, many of which are aggressive. Perhaps the ants, by their aggressiveness, somehow protect the plants, which repay the ants in nectar and shelter. This hypothesis, rather startling at the time, was termed the *protectionist hypothesis*, and it envisioned the relationship between the plants and ants as mutualistic. The alternative idea, called the *exploitationist hypothesis*, argued that the ants merely fed on the sugary nectaries but provided no actual service to the plants. An early (1910) description of the exploitationist hypothesis suggested that the plants "have no more use of their ants than dogs do their fleas" (quoted in Bentley 1977).

Cecropias are among the most common plants with extrafloral nectaries and bead bodies. Cecropia nectaries are termed *Mullerian bodies* and are located at the base of the leaf petiole, where the large leaf attaches to the stem. Ants of the genus *Azteca* live in domatia within modified hollow pith of the stem and feed on the Mullerian and bead bodies. I have frequently encountered these ants, and they are not nice ants. Cutting cecropias for use as net poles (to catch and band birds), I have been attacked vigorously by Aztec ants, and I have little sympathy for the exploitationist hypothesis. The ants of a cecropia are pugnacious and thus protective of their tree. If I were a cecropia, I'd want some of these ants living on me. Cecropia have additional structural properties that suggest a coevolution with Aztec ants. The underside of the wide, palmate leaves is velvetlike, with a carpet of tiny hairs and hooks that allow ants to gain purchase and so move easily across the leaf. Cecropia species that are not mutualistic with Aztec ants have leaves with smooth undersides (Benson 1985).

Janzen (1966) settled the controversy at least for one ant species. Janzen studied *Pseudomyrmex ferruginea*, which occurs on five species of *Acacia* tree. Commonly called the bull's horn acacia, the tree has pairs of large hollow thorns on its stem that serve as homes for the ants. A single queen ant burrows into a thorn of a sapling acacia to begin a colony that can increase to as many as 12,000 ants by the time the tree is mature. Janzen noted that by the time the fast-growing tree was seven months old, 150 worker ants were "patrolling" the stem. The acacia ants attack any insects that land or climb on the tree, including beetles, bugs, caterpillars, and other ants. Ants also clip plants that begin to grow nearby or overtop and shade the acacia (thus taking its sunlight), and they attack cattle or people if they brush against the tree. Ants become very active, swarming out of the thorns and over the foliage even if merely exposed to the odor of cattle or people. I was once attacked on the neck by a single acacia ant, and the formic acid irritated me for over a day. I can well imagine the discomfort that would have occurred if I had been stung by many of these ants.

Why do ants live on acacias? They obtain shelter within the thorns, but they also obtain nutrition from two kinds of extrafloral nectaries. One type is termed *Beltian bodies*, which are small orange globules growing from the tips of the leaflets of the compound leaves, and the other type is called *foliar nectaries*, located on the petioles.

Janzen performed a field experiment that discriminated between the protectionist and exploitationist hypotheses. He treated some acacias with the insecticide parathion, and he also clipped thorns to remove all ants from the treated trees. The antless trees did not survive nearly as well as control trees, which were permitted to keep their ants. Janzen estimated that antless acacias were not likely to survive beyond one year, either falling prey to herbivores or being overtopped by other, competing species of plants. Ants are needed to attack herbivores and clip other plants. An antless acacia is doomed. Janzen concluded that the ants and acacias are obligate symbionts, depending entirely upon each other. The protectionist hypothesis is correct, and an impressive mutualism has coevolved between acacias and *Pseudomyrmex*.

Though both *Azteca* and *Pseudomyrmex* are aggressive ants, many ant plants harbor less obviously aggressive species. However, these more docile ants may nonetheless protect the plant by eating tiny mites or insect eggs (Benson 1985).

Ant-plants may obtain more than just protection from ants. Some may benefit by securing nutrient-rich substrate as a byproduct of ant colony construction and refuse. For example, some epiphytes are known to grow only on ant nests (Benson 1985).

Extrafloral nectaries occur also on temperate zone plants (Keeler 1980) but seem more abundantly represented among tropical species (Oliveira and Leitao-Filho 1987). In a survey of riparian and dry forests in Costa Rica, plants with extrafloral nectaries ranged in percentage cover from 10 to 80% (Bentley 1976). In the Brazilian cerrado, cover by woody plants with extrafloral nectaries ranged from 7.6 to 20.3% (Oliveira and Leitao-Filho 1987). Many plants with extrafloral nectaries house ants, but the degree to which they protect their hosts varies (Bentley 1976, 1977).

But It Can Get Complicated

Extrafloral nectaries, at least in some cases, undoubtedly function as effective bribes from plants to ants, the plant "buying" protection from the ants. But if ant protection can be bought, why not, in an evolutionary sense, sell to the highest bidder? Apparently a few species of lepidopterans from the families Riodinidae and Lycaenidae have discovered this basic tenet of economics. In a remarkable example documented in Panama by DeVries (1990, 1992) and DeVries and Baker (1989), caterpillars of the butterfly *Thisbe irenea* entice ants to protect them, rather than the ants' host plant (*Croton*), and then the rapacious caterpillars eat the leaves from the very plant the ants were once protecting! These caterpillars, termed *myrmecophilous* for their "ant-loving" habits, have evolved at least three separate organs that act to attract and satisfy ants: nectary organs that produce protein-rich ant food; unique tentacles that release chemicals mimicking those of the ants themselves and signaling them to defend; and vibratory papilla, which, when the caterpillar moves its head vigorously, make sounds that travel only through solid objects, but which immediately attract ants. Ants apparently have a much stronger preference for protein-rich caterpillar nectar droplets than for the carbohydrate-rich *Croton* nectaries (why settle for a sugary soda when you can have a burger?), and the ants are essential in protecting the otherwise vulnerable caterpillars from marauding predatory wasps. Thus, by "bribing" the ants, the caterpillars have succeeded in both averting the main protective adaptation of the plant and ensuring their own relative safety from their major predators, wasps.

Fungus Gardens and Leafcutting Ants   *Figures 90, 91, 92*

No one can visit the Neotropics without encountering leafcutting ants. Throughout rainforests, successional fields, and savannas, well-worn narrow trails are traversed by legions of ants of the genera *Acromyrmex* and *Atta* as they travel to and from their underground cities, bearing freshly clipped leaves. Their trails take them up into trees, shrubs, and vines where they neatly clip off rounded pieces of leaves, which they carry back to their colony. The ants live in underground colonies of up to eight million individuals, consisting of a single large queen and myriad worker ants, most of which remain subterranean. Workers are of several size classes: very small (minimas), medium-sized (medias), and large (maximas). Soldiers, the principal defense class, are large and well armed with formidable pincer jaws. You'll meet them if you dig into a colony. *Atta* colonies are underground, but the bare mounds of soil that mark their multiple entrances spread widely and obviously on the surface. These abundant ants make no secret of their presence. The sight of thousands of leafcutters marching along, most of which are bearing neatly clipped leaf fragments, is unique to the Neotropics. Leafcutters are not to be found in the Old World tropics.

The impact of leafcutting ants may prove enormous. On Barro Colorado Island in Panama, leafcutter ants have been estimated to consume 0.3 ton of foliage per hectare per year, equal to the combined effects of all vertebrates in the forest (Leigh and Windsor 1982). Indeed, estimates suggest that somewhere between 15 and 20% of net primary productivity within a rainforest is consumed by the leafcutters, and the ants may have a devastating effect on local agriculture. Leafcutters are generally selective as to which species they clip (Wetterer 1994). For example, in Guanacaste, Costa Rica, one *Atta* species clipped mature leaves from only 31.4% of the plant species available. Another species used leaves from only 22% of the available plant species, indicating a strong selectivity of both leafcutter species (Rockwood 1976). The commonness or rareness of a plant species has no correlation with *Atta* preference. The ants often travel far from their nest to seek out a certain plant species. Rockwood hypothesized that internal plant chemistry strongly influences *Atta* diet, a suggestion borne out by subsequent research (Hubbell et al. 1984). Ants seem to concentrate on plants with minimal amounts of defense compounds in their leaves (see chapter 6).

Leafcutter ants are part of a larger ant group called the fungus garden ants (Myrmicinae: Attini), each of which, remarkably, cultivates a particular species of symbiotic fungus that makes up its principal food source (see below). Some fungus/ant relationships may be as old as 50 million years. There are approximately 200 fungus garden species, of which 37 are leafcutters. The remaining species, most of which are inconspicuous, cultivate their fungus on some combination of decaying plant or animal organic matter (Holldobler and Wilson 1990). Though most abundant in the tropics, fungus garden ants also occur in warm temperate and subtropical grasslands. One enterprising species even occurs as far north as the New Jersey pine barrens (Wilson 1971).

Leafcutting ants taste and may ingest the sap of the leaves they cut, perhaps using the sap as an additional food source (Wetterer 1994). They do not, however, consume any leaves but rather clip and carry leaf fragments to their colonies. There they use the leaves to make media to culture a specific fungus. This odd fungus, which is never found free-living outside fungus garden ant colonies, is the ants' only food. Leaves brought to the colony are clipped into small pieces and chewed into a soft pulp. Before placing the pulpy mass on the fungus bed, a worker ant holds it to its abdomen and defecates a fecal droplet of liquid on it. The chewed leaf is then added to the fungus-growing bed and small fungal tufts are placed atop it. Other ants sometimes add their fecal droplets to the newly established culture. Detailed photographs from inside an *Atta* colony can be seen in Moffett (1995b).

Worker ants collecting leaves avoid those that contain chemicals potentially toxic to the fungus (Hubbell et al. 1984). One tree, *Hymenaea courbaril*, a legume, has been shown to contain terpenoid (see chapter 6), which is antifungal (Hubbell et al. 1983). *Atta* ants must obviously avoid clipping leaves from this species. The tree has evolved a protection from *Atta*, not by poisoning the ant, but by poisoning its fungus!

The relationship between ants and fungus is unique. The ants culture only a few fungal species, all of which are members of one group, the Basidiomycetes (family Lepiotaceae), a group whose free living members include the familiar parasol mushrooms. The fungi are always in pure culture, protected from contamination from other fungal species by constant "weeding" by ants. Without the attention of the ants, the fungus is quickly overtaken by other fungal species. Both ants and fungi are totally interdependent, an example of a complete obligatory mutualism (Weber 1972). Ant and fungus are inextricably linked by evolution: only the queen reproduces, and when a queen ant founds a new colony she takes some of the precious fungus with her inside her mouth. Fungus and ants disperse together.

Detailed studies of the fungus-ant relationship at the biochemical level has revealed the multiple roles that ants play in culturing the fungus (Martin 1970). The ants clean the leaves as they chew them to make the culture bed pure. Ant rectal fluid contains ammonia, allantoic acid, the enzyme allantoin, and all twenty-one common amino acids. These compounds are all low molecular weight nitrogen sources, and they are the key ingredients in making the culture optimal for the fungus. The fungus lacks certain enzymes that break down large proteins (all of which are made up of chains of amino acids). Thus it depends totally on the ant rectal fluid to supply its amino acids. Experiments attempting to grow the fungus in a rich protein medium failed. It can only grow in a medium of small polypeptides and amino acids. Ants also supply enzymes necessary to aid in breaking down protein chains.

Martin (1970) summarized the functions of the ants as (1) fungal dispersal, (2) planting of the fungus, (3) tending the fungus to protect it from competing species, (4) supplying nitrogen in the form of amino acids, and (5) supplying enzymes to help generate additional nitrogen from the plant medium.

The fungus garden ants are the expert gardeners of the insect world, and their labors pay off handsomely. The fungus symbiont digests cellulose, an energy-rich compound that ants cannot digest. Not only that, but the fungus is unaffected by many, if not most, of the defense compounds contained within leaves of many plant species (see chapter 6). By eating the nutritionally rich fungi, ants circumvent the numerous and diverse defense compounds typical of Neotropical plants, while at the same time tapping into the immense abundance of energy in rainforest leaves. Fungus garden ants owe their remarkable abundance to their unique evolutionary association with fungi.

How such a sophisticated and apparently ancient evolutionary relationship began is difficult to know, but one possibility is that the ant-fungus relationship was initiated simply through predation by ants on fungus. What began as predation evolved over the millennia into mutualism.

Some suggestion has been made that mutualistic associations such as is typified between ants and fungus represent a clear example of how nature can be cooperative rather than competitive, a kind of anti-Darwinian view of nature. But cooperation, however real, is not at all contrary to predictions of natural selection theory. Any obligate or faculative mutualism can just as easily be described as mutual or reciprocal parasitism, where one party is exploiting the other and being exploited in return (though to the ultimate benefit of each). Each party in the mutualism acted and continues to act as a selection pressure on the other. There is nothing anti-Darwinian about that.

For more on the intricacies of coevolution, see Futuyma and Slatkin 1983. For more on ants of all sorts, see Holldobler and Wilson 1990.

The Importance of Fruit in the Tropics

Fruit is both abundant and (relatively) constantly available throughout the year in the Neotropics, making it an important resource for birds, mammals, certain reptiles, occasional fish (chapter 8), and all manner of arthropods (Levey et al. 1994). Where can you go to see a spider monkey, a spangled cotinga, and a common iguana, all in the same tree? Try a fig tree, when it has mature figs. Many Neotropical animals are considered to be *frugivores*, creatures whose diet includes more than 50% fruit (Levey et al. 1994).

In the temperate zone, fruit is a distinctly seasonal resource, occurring from midsummer through autumn. Many birds migrating to winter in the tropics feed heavily on fruit in the fall, but, because fruit is ephemeral in the temperate zone, no bird families have specialized as frugivores (Stiles 1980, 1984). In the tropics, entire families of birds, such as the manakins, cotingas, toucans, parrots, and tanagers, depend heavily on fruit, and some species are almost exclusively frugivores (Snow 1976; Moermond and Denslow 1985). In addition, mammals ranging from bats to agoutis to many monkey species utilize fruit as a major component of their diets. Fruits provide a calorically rich, nontoxic, and easily acquired resource. But there are downsides to a diet of fruit. Protein is usually sadly lacking, thus an all-fruit diet, while rich in calories, is typically nutritionally deficient. Also, fruiting patterns vary, often significantly, in both time and space. In other words, two fruiting fig trees may be widely separated, necessitating searching by frugivorous animals. For example, the great green macaw (*Ara ambigua*), a large parrot species, must make extensive, irregular movements throughout its range in Costa Rica, searching for satisfactory fruiting plants (Loiselle and Blake 1992). Seasonal changes in abundance of most fruits are common throughout the tropics (Fleming et al. 1987). Some montane frugivores must undergo regular seasonal migration to lower elevations in search of favored fruits (Wheelwright 1983; Loiselle and Blake 1992).

The function of a fruit is to advertise itself to some sort of animal so that it will be eaten. The seed(s) contained within the fruit then passes through the alimentary system of the animal (or is regurgitated), and, because the animal is mobile, seeds are deposited away from the parent plant. Fruits evolved as adaptations for seed dispersal. The animal derives nutrition from the fruit but also disperses seeds; thus, the relationship between animal and plant is mutualistic, in some ways rather like pollination: the animal is bribed in return for its mobility (but see Wheelwright and Orians 1982, and below). It is to the plant's ultimate advantage to invest energy to make fruit as well as to the animal's immediate advantage to eat the fruit. Some parrot and pigeon species, however, digest the seed as well as the pulp of the fruit (or else they injure the seed, and it does not germinate). These "seed predators" are not useful as dispersers and must be considered parasites rather than mutualists.

Some plants do not "invest" in expensive fruits to attract animals as seed dispersers. Some rainforest canopy trees, vines, and epiphytes utilize wind or water dispersal of seeds, though many other species are animal-dispersed (Kubitzki 1985). Wind dispersal is most common at the canopy level or in deciduous forests, where leaf drop can help facilitate wind movement of seeds (Janzen 1983a). A study in Costa Rica surveyed 105 tree species in a deciduous forest and found that 31% were clearly wind-dispersed (Baker and Frankie, cited in Janzen 1983a). In dense interior rainforests animals are far more important than wind for seed dispersal.

Ornithologists David Snow (1976) and Eugene Morton (1973) have considered the evolutionary consequences to birds of a diet mainly of fruit. Fruit typically is temporally and spatially a *patchy resource*, meaning that it may be abundant on a given tree (for instance, a fruiting fig tree), but trees laden with mature edible fruits can be widely spaced in the rainforest. And for most of the year, a fruiting tree may be barren of fruit. Such a resource distribution selects for social behavior rather than individual territoriality. Flocks of fruit eaters can locate fruiting trees more efficiently than solitary birds, and there is no disadvantage to being in a flock once the fruit is located since there is usually more than enough fruit for each individual. Even if not, it is extremely difficult for one individual to defend the resource, excluding all others. Therefore it is not surprising to encounter flocks of parrots or groups of toucans.

The purple-throated fruit crow (*Querula purpurata*), like many species, feeds on both insects and fruits, but its frugivorous habits may have been instrumental in the evolution of its social behavior (Snow 1971). These birds live in small communal groups of three or four individuals that roam the forest together in search of preferred species of fruiting trees. Within the social group there is virtually no aggression, and all members of the group cooperate in feeding the nestling bird (they have clutches of only one). The nest is in the open and is vigorously defended by the entire group. Fruit-eating mammals also tend toward sociality. Pacas, coatimundis, and pecarries are organized in bands and herds.

Snow points out that frugivorous birds tend to have much "free time," since fruits are generally easy to locate and require virtually no capturing time and effort. The male bearded bellbird, a frugivore that I will discuss in more detail in chapter 12, spends an average of 87% of its time calling females to mate. Another frugivore, the male white-bearded manakin, spends 90% of its day courting females!

Frugivorous birds are generally abundant compared with insectivorous species, again, because fruit is abundant and readily available. Insects, on the other hand, are widely dispersed, often difficult to find and capture, and represent far less overall biomass. In one area of rainforest in Trinidad, Snow (1976) netted 471 golden-headed manakins and 246 white-bearded manakins, a total of 717 birds. In this same area he caught eleven species of tyrant flycatchers, but their combined total did not equal that of the two frugivorous manakins. Snow's wife, Barbara Snow, made a detailed study (Snow and Snow 1979) of the ochre-bellied flycatcher (*Pipromorpha oleaginea*) on Trinidad (page 106). As mentioned earlier, this species is undergoing an evolutionary diet shift. Though it is a tyrant flycatcher, it feeds almost exclusively on fruit. The Snows found the ochre-bellied flycatcher to be the most abundant forest flycatcher in Trinidad and attributed its numerical success to its diet of fruit. Fruit will support more of a given species than will a diet of insect food. In Belize, my colleague William E. Davis, Jr., and I have found the ochre-bellied flycatcher to be by far the most abundant of the 23 flycatcher species we encountered. We netted 102 ochre-bellies compared with 14 sulfur-rumped flycatchers (*Myiobius sulphureipygius*), the next most frequently netted flycatcher species. Studies throughout the Neotropics confirm the high biomass of frugivorous bird and mammal species in lowland rainforest (Fleming et al. 1987; Gentry 1990a). This is not surprising given that 50–90% of tropical trees and shrubs (depending on site) have seeds that are normally dispersed by frugivorous vertebrates (Fleming et al. 1987). Frugivores are not as species-rich as insectivores, but estimates are that 80–100 species of primarily frugivorous primates, bats, and birds typically occur in forest sites ranging from Central America to Amazonia (Fleming et al. 1987).

As noted above, a diet of fruit is not without potential problems. Interspecific (especially diffuse) competition may occur due to the localized nature of the resource attracting many potential feeders (Howe and Estabrook 1977). Nutritional balance may be lacking. Seeds are usually undigestible, and fruits tend to be highly watery and contain little protein compared with carbohydrate (Moermond and Denslow 1985). Small birds tend to eat small, carbohydrate-rich fruit, and many diversify their diets to include arthropods. Large frugivores, such as toucans, eat many different-sized fruits, including those rich in oil and fat (Moermond and Denslow 1985). Many of these species also supplement their diets to include animal food. For a thorough review of frugivory patterns at La Selva, Costa Rica, see Levey et al. (1994).

Who Gathers at the Fruit Tree?

To put it mildly, fruiting trees attract many species. One of the best ways to see both birds and mammals is to locate a tree laden with fruit and watch what comes by to feed. On January 14, 1982, over a one-hour period in mid-morning, I observed the following seventeen species of birds at a single fruiting fig tree in Blue Creek Village, Belize: masked, yellow-winged, and blue-gray tanagers; masked tityra; clay-colored robin; black-and-white warbler; streak-headed woodcreeper; yellow-throated euphonia; red-legged honeycreeper; northern and orchard orioles; buff-throated saltator; collared aracari; aztec parakeet; lovely cotinga; social flycatcher; and black-cheeked woodpecker. Not all of these species fed directly upon the fruits because insects are also attracted to fruit. Both frugivores and insectivores benefit from a visit to a fruit tree. Leck (1969) observed sixteen species of birds ranging over eleven families on a single tree species (*Trichilia cuneata*) in Costa Rica, of which eleven were observed directly feeding on fruit.

Fruit is an important resource in both montane and lowland forests. A comprehensive survey of fruiting trees and fruit-eating birds at Monteverde cloud forest in Costa Rica found that 171 plant species bore fruit that was fed upon by 70 bird species (Wheelwright et al. 1984). Some birds depended heavily on fruit, others casually. Among the birds were 3 woodpecker species, 9 tyrant flycatchers, 8 thrushes, 8 tanagers, and 9 finches, as well as toucans, pigeons, cotingas, and manakins. Though some birds were observed to feed on fruit only rarely, it was clear from this study that fruit represents an important resource for a very large component of the avian community. In comparison, at the lowland forest at La Selva, of 185 tree species studied, 90% produce fleshy fruits, of which approximately 50% of the species are primarily bird dispersed, 13% bat dispersed, and the remainder dispersed by other mammals such as monkeys and agoutis. Of 154 species of shrubs and treelets at La Selva, 95% bore small fleshy fruits, most of which were dispersed by birds or bats (Levey et al. 1994).

A study of fruit dispersal of *Casearia corymbosa* in Costa Rica recorded twenty-one species feeding on the fruits, none of which really contributed to seed dispersal (Howe 1977). These species failed to move significantly away from the fruiting tree. Only one bird species, the masked tityra, *Tityra semifasciata*, was judged to be an efficient seed disperser. This robin-sized black and white bird fed heavily on *Casearia* fruits and regurgitated viable seeds at considerable distance from the parent tree. The tityra is common and has a high feeding rate, characteristics enhancing its efficiency as a seed disperser. The tree *Tetragastris panamensis* on Barro Colorado Island attracted an assemblage of twenty-three birds and mammals that fed on the fruits, which were produced in "spectacular displays of superabundant but (nutritionally) mediocre fruit" (Howe 1982). But interestingly, actual seed dispersal of this species was less effective than another species, *Virola surinamensis*, which produced fewer but nutritionally better fruits. Fewer bird species fed on *Virola*, but those that did regurgitated seeds one at a time, well away from the plant where they were ingested (Howe 1982). In a study performed in the cloud forest at Monteverde, Costa Rica, seeds of three species of gap-dependent plants were heavily consumed by six bird species. However, only three of the six were useful to the plants as seed dispersers. The other three dropped seeds at the site of the parent plants or else destroyed the seeds in their guts (Murray 1988).

As the examples above demonstrate, the efficacy of seed dispersal by frugivores is difficult to generalize, and thus there is less evidence of finely tuned
mutualisms such as are common in pollination systems (Wheelwright and Orians 1982; Levey et al. 1994). Different assemblages of frugivore species can
exist throughout the range of a given plant species (Wheelwright 1988), so the
important dispersal species in one location will not be the same as those at
another locale. Some frugivore species which otherwise seem to be specialists
on a given plant species may be ineffective seed dispersers. The resplendent
quetzal (*Pharomachrus mocinno*) feeds largely on fruits from the laurel family
(Lauraceae) but, because of its habit of perching for long time periods as it
feeds (thus defecating seeds beneath the parent plant), it is not a very efficient
seed disperser (Wheelwright 1983).

Birds are selective about the size of the fruits they eat (Wheelwright 1985;
Levey et al. 1994). Species such as toucanets pluck fruit, juggle it in their bills,
and then reject it by dropping it. Large fruits are particularly at risk of rejection and may be found scarred by bill marks. Wheelwright hypothesized that
plants are under strong selection pressure to produce small to medium-sized
fruits, as larger ones are rejected by most bird species except those with the
widest gapes. Since wide-gape species were observed to feed just as readily on
small as on large fruits, plants that produce large fruits must depend upon
there being many large-gape species to insure seed dispersal. Large fruits do
permit more energy to be stored in the seeds, an advantage once dispersal and
germination have occurred.

Studies in Costa Rica indicated two patterns in methods by which birds take
fruit (Levey et al. 1984; Levey 1985; Moermond and Denslow 1985; Levey et al.
1994). Calling them *mashers* versus *gulpers*, the researchers noted that some
birds mash the fruit, dropping seeds, while others gulp the fruit, ingesting
it whole, and either regurgitate or defecate seeds. Mashers are finches and
tanagers, and gulpers are toucans, trogons, and manakins (see chapter 12).
Mashers are more sensitive to taste than gulpers. Levey (1985) found a distinct
preference among masher species for 10–12% sugar solutions as opposed to

Golden tanager, a masher

Cuvier's toucan, a gulper

lower concentrations. Gulpers, which swallow fruit whole, are taste insensitive. Gulpers' wide wings aid them in hovering at the fruit tree, and they have wide, flat bills useful in plucking and swallowing the fruits.

Precise measurements of fruit volume and seed dispersal were reported from a single fig tree in a lowland deciduous forest in Costa Rica (Jordano 1983). The estimated total fig crop was approximately 100,000, all of which were taken within five days. During the first three days alone, 95,000 were devoured! Birds were the principal feeders, eating 20,828 figs each day, about 65% of the daily loss. Mammals and fruits falling to the ground were the other source of loss. Parrots, which are more often than not seed predators, accounted for just over 50% of the daily total of figs. The most efficient seed dispersers—orioles, tanagers, trogons, and certain flycatchers—took only about 4,600 figs per day. Approximately 4,420,000 seeds were destroyed each day, mostly from predation by wasps and other invertebrates. Parrots were estimated to account for 36% of the seed loss. Only 6.3% of the seeds lost from the tree each day were actually dispersed and undamaged, indicating the high cost of seed dispersal.

The Oilbird—A Unique Frugivore

Perhaps the most unique of the many Neotropical frugivores is the oilbird, *Steatornis caripensis*, often called *guacharo*. Only one species exists in this odd family (Steatornithidae), and it ranges from Trinidad and northern South America to Bolivia (Hilty and Brown 1986). The oilbird is a large nocturnal bird, its body measuring 18 inches in length and its wingspread nearly 3 feet. Its owl-like plumage is colored soft brown with black barring and scattered white spots, and its head is punctuated by a large hooked bill and bulging, wide, staring eyes. Oilbirds are fascinating enough as individuals, but they come in groups. Colonies are widely scattered throughout the range of the species, as the birds live in caves, venturing out only at night to feed on the fruits of palms, laurels, and incense, often obtained only after flying long distances from the cave (Snow 1961, 1962b, 1976; Roca 1994). Fruits are taken on the wing as the birds hover at the trees, picking off their dinners with their sharp hooked beaks. Oilbirds probably locate palms by their distinct silhouettes, but they are thought to home in on laurels and incense through odor. Fruits from these trees are highly aromatic.

Enter an oilbird cave and be greeted by a cacophony of sound, a chaotic chorus of anxious growls and angry screams. In the dark, dank cave the flapping wings of the disturbed, protesting host conjure up thoughts of tropical demons awakened. Soon, however, the birds calm and flutter restlessly back to their nesting ledges, snarling as they resettle. Among the din you hear some odd, clicklike noises.

These vocalizations are one of the reasons why oilbirds are unique. They are the only birds capable of echolocation, the same technique by which bats find their way in the dark. The clicks are their sonar signals, sent out to bounce off the dark cave walls and direct the birds' flight. Oilbird echolocation lacks the sophistication found in bats, but it serves adequately to keep the birds from crashing against the cave walls as they fly in total darkness.

Oilbirds are closely related to the diverse family of nightjars, of which the whip-poor-will (*Caprimulgus vociferous*) is a common example. However, nightjars differ from oilbirds in that are all insectivorous, none are colonial, nor do any live in caves. How did the oilbird evolve frugivory, sociality, and its cave-dwelling habit? Snow (1976) has suggested a scenario for oilbird evolution that begins with a critical diet shift from insects to fruit. Snow hypothesizes that oilbirds were originally "normal" nightjars feeding on insects. However, large fruits offer a potentially exploitable high calorie resource, especially to a night bird, since there are few large bats and no other nocturnal birds to offer competition. Oilbirds thus became oilbirds when their ancestors shifted to a new diet of fruit.

Oilbirds

Frugivory selected for several adaptations. First, the birds' olfactory sense became enhanced as it provided an advantage in locating aromatic fruits. Second, social behavior began to evolve because fruits are a patchy resource and birds would tend to come together at fruiting trees. More important, however, is that a diet of fruit, though rich in calories, is nutritionally unbalanced, and, consequently, nesting time is prolonged. Incubation of each egg lasts just over one month. Once oilbirds hatch, they fatten up immensely in the nest due to a buildup of fat from the oily fruits, though they take a very long time to acquire sufficient and proper protein to grow bones, nerve, and muscle. By the time a juvenile is two months after hatching, it has still not left its nest but may weigh 1.5 times what either of its parents weighs! The name *oilbird* refers to the fact that juveniles put on so much temporary fat that they can be boiled down for the oil. Native people also occasionally use them for torches since they burn so well! The total time it takes an oilbird to go from newly hatched egg to fledging and independence is nearly one hundred days. For a full clutch of four eggs to develop requires approximately six months. Such an extended development time makes it very risky (even absurd) to nest on the forest floor, the traditional nightjar nesting site. Also, Snow notes that the defecated seeds that would surround a ground nest would serve to bring attention to it. Cave dwelling offers much more protection for the nest, but caves are also very patchily distributed resources. Again, sociality is selected for, and oilbirds became colonial cave dwellers.

Cave dwelling selected for the development of echolocation and also for a larger clutch size. Most tropical birds lay very few eggs in a given nest, often only one (like the purple-throated fruit crow mentioned above). Predator pressure is likely to be a major reason for the small clutch sizes, since nests can be more secretive if there are fewer mouths to feed. Given the safety of caves, however, oilbird nests are not under severe predator risk, and oilbird clutch size is normally four eggs. The nest is built up using droppings and is located on a cave ledge. Thin, yellowish, light-starved seedlings sprout from defecated seeds around the nest. Birds are thought to pair for life.

The scenario given above, plausibly accounting for all major oilbird adaptations, may or may not be true. Because it is not possible to go back in time, it is not possible rigorously to test Snow's hypothesis.

Research on mitochondrial DNA (mtDNA) has suggested that the oilbird is an ancient species that actually originated in North America perhaps as long as 50 million years ago (Gutierrez 1994). At that time, climate was much more tropical at northern latitudes, so fruit would have been readily available throughout the year. Oilbirds may have invaded the present range by crossing the Pleistocene land bridge (page 113) during the recent glacial periods. Such a migration could have seriously reduced the population, creating what evolutionists call a "genetic bottleneck," where surviving individuals bear a strong genetic similarity, one to another. Oilbird mtDNA shows little variation among widely scattered colonies, evidence that the separated populations are not becominig genetically distinct. Such similarity may also be due, of course, to dispersal of young birds among colonies, thus promoting large amounts of gene flow (Gutierrez 1994).

The oilbird is an important seed-dispersing species. In an exhaustive study centered in Cueva del Guacharo (Guacharo Cave) near the town of Caripe, Monagas, Venezuela, a mountainous, heavily forested region in northeastern Venezuela, Roca (1994) demonstrated by radio tagging that oilbirds have home ranges that encompass up to 96.3 km^2 (37 mi^2) and may have to fly up to 150 km (93 mi) between feeding sites. Indeed, dispersing individuals fly even farther in search of food, up to 240 km (150 mi) in a single night. Given that an adult oilbird requires approximately 50 fruits daily, Roca calculated that the entire colony he studied collectively regurgitated approximately 15 million seeds each month, a biomass of about 21 tons of seeds! Roca estimated that about 60% of the seeds were dispersed in forest. Oilbirds would seem to qualify as important species in maintaining the plant biodiversity of the forests in which they forage and, as such, merit strict conservation measures, especially around their caves.

6

The Neotropical Pharmacy

ROPICAL rainforests are green. Myriad leaves, large, small, simple, and
compound, adorn trees and vines from ground to canopy. Careful examination of leaves reveals that most show little if any damage from
insects or other herbivores. But why?

At least in theory, tropical rainforest leaves are much like a lake full of pro-
verbial sitting ducks: they are potential food sources for insects, other herbivores, and pathogens year-round. They can't run away. There is no cold winter
season when most arthropods are inactive or outright dead. Yet tropical leaves,
for the most part, remain largely intact, albeit with a hole or two here and
there. In one Amazon rainforest, for example, biomass of living vegetation was
estimated at approximately 900 metric tons per hectare. The mass of animals,
all mammals, birds, reptiles, insects, and other creatures, totaled a mere 0.2
metric ton per hectare (0.5 ton per acre), thus the biomass of animals was only
0.0002% of the plant biomass! Of the total number of animals in this Amazonian forest, only 7% ate living plant material such as leaves and stems; 19%
(mostly termites) depended on eating living or dead wood; and fully 50% ate
only dead vegetation. The remainder were carnivores (Fittkau and Klinge
1973). Such an imbalance between plant and animal biomass is noticeable to
anyone with experience in rainforest. In general, herbivores in the tropics take
only a small percentage of what is potentially available to them (Coley et al.
1985). How do tropical plants protect themselves against the herbivorous
hordes as well as against invasive pathogenic bacteria and fungi?

Plant Defense Compounds   *Figure 37*

The answer is drugs. Leaves of both tropical and temperate zone plants are
abundantly laced with noxious chemicals, some of which are very familiar to
humans. Daniel Janzen (1975) puts it well: "The world is not coloured green
to the herbivore's eyes, but rather is painted morphine, L-DOPA, calcium oxalate, cannabinol, caffeine, mustard oil, strychnine, rotenone, etc." Many well-
known poisons and stimulants in human culture, drugs ranging from curare to
cocaine, come from tropical plants. These chemicals, along with many others
present in plants, are often termed by plant physiologists as "secondary compounds" because most seem to lack any direct metabolic function, such as
involvement with photosynthesis. But secondary compounds do have a function, and a vital one at that.

Plants contain many different kinds of secondary compounds, and any given plant may have dozens. Because they collectively help protect plants, they are commonly called defense compounds or *allelochemics*. They may have originated as genetically accidental metabolic byproducts or chemical wastes that, by chance, conveyed some measure of protection to the plant from attacks by either microbes or herbivores. Such mutations would confer high fitness, and thus natural selection would favor their rapid accumulation (Whittaker and Feeny 1971). Most researchers attribute no direct metabolic function to secondary compounds, arguing instead that they serve entirely for defense (Ehrlich and Raven 1967; Janzen 1969b, 1985). However, some evidence exists that some secondary compounds may be directly functional (Seigler and Price 1976; Futuyma 1983). Nonetheless, there is ample evidence favoring the argument that many secondary compounds function primarily for plant defense. Most plant species contain an impressive variety of defense compounds representing several major chemical groups. Some defense compounds function principally to protect against herbivores, some to protect against bacteria and fungi.

Alkaloids

Alkaloids are among the most familiar and addictive drugs known. Such familiar names as cocaine (from coca), morphine (from the opium poppy), cannabidiol (from hemp), caffeine (from teas and coffee), and nicotine (from tobacco) are all alkaloids. Taken together, there are over 4,000 known alkaloids that are globally distributed among 300 plant families and over 7,500 species, and a single plant species may contain nearly 50 different alkaloids (Levin 1976). By no means do all alkaloid-containing species occur in the tropics. Although tropical species contain many alkaloids (see below), alkaloids are also well represented among temperate zone plants. It is estimated that approximately 20% of all vascular plant species contain alkaloids (Futuyma 1983). Alkaloids are found not only in leaves but almost anywhere in the plant, including seeds, roots, shoots, flowers, and fruits.

Most alkaloids have a bitter taste. In mammals, depending on dosage level, they tend to interfere with both liver and cell membrane function. They may also cause cessation of lactation, abortion, or birth defects. Many are addictive. Of all of these characteristics, it is probably the bitter taste combined with the difficulties of digestion and liver function that discourage animals from munching alkaloid-rich vegetation.

As an example, caffeine has been experimentally shown to discourage insect feeding (Nathanson 1984). Caffeine is a type of alkaloid called a methylxanthine, and both it and synthetic methylxanthines can seriously interfere with enzyme systems of tobacco hornworms (*Manduca sexta*). Damage occurred at concentrations noted to be normal for field plants, thus caffeine, though a stimulant to humans, is in reality a form of insecticide. However, most alkaloids do not seem to function as insect antifeedants (Hubbell, pers. com. 1987). Further, some alkaloids may serve not primarily as defense compounds but rather to store carbon and/or nitrogen (Futuyma 1983).

Phenolics and Tannins

Phenolic compounds are often abundant in plants (Levin 1971). One group adds the pungency to many of the most well-known spices, and another, known as the tannins, provides the basic compounds used in tanning leather. Tannins are particularly abundant in temperate oak leaves as well as in many tropical plant species such as mangroves.

Work performed on *Cecropia peltata* on Barro Colorado Island, Panama, indicated that tannins were heavily concentrated in young trees but declined in concentration in older plants (Coley 1984). Tannin levels were lower in plants grown in the shade, indicating that access to high levels of photosynthesis may be necessary for tannin production. In field experiments, low tannin plants experienced twice the level of herbivory as those with high tannin levels. However, leaf production was inversely correlated with tannin levels. The more leaves on the tree, the lower the tannin per leaf, indicating that tannin production, though perhaps protective, is clearly costly to the plant. Trees like *Cecropia*, experiencing intense competition for light, are probably best served by limiting tannin protection in favor of rapid growth.

Phenolics are small proteins stored in cell vacuoles that are broken when an insect or other herbivore bites the leaf. Upon release, the phenols combine with various proteins, including those enzymes necessary for splitting polypeptides (parts of proteins) in digestion, perhaps making it more difficult for a herbivore to digest protein. Leaf damage by insects or pathogens may stimulate production of phenolics (Ryan 1979).

By no means are phenolics or tannins effective in discouraging all insect herbivores. Leafcutting ants, major herbivores throughout the Neotropics (page 133), seem undeterred by them (Hubbell et al. 1984). The role of phenolics, and tannins in particular, as antiherbivore adaptations is rather questionable. There is little direct evidence, for instance, that tannins serve as a general defense mechanism (Zucker 1983). Tannins may discourage some insects initially, only to stimulate natural selection for resistance to them, as has happened, for instance, with the use of insecticides like DDT. Some insects, especially those adapted to feed on a diversity of plant foods, have enzymes that detoxify some defense compounds (Krieger et al. 1971). Some kinds of tannins may be effective principally against insects, while others may function principally against microbes and pathogens (Zucker 1983; Martin and Martin 1984).

Saponins

Saponins are soaplike compounds that are relatively common in tropical plants and act to destroy the fatty component of the cell membrane. Some indigenous people utilize saponins to poison and capture fish. Saponins, leached from leaves, act on fish gills, interfering with respiration.

Cyanogenic Glycosides

Everyone knows that cyanide can be a deadly poison. Many species of tropical plants contain compounds called cyanogenic glycosides consisting of cyanide linked together with a sugar molecule. When combined with enzymes from either the plant or a herbivore's digestive system, the sugar is released, leaving hydrogen cyanide. Needless to say, these highly potent plants discourage herbivores. It is interesting to note that some strains of manioc, from which cassava is prepared (page 162), contain high concentrations of cyanogenic glycoside in the root, the part that is eaten. Thus the root must be washed extensively with water to eliminate the cyanide before it is eaten (see below).

Cyanogenic glycosides are well represented in passionflowers (*Passiflora* spp.), fed upon by caterpillars of *Heliconius* butterflies (see below). Each *Heliconius* species is able to detoxify one or two cyanogenic glycosides, and thus these butterflies are selective, each species feeding only on select passionflower species (Spencer 1984). The butterfly caterpillars have hydrolytic enzymes specific for detoxifying certain cyanogenic glycosides.

Cardiac Glycosides

Cardiac glycosides interfere with heart function. Digitalis, from the temperate zone plant foxglove (*Digitalis purpurea*), is a cardiac glycoside commonly used as a heart stimulant. Cardiac glycosides in high concentration can be fatal to normal hearts. Among the many types of plants with cardiac glycosides are members of the milkweed family.

Terpenoids

Terpenoids are a complex group of fat-soluble compounds that include monoterpenoids, diterpenoids, and sesquiterpenoids. Some are used in the synthesis of compounds that may mimic insect growth hormones (preventing rather than promoting growth of the insect) or can be modified into cardiac glycosides (Futuyma 1983). Some terpenoids discourage both insects and fungi. One terpenoid in particular, caryophylene epoxide, has been shown to repel completely the fungus garden ant *Atta cephalotes* from clipping leaves of *Hymenaea courbaril.* This terpenoid was shown to be highly toxic to the fungus that the ants culture (Hubbell et al. 1983). In a survey of forty-two plant species from a Costa Rican dry forest, three-quarters contained terpenoids, steroids, and waxes that strongly repelled leafcutting ants (Hubbell et al. 1984).

Toxic Amino Acids

Some tropical plants, especially members of the bean and pea family (legumes), contain amino acids that are useless in building protein but instead interfere with normal protein synthesis. Canavanine, for example, mimics the essential amino acid argenine. Perhaps the best known of the toxic amino acids is L-DOPA, which can be a strong hallucinogen. Both canavanine and L-DOPA have been found to be concentrated in the seeds of some tropical plants. In general, the major function of nonprotein amino acids, at least in legumes, seems to be to discourage herbivores (Harborne 1982).

Calcium Oxalate

Among of the most familiar tropical plants are the large-leafed philodendron vines of the arum family. The widespread skunk cabbage (*Symplocarpus
foetidus*), which grows throughout swamps and streamsides in eastern North
America, is among the relatively few arums to reach the northern temperate
climes. Skunk cabbage, like most of the other arums, contains crystals of calcium oxalate, a caustic substance that makes the delicate tissues of the mouth
burn. Most tropical arums have very large leaves, tempting targets for herbivores. Given their obvious large leaf size, it is easy to see how successfully protected arums must be. Very few of their huge leaves are significantly chewed.

Other Types of Defense

We drive around every day on a defense strategy of tropical plants, namely,
rubber tires. The rubber tree (*Hevea brasiliensis*), which can reach heights of
36.5 m (120 ft), is but one of many tropical trees that produce latex, resins, and
gums, all of which act to render the trees less edible. Rubber tree sap is a milky
suspension in watery liquid contained in ducts just below the bark, external to
the cambium and phloem (Janzen 1985). It congeals upon exposure to air and
may aid in closing wounds, protecting against microbial invasion, and hinder-
ing herbivores. Indeed, imagine an insect such as a weevil larva or termite that
drills through bark only to chew into sticky, congealing goo. Latex is present
in plants of many families (Euphorbiaceae, Moraceae, Apocynaceae, Cari-
caceae, Sapotaceae, others), a case of a convergent defense adaptation among
many species of tropical plants (Janzen 1985). The chicle tree, *Manilkara zapota*, which grows in rainforests throughout Central America, produces a natural latex called chicle from which chewing gum is made.

In those trees whose leaves are chemically nonrepellent to leafcutting ants,
sap adhesion to the insects' mouthparts and appendages discourages ants
from clipping the leaves (Hubbell et al. 1984). Leafcutter ants will take leaves
that are removed from *Euphorbia leucocephala* trees but will not clip intact
leaves, presumably because clipping induces copious sap flow from intact
leaves (Stradling 1978). However, some insect herbivores have adapted to
latex defense. For example, one caterpillar species clips leaves of papaya in
such a way as to cause the defensive latex to flow away from where the insect is
feeding (Janzen 1985).

In addition, tropical trees may be spiny or thorny or have leaves coated with
diminutive "beds of nails" called trichomes that sometimes literally impale
caterpillars (Gilbert 1971). Experiments have shown that sharply toothed leaf
edges can reduce caterpillar grazing. When teeth are experimentally removed,
caterpillars feast (Ehrlich and Raven 1967). In Belize, I am careful to avoid
grabbing the trunk of a young warree palm, *Acrocomia vinifera*. This common
understory species has 5-cm (2-in)-long spines lining its lower trunk. Each of
these sharp, jagged stilettos is coated with lichens and various microbes and
can cause infection easily if they penetrate the hand. Warree palms also have
long, sharp spines on the undersides of leaf midribs. Warrees are not alone in

their sharp defense, as many species of understory palms throughout the tropics are also generously endowed with spines. Even the actual wood of tropical trees is often quite hard, a possible adaptation to discourage the ever-present hordes of termites.

Leaf toughness, nutrition value, and fiber content also affect ability to resist herbivores. Coley (1983) examined rates of herbivory and defense characteristics of forty-six canopy tree species on Barro Colorado Island. She compared young with mature leaves and gap-colonizing species with shade-tolerant species. In general, young leaves were grazed much more than mature leaves even though many of the young leaves were loaded with phenols (indicating that phenols do not discourage herbivory). Young leaves were, however, richer in nutrients and lower in fiber and toughness. Mature leaves of gap-colonizing species were grazed six times more rapidly than leaves of shade-tolerant species, and, overall, gap-colonizing trees had fewer tough leaves with lower concentrations of fiber and phenolics than shade-tolerant species. Gap tree leaves grew faster but had shorter lifetimes. Coley concluded that leaf toughness, fiber content, and nutritive value were more influential than defense compounds in affecting patterns of herbivory. Studies on nutritional choices of howler monkeys (see below) resulted in a similar conclusion (Milton 1979, 1981).

The typical tropical plant combines various defense compounds with mechanical defenses. Some plants contain up to ninety different defense compounds, ranging from carcinogens to stimulants, laced throughout the plant's tissues. Spines may line the leaf edge, and thorns help protect the bark. There may be latex chemicals, and even the wood may contain crystals of silicon, a sort of "ground glass" for protection (Janzen 1975, 1985). Some plants utilize insects in their defense. The bull's horn acacia, described in chapter 5, is protected by its resident *Pseudomyrmex* ants. For these acacias, ants ecologically take the place of defense compounds. Acacia species without ants contain cyanogenic glycosides, which ant acacias lack (Janzen 1966). Ants also are implicated in the defense of other species such as cecropias and passionflowers.

The Evolutionary Arms Race

Given that many species of Neotropical plants have numerous kinds of defense compounds as well as mechanical defenses, it may come as a surprise that any kind of herbivore is able to eat them. But some do. Evolution works in both directions. Bear in mind that defense compounds evolved in response to selection pressures presumably exerted by various herbivores and pathogens. Once a kind of defense compound evolves, it, in turn, acts as a selection pressure on herbivores and pathogens to evolve resistance, or some other strategy to circumvent the plant defense (recall the coevolution discussion in chapter 5). Once again vulnerable, the plant is then under enhanced selection pressures to evolve yet another defense compound, which, should that occur, merely acts as yet another selection pressure on the herbivore. And so it goes, the evolutionary treadmill. One evolutionist (Van Valen 1973) has likened this process to the fabled Red Queen of *Alice in Wonderland*, always having to run to stay in place. Because of the realities of coevolution, and because of the many millennia in which coevolution has been ongoing in the Neotropics, plants have accumulated a substantial reservoir of defense compounds as they run through evolutionary time to stay in place in the rainforest.

Latitudinal Trends

As I mentioned earlier, defense compounds are well distributed in the temperate zone and are even in the arctic, as well as in the tropics. Lemmings, the small (and cute) tundra-inhabiting rodents whose populations sometimes swell enormously, are periodically affected by defense compounds in some of their food plants. It is true, however, that tropical ecosystems are habitats where defense compounds are abundantly represented. Just as species diversity increases as latitude decreases (page 86), so does the presence of defense compounds.

Levin (1976) made a detailed study of the geographic distribution of alkaloid-containing plants. Groups of herb, shrub, and tree species each contained a significantly greater percentage of tropical species with alkaloids. In all, 27% of temperate species tested contained alkaloids compared with 45% of the tropical species tested. The evolutionarily more primitive plants, such as the magnolias, many of which are tropical species, contained more alkaloid-bearing representatives than species from the more recently evolved families.

In Kenya, 40% of the plant species tested contained alkaloids compared with only 12.3% in Turkey and 13.7% in the United States. In Puerto Rico, a tropical but much smaller geographic area than the United States, 23.6% of the plants tested contained some alkaloids. Levin's analysis revealed a relatively steady decrease in the percentage of plants bearing alkaloids from the equator northward, a trend termed a *latitudinal cline*. One possible explanation for this cline is that nontropical species are less subject to pest pressure, and thus high alkaloid content has not been selected for in many of these species. Another is that pest pressure may differ substantially between tropical and temperate areas, selecting for different defense chemicals.

Ecological Trends

Figure 37

Defense compounds are found in plants from virtually all habitats, but some trends are apparent. Tropical lowland forests, mangrove swamps, deserts, and mountain rain and cloud forests are all habitats where defense compounds are abundant. Alpine forest and grassland as well as disturbed areas (see below) have fewer plants containing defense compounds.

Defense compounds are abundant in lowland forest occurring on nutrient-poor white, sandy soils in the northern Amazon region (page 56). Leaves from the vegetation of white soil forests, which are very expensive to replace (Janzen 1985), are long-lived and have such high concentrations of defense compounds that even after the leaf drops from the plant nothing can really eat it. The leaf must be leached of its defense compounds by rainfall before it can be broken down and its minerals recycled. The blackwater rivers that are characteristic of white, sandy soil regions are black because of the leached phenolics (e.g., tannins) in the water. The probable reason for the high concentration of defense compounds in white, sandy soil forests is that it is more efficient for the plants to manufacture leaves packed with defense compounds than to replace leaves ravaged by microbial pathogens, fungi, or herbivores (Janzen 1985). Given the shortage of minerals in the soil, the replacement of leaves is a more costly enterprise than the synthesis of defense compounds that aid in leaf longevity.

Areas undergoing ecological succession tend to have species that invest in defense compounds differently from those on poor soil. Most successional species are racing to maximize their rates of growth. They seem to synthesize alkaloids, phenolic glycosides, and cyanogenic glycosides, all present in low concentration, collectively representing a relatively low energy cost. In contrast, plants on nutrient-poor soils invest in metabolically "expensive" defense compounds such as polyphenols and fiber (e.g., lignin) that are retained as immobile defenses in the leaves and bark. Trees grow more slowly but are well protected as they grow. These contrasting patterns in plant defenses may be related to resource availability (Coley et al. 1985). On sites where resources are poor, "expensive," long-lasting defense compounds are favored. On resource-rich sites, "cheaper," shorter-lasting defense compounds are favored because the tree is able both to devote sufficient energy to rapid growth and to replace defense compounds as needed.

Many successional species are subject to herbivore damage (Brown and Ewel 1987). Cecropias are fast growing, and I have seen many with extensively damaged leaves. Perhaps cecropias "trade off" protection by defense compounds for rapid growth. I have also seen vines and herbs thoroughly grazed in successional areas adjacent to mature forests where understory leaf damage was far less. Areas undergoing early succession are resource-rich. As more and more species invade and grow on the site, competition among individuals reduces the overall availability of resources per individual and selects for those species adapted to grow more slowly but persist. Janzen (1975) has estimated that insect density is five to ten times greater in successional areas than in the understory of a mature rainforest, presumably because successional species are more palatable to insects.

The battle between insect herbivores and a plant species was well documented for the understory species *Piper arieianum* at La Selva Biological Station in Costa Rica. Insect damage to this piper species varied greatly from plant to plant, and some plants were genetically more resistant to herbivores than others (Marquis 1984). Herbivory likely has had a very strong selective influence on the evolution of piper defenses. Perhaps some individual pipers "risk" investing little in defense compounds but more in rapid growth and reproduction, while others grow more slowly but are better protected. Plants with few defenses may be partially protected by occurring among well-defended plants, where herbivores are scarcer.

Plants versus Monkeys

Consider what it would be like to be a monkey living in a rainforest. Neotropical monkeys are essentially vegetarians and live in what would seem an idyllic habitat burgeoning with leaves and fruits. Appearances can be deceiving, however. Given the abundance of defense compounds packed into all tissues of tropical plants, how do monkeys cope with a world green on the outside but poison on the inside?

Sometimes science begins with serendipity. A casual observation, perhaps unimportant or only vaguely interesting to most people, becomes a starting point for a study that yields new and exciting perspectives on how nature functions. For Kenneth Glander (1977), the serendipitous observation happened while he was watching mantled howler monkeys, *Alouatta villosa* (page 300), in Costa Rica. A female animal with her juvenile clutching on apparently became disoriented and fell 10.5 meters (about 35 ft) out of a carne asada (*Andira inermis*) tree. Monkeys, understandably, do not often fall from trees. For a monkey, the tree is everything: it is food, it is shelter, it is security. Any monkey that falls from a tree is either ill or wounded. The female was dazed from her fall but recovered sufficiently to climb back up the tree. Her baby was not injured. But why had the monkey lost its balance and fallen out?

Glander became curious about the monkey that fell from the tree as well as about several dead howlers he found in his study area. He wondered if these animals had perhaps been poisoned by defense compounds present in the foliage they had eaten. Glander was aware of the fact that local people use the crushed leaves of the common madera negra tree, *Gliricidia sepium*, to obtain the poison rotenone, used against rats. Madera negra leaves also contain various alkaloids. Howlers eat madera negra leaves.

For Glander, the work had begun. Nature does not relinquish information easily, and Glander spent 5,000 hours in the field observing howlers. He marked each of 1,699 trees in the study area so that he could document the exact trees in which monkeys fed.

Glander learned that howlers were extremely selective in their feeding choices. Of a total of 149 madera negra trees in the study area, the howler troop only fed in 3, and they were always the same 3 trees! These trees were found to have leaves free of alkaloids and cardiac glycosides. Other madera negra trees had high concentrations of defense compounds. The howlers had apparently learned which trees were safe to eat. Imagine living in a town with 149 restaurants, where most put poison in the food they serve, but three serve safe, healthful meals. The people of the town would learn quickly by sad experience which restaurants were safe. So it is with the howlers in Costa Rica. Glander speculated that defense compounds may have been one factor affecting evolution of intelligence in primates. Poisons provide a selection pressure to remember which trees are safe and which are not and to communicate the information to others in the troop. The very social structure of the howlers and other monkey species may be evolutionarily related to stresses imposed by living in a world of toxic trees.

Glander found that mantled howlers tended to favor young leaves that are relatively high in nutritional value but have not yet become loaded with defense compounds. When given no choice but to eat mature leaves, they ate only a little and then moved to a different tree. This behavior would help avoid too high a dose of one type of defense compound. Sometimes they ate only the leaf stalk or petiole, ignoring the blade. The petiole has the least alkaloid content. The total picture is not simple, however. Katherine Milton (1979, 1982), who studied mantled howlers on Barro Colorado Island in Panama, found that protein and fiber content, not secondary compounds, seemed to be more important factors affecting leaf choice. Fiber makes leaves more difficult to digest and is least concentrated in the preferred younger leaves. Protein is proportionally higher in young than older leaves. The more protein to fiber, the more desirable the leaf was to howlers.

Milton (1981) learned that howlers have long intestinal systems, especially the hindgut. It takes food up to 20 hours to pass through a howler's digestive system. Spider monkeys, *Ateles geoffroyi*, eat mostly fruits, much easier to digest because they contain more protein and fewer defense compounds than leaves. It takes food only about 4.4 hours to move through the shorter gut of a spider monkey. Howlers, with their long hindguts, are able to digest leaves efficiently, coping to a reasonable degree with both high fiber and defense compounds. Howlers and spider monkeys do not compete intensively with each other for food because each focuses on a different primary food source.

Plants versus Insects

Given that tropical plants contain large quantities of defense compounds, how do insects manage to find anything to eat? One answer, at least for some, appears to be evolutionary specialization, adapting to eat relatively few kinds of plants. Unlike most monkeys and other large herbivores with broad diets, many insects, both tropical and temperate, have tendencies to feed exclusively on one type of plant rather than upon a wide array of species. For instance, heliconid butterfly caterpillars, which I discuss below, feed exclusively on vines in the genus *Passiflora*, the passionflowers. The colorful caterpillar of *Pseudosphinx tetrio* (page 83) feeds only on the leaves of *Plumeria rubra*, the frangipani tree. In Trinidad, I observed at least a dozen of these large, boldly patterned caterpillars totally defoliate a frangipani, while all surrounding trees were untouched.

Toxic defense compounds are undoubtedly a major selection pressure in the evolution of specialization among insect herbivores. Those insects that evolve enzyme systems that detoxify defense compounds or somehow sequester them (see below) are able to focus their appetites on specific plant species. However, other factors can also select for diet specialization (Futuyma 1983). For instance, plant compounds may be repellent but not actually toxic. Insects may overcome the repellency and adapt to recognize a host plant by its repellent compounds. Interspecific competition may also select for host specialization among insects, as might avoidance of certain parasitoids (which are attracted to certain plant-specific compounds). Insects might also be practicing optimal foraging behavior, searching for the food that provides the highest reward for the smallest effort.

Even when feeding on a specific host species, insects may employ behaviors that minimize exposure to defense compounds. For example, the caterpillars of the butterflies of the genus *Melinaea* feed on plants in the Solanaceae family, especially in the genus *Markea* and *Juanaloa*. The caterpillars, like those mentioned earlier that avoid latex from papaya, are well known for cutting the leaf veins of their host plants, thus preventing the defense compounds from reaching the leaf blade, where the caterpillars are feeding (DeVries 1987).

Many tropical insects are not specialized to feed on but a single type of host but are generalist feeders. For example, leafcutter ants are among the least specialized herbivores in the tropics, typically foraging on a wide diversity of plant species (Rockwood 1976; Hubbell et al. 1984). The leafcutters accomplish this foraging diversity by virtue of the fact that it is their cultivated fungus, and not them, that digests the plant tissue, and the fungus is in general far more tolerant of defense compounds than the insects. Because of this remarkable feeding adaptation, leafcutters are abundant, a real evolutionary success story.

Bruchid Beetles

Beetles in the family Bruchidae provide examples of both feeding specialization and adaptation to host defenses. Bruchids are seed predators, especially on legumes. Females lay eggs on seed pods or directly in seeds. Larvae enter pods upon hatching (or are hatched inside pods) and feed on seeds before pupating. One bruchid species, *Merobruchus columbinus*, killed 43% of the seed crop from its host tree *Pithecelobium saman* in Guanacaste, Costa Rica (Janzen 1975). In Costa Rica, 102 of 111 species of bruchid beetles from a deciduous forest fed on only one host plant (Janzen 1975), a very high degree of specialization. In a survey of the beetle families Bruchidae, Curculionidae, and Cerambycidae, about 75% of the species fed on a single plant species. Only 12% fed on three or more host plants (Janzen 1980c).

Bruchid species are widespread in the tropics, and a single species of bruchid may feed on several different host plants throughout its range. However, Janzen found that within any local area a bruchid species will feed only on one host species. In other words, local races exist within some bruchid species, each of which is specialized. Bruchids have a diverse array of adaptations for dealing with the defenses of their host plants (Janzen 1969b).

Center and Johnson (1974) have documented the complex game of coevolution that illustrates the evolutionary arms race between bruchids and their host plants. Some host plants have evolved toxic seeds containing hallucinogenic alkaloids, saponins, pentose sugars, and free amino acids. Bruchids either avoid these toxins or are physiologically resistant to them (see below). Some seed pods produce sticky gum following penetration of bruchid larvae. Bruchids evolved a period of quiescence in embryonic development, until seeds mature and it is too late for pods to produce the gum. Some pods fragment or explode, scattering seeds and thus avoiding larvae that enter after hatching on the pod wall. Bruchids evolved the habit of ovipositing directly on the seeds, after they are scattered. Some seed pods flake, thus helping remove eggs from their surface. Bruchids oviposit beneath the flaking substance, thus avoiding it, or else they have an accelerated embryonic development, hatching and entering the pod before it flakes. Some seeds remain very small (thus not supplying much bruchid food), then growing quickly just prior to dispersal. Bruchids nonetheless enter and eat immature seeds, or else delay their maturation until seeds are mature and thus larger. Finally, some legumes make very tiny seeds, too small for bruchids to mature within them. Bruchids eat them anyway, devouring several seeds during development, not just one. Bruchids adapt sufficiently well that most host plants are rarely free from bruchid seed predation (Janzen 1969b; Center and Johnson 1974). There can be little doubt that bruchids are doing quite well in the coevolutionary arms race, running well to stay in place.

The physiological ability to deal with plant toxins may be fortuitous, at least in the case of bruchids and other beetles that digest seeds containing the toxic amino acid L-canavanine. Certain species of beetles that specialize on eating seeds high in L-canavanine contain an arginase enzyme that can efficiently degrade the toxic amino acid. However, many beetles that do not feed on plant seeds high in L-canavanine also have arginase. Conceivably, these species could also detoxify L-canavanine. Why don't they? Degradation of L-canavanine may still result in its incorporation into aberrant proteins, called canavanyl proteins. Other metabolic systems are also necessary to deal efficiently with L-canavanine, which thus far these other insects have probably not evolved (Bleiler et al. 1988).

An analysis of larvae of two beetle species that dine on L-canavanine-containing seeds found that both could cleave L-canavanine to L-canaline and urea. Rather than excrete the urea, which would represent a significant loss of nitrogen to the developing insect, the insects utilize another enzyme, urease, to synthesize ammonia from urea. The ammonia is incorporated into synthesizing various amino acids used in growth, and the L-canaline is not incorporated into aberrant amino acids (Bleiler et al. 1988).

Heliconid Butterflies and Passionflowers

Butterflies in temperate and tropical areas are known for their affinities for feeding on specific plant families. The pipevine swallowtail caterpillar (*Battus philenor*) of temperate woodlands feeds on nothing but plants of the birthwort family (Aristolochiaceae). The painted lady caterpillar (*Vanessa virginiensis*) is evolutionarily committed to composites (daisies, etc., Compositae), and the monarch caterpillar (*Danaus plexippus*) specializes on milkweeds (Asclepiadaceae). Caterpillars are more selective than adults because adult butterflies feed on nectar, aiding in pollen dispersal. Their involvement with plants is fundamentally mutualistic, and defense compounds would be of little selective importance. Caterpillars, however, are voracious herbivores, and, being folivores, they are the enemy of plants. Thus they encounter high concentrations of defense compounds as they chew on leaves. Since different families of plants produce different combinations of defense compounds, natural selection has acted on caterpillars in such a way that various caterpillar species have evolved tolerance for defense compounds associated with different plant families.

Heliconid butterflies (Heliconiinae) are a diverse and colorful group, almost all of which are Neotropical (DeVries 1987). They are usually considered part of the brush-footed butterflies (Nymphalidae), although some taxonomists place heliconids within their own distinct family. Nymphalids number nearly 3,000 species globally, but heliconids are represented by only about 50 species, with many local races throughout tropical America (see below).

Commonly called longwing butterflies, only two species, *Heliconius charitonius* and *H. erato*, regularly reach the United States, and both of these are found only in the extreme southern part of the country, such as the Everglades in southern Florida.

Heliconid caterpillars feed almost exclusively on species of *Passiflora* or passionflower (Passifloraceae), a common vine numbering approximately 500 species. Like heliconids, passionflowers are largely Neotropical, and another name for heliconids is "passionflower butterflies" (DeVries 1987). Not many kinds of herbivores eat passionflower vines, probably because these vines contain effective defense compounds. Most passionflowers contain various cyanogenic glycosides and cyanohydrins, and there is a strong correlation between preferences by heliconid caterpillars and specific defense compounds that may be present (Spencer 1984). The high diversity of cyanogens among passionflower species may be an evolutionary response to herbivory by heliconids. Heliconids, however, seem able to adapt to the ever-changing cyanogen regime by evolutionary changes in their hydrolytic enzymes and by sequestration of cyanogens. Heliconid caterpillars have largely solved whatever toxic challenges may be presented by passionflower and are capable of defoliating passionflower with little difficulty. To passionflower, these caterpillars are the enemy.

Heliconid butterflies lay small numbers of eggs in globular yellow clusters directly on passionflower leaves, favoring young shoots. When the eggs hatch, the little caterpillars are sitting on their food source. The trick for the plant, therefore, is to prevent the adult female butterfly from locating, selecting, and laying eggs on its leaves.

Detailed studies (Gilbert 1975, 1982; Benson et al. 1976) of passionflower versus heliconids have shown that the plant's defenses seem to be both an attempt to repel and to fool the caterpillars. Passionflower produces extrafloral nectaries (page 130) that attract various species of ants and wasps. These aggressive insects help repel heliconid caterpillars. Certain amino acids such as proline, tryptophan, and phenylalanine are very common constituents

Heliconius butterflies

of nectaries (Horn et al. 1984). Proline is important for insect flight, and the other two amino acids are important in insect nutrition. The nectaries attract flying hymenopterans (bees and wasps) as well as ants. Some passionflowers are protected exclusively by ants, some exclusively by wasps, some by both. At least one study has shown that caterpillar survival is much lower on *Passiflora* with attending ants (Smiley 1985). Caterpillar mortality rate was 70% on ant-attended plants compared with 45% on nonant plants, suggesting strongly that ants do protect passionflowers from heliconids.

Some *Passiflora* extrafloral nectaries appear, to the human eye, to mimic heliconid egg clusters (Gilbert 1982). Passionflower vines typically have young leaves spotted with a few conspicuous yellow globs, the egg mimics. Female heliconids will not lay eggs on a leaf already containing egg masses. The mimic egg masses presumably prompt the female to keep searching and find another victim. Gilbert believes the mimic eggs to be a recent evolutionary development in the plant-insect battle. Currently only 2% of passionflower species have mimic eggs.

Leaf shape varies within a species of passionflower, and passionflower leaves often resemble other common plant species growing nearby (Gilbert 1982). Perhaps heliconids may be tricked by the similarity of appearance (leaf mimicry) and thus overlook a passionflower. This speculation depends, of course, on the butterfly using visual cues to locate passionflower vines. If the insect depends principally on scent, such leaf mimicry would seem useless. More research is required to answer the question.

For the time being, at least one passionflower species, *Passiflora adenopoda*, may have actually won the coevolutionary battle between insect and plant. Its leaves are covered by minute hooked spines called trichomes. Resembling a Hindu's bed of nails, trichomes impale the soft-skinned caterpillars. Once a caterpillar is stuck, it starves (Gilbert 1971). Does the future hold in store thick-skinned caterpillars? Time will tell. At least one butterfly species, *Mechanitis isthmia* (an ithomiid, not a heliconid), has adapted to thwart trichome defense. Caterpillars feed on plants of the tomato family and succeed in countering trichomes by spinning a fine web covering them, enabling the caterpillars to move easily over the leaf surface to feed on the leaf edges (Rathcke and Poole 1975).

Why Are Heliconius Butterflies Pretty?

Heliconid butterflies are neither rare nor inconspicuous. In fact, they are among the most obvious and beautiful butterflies of the tropics. They fly slowly, almost delicately, and are very easy to see along forest edges as well as in interior rainforest. When atop a plant, the butterfly appears brilliantly colored, many almost iridescent. Why are they so conspicuous? Consider the potential risk to the insect. Most tropical bird species feed heavily on insects. Remember, there are over 300 species of flycatchers alone! It would seem suicidal for a butterfly group to be colored like neon signs, seeming to say "eat me."

Remember, however, that in the tropics, obviousness may serve as a warning (page 81). If heliconid butterflies are distasteful to would-be predators, it may be to the butterflies' potential advantage to be obvious. Once a bird has eaten an unpalatable insect, it may remember its unpleasant experience and resist trying to devour other insects in the group.

Heliconid butterflies are only one group among many that are brilliantly colored, including one well-studied example from the temperate zone. The conspicuous orange monarch (*Danaus plexippus*), familiar to anyone even vaguely interested in lepidopterans, is another example. Lincoln Brower (1964, 1969) tested the palatability of monarchs to blue jays (*Cyanocitta cristata*) and learned that monarchs are, indeed, poisonous to the jays. Minutes after a jay ate a monarch, it vomited up the remains. From that point on, the bird would not touch a monarch. The bird easily remembered the bright color pattern of the butterfly.

Monarch caterpillars feed on milkweeds, many of which contain high concentrations of cardiac glycosides. Caterpillars sequester glycosides in their tissues, and after metamorphosis the adult butterfly contains glycosides, rendering it unpalatable. Monarchs have turned a neat evolutionary trick. They have adapted to the milkweed's defense compounds, cardiac glycosides, and applied them to their own protection. Not only do they obtain virtually exclusive use of milkweed (few other insects can eat it), but they also are protected by using milkweed's defense compounds. Heliconid species apparently do the same thing (Brower et al. 1963). Butterflies adorned by warning coloration are likely to store plant defense compounds (or to use them as precursors to synthesize their own unique defense compounds) and thus to be able to sicken their predators (Brower and Brower 1964).

Curiously, in the Neotropics, some monarchs are palatable to birds during at least part of the year. In Guanacaste, Costa Rica, DeVries (1987) observed that captive magpie jays (*Calocitta formosa*) ate monarchs taken during early wet season but not those taken from populations later in the year. DeVries points out that the three species of *Danaus* butterflies that occur together in Costa Rica do not appear to form part of a mimicry complex (see below) and that it remains an unanswered question as to whether or not adults of any of these species have the same distasteful properties that have been so well documented for North American monarch populations.

In Costa Rica, Woodruff Benson (1972) altered the coloration pattern of *Heliconius erato*, a common, brightly colored, unpalatable butterfly. He marked the butterflies to alter their wing patterns such that predatory birds would not have seen such patterning before. He also established a control group, identical to specimens normally occurring. He released equal numbers of normal, control butterflies and altered, uniquely patterned individuals. Significantly fewer of the altered individuals were recaptured, an obvious indication that fewer survived. Some altered butterflies that were recaptured showed damage from bird attacks. Benson was even able to identify one bird species, the rufous-tailed jacamar (*Galbula ruficauda*), by the shape of wounds it left in a butterfly's wing. Benson concluded that wing pattern does indeed serve as protection, once predators learn the pattern and associate it with unpalatability.

There is an obvious cost associated with warning coloration. At each generation some individuals are sacrificed in order to educate the predators. Why should a butterfly give up its life for the good of its species? The answer is that it probably doesn't but, rather, gives up its life only for the good of its own genes. Heliconid butterflies have a limited home range, and a single female can live for up to six months, a virtual Methuselan age for a butterfly. Because the basically sedentary females lay many eggs over the course of their long lives, the local population in any given area is likely to consist of close relatives: brothers, sisters, cousins, second cousins, etc. If a single individual is lost to a predator, but the predator subsequently avoids other members of the group, the other group members each benefit. And, if most of the members of the group have genes in common with the sacrificed individual, then, in an evolutionary sense, the individual has indirectly promoted its own genetic fitness. It has acted to protect *copies* of its own genes. This process, called *kin selection*, may help explain how warning coloration evolved in heliconids. Thus far, it has not been rigorously tested and so must remain speculative. Interestingly, heliconid butterflies are among the only butterflies to have communal roosts, aggregations that may be "extended families."

Mimicry Systems   *Figure 94*

Although heliconids and many other tropical butterflies are striking in appearance, many are often difficult to identify because different species look remarkably alike: they mimic each other. When you try to identify a tropical butterfly, look very closely because it may not be what it appears to be. It may be another species entirely, a mimic of what you think it is.

Batesian Mimicry

Henry Walter Bates (1862), one of the first naturalists to explore Amazonia, was amazed to discover that some unrelated species of butterflies look alike. He suggested that a palatable species may gain protection from predators if it closely resembles a noxious unpalatable species, a phenomenon now termed *Batesian mimicry*. Bates was correct. The unpalatable species, termed the model, is essentially parasitized by the palatable species, termed the mimic. Because it closely resembles an unpalatable species, the mimic enjoys the umbrella of protection provided by the presence of the model. For the model, the presence of the palatable mimics makes the education of predators more difficult. Suppose a predator encounters one or even two palatable mimics *as its first experience*. It may be subsequently more difficult for the predator to believe that the noxious model is always noxious. Batesian mimicry is therefore most effective when the mimic is not very abundant. If, for instance, it were as abundant as its model, the entire system would be relatively unprotective, because predators would encounter palatable mimics as readily as unpalatable models.

A classic example of Batesian mimicry is the North American viceroy butterfly, *Basilarchia archippus*. The viceroy, a member of the family Nymphalidae, bears a strong resemblance to the monarch, a Danaid. Both are orange with black wing veins. As I discussed above, monarchs, by virtue of a diet tainted by cardiac glycosides from milkweeds, are unpalatable. Viceroys feed mostly on plants of the willow family and are quite palatable. Any lepidopterist will tell you that it's much harder to find a viceroy than a monarch. The mimicry works but only if the mimic's population is small in any given area. (One interesting twist on mimicry is that some monarchs "mimic" themselves! There are milk-weeds that contain little or no cardiac glycosides and are fed upon by monarch caterpillars. Butterflies from these caterpillars are palatable but still gain pro-tection by occurring with monarchs that are unpalatable, a phenomenon Lin-coln Brower (1969) terms *automimicry.*)

Butterflies exhibit many examples of Batesian mimicry in the Neotropics (Gilbert 1983). However, they are not the only Batesian mimics. Paul Opler (1981) unraveled a Batesian mimicry system in which a single innocuous insect species, *Climaciella brunnea,* a predatory mantispid in the order Neuroptera (to which belong the lacewings, alderflies, dobsonflies, and others), has evolved an uncanny resemblance to five different wasp species. The stingless neurop-terans, like viceroys and other butterflies, mimic an array of models, and all gain protection by appearing to be what they are not.

Mullerian Mimicry

Although Batesian mimicry is well represented in the tropics, recall that most plant species there have some combination of defense compounds. Therefore, any caterpillar species will likely have to cope with defense com-pounds in adapting to a food source. Unpalatability of caterpillars should be relatively common in the tropics, because so many of the food plants encoun-tered by the larval insects have defense compounds that, if stored or metaboli-cally modified by the insect, could render the creature unpalatable. In 1879, Fritz Muller suggested that two or more unpalatable species could benefit by close resemblance. If two unpalatable species look alike, the would-be preda-tor needs to be educated only once, not twice. The closer the resemblance, the greater the advantage to both species. This concept of convergent patterns among unpalatable species is called *Mullerian mimicry.* Unlike Batesian mim-icry, Mullerian mimicry is basically mutualistic, because individuals of both species benefit from the mimicry.

Both *Heliconius erato* and *H. melpomene* are unpalatable, both are brilliantly colored, and they look remarkably alike (DeVries 1987). What is even more remarkable is that there are eleven distinct races of *H. melpomene* in the Amer-ican tropics ranging from Mexico to southern Brazil. These races do not look the same. John R. G. Turner (1971, 1975, 1981) learned that for every local race of *H. melpomene,* there is a virtually identical local race of *H. erato.* Both species have converged in *racial variation* throughout their ranges! Only one race of *H. erato,* which is restricted to a small range in northern South America, lacks a *H. melpomene* counterpart. These two species provide a clear example of Mullerian mimicry.

Evolution of Mimicry

It may seem odd that two different species, with presumably different genet-ics, could somehow evolve to look essentially identical. But such an evolution-ary concordance is not really unusual. There are many well-known examples of what evolutionists call convergence, or parallel evolution, where genetically distinct organisms evolve close resemblance under similar selection pressures. Many Australian marsupials, for instance, bear close anatomical resemblance to certain North American placental mammals, though they remain genetically quite distant. It should not be surprising, therefore, that various species can become mimics of others or of each other. It has been suggested that the dramatic racial variation apparent in *Heliconius melpomene* and *H. erato* is an example of rapid evolution, a form of punctuated equilibrium (Gould and Eldredge 1977), but one in which dramatic racial variation over a species' range is not accompanied by subsequent speciation (Turner 1988).

Recent research into mimicry has shown that some mimicry patterns appear to be controlled not by hundreds or dozens of genes but rather by one or a few genes, each with large effects on the organism's phenotype. For example, in *Heliconius melpomene* and *H. erato* there are about thirty-nine genes that appear to affect color pattern, but the racial differences among the two species, as well as the racial concordances, appear to be influenced by only four or five genes (Nijhout 1994). Additional research focused on mitochondrial DNA (mtDNA) has produced the surprising result that *Heliconius melpomene* and *H. erato* have not shared a recent common evolutionary history. In other words, these species are not very close evolutionary relatives, a real surprise considering their overall phenotypic similarity. The diverse races of both species appear to have evolved within the past 200,000 years, and, again a surprise, convergent phenotypes have evolved independently both within and between species (Brower 1996). However, given the possibility that color pattern could be determined by only a few regulatory genes, perhaps the wider evolutionary separation between the two heliconids, even in light of their uncanny similarity of wing patterns across their ranges, should not be so surprising.

Mimicry Complexes

Both Batesian and Mullerian mimicry have led to the evolution of extensive convergences among large groups of butterflies in the tropics. Dozens of distinct species converge in appearance, making butterfly identification a taxonomist's nightmare. Christine Papageorgis (1975), working in Peru, identified five distinct color complexes of butterflies, each of which contains many mimicking species. She also found that each *mimicry complex* tended to occupy a distinctly different height in the rainforest.

From the forest floor to 2 meters (6.5 ft) is the *transparent complex*, butterflies with transparent wings with black wing veins. From 2 to 7 meters (23 ft) is the *tiger complex*, consisting of yellow, brown, black, and orange striped butterflies, most of which are Mullerian mimics. From 7 to 15 meters (49 ft), the *red complex* dominates. These butterflies include both *Heliconius erato* and *H. melpomene*. From 15 to 30 meters (98 ft) is the *blue complex*, all of which are heliconids with irridescent blue on the hindwing and yellow bands on the forewing. Finally, from 30 meters to the canopy and above is the *orange complex*, another group of heliconids with bright orange wings with black wing veins.

Papageorgis wondered why these five complexes were vertically stratified in the rainforest. After some experimental work, she concluded that each complex may be relatively cryptically colored in flight at the height in the rainforest where it normally flies. She related the coloration pattern of the complex to the pattern of light penetration in the rainforest and suggested that each color complex is most difficult for predators to see *in flight* at the height normally occupied by the complex. Papageorgis argued that brightly marked butterflies do not necessarily attract predator attention but, "on the contrary, can be cryptic in flight, thus giving a butterfly pursued by an un-informed or forgetful predator a second chance of escaping into the foliage." If Papageorgis is correct, tropical butterflies may employ both cryptic and warning coloration within the same species, depending upon where exactly they are flying, or if they are perched.

Ethnobotany/Ethnozoology: Multiple Uses for Multiple Species

Native peoples in the tropics have many generations of experience in deal-ing with defense compounds. For example, the bitter-tasting strains of manioc (page 184), whose meter-long, thick roots provide a basic source of carbohy-drate, are protected from herbivores by cyanogenic glycosides. It is only in the poorest, least fertile soils that one finds bitter, high-cyanide manioc. Just as with most plants on infertile soils, manioc apparently requires powerful de-fense compounds because the cost of replacing herbivore-damaged tissue is too great relative to the potential for productivity (page 144). Because of its cyanide protection, manioc is easy to grow, one of the few crops that people can successfully farm on poor soils. But what to do about the cyanide? After the root is harvested, it must be first grated and then thoroughly washed, soaking throughout the night so that the toxic compounds become soluble in water. The doughlike mash is then placed into an elongate, flexible cylinder, nor-mally woven from palm fronds, that constricts when pulled and is used to wring out dough. The cylinder is hung from a tree branch and a horizontal pole is attached to the base. Usually two women sit on the pole, one on either side of the cylinder, and their combined weight elongates, squeezes, and com-presses the cylinder, which forces out the cyanide-containing liquid. The mash, now essentially free of cyanide contamination, is wrung out and baked on a flat stone (Schultes 1992). In western Amazonia, the cylinder is called a tipi-tipi. In Belize, it is a whola, the local name for the boa constrictor.

Given the abundance and diversity of plants and animals in the Neotropics, coupled with the long history of human occupation of the region, it is not surprising that the indigenous people have found multiple uses for the diverse array of chemicals contained within the many species of native flora and fauna. To name but the broadest categories, chemicals have been extracted for use in arrow (dart) poisons, hallucinogens, fish poisons, drugs for medical and re-lated use, stimulants and spices, essential oils, and pigments (Gottlieb 1985).

The science of ethnobotany, which has attracted a great deal of interest in recent years (Cox and Balick 1994; Joyce 1992; Plotkin and Famolare 1992), is, as the name implies, the study of how indigenous people have learned how to use ambient vegetation for a diversity of pragmatic purposes. It is broader in scope than merely the extraction and subsequent usage of chemicals contained within the plants. Ethnobotany also includes a consideration of all uses of plants, including for food and fiber. It is an interdisciplinary field involving botany, anthropology, archaeology, plant chemistry, pharmacology, history, and geography (Schultes 1992). In this chapter, however, I will discuss only some of the pharmacological, narcotic, and hallucinogenic uses of drugs extracted from various Neotropical plants. In the following chapter, I will provide additional examples of other applications of ethnobotanical knowledge. For a general introduction to ethnobotany, see Balick and Cox (1996). The most distinguished student of Neotropical ethnobotany, indeed, the man who pioneered the field, is Richard Evans Schultes, whose remarkable career has spanned nearly five decades and has taken him throughout tropical America. For an encyclopedic and insightful treatment of Amazonian ethnobotanical information, see Schultes and Raffauf (1990). For a fascinating popular account of hallucinogenic drugs, see Schultes and Hofmann (1992).

Ethnobotany is not confined to the tropics. A perusal of old herbal manuals will quickly reveal that numerous North American plant species were relied upon for pharmaceutical applications in years past, until the advent of modern medicines. Some still are used. As one example, resin from the mayapple (*Podophyllum peltatum*), a common understory spring wildflower throughout eastern forests, was commonly employed by Native Americans to remove warts. It is still used today to treat venereal warts (Schultes 1992). In the Neotropics, many make the assumption that ethnobotany applies only to isolated indigenous tribes such as the Yanomamo, but this is false. Modern populations of mixed-heritage humans such as the *mestizos* and *riberenos* of Peru or the *caboclos* of Brazil, all of which are linked to native Amerindian cultures, but also under strong modern influence, make heavy use of ethnobotanical knowledge (Phillips et al. 1994).

Ethnobotanical insight is gained culturally through the generations, essentially by trial and error. Not all indigenous groups possess equally sophisticated ethnobotanical understanding. With regard to extraction and preparation of various drugs and drug combinations, the knowledge is often housed in the mind of but one revered individual, the shaman or medicine man. Information is never written down but is instead passed on from one generation to the next by the shaman, who is both a teacher and a practitioner, a person of substantial power in the community. Illnesses are rarely, if ever, blamed on organic causes but are usually assumed to have been caused by evil spirits or curses (Schultes 1992). It is the shaman who communicates with the spirit world—and who cures headaches, back pain, bug bites, and constipation. Unfortunately, a shaman may die of old age before passing his knowledge to the next generation. There is widespread fear among ethnobotanists that much knowledge is currently being lost as traditional tribes experience the impact of modern culture, and fewer young people study to be shamans. Loss of such slowly accumulated, essential knowledge would be a pity.

Ethnobotanist Mark Plotkin (1993), who studied with shamans in northeastern South America, describes many intriguing examples demonstrating the sophisticated knowledge of local people in the use of tropical defense compounds. Alkaloid-containing sap from a common liana is used to help cure fever in children. Rotenone, a potent vasoconstrictor, is extracted from another common liana and employed to kill fish, a critical protein source. Plants that even a skilled botanist has difficulty telling apart are easily recognized as separate species by the shaman. Equally intriguing is Plotkin's vivid descriptions of how he was tutored in this knowledge (including the use of hallucinogens) by various shamans whose trust and respect he patiently won.

I met Piwualli, a shaman, whose circuit included the villages along a section of the Napo River in Ecuador. Said to be seventy-three but looking considerably younger, Piwualli ushered my group to a dilapidated wooden table outside his house. The table was piled high with clumps of dried herbs, part of the bush doctor's pharmacopoeia. Piwualli spoke no English and only a little Spanish, but our guide knew Piwualli's language and acted as translator. Piwualli described the multiple uses of the various plants. One sounded remarkably like it was used for symptoms characteristic of certain severe nervous disorders such as Parkinson's disease. Interestingly, Schultes and Raffauf (1990) list three species of unrelated plants used by indigenous people in this region to treat "palsy-like trembling."

The suggestion that many serious ailments may be helped by potent compounds from the tropics is both intriguing and promising. For many years the alkaloid quinine, from the Neotropical shrub/small tree genus *Cinchona*, has been reasonably effective in combating certain malarias. Resin extracted from plants of the genus *Virola*, used as powerful hallucinogens (see below), may also prove to be very effective in controlling or even curing chronic fungal infections, which currently can only be suppressed by modern medicines (Schultes 1992). Plotkin (1993) notes that only about 5,000 of the world's 250,000 species of plants have been thoroughly investigated as to pharmacological properties and that the 120 plant-based prescription drugs are derived from only 95 species. Gottlieb (1985) points out that as of 1977, only 470 of the estimated 50,000 Brazilian flowering plant species, or about 1%, had been examined for the existence of chemical compounds. Cox and Balick (1994) assert that "less than half of 1 percent [of flowering plant species] have been studied exhaustively for their chemical composition and medicinal value." There is obviously so much to learn.

Surveys are being conducted in an attempt to evaluate the pharmacological potential of Neotropical plant species. Schultes (1992) describes how scientists aboard the research vessel *Alpha Helix*, normally an oceanography ship, sailed the Amazon for a year and collected 960 plants representing 3,500 specimens, conducting biochemical analyses of these plants using the sophisticated shipboard laboratories. One innovative approach was created by Thomas Eisner of Cornell University. He was catalytic in convincing the Merck Pharmaceutical Company to enter into an agreement with the government of Costa Rica, such that Merck would provide one million dollars for inventory and conservation purposes in exchange for exclusive rights to survey the flora for compounds of potential medical use (Cox and Balick 1994). Eisner calls his approach "chemical prospecting," and, though the eventual outcome of such searching is unknown at present, it is undeniable that there is potential for finding medically useful drugs within tropical flora. Shaman Pharmaceuticals, which was created in 1988, is a pioneer company attempting to discover and apply ethnobotanical knowledge to the needs of modern medicine (Joyce 1992).

Plotkin (1993) emphasizes the obligation to share any benefits, both monetary and otherwise, that may be derived from ethnobotanical studies with the indigenous people who, in fact, obtained the knowledge in the first place. Such a policy not only is morally compelling, it has strong conservation potential. For example, the Terra Nova Rain Forest Reserve in Belize was established in 1993 by a group named the Belize Association of Traditional Healers, an assemblage that includes people from most of the cultural and ethnic groups in Belize, a country in which about 75% of the people are estimated to be dependent on plant medicines for their primary health care needs (Balick et al. 1994). The reserve, a 2,400-ha (5,928-acre) area of lowland forest, will be managed to accomplish the following: cultivation and documentation of medicinally useful plants and protection of the plants from overharvesting; ethnobotanical and ecological research; encouragement of ecotourism, with walks and seminars designed to teach about the uses of the plants.

Schultes and Raffauf (1990), in their book *The Healing Forest*, discuss approximately 1,500 species and variants of plants from 596 genera and 145 families, all of which have medicinal or toxic uses by indigenous peoples in northwest Amazonia. It is facinating to see the range of symptoms that are treated as well as the diversity of plants that are applied to certain common ailments or conditions. For example, there are 38 plants that can be used for diarrhea, 25 for headache, 18 for muscular aches and pains, and 38 for toothache. There are many plants that can be used for various insect bites (including 16 for ant bites), 36 for intestinal parasites, and 29 for snakebites. There are 26 plants listed for use as contraceptives. In addition, there are plants alleged to have use in treating such conditions as sinusitis, stiff neck, bleeding gums, stomach ulcers, cataracts, asthma, swollen breasts, epilepsy, testicular swelling, tumors, boils, blisters, mange, and baldness, a selection that is by no means comprehensive. Of course, you should bear in mind that the degree to which these diverse applications achieve success is debatable. I would, for instance, be very dubious that any shaman could cure someone envenomated by a bushmaster. I would be equally dubious about the success rate in curing various serious ailments, such as tuberculosis. Where available, many indigenous groups readily accept modern medicine (though it can be argued that part of the reason for such acceptance is that these peoples have been afflicted with various ailments from exposure to settlers). Still, the efficacy of ethnobotanical treatments for many afflictions seems undeniable, and, as mentioned above, there is still so much to learn.

Besides medicinal uses, many plants are used to extract various poisons for hunting, and many other plants are used for hallucinogenic or narcotic purposes. I will close this chapter with a brief look at some of the best known of these.

Curare

Charles Waterton, who first journeyed to Amazonia in 1812, was undoubtedly a wonderfully entertaining dinner guest. What stories he must have told. This aristocratic, eccentric explorer of the Amazon demonstrated uncommon skill at taxidermy as well as intrepid drive for exploration and discovery. And one of his discoveries was curare. Waterton (1825) describes a vine, called Wourali, that supplies the primary ingredient for arrow poison and the "gloomy and mysterious operation" in which the poison is extracted and prepared, only by certain skilled individuals. He details how a large ox, estimated to weigh nearly 1,000 pounds, died within twenty-five minutes of being shot in the thigh with three poisoned arrows. The poison, said Waterton, produced "death resembling sleep."

Curare has such a powerful effect of relaxing muscles that it induces paralysis. And that's the basic idea. Curare is added to the tips of arrows and darts and then used by skilled hunters to bring down various species of mammals and birds. If you look at the small darts that are the ammunition of blowguns, you will see immediately that these weapons would do little more than make a pinprick in their intended prey were it not for the presence of the poison. The arrow or dart doesn't bring the creature down—the curare does. Curare and its derivatives are well known by practitioners of modern medicine, as they are commonly employed during certain surgical procedures.

Curares are extracted from many different kinds of plants from an array of different families. Indeed, over seventy-five plant species are utilized for this purpose in the Colombian Amazon. Most curares are mixtures of several plant species (often prepared specifically for the kind of animal sought), with much variation not only from tribe to tribe but from one shaman to another (Schultes 1992). The art of preparing curare requires careful attention to detail. It is a dangerous substance. It is remarkable that so many different combinations of curare poisons have been discovered and utilized by the indigenous Amazonian peoples (Schultes 1992; Gottlieb 1985).

The plant group from which curare takes its name is the genus *Curarea*, formerly *Chondrodendron*. *Curareas* are lianas, beginning as rooted shrubs that eventually become climbers. *Curarea toxicofera* is a species that is widely used by many tribes. The curare is extracted from the bark and wood of the stem and is often mixed with other species, particularly those in the genus *Strychnos* (Schultes and Raffauf 1990).

Cocaine

Cocaine is, without question, an addictive and powerful narcotic drug and a major societal problem in parts of Western society. Chemically, it is a powerful alkaloid extracted principally from a small, unpretentious shrub, *Erythroxylum coca*, var. *ipadu.*, commonly called coca (Balick 1985). A second species, *E. novogranatense*, is cultivated along the eastern slopes of the Andes and does not occur in lowland areas. Coca contains numerous alkaloids, but cocaine is the one in greatest concentration. Though cocaine is considered a scourge of society in North American culture, it has important traditional uses by South American indigenous peoples: as medicine, in certain rituals, for chewing, and for nutrition (Balick 1985). Studies cited by Balick show that ingestion of 100 grams of coca leaves is sufficient to supply one's daily needs for calcium, iron, phosphorus, and vitamins A, B2, and E. Chewing wads of coca leaves is also important in suppression of fatigue, providing added endurance for people in the rarified air of the high Andes. It should be emphasized that a leaf contains only about 1% cocaine, and even its effects are modified by other compounds in the leaf (Boucher 1991), so chewing coca leaves is not the same as snorting crack cocaine (which affects the brain in as little as five seconds). Coca leaves are also applied to wounds or boiled to make a tea. I can assert from personal experience that coca tea helps to attenuate the unpleasant effects of high-altitude sickness common to visitors in the Andes.

Most coca that is grown to be used as a narcotic is from Peru and Bolivia (though it is purified and shipped from Colombia, which produces about 80% of the world's cocaine), especially the Upper Huallaga Valley in Peru (along the east slopes of the Andes), where it is estimated that 60% of the world's coca is grown (Boucher 1991). Unfortunately, it is very lucrative to farm coca for cocaine. For example, a hectare of coca in Bolivia can yield $6,400, compared with $1,500 for coffee, $600 for bananas, and $300 for corn (Boucher 1991). Such profitability, coupled with the reality that coca has many traditional uses, suggests that the eradication of the cocaine trade is at best highly problematic.

One historic note of interest is that the soft drink Coca-Cola was at one time really COCA-Cola. In 1903, based on the recommendations from a report by the U.S. Commission on the Acquisition of the Drug Habit, the producers of Coca-Cola eliminated the minute amount of cocaine that, up to that time, had been included in the recipe (Moeser, in Boucher 1991). The report asserted that cocaine was being used mostly by "bohemians, gamblers, prostitutes, burglers, racketeers, and pimps."

Intoxicants and Hallucinogens

Perhaps the best-known hallucinogen in the Neotropics is the genus *Virola* in the family Myristicaceae (nutmegs). There are 62–65 species of these understory trees throughout the Neotropics, and a few of them are widely used throughout western Amazonia and much of the Orinoco Basin to achieve rapid and extreme intoxication and subsequent hallucinations. Such a practice serves multiple functions, ranging from spiritual divination to ritualistic diagnosis and treatment of disease (Schultes and Hoffmann 1992). In many tribes only the shaman takes epena, ebena, or nyakwana, as the *Virola* preparation is known, while in others, such as the Yanomama, all male members of the group participate. The drug itself is obtained from cambial exudate on the inner bark of the tree, which is boiled, simmered, and refined into a reddish powder. In most cases the drug is taken as a powdered snuff, blown with great force into the nostrils and sinuses, using an elongate pipe made from a plant stem. In some cases, however, the drug is administered orally, in the form of a pellet. Once administered, the drug, which is a combination of strong indole alkaloids (Gottlieb 1985; Schultes and Hoffmann 1992), causes immediate tearing and mucous discharge followed soon by a restless sleep during which the person is subject to extreme visual hallucinations described as "nightmarish." Details of this experience can be found in Schultes and Hoffmann (1992), Schultes and Raffauf (1990), and Plotkin (1993). Besides use as a hallucinogen, *Virola* is used for an array of medical problems (Schultes and Raffauf 1990).

Aztecs and Mayans of Central America routinely used mushrooms and various "psychedelic fungi" in their religious rituals. One mushroom in particular was said to provide its users with "visions of Hell" (Furst and Coe 1977). Peyote, made from the cactus (*Lophophora* spp.), is a widely used alkaloid hallucinogen throughout Middle America.

Balche was an intoxicant that Mayans used in many of their religious rituals. It was made from fermented honey plus a powerful bark extract from a tree, *Lonchocarpus longistylus*. When ingested by mouth, balche, as well as other potent drugs, often caused the user to become violently ill with nausea. Furst and Coe (1977) have suggested that Amerindians may have routinely taken these drugs by rectal injection, a practice they refer to as ritual enemas. Illustrations on Mayan vases depict administration of enemas, even including the Mayan gods. It is unlikely that such an event would have been singled out by an artist if the enema were being taken to relieve constipation. Rather, the practice allowed people to avoid the inevitable nausea that would occur if the balche had to pass through the stomach.

It should be emphasized that the widespread usage of potent hallucinogens is probably a cultural result of the abundant availability of defense compounds, especially alkaloids. The effect has been that throughout the Neotropics, indigenous groups have developed cultures in which the spirit world and literal world are often hard to separate.

Long-billed gnatwren, typical arthropod feeder that likely avoids Heliconid butterflies and their mimics.

7

Living Off the Land in
the Tropics

THE NATIVE peoples of tropical America thrived in jungle and rainforest for several millennia before the Spaniards arrived. The origin of the native peoples of the Americas can be traced to presumed migrations from Asia across the Bering Strait as long ago as 20,000 years (some say as much as 42,000). It is estimated that people first crossed the isthmus of Panama to colonize South America about 15,000 years ago (Meggers 1988). Though recent geologically, such a time span is certainly ample for the development of human culture well attuned to the resources of the rainforest and surrounding ecosystems. However, human prehistory in Amazonia is still poorly known (Meggers 1985). Recent work has suggested that human prehistory in Amazonia may be much older, perhaps as old as 32–39,000 years (Goulding et al. 1996).

Until recently it was generally believed that South American Amerindian civilization probably originated in the Andes and spread eastward into Amazonia. That view has been challenged by artifacts that apparently predate Andean artifacts. Middens have been discovered in the Santarem region of Brazil, near the confluence of the Tapajos and Amazon rivers, containing pottery and other artifacts that date from about 8,000 to 7,000 years before the present (Roosevelt et al. 1991). The pottery is apparently at least 1,000 years earlier than that found in northern South America and 3,000 years earlier than Andean and Mesoamerican pottery. Archaeologists investigating this site suggest that by 2,000 years ago, a large and agriculturally sophisticated population could have been supported on the rich alluvial soils deposited by the annual flood cycle along the *varzea*. There is also evidence for the existence of complex societies from about A.D. 500 to A.D. 1400 from Marajo Island, a large island at the mouth of the Amazon River (Bahn 1992; Gibbons 1990). On Marajo, mounds have been found to contain multiple house foundations, one upon another, plus an abundance of pottery, indicating a long human occupation. The artifacts found both at Santarem and Marajo are elaborate, consisting of burial urns and other sophisticated pottery, finely carved jade, and large statues of presumed chieftans. Prior to the recent work at Santarem and Marajo, the only truly urbanized civilization ever to occur in a Neotropical rainforest environment was that of the Classic Mayans. The evidence now suggests the possibility that comparable urbanization, population concentration, and widespread agriculture existed among Amazonian tribes. *Varzea* regions, where soil fertility is annually renewed during the flood cycle, seem to have supported large and permanent settlements from about A.D. 500 until the European conquest (Meggers 1988). Francesco de Orellana, European discoverer of the Amazon River (page 198), observed dense human populations along much of the river when he navigated it in 1542 (Goulding et al. 1996). Still, nothing comparable to the physical remains of Tikal or Machu Picchu has as yet been unearthed in Amazonia.

As a generality, the culture of any people is deeply interwoven with their pragmatic knowledge of the land from which they extract food and fiber. Many cultural anthropologists now approach their studies from the perspective of human ecology, with the focus on the ways culture develops in response to environomental opportunities and challenges. Aboriginal Americans became expert hunters and learned how to create and farm their own forest gaps without the soil permanently losing fertility, even in such areas as the Guianan Shield, where soils are extraordinarily poor in nutrients.

Successful exploitation of rainforest resources requires high levels of skill and local knowledge. For example, in northeastern South America, people plant between seventeen and forty-eight varieties of manioc (cassava or yuca; see page 184) together in the same plot (Dufour 1990). Such a practice acts to minimize damage by insects, which tend to specialize on one variety. Manioc is also planted close together to form a canopy of shade to keep the ground cooler, and some weeds are permitted to remain in crop rows to guard against rapid soil erosion (Plotkin 1993). Once the manioc root is harvested, the cyanide-yielding prussic acid is skillfully eliminated (page 162), rendering the root edible.

It is amazing to note the diversity of uses for various palm species. For example, throughout its extensive range, native peoples make diverse use of the moriche palm (*Moriche flexuosa*). Called the *koi* in Suriname and the *buriti* in Brazil, this species, which has been called the "tree of life" (Carneiro 1988), provides wood for canoes and houses, as well as thatch and material for weaving. Its fruit is used for oil, and its unopened flowers are used to make wine or flavoring for ice cream (Plotkin 1993; Goulding et al. 1996). It is also used for making bowstaves, spears, arrow shafts, and manioc strainers (Carneiro 1988). Indeed, fruit from moriche palm is reported to be the third most important fruit, after bananas and plantains, sold at the markets in Iquitos, Peru (Goulding et al. 1996). Some other palm species are equally as important as moriche (Balick 1985, and see below).

Some aboriginal groups have learned to extract potent poisons ranging from batrachotoxins in frog skin (page 81) to curare from various plants (page 165). The spiritual world is extremely important in Amerindian culture, hardly surprising when one considers how many hallucinogenic drugs (chapter 6) can be extracted from tropical plants, ranging from mushrooms to legumes (Schultes and Hoffmann 1992). Consequently, one of the most important members of Indian society is the shaman, the person who holds the knowledge about the varied uses of local plants and animals, and who, it is believed, is able to communicate with the spirit world. Amerindian hunting skills have been widely documented, from the stealth and speed with which they move through the forest, to their often uncanny accuracy with a bow and arrow or blowgun.

Different peoples use the land in different ways. Some tribes are apparently much better at certain skills (such as plant identification and pharmacological use) than other tribes. Knowledge differs from region to region, tribe to tribe. People living on *varzea* have different cultural adaptations for using the land from those living on *terra firme* (Meggars 1985, and see below).

Hunting and Gathering

Hunting and gathering in its most pristine form includes no element of farming. Hunter-gatherers take naturally occurring resources with only small temporary alteration of the ecosystem. They do essentially nothing to redirect the energy flow through the ecosystem to themselves by the simplification of food chains or elimination of competing species. Their collective impact on the ecosystem is therefore small. Because they attain such a tiny percentage of the forest's net primary productivity (page 44), hunter-gatherer populations must be low density because there is very little actual energy directly available to the people. Populations remain small. Most trees, epiphytes, vines, and animals are not used for food or fiber. Forest is not cut or is very minimally cut. The people forage and hunt, taking only what they find and catch. The average size of a hunter-gatherer population is only about 1 per 2.6 km^2, though this figure varies somewhat from one region to another.

Because of the variety of hunter-gatherer tribes throughout the Neotropics as well as differences in habitat from one region to another, cultures vary. It is not possible to describe one culture as typifying all. Nonetheless, within certain wide limits, general patterns of culture emerge. Hunter-gatherers are usually relatively nomadic, living in a small temporary village or encampment for some time and eventually moving on when they have exhausted the game or essential plants from a given locality (though some villages are fairly permanent and people walk long distances to hunt and gather). Hunting is accomplished by careful, quiet stalking, using a blowgun, bow and arrow, or spear to bring down essential protein: large birds, monkeys, sloths, agoutis, pacas, tapirs, and others. Often, but not always, arrows or darts are tipped with poison. Rifles and shotguns are becoming more frequent. Protein is also supplied by certain arthropods, especially large grubs, and, where tribes are living along rivers, by hunting (also sometimes using poison) for fish, turtles, capybara, and crocodilians.

Some Neotropical hunter-gatherer tribes are highly territorial and are known, at least in the past, for high levels of aggression (shrunken human heads are part of the cultural artifacts of some Amazonian aboriginal groups). In Brazil, Mundurucu headhunters in the recent past made no distinction between people from different tribes and animals such as peccaries and tapirs—all were hunted. Tribal warfare was probably a response to the need to protect areas of forest for the exclusive use of a single tribe, thus increasing the forest's yield (Wilson 1978). Tribal raids were also done to procure women, ensuring genetic outbreeding through aggression, a custom inculcated within the culture. Given those explanations, it should not be lost on the reader that aggression in extreme forms has been a major component of many Amazonian cultures, including groups such as the Mundurucu and Yanomamo.

There are few "pristine" hunter-gatherers in the Amazon Basin, and essentially none in Central America. Most people now use agriculture of some sort to supplement their diets, and most have some periodic direct contact with the "modern" world. Shotguns are rapidly replacing blowguns. When Europeans first began coming to South America around 1500, the estimated total population of aboriginal humans throughout Amazonia was about 6.8 million (Denevan 1976), though some estimates suggest a number half that size. Most people were settled along riverine areas where varzea-type floodplains assured annual renewal of soil fertility. Interior *terra firme* forests and savanna areas apparently were far less populated. By the early 1970s, the indigenous population was only a half million, and in Brazil alone the number dropped precipitously, from about a million to about 200,000 during the twentieth century (Dufour 1990). Today the estimate for all of Amazonia is only about 250,000, a twenty-four-fold decrease from an estimated six million (Carneiro 1988). Amerindian populations were reduced by a lethal combination of outright conquest and genocide, slavery, and, probably most significant, the introduction of European diseases to which the Amerindians had little natural resistance. In Amazonia, tribes that inhabited varzea, who represented the largest Amerindian populations, fared worst, being essentially decimated by Europeans. Only those tribes such as the Yanomamo (see below), the Javari, and the Xingu, which inhabitated remote and inaccessible forest, survived the conquest period, and even they suffered reductions in population whenever there was European contact.

Today, most Amerindian tribes live on anthropological reserves or "indigenous areas," called *resguardos*, which are lands in which aboriginal groups are permitted to follow their traditional lifestyles. In Brazil, Indian lands are administered by the government agency FUNAI (Fundcao Nacional do Indio), the National Indian Foundation. This agency governs a huge area representing 100.2 million ha (247 million acres) in 371 reservations in the Brazilian Amazon, or roughly 20% of Brazilian Amazonia (Peres 1994). In the northern state of Roraima, about 42% of the land area is reserved for use by Indians, even though Indians represent only about 15% of the population of Roraima (Brooke 1993).

Indigenous tribes face many issues. Amerindian populations continue to be forced into retreat in some areas as people with different cultural backgrounds, often from overpopulated, extremely poor urban areas, migrate to the new frontier of the rainforest, some to begin subsistence farming, some in search of gold. This trend has greatly accelerated in Amazonian Brazil due to the continually expanding Transmazonian highway system (see chapter 14). Equally distressing, in some areas aboriginal tribes (for example, the Nambikwara tribe of Mato Grosso, Brazil) are deliberately exploiting their own lands for short-term profit, granting permission to outsiders for logging and gold mining (Peres 1994).

The Yanomamo group of the state of Roraima in northern Amazonia, who inhabit the highland rainforest near the border between Venezuela and Brazil, represent an aboriginal population who combine hunting and gathering with simple agriculture. Approximately 15,000 Yanomamo remain, the largest of the remaining forest-dwelling Amazonian tribes (though they are much reduced from previous times), and they are scattered in groups throughout an overall territory of about 40 km^2 (15,400 sq mi). In the past, the intense level of aggression manifested among local villages earned the Yanomamo the description of being "fierce people" (Chagnon 1992), though it has been argued that their level of aggression is not significantly different from that of modern societies (Plotkin 1993). Yanomamo live in scattered villages, hunting local game, gathering foods and fiber from the forest, and clearing small plots for the cultivation of plantain. They continue to practice rituals such as use of powerful narcotics to interact with the spirit world (page 167) and the consumption of the ashes (mixed into a soup) of their dead. Today the Yanomamo are under threat from highway construction through their territory, bringing an influx of gold miners (Collins 1990; Brooke 1993).

Examples of Hunter-Gatherer Lifestyle

The hunting habits and caloric intake of the Ache in eastern Paraguay form the subject of a study that illustrates the adaptiveness of the hunter-gatherer lifestyle (Hill and Hurtado 1989). The Ache are a society of four independent groups, with each group typically made up of ten to fifteen small bands (of about forty-eight people each) that roam throughout a territory of about 18,500 km^2 (7,000 sq mi), most of which is semideciduous evergreen forest. On average, an Ache forager consumes 3,700 calories per day, which, when compared with 2,700 cal/day for an active adult North American, is certainly a large intake. Further, an average of 56% of the Ache calories comes from mammalian meat, with honey providing 18% and plants and insects together providing 26%. Of course these averages vary from day to day and with season. A typical Ache foraging trip involves men leading, carrying bows and arrows, and women and children following, carrying supplies in baskets. The group "bushwhacks" through forest without following existing trails. Eventually the sexes separate as the men move more rapidly in search of prey. A simple camp is made at day's end.

The Ache hunt approximately fifty species of vertebrates, including fish, a far smaller number than the total potentially available to them. They exploit only about forty species of edible fruits and insects, though, again, there are many more potentially available. Only seventeen different resources are estimated to account for an astonishing 98% of the total caloric intake, showing that the people are highly selective in their choices, ignoring most resources (Hill and Hurtado 1989). Like numerous other hunter-gather societies that have been studied, the Ache practice extensive food sharing, a habit strongly encouraged by cultural reinforcement from childhood through adulthood. Hunters readily share their kills to the point where a hunter rarely eats any of his actual kill. Meat and honey are shared the most, with less sharing of plant and insect food. This is hardly surprising and is clearly adaptive to the individual as well as well as to the society as a whole because meat and honey are highly scattered, rather unpredictable resources compared with insects and plants. A hunter cannot be successful every day but might be very successful on any given day. But with food sharing as the norm, a hunter who fails to make a kill does not go hungry, nor does his family. Food sharing greatly reduces the day-to-day caloric variability that would otherwise prevail from person to person if sharing did not occur (Hill and Hurtado 1989). Like food sharing among vampire bats (page 100), it is a clear example of reciprocal altruism.

The Ache selectivity with regard to prey choice is a probable example of what ecologists call "optimal foraging." In other words, does a hunter kill whatever he finds, regardless of the amount of energy required to stalk and make the kill (relative to that received), or does he ignore certain prey, preferring to invest more energy searching for prey with a higher payoff? Is it better to stalk and shoot a small monkey than to ignore it and keep searching for a peccary, which will provide far more meat? In the Ache study, return rates for a whole day of foraging (the actual work of finding and procuring food) were calculated to be 1,250 calories per hour for men and about 1,090 cal/h for women, but the average caloric value for food acquired was about 3,500 cal/h for men and 2,800 cal/h for women. Obviously, the selectivity results in substantial caloric benefit over cost (Hill and Hurtado 1989).

One prevalent belief is that hunter-gatherer peoples have relatively large amounts of downtime, when they are not hunting and gathering. They do not, after all, need to labor over clearing forest or plant and weeding crops. The Ache study showed that men spend about 6.7 h/day searching, making a kill (or finding honey), and processing food, plus another 0.6 h/day working on their hunting tools, rates of labor comparable to the average working day of a North American. Ache men spend about 4.5 h/day for socializing. For women, about 8 h/day is spent in light work or childcare, 1.9 h in subsistence activities, and 1.9 h moving camp. In total, men supply about 87% of the energy in the Ache diet and almost 100% of the protein and fat (Hill and Hurtado 1989).

Birth and death rates among the Ache are generally reflective of most hunter-gatherer peoples, though there is certainly variability among groups. About one in every five Ache children dies before reaching one year of age, and about two out of every five fail to survive to age fifteen, so the survivorship from birth to puberty is about 60%. Only about 32% of child mortality is due to illness; fully 31% is from homicide! Among Ache adults, most (73%) die from warfare or accidents (including snakebite and jaguar attacks, the occupational hazards of hunting). Illness accounts for only 17% of adult mortality. On average, Ache women give birth every thirty-eight months (Hill and Hirtado 1989).

It is often assumed that hunter-gatherer nomadism results from hunting efforts that eventually deplete local game populations. Without question, tribes are dependent on local game sources, and there is much documentation of wildlife depletion due to hunting (see below). But this need not always be the case. A study of the Siona-Secoya Indian community in terra firme rainforest of northeastern Ecuador documented 1,300 kills representing 48 species, including various mammals, large birds, and reptiles (Vickers 1991). The average number of kills per 100 man-hours of hunting was only 21.16 (with sample size of 802 man-days and 6,144 man-hours), and the mean number of kills per man-day of hunting was only 1.62. These figures do not suggest that hunting pressure was sufficient to deplete the animal populations and are consistent with other studies (Ruddle 1970; Redford and Robinson 1987). In the total hunting area of 1,150 km² (440 sq mi), only one bird species, a currassow (*Mitu salvini*), was depleted due to hunting, though in the area immediate to the village, two species, a bird (trumpeter, *Psophia crepitans*) and a monkey (woolly, *Lagothrix lagotricha*), were depleted. Other animals did not suffer population declines. The human population was low, approximately 0.2 person/km² (.12/sq mi), which is considered typical of aboriginal inhabitants of Amazonian lowland forest. As with the Ache, hunters were highly selective, concentrating on large game such as peccaries, tapir, and woolly monkeys. It has even been argued that human modification of rainforest (through the planting of selected fruiting species) can increase animal populations such as tapir, deer, and peccary (Posey 1982). In Vickers' study, factors such as depletion of cedar trees (*Cedrela odorata*) for canoes and cutting palms for thatch were more important than loss of game animals in forcing the settlement to move. In the ten years that the settlement remained in a given area, game was not significantly depleted by hunters (Vickers 1988).

In contrast, if one looks at the entire rural population of Amazonian Brazil, including colonists and aboriginals, subsistence hunting has much greater impact. Given a presumed population of nearly 3 million persons (living outside of cities and the vast majority of whom are not aboriginal people living in tribes) in an area of about 3.6 million km² (1.4 million sq mi), and considering the average per annum consumption rate, about 14 million mammals are killed each year, a very significant number. Adding birds and reptiles, the total number of animals killed by subsistence hunting jumps to about 19 million annually. Add to that figure the number of animals fatally wounded but not taken, and the estimate jumps to a staggering 57 million per year (Redford 1992). There are numerous examples of hunting pressure being responsible for the local depletion of animals (Robinson and Redford 1991). Hunting impact is a complex issue, as it is frequently done not only for subsistence but also for commercial profit.

Slash-and-Burn Agriculture   *Figures 63, 64*

Agriculture, which began about 7000 B.C., involves the simplification of nature's food webs and subsequent rechanneling of energy to humans. Instead of the sun's captured energy being dissipated among many herbivores and carnivores with just a meager amount going to humans, it is refocused on but a few crop species that humans have planted for their own use. Consequently, much more energy is available to humans, so the human population can be ten to a hundred times larger than is the case with a hunter-gatherer lifestyle. There is, however, a substantial cost to agriculture. To simplify food webs, people have to perform labor. It takes work to farm because not only must the habitat be radically changed, but the needs of the crop plants must be met as well. Farmers must work to protect their investments, diligently weeding to remove competing species, defending crops against the continuous onslaught of herbivorous insects and rodents, and somehow insuring that adequate nutrients from the soil remain available. Finally, activities such as planting and harvesting must be carefully planned with regard to seasonal cycles, and efforts must be made to grow sufficient food to carry the people through seasonal shortages. Work done by nature in hunter-gatherer societies essentially for free comes with a big price tag in agricultural societies. But of course the dividends are also much larger. Agriculture basically represents a disturbance, an anthropogenic gap, created in the ecosystem. The farmer works to prevent the normal successional processes from closing that gap so that crops can be grown for exclusive use by humans and/or their domesticated animals. Just as successional areas are temporally unstable, eventually succeeding to forest, so agricultural systems are intrinsically unstable systems. The farmer's labor provides the stability against nature's tendency to diversify the system. I once heard it said that if an interstellar intelligence from some distant point in the universe should perchance visit Earth, it might well conclude that the dominant species were corn, wheat, and rice, which had combined to enslave a group of odd bipedal creatures who tend to their every need. Such is agriculture.

There is now ample historical evidence suggesting that human populations have made extensive use of fire in Amazonia (Roosevelt 1989) and throughout Central America. For example, charcoal fragments contained in soil from lowland rainforest at La Selva, Costa Rica, date to as long ago as 2,430 years before present (Horn and Sanford 1992). Similarly, charcoal fragments suggest extensive burning activity by humans in the high-elevation paramo zone of Costa Rica (Horn and Sanford 1992; Horn 1993). While charcoal fragments could be due to natural fires, it is thought more likely that the fires were set by humans, probably to facilitate hunting, open understory, or prepare for agriculture. Indeed, fire is the key ingredient in making it possible to farm the Neotropics sustainably.

Tropical peoples face a challenge in attempting to farm nutrient-poor rainforest soils. Most of the minerals and nutrients are not in the soil but in the biomass: the trees, lianas, and epiphytes. The problem is that to clear an area for farming, it is obviously necessary to remove the mass of vegetation. But to do so seems to doom the farming effort because the often mineral-poor soil will not sustain very much in the way of crops. The way out of this dilemma is fire, applied in a practice what has come to be termed *swidden* or *slash-and-burn* agriculture (Beckerman 1987; Dufour 1990).

A small plot of land (usually 0.4–0.6 ha) is chosen and machetes and axes are used to cut down all of the vegetation. Trees too large to be cut are girdled, killing them. The tangled pile of vegetation is then set on fire rather than removed. Fire eliminates the leaves and wood while at the same time releasing the nutrients and minerals contained within. The soil surface is fertilized by the ash from the biomass, and the ash tends to be alkaline, raising the pH, making the soil less acidic. The farmer can then plant crops, and the soil will be fertile, but only for a few years. Rainfall will still act to erode the now exposed soil and leach minerals. The crops themselves are removed, of course, and with them go more of the minerals. The result is that fertility and yield decline steadily. Typically, staple crops include manioc, plantains, bananas, sweet potato, pineapple, chili peppers, and others (Dufour 1990). Plots are normally planted as polycultures rather than monocultures, a practice that helps with pest control and slows the rate of natural succession (see below).

Crop losses from predation by such creatures as agoutis are actually antici-
pated, and "extra" sweet manioc is often planted for rodent consumption. In
a detailed study of the Tukanoan Indians of Colombia, it was learned that most
intensive farming of the swidden plots occurs from the twelfth through the
twenty-fourth month after burning (Dufour 1990). At least two manioc crops
are harvested over this time, and manioc harvesting continues until the thrity-
sixth month after cutting. Individual households may have several swidden
plots of differing ages, including some older plots that are used for other products such as fruit trees, medicinal plants, or plants that produce fish poisons
(Dufour 1990).

In one Amazon region studied, the yield for a single village was 18 tons/
hectare the first year, 13 tons/ha the second year, and only 10 tons/ha the
third year, a reduction from the first year of 45% (Ayensu 1980). Within a few
years the plot will be abandoned, allowing natural succession to occur, closing
the gap. The typical time sequence for swidden agriculture is to farm the plot
for two to five years, sometimes only for one year, sometimes for as long as
seven years, and then abandon it for at least twenty years. Ideally (but rarely),
an area just abandoned will not be recut for nearly a hundred years or so,
permitting substantial recovery of the system. Slash-and-burn agriculture requires constant rotation of sites and often results in a nomadic population who
must move around in the rainforest to find suitable plots to farm. Because of
soil nutrient limitation and therefore the need to allow forest regeneration,
total human population remains generally low.

In some areas, a practice known as swidden fallowing significantly extends
the usefulness of the plot. A swidden fallow is somewhat akin to agroforestry in
that people plant longer-lived species such as peach palm, guava, coca, various
tuber crops, breadfruit, and copal, along with encouraging the growth of various shade trees useful as firewood, thatch, or medicinal plants. This practice
is also common around permanent campsites and trails, resulting in a significant alteration of the plant species richness of the local area, forming a complex mosaic of agricultural and agroforestry plots as well as natural but disturbed forest in various stages of succession. Given the extensive and varied
use to which indigenous people have put the land, it is probably hard to know
exactly what "natural forest" is throughout floodplain areas in Amazonia
(Dufour 1990).

An experimental study in Costa Rica demonstrated that slash and burn does
not, in the short run, degrade the soil (Ewel et al. 1981). Researchers cut,
mulched, and burned a site that contained patches of eight- to nine-year-old
forest and seventy-year-old forest. Before the burn there were approximately
8,000 seeds per square meter of soil, representing 67 species. After the burn
the figure dropped to 3,000 seeds/m^2, representing 37 species. Mycorrhizal
fungi survived the burn, and large quantities of nutrients were released to the
soil following burning. The remaining seeds sprouted, and vegetation regrew
vigorously on the site. Studies in Amazonia also indicate that ecosystem function is in no way permanently impaired by swidden practices (Uhl 1987). Studies on bird communities also support the idea that swidden agriculture mimics
natural disturbance processes. In an Amazonian study site, shifting cultivation
had strong impact only from the time the plot was cut until about ten years

later. Thereafter, the bird communities that inhabited the regenerating sites were essentially the same as those that occurred following natural disturbance (Andrade and Rubio-Torgler 1994).

Kekchi Mayan Life in Blue Creek Village, Belize   *Figure 62*

Not all slash-and-burn farming requires a nomadic lifestyle. For some peoples, especially where the soils are young and thus fertile, it is possible to establish a stable village and farm the surrounding forest, rotating plots every few years but not leaving the basic area. Blue Creek, in the Toledo District of southern Belize, is such a place. Sitting at the base of the limestone Maya Mountains near the Guatemalan border, the region receives about 4,600 mm (180 in) of precipitation annually, most of it from June through December. The 4,000 people who inhabit the district are Kekchi Mayan, and their ancestors lived in highland and lowland areas of Alta Verapaz Department in eastern Guatemala until the time of the Spanish conquest. The Kekchis migrated to Belize some years after the conquest and live there today under the protection of the Belizean government in scattered villages throughout the district. The people speak their own Kekchi language, plus English and Spanish. Blue Creek Village, typical of Kekchi villages, supports about 160 people and consists of several dozen wooden, dirt-floored houses with palm-thatched roofs. Small lawns are kept in front of each house, the grass cut by machete. In and around the village are planted bananas, plantain, breadfruits, cacao, pineapple, and various kinds of citrus trees. Pigs and chickens wander among dogs, cats, and numerous children. Blue Creek flows through the village, and all bathing and washing of clothes is done in the creek. At night, for amusement, men sometimes play the xylophone-like marimba, their music distinctly Hispanic-sounding, probably quite different from the musical sounds of the classic Mayans.

Men clear communal areas called *milpas*, a term for Mesoamerican slash-and-burn plots of a few acres. Usually the milpas will be a considerable distance, often miles, from the village, and thus the impact of farming is far from obvious at first. Mayans, one learns quickly, are good walkers. Crops include corn and beans, manioc, sweet potatoes, and yam beans, along with tomatoes, squash, and peppers. Crops are planted and periodically tended until they can be harvested.

A strong sexual division of labor exists in the Kekchi village. Women keep house, tend children, prepare food, and weave, although weaving is giving way to buying ready-made clothing. Rising well before dawn, women grind corn to prepare tortillas, bake cassava bread, and kill and dress chickens or other animal food. During the day women care for children and do such chores as laundry. The men build and repair the homes, tend the milpas, or go into the forest to hunt, using rifles and shotguns, not bow and arrows.

Blue Creek represents a late-twentieth-century version of slash-and-burn agriculture. It may appear that the Indians are quite traditional in their culture, but this view has been strongly challenged (Wilk 1991). Most important, the use of communal land and shifting milpa agriculture (which appears very traditional) actually developed relatively recently, after the Kekchi Mayans migrated from Guatemala to Belize, and thus does not represent a precolonial culture. Today the Kekchi Mayan culture continues to change rapidly. North American farmers, many of them Mennonites, have moved near the village and into the surrounding area, bringing with them the techniques and equipment of intensive farming. Tractors are replacing machetes. The Belizean government is attempting to initiate an individual land-tenure system to replace the communal system. In addition, subsistence farming is being supplemented by cash-crop production, and many of the men are now taking jobs outside of the village.

Central American Successional Crop System

Slash-and-burn agriculture has much in common with ecological succession in that it uses the successional process to restore the soil after use for farming. Polycultures, in particular, can be effectively planted to maximize the productivity of a given plot. Instead of only one crop, several surface crops, such as corn and beans, share the same plot with root crops (such as manioc or sweet potatoes) while the border of the plot may be planted in peppers and tomatoes. Polycultures are more resistent to insect attack because the crop biodiversity provides habitat for herbivore predators and reduces the competitive effects of undesirable weeds.

Comparisons of various monoculture crops with mixed-species plots have shown that the more diverse plots had significantly more root surface area, and thus increased species diversity may enhance an ecosystem's ability to capture nutrients (Ewel et al. 1982). Ewel and his coauthors conclude,

> The maintenance of root systems having high surface area of absorbing roots well distributed in the soil profile may be one of the most important features to strive for in designing agroecosystems appropriate for the humid tropics, where soil-nutrient storage is often low and leaching rates are high. Such root systems can be achieved by designing systems that are diverse and long-lived.

Robert Hart (1980), working in Costa Rica, has suggested that farming can be directly analogous to succession. He presents a scheme, which appears to be a modification of the swidden fallow system described above, whereby crops are rotated into and out of plots on the basis of their successional characteristics. Using such a system, Hart claims that it would be possible to utilize a plot of land continuously and productively for at least fifty years or more. To quote Hart:

> Early successional dominance of grasses and legumes can be assumed to be analogous to maize (*Zea mays*) and common bean (*Phaseolus vulgaris*) mixtures. Euphorbiaceae, an important family in pioneer stages of early succession, can be represented by cassava (*Manihot esculenta*), a root crop in the same family. In a similar replacement, banana (*Musa sapientum*) can be substituted for *Heliconia* spp. The Palmae family can be represented by coconut (*Cocos nucifera*). Cacao (*Theobroma cacao*) is a shade-demanding crop that can be combined with rubber (*Hevea brasiliensis*) and valuable lumber crops

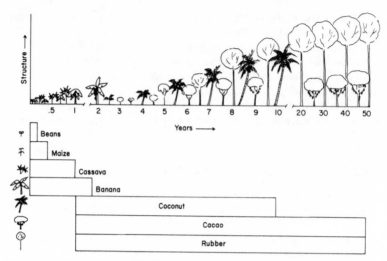

Sequence of crops that would mimic successional patterns. See text for explanation. From R. D. Hart (1980). Reproduced with permission.

such as *Cordia* spp., *Swietenia* spp., or other economically valuable members of the Meliaceae family to form a mixed perennial climax.

One example of the kind of system described above is found in Selva Lacandona in Chiapas, Mexico, the home of 450 remaining Lacandon Maya Indians (Nations 1988). These people combine agriculture with forestry, growing up to seventy-nine varieties of food and fiber crops in one-hectare plots in a manner that closely mimics succession. Trees are felled and burned, and a combination of fast-growing tree and root crops such as plantains, sweet potatoes, bananas, papaya, and manioc are immediately planted. Once these crops become established, others are added throughout the year, depending upon appropriateness of the season: corn, chilies, limes, coriander, squash, tomatoes, mint, rice, beans, sugar cane, cacao, and others. Certain natural species, such as sapodilla, wild pineapple, and wild dogbane, which invade as part of normal succession, are also permitted to grow among the planted crops. These are also eventually used for various purposes. As the plot diversifies, animal species such as peccaries, brocket deer, and pacas are attracted, all of which are valuable game animals. The Lacandon system permits a single hectare to remain productive for five to seven consecutive years before rotation is necessary. Eventually, trees such as rubber, cacao, citrus, and avocado are planted, and the yield from these species adds yet another five to fifteen years of productivity to the plot.

Amazonian Life on *Terra Firme* and *Varzea*   *Figures 162, 163, 164*

Because of the greater availability of fish and other aquatic game plus the renewal of soil fertility by the annual flooding cycle, Amazonian varzea populations have been estimated to be about seventy times greater than those of indigenous people on terra firme ($37.8/km^2$ compared with $0.5/km^2$) (Denevan 1976). Meggers (1985) has compared cultural adaptations on terra firme and varzea, and there are some striking differences.

Peoples on terra firme rely heavily on sophisticated slash-and-burn shifting cultivation, with manioc being the principal crop. Crop yield drops annually, and thus plots are typically abandoned within about three years. For productivity to continue, new plots must be put into cultivation each year, and at least twenty years must pass after plot abandonment in order that soil fertility be restored. Manioc is an ideal crop, as it grows well in poor soil, is easily harvested as needed, and is a fine source of continuous carbohydrate. Protein is obtained from animals. In addition, many kinds of naturally occuring plant materials (nuts, fruits, seeds, etc.) are consumed. Villages must move after about five to seven years due to depletion of local resources. Interestingly, among some aboriginal cultures the motivation for abandoning a village is based not on a direct measurement of resource availability but rather on the cultural belief that the death of an adult in the community is a sign that the people must move, a belief that assures nomadism. Sorcery and warfare are common, which means that as populations increase, violence becomes more common, and populations are then reduced.

Varzea populations enjoy a continuous and abundant source of protein in the form of fish, caiman, birds, manatees, and turtles, as well as mammals such as capybara (though such creatures have often been seriously depleted by too much hunting pressure). Wild rice is harvested and crops (beans, peppers, cocoa, bananas) are planted to make maximal use of the annual flooding cycle. The variety of manioc that is planted matures in six rather than twelve months. Maize is grown as well. The period of most stress is the three months during full flood, when food from the previous harvest must be consumed, as none can be grown at that time. Because of year-to-year variability in the flooding cycle, crop yields can vary dramatically. Cultures reflect this reality: practices such as female infanticide are common and, of course, act to lower the population.

Nonindigenous Farmers in Amazonia

Should you travel anywhere along the Orinoco, Amazon, or the major river tributaries, you will notice immediately that riversides are abundantly inhabited by humans, particularly in varzea areas. When Europeans colonized Amazonia, they bred with Amerindians and their descendants became the people who today make their living by farming and fishing the floodplains. The riverine peasantry are called *caboclos, riberenos, mestizos,* or *campesinos,* depending upon region. In many ways, these people work the floodplain as described above for traditional varzea populations, with the exception that they make much more use of market economy rather than rely entirely on subsistence. They cash-crop rice, for instance, and sell fish at market. Indeed, the largest and most diverse fish market in Amazonia is at Manaus, Brazil, where between 30,000 and 50,000 tons of fish are marketed annually (Goulding et al. 1996). Riverine people also harvest such things as Brazil nuts, palm fruits, and rubber for commercial sale (Padoch 1988; Dufour 1990). Note that such usage is not necessarily environmentally damaging, unless overharvesting occurs (and it often does, with such creatures as tapir, manatee, turtles, and capybara, though not so much with plants).

Mestizo Life along the Tambopata River in Southeastern Peru

A detailed study by Phillips et al. (1994) demonstrates the importance of diverse floodplain habitat to the local mestizo population along the Tampopata River in southeastern Peru. Mestizos are relatively recent colonists who live along riverine floodplain and who are of diverse heritage (see above). Interviews conducted with men, women, and children ranging in age from five to sixty-seven indicated that the vast majority of plant species occurring in the region were of at least some use to the people (though some species much more so than others). This, of course, demonstrates that the people have a strong pragmatic knowledge of the ecology of the region as well as a dependency on it.

Mature forest tended to have the broadest range of usage because it supplied both construction materials and food, but lower floodplain forest was most useful for medicinal purposes, and swamp forest was important for commercial harvesting. The people used an estimated fifty-seven woody plant families and 87.2% of the tree and liana species found on the inventory plots. Remarkably, in one 6.1-ha (15-acre) plot, 94% of the woody stems were regarded as useful to the mestizos. Not surprisingly, palms were of particular importance. Studies from other riverine areas cited by the authors indicate a generally similar broad usage of the biodiversity of the floodplain (see also page 195).

Prehistoric Intensive Agriculture

In Amazonia, prehistoric intensive agriculture was undoubtedly restricted to varzea areas where soil fertility could be maintained. Otherwise, only the Incas, who practiced terraced farming on the Andean slopes, could be described as having intensive agriculture. In Central America, there is much archaeological evidence suggesting that Mayans supported their vast population through techniques of intensive agriculture and silviculture (Hammond 1982). Such practices would result in large-scale landscape alterations, with much forest cutting.

Intensive agriculture in Central America was accomplished largely by two methods, hill terracing and raised fields in swamps and marshland. Hill terracing involves the construction of stone walls along hillsides, the walls acting to retard erosion and trap soil washed by rains. (This method was also widely practiced by Incas of the Andes.) Hill terracing permitted Mayans to cultivate a given plot for much longer than ordinary slash-and-burn techniques because the soil fertility was preserved. Raised fields involve the excavation of drainage canals to reduce water levels and thus raise dry fields from what was previously swampland. Ancient Mayans not only used the raised fields for agriculture but probably also used the canals for keeping fish and turtles, both important protein sources.

Past raised-field agriculture is revealed in part by patterns evident in aerial photographs of the landscape today. A large area in northern Belize (Pulltrouser Swamp) was under intensive wetland cultivation by Mayans between 200 B.C. and A.D. 850 (Turner and Harrison 1981, 1983). Mayan techniques succeeded in preserving soil fertility and may have permitted uninterrupted farming throughout the year. Using both aerial photography and remote-sensing, side-looking airborne radar imagery, a pattern was exposed showing that vast areas of northern Guatemala and Belize contain canals that were probably constructed by early Mayans and used in connection with agriculture (Siemans 1982). Modern attempts to model early Mesoamerican intensive agriculture using raised fields and canal systems (called *chinampas*) can, indeed, produce impressively high yields (Gomez-Pompa et al. 1982; Gomez-Pompa and Kaus 1990).

The Mayans of Tikal probably cultivated the *ramon* (or breadnut) tree (*Brosimum alicastrum*). This tree is today abundant throughout the Guatemalan Peten region, and a single tree has the potential to yield 2,200 pounds of edible nutritious seeds (Hammond 1982). In addition, the breadnut tree is tolerant of many soil types and grows rapidly, an ideal tree for cultivation. Its fruits and seeds would have provided sources of nutrition for both humans and domestic animals, its leaves would have been used for animal forage, and its wood for construction (Gomez-Pompa et al. 1982). Many researchers believe that the present abundance of *ramon* throughout areas formerly densely populated by Mayan civilization resulted from Mayan silviculture. Evidence cited by Hammond suggests that Mayans preserved these seeds in underground chambers called *chultunobs*. Breadnuts probably served as a "famine food," to be used when times were difficult.

Major Neotropical Crops: Food, Fiber, Oil, and Rubber

Vegetables   *Figures 69, 70*

Two of the world's three most important food crops, corn and rice, are from the tropics. They are important carbohydrate sources, and both are well-known members of the grass family. The third major crop that feeds the world's peoples, wheat, is a temperate grass.

Corn, or maize (*Zea mays*), is farmed virtually everywhere in the American tropics. It is undoubtedly native to the Americas, and evidence of its ancestry from a very small area of southern Mexico has been found (Iltis et al. 1979). An annual, it must be replanted every year. Corn, to give but one example, is widely used to make tortillas, baked pancake-like patties, a staple food throughout much of Central and South America.

Rice (*Oryza sativa*) originated in Asia and was brought to the American tropics by the Spanish. Both upland rice and paddy rice are grown in various places, depending upon site conditions. Native wild rice grows along the Amazon floodplain (*varzea*) and is widely cultivated there as well. Rice grains are cooked and eaten as is or sometimes ground up and made into a bread. Rice can also be fermented for wine.

Beans (*Phaseolus vulgaris*) are the most important vegetable protein source. Several varieties are grown throughout the tropics. Rice and beans are a ubiquitous dish, served in essentially every home and restaurant, so it is not possible to visit anywhere in the Neotropics without the opportunity to partake of rice and beans. And you should. When rice is cooked in coconut milk, as it commonly is in Belize, and the rice and beans spiced with hot pepper sauce, the combination is worthy of any tropical gourmet's palate. In areas where corn is more common than rice, the staple is tortillas and beans.

Manioc (*Manihot esculenta*) is also referred to as cassava or yuca. It is a tuber crop whose root is very rich in carbohydrate and is probably the single most important crop in South America. A perennial, manioc originated in the American tropics (though it has been planted throughout the world's tropics) and will grow annually without replanting. It is an essential staple throughout the Neotropics, usually ground into a paste and made into hard bread. The plant itself is a small, spindly tree with palmate, compound leaves, most unpretentious in appearance. But the thick root is long, often more than a meter. On nutrient-poor soils, manioc contains prussic acid, a cyanogenic glycoside that protects the root from herbivores and, as described on page 162, must be removed prior to eating (Hansen 1983b). Many varieties of manioc exist with variable levels of prussic acid concentration. There are sweet manioc varieties with essentially no prussic acid and bitter manioc, which has high concentrations of cyanide compounds. The sweet-tasting manioc varieties grow only in the most fertile soils, while the bitter strains are found in soils of low fertility (Hansen 1983b).

Sugar cane (*Saccharum officinarum*) is throught to have originated in New Guinea. A tall perennial grass, it is grown principally as an export crop in the American tropics. Often grown in a large-scale monoculture, it is subject to many diseases as well as insect and rodent pests.

Other important vegetable crops derived from the tropics include tomatoes (*Lycopersican esculentum*), squash (*Cucurbita* spp.), peppers (*Capsicum* spp.), and potatoes (*Solanum tuberosum*).

The Tropical Fruit Basket   *Figure 71*

Bananas had their origins in India, Burma, and the Philippines, developing through a series of hybridizations from two originally wild species (*Musa acuminata* and *M. balbisiana*) (Vandermeer 1983b). The first bananas in the American tropics were introduced in 1516 in Hispañola. The odd, drooping arrowhead flower and the immense, elongate leaves make the entire plant as distinctive as its fruit. Cultivated bananas, because they are hybrids, must be artificially propagated and cannot pollinate themselves. The tiny black spots inside the fruit, which many people believe to be seeds, are actually aborted eggs. Bananas are often grown on large-scale plantations, but most slash-and-burn farmers have a few plants around their houses. Eaten raw or fried as plantains, bananas are very important to tropical peoples.

Papaya (*Carica papava*) is a small tree that produces cauliflorous fruits. An early successional species, the tree has very soft wood and a brief life cycle, but the domesticated species has never been found growing in the wild. Its origins are thought to have been from southern Mexico or Costa Rica. Pollination probably occurs by wind, though insects may also be involved. The fruits turn from green to yellow to orange when ripe and are a favorite food throughout the tropics.

Pineapple (*Ananas comosus*) is a terrestrial bromeliad originally from South America. The spikey, sharply spined leaves protect the plant, whose single flower cluster grows in the center of the leaf rosette. Domestic pineapples must be artificially propagated, though some pollination by insects can occur. Most agricultural families have a few pineapples as part of their "dessert" crops.

Mango (*Magifera indica*) is a native of Indian and Burmese rainforests that was brought by either the Spanish or Portuguese to the New World. It usually grows as a small tree with leathery, waxy green leaves that are produced in clumps toward the branch tips. The tiny flowers, which number up to 5,000 on a single branch tip, are yellowish to pinkish, fragrant, and attract flies as pollinators. The fruit is variable with respect to both its size and color (ranging from yellowish to green to reddish). Mango fruits abound with terpenoids (chapter 6), which give the fruit its unique flavor.

Like mango, the breadfruit tree (*Artocarpus altilis*) is not native to the American tropics, though it is very widely propagated and is a common sight throughout the region. A striking tree in appearance, the breadfruit has large, deeply lobed, dark green leaves. The green, knobby-skinned fruit, which gives the tree its name, is very large—15–20 cm (6–8 in) in diameter—and weighs about 6 pounds. Very rich in carbohydrate, the fruit can be fried, boiled, or baked. An infamous tree historically, the breadfruit was first discovered in Polynesia by Captain Cook. The botanist Joseph Banks, who accompanied Cook, judged it a suitable source of starch for British slaves in the Caribbean. Captain William Bligh was ordered to transport healthy breadfruits from Tahiti to Jamaica, but the well-documented mutiny on the *Bounty* prevented the success of Bligh's mission. Undaunted, Bligh returned on a second mission and succeeded, bringing 1,200 trees to Port Royal aboard the *Providence* in 1793 (Oster and Oster 1985). Initially the fruits served merely as a food source for pigs, and it was only after the emancipation of slaves in 1838 that the people found breadfruits to be respectable as human food.

The Little Tree That Tastes So Good
Figures 74, 75

Cacao is used to make chocolate and comes from the tree *Theobroma cacao*. The Latin name means "food of the gods," a reference to the belief that the gods had been the original suppliers of the delicacy (Hansen 1983a). There are twenty-two species in the genus *Theobroma*, all from the Neotropics. Most of the cultivated forms, including cacao, are from Central America, where it was widely cultivated throughout the region well before the arrival of the conquistadores. Indeed, the Spanish first experienced cacao in the court of Montezuma (Balick 1985). Cacao is a small tree, normally part of the rainforest understory. It periodically drops its shiny, oblong leaves and grows new leaves, which are initially reddish. A cauliflorous tree, the flowers (and therefore the oblong fruits) grow directly out of the trunk. Pollination is accomplished by a small midge that breeds on the surface of open water. Unfortunately, pollination in many areas may be relatively inefficient, presumably due to a loss of insect pollinators (Young 1994). Ancient Mayans grew cacao on raised fields surrounded by canals. The pollinating midge thrived in such areas. Cacao fruits can be produced throughout the year but are most abundant at the end of wet season. Fruits are variable in color and texture. Mammals are the principal seed dispersers. People first dry and then roast the seeds to make cocoa. For a complete account of the ecology and history of cacao, see Young (1994).

The Little Caffeinated Shrub

Figures 72, 73

Coffee (*Coffea arabica*), originally from Ethiopia, is an understory shrub with many widely spreading, horizontal branches. A single plant can be covered with small white flowers. Wind-pollinated, the plant often self-fertilizes. Leaves are very shiny and dark green. Though coffee can be grown throughout the tropics, the best beans are from regions at low to middle elevations, between 250 and 450 m (about 800–1,500 ft.), that have a dry season (Boucher 1983). Coffee is traditionally grown on plantations beneath tall trees that provide shade. However, there is a new coffee variety, already planted heavily in virtually all coffee-growing areas throughout northern Latin America, that grows well in full sunlight (and thus is called "sun coffee"). This is most unfortunate, because the agricultural practice of sun-grown coffee destroys the forest overstory, vastly reducing the plant and animal species richness of the ecosystem, as well as creating other potentially damaging ecological effects. Shade-grown coffee, on the other hand, maintains a basic forest structure and thus is compatible with maintenance of biodiversity and normal ecological functioning of the habitat. For a complete discussion of the issues surrounding shade and sun coffees, see Perfecto et al. (1996).

Palms

Many products of palms are important at least to some degree for their commercial value. Indeed, the commercial harvest of palms has been suggested as a potentially important tool in saving portions of forests from being cut for agriculture. Some palms are utilized principally for fiber. For example, the species *Aphandra natalia* is widely used in Ecuador to make brooms. Another species, *Astrocaryum chambira*, is used to fashion hammocks, nets, and *chigras*, the string bags heavily used throughout the tropics and a favorite purchase of tourists (Pedersen and Balslev 1992). Other palms are used mostly for oil. *Astrocaryum aculeatum*, known as murumuru or tucum (depending on region), compares well with coconut oil in terms of fatty acid composition. The fruit of both this species and moriche palm (page 170) is used as a food and is reported to have three times the vitamin A of carrots, in addition to other nutrients (Balick 1985).

Coconut palm (figure 16) (*Cocos nicifera*) is grown in coastal areas (and occasionally in lowlands) throughout the tropics. Because of its remarkable dispersal ability (coconuts float!), it is difficult to pinpoint the coconut palm's place of origin, but it is probably not the western hemisphere (Vandermeer 1983c). The tree was not widely distributed in the Americas before Columbus arrived. The coconut tree is unmistakable with its spreading fronds and clusters of brown coconuts below. It is fun to watch a boy climb the tree, shake loose a ripe coconut or two, and adeptly remove the husk with his machete, making a small hole so you can drink the milk. The coconut is then shattered and its white "meat" eaten. Coconut can grow in any wet tropical area in almost all soil types and is by no means confined to coastal beaches. In addition to growing wild, coconut is an important plantation species, where it is grown for copra, made from the dried seed. The copra contains 60–68% oil, which is extracted and widely exported.

Another imported plantation species is African oil palm (*Elaeis guineensis*), which was brought to the Neotropics during the slave trade (Vandermeer 1983a). This tree is widely grown on plantations throughout the Neotropics, and acreage devoted to producing oil palms is increasing rapidly, a mixed blessing since natural habitat is destroyed to make room for this nonnative species. Each seed contains 35–60% oil, the commodity for which the tree is grown.

A Big Tree, a Tasty Nut

The Brazil nut, which itself is not small as nuts go, comes from a Neotropical giant, *Bertholletia excelsa*, a canopy emergent that can reach heights of up to 40 m (about 130 ft). The clusters of dense yellow flowers mature into large, hardened fruits, each of which can harbor between one and two dozen seeds, the tasty nuts. Fruits are split open by harvesters and the nuts removed for eventual sale. The tree is found throughout most of Amazonia, and nuts are harvested throughout the region. Nuts can be eaten raw or roasted, and, given that a Brazil nut is about 65–70% lipid (oil) by weight (Balick 1985), one should eat rather few. They are good but fattening. Brazil nut trees grow on both floodplain and terra firme but tend to be widely spaced.

Rubber, Past and Present

Rubber is a yellowish-white latex contained in phloem cells of the emergent tree *Hevea brasiliensis*. The rubber tree grows throughout Amazonia (where it is native), on both floodplains and uplands. Like the Brazil nut, it can reach 40 m in height. Though there are nine species of *Hevea*, it is *brasiliensis* from which about 99% of the world's rubber is obtained (Balick 1985). The latex is typically obtained by the process of tapping, meaning that a knife with a V-shaped edge is used to pierce the cambium carefully without cutting through it. Cuts are angled progressively from top left to bottom right.

Though rubber is from a Neotropical tree, most rubber today is produced by trees grown in the Asian tropics, a story brieflly worth telling. European colonists in the mid-eighteenth century quickly learned from the various aboriginal tribes that latex from *Hevea* could be used to waterproof articles such as boots. The rubber trade began. In 1839 Charles Goodyear learned how to vulcanize rubber, making it resistant to cracking from cold or melting from heat. This, plus the introduction of steamboat traffic in the Amazon, made big-time rubber export possible and economically compelling, and the so-called South American rubber boom began in earnest. Manaus, Brazil, expanded from a small city of about 5,000 persons in 1870 to a population of 50,000 by 1910. As the city grew, so did its sophistication, including the construction of a world-class opera house (which was recently restored). Iquitos, Peru, followed a similar course. But the boom eventually went bust some years after a Britisher, Henry Wickham, in 1876 shipped about 74,000 *Hevea* seeds to Kew Gardens in London, where they were cultivated and eventually shipped to plantations scattered among the British colonies in Asia. Plantation-grown rubber from Asia quickly outcompeted South American rubber on the open market. Though the huge rubber boom ended for South America, rubber has always been harvested there, and, during World War II the South American rubber industry experienced a brief period of growth. Today South American rubber comes from modest efforts by local rubber tappers, some of whom are responsible for the idea of using natural rainforest as extractive reserves (Fearnside 1989). Extractive reserves will be discussed in chapter 14. For a more detailed account of the story of the rubber industry, see Goulding et al. 1996.

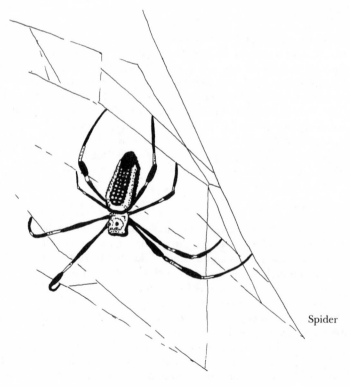

Spider

8

Rivers through Rainforest

RIVERS contribute to biodiversity by providing varied habitats for many species of plants and animals. Because of the seasonality so prevalent throughout much of the Neotropics, rivers experience an often dramatic annual flood cycle (ranging, for example, approximately 15 m (50 ft) annually at Iquitos, Peru) that exerts a significant impact on bordering ecosystems, especially gallery forests. In addition, rivers provide a cornucopia of habitats: swamps, marshes, streams, oxbows, river islands. Each of these habitats hosts species that otherwise would not be present.

Two South American riverine systems are of obvious importance because of their vast areas and locations within the heart of Amazonia. These are the Orinoco and Amazon basins.

The Orinoco

The Orinoco River, nearly 2,560 km (1,600 mi) long, flows northeast from the Rio Guaviare in eastern Colombia, bisecting Venezuela before exiting to the Atlantic Ocean. Considering average annual discharge, it ranks as the third largest river in the world. The Orinoco Basin, though large, is but one-sixth the area of the Amazon Basin, to which it connects via the Rio Casiquiare, which flows into the Rio Negro, itself a part of the Amazon Basin. Much of the Orinoco Basin drains the ancient Great Guianan Shield, located southeast of the main river (see below).

The Orinoco begins at an elevation of 1,074 m (3,523 ft) in the Parima Mountains close to the border between Venezuela and Brazil. It soon bifurcates into the southern and northern streams, the former of which flows south, eventually joining the Rio Negro and flowing into the Amazon, the latter flowing north and east, joining major tributaries such as the Rio Meta, Rio Arauca, and Rio Apure. The major city located along the Orinoco is Ciudad Bolivar (in Venezuela), where the river is typically about 244 m (800 ft) wide. Ships can navigate the Orinoco for about 1,120 km (700 mi) from its mouth to the Cariben Rapids, 9.6 km (6 mi) from the Rio Meta.

Like the Amazon and its tributaries, the Orinoco is a seasonal river. At Ciudad Bolivar the annual variation in water level is normally between 15 and 18 m (approx. 50 to 60 ft). Rainy season in the Orinoco Basin is generally from April to October, and dry season is from November through March. The discharge rate varies dramatically with seasonality, the lowest flow during dry season being but 1/25 to 1/30 of the highest flow during wet season (Meade and Koehnken 1991).

The Orinoco River bisects two distinct geological areas. The right (south) bank of the Orinoco borders Precambrian bedrock from the Guianan Shield, among the oldest geological formations on the continent and, for that matter, on Earth. In contrast, the land bordering the left (north) bank of the main river is geologically young, formed only a few centuries ago from sediments washed from the Andes and transported across the flattened Llanos (Meade and Koehnken 1991). The effect of these differing geological histories is reflected in the differing characteristics of the tributary rivers that drain into the main Orinoco. The right-side tributaries typically are stable, constrained by crystalline bedrock and, especially within the Guianan Shield, abundantly supplied with rapids and waterfalls. The left-side tributaries are unstable, with shifting channels formed by alluvial deposits from the river (Meade and Koehnken 1991).

The Orinoco flows west, then north, before beginning its major eastward flow. The river itself has had a strong influence on the geology of the region, having helped cut channels through parts of the the Guianan Shield, thus contributing to the isolation of a unique series of flat-topped tablemountains called *tepuis*, some up to 1,500 m (approx. 5,000 ft) tall (page 225). The highest waterfall on the planet, Angel Falls, drops about 1,000 m (3,281 ft) from the top of one of these tepuis. Much of the Orinoco flows quietly and slowly through the vast, marshy Llanos region of Venezuela (see page 229), where wildlife is abundant.

The Orinoco meets the Atlantic Ocean at the Amacuro Delta and Gulf of Paria, an area of extensive mangrove forests that Columbus first explored in 1498, calling it a "gateway of the Celestial Paradise." The first European to navigate the length of the Orinoco was a Spaniard named Deigo de Ordaz, whose expedition took place in 1531–32. Like most Spanish explorers, Ordaz's motivation was the quest for El Dorado, the mythical "gilded chieftain" who covered himself with gold (see below).

The Amazon

Figures 155, 156, 157, 166, 167

The name "Amazon River," or Rio Amazonas, refers to the vast river that forms at the confluence of the Maranon, Ucayali, and, to a lesser degree, Tigre, rivers just west of Iquitos, Peru, and flows eastward several thousand miles to the sea. The name "Amazon" is used in Peru (and in general language) from the confluence point eastward. In Brazil the name is formally used from Manaus eastward. West of Manaus the giant river is frequently called the Rio Solimões.

The headwaters of the immense river can be traced to a small, unremarkable tributary, the Carhuasanta, at an elevation of 5,598 m (18,363 ft) in the cold, wind-swept Peruvian Andes only about 192 km (120 mi) from the Pacific Ocean. The Carhuasanta flows into the Hornillos, which in turn joins the Apurimac, a major tributary that eventually joins the Ene, the Tambo, and finally the Ucayali. The Amazon system plunges in elevation initially but drops only about 1.5 cm/km (2 in/mi) once outside the Andes, eventually flowing its full course of 6,696 km (4,185 mi) to the Atlantic Ocean.

Though the Nile ranks as the world's longest river, just exceeding the Amazon, the Amazon carries by far the world's largest volume of water. It is estimated that 16% of all river water in the world passes through the 320-km (200-mi)-wide delta of the Amazon (Muller-Karger et al. 1988), which daily discharges about 4.5 trillion gallons, or about 7.1 million cubic feet of water per second (200,000 m^3 per second). This represents a discharge of about 4.4 times that of the Congo (Zaire) River, the next most voluminous river. The plume of muddy water from the Amazon can be seen as far as 100 km (about 60 mi) out to sea and has been traced by NASA's Coastal Zone Color Scanner as it moves toward Africa between June and January and toward the Caribbean from February through May (Muller-Karger et al. 1988). The river itself is over 10 km (about 6 mi) wide as far as 1,600 km (1,000 mi) upriver, and large ships can navigate for over 3,700 km (2,300 mi), eventually docking at Iquitos, Peru (Dyk 1995; Bates 1964). Two Amazonian tributary rivers, the Negro and the Madeira, rank as the fifth and sixth largest rivers in the world with regard to annual discharge. By comparison, the Mississippi River ranks about tenth and has only about one-twelfth the annual discharge of the Amazon (Meade et al. 1991).

The Amazon River once flowed in the opposite direction, draining into the Pacific Ocean near what is today the port city of Guayaquil, Ecuador. It is now believed that the river changed to its present west-to-east course as recently as 15 million years ago, when the Andean uplift profoundly altered the course of the river as well as patterns of biogeography throughout the Amazon Basin. Initially the uplift of the Andes created a gigantic lake, bordered on the west by the newly arisen mountain chain and on the east by the extensive Guianan and Brazilian shields. The Amazon finally made its way to the Atlantic during the Pleistocene, cutting through its eastern barrier in the vicinity of Obidos, Brazil (Goulding 1980). Many widespread trees that can be found along the river were probably dispersed eastward by the altered course of the river water (Goulding 1993).

The Amazon Basin, drained by the Amazon River and its gigantic tributaries, covers an area of about 6.5 million km^2 (2.5 million sq mi), essentially 40% of the total area of South America. Approximately 1,100 tributaries service the main river, and some of them, like the Rio Negro, Napo, Madeira, Tapajos, Tocantins, and Xingu, obviously rank as major rivers. Amazon tributaries vary in color from cloudy yellow to clear black depending upon where they originate and their geological and chemical properties (page 56). Examples of major blackwater rivers include the Rio Negro and Rio Urubu. Clearwater rivers include the Rio Tapajos, Rio Trombetas, Rio Xingu, and Rio Curua Una. Whitewater rivers include the Rio Jutai, Rio Jurua, Rio Madeira, Rio Purus, Rio Napo, and the upper Amazon itself. Before the ever-increasing numbers of roads and airstrips were built, these tributaries served as "interstate highways" providing the only access to the interior. Cities such as Iquitos, Peru, are, even now, approachable only by boat or airplane. There are no access roads. Where rivers flow there is continual adequate soil moisture, and evergreen gallery forest lines the banks.

The Amazonian Flood Cycle

Standing in the downpour watching the trails turn into mud, I was at first surprised that the Amazon itself was dropping nearly a meter a day, even though it was raining heavily throughout the region. But the river's depth was not closely related to the rain falling on this rainforest near Iquitos, Peru. The river was dropping because snow had stopped falling in the Andes and the meltwater had already drained and been absorbed by the Amazon. Now the huge river was receding from its peak flood.

The timing of floods and the distribution of floodwaters are the result of the complex pattern of seasonal precipitation, much of it in the distant mountains. Because of the vast area of the Amazon Basin, at any given time some regions will be experiencing flood while others will be low water. This is because the Equator divides Amazonia (though it is north of the river itself), and many of the major Amazonian tributary rivers are either partly north or entirely south of the Equator. Rainy season generally occurs in southern Amazonia from October to April. Rainy season in Manaus, in the middle of the basin, is from November to May, and rainfall is highest in the northern part of the basin from April through June (Junk and Furch 1985). The wettest months in Iquitos (which receives 118–157 m [3,000–4,000 mm] of annual rainfall) are February through May, though there is much variability. In fact, according to meteorological records kept at Iquitos Airport, every month of the year except May has, in one year or another, been the low water month (P. Jensen, pers. comm. 1996), so seasonality in Iquitos is, if anything, poorly defined.

In general, flooding in the northern waters occurs as southern waters are low and vice versa. In areas fed by one major tributary there is a single annual flood. But in those regions fed by both southern and northern tributaries, there are two annual flood periods, which may vary from one another in intensity. As rainy season proceeds, flood waters build such that the peak of the flood cycle usually occurs at the onset of dry season. Because some parts of the Amazon are in flood while other parts are in low water, there is little difference between the minimum and maximum annual discharge rates, which vary by a factor of only two or three (Nordin and Meade 1985). This is in marked contrast to the Orinoco River, all of which lies north of the Equator (see above).

The low, flat geomorphology of the Amazon Basin also contributes strongly to its propensity for flooding. Though sediment has a strong tendency to build up along river banks, forming levees, Amazonian rivers will routinely overflow their banks at full flood. The general floodplain is characterized by land that is not uniformly flat, creating habitats such as temporary lakes and swamps. Floodplain forest is estimated to occupy approximately 100,000 km^2 (38,000 sq mi) within the total Amazon Basin (Goulding 1980).

The importance of the flood cycle should not be underestimated. Amazonian rivers experience an annual fluctuation that averages between 7 and 13 meters (23–43 ft), which can result in a floodplain forest innundated up to 10 meters annually, a water fluctuation that can bring river water up to 20 km (12.5 mi) into the neighboring forest (Goulding 1993). Flooding is essential in dispersing sediment, fertilizing varzea floodplain, and enabling fish and other

organisms to make use of gallery forest during high water. Zooplankton reproduction peaks during high water, and this resource, which washes into neighboring rivers as the flood recedes, provides an invaluable resource for fish,
especially, but not exclusively, during their juvenile life cycle stages (see
below). Until recently, the extensive damming activities that are routinely seen
in countries such as the United States were unknown in South America. Damming changes the flood cycle, isolating previously flooded areas from the annual flood. Doing so throughout much of Amazonia would cause a substantial
disruption of many ecological relationships and interdependencies (Goulding
et al. 1996).

Overview of Riverine Habitats

There are many kinds of habitats that one can consider part of an overall
riverine ecosystem. A river is dynamic: it varies seasonally and, with time, turns
and twists within its floodplain, creating a diversity of habitats. Below is a brief
summary of some of the most important of these riverine habitats (see Remsen
and Parker 1983; Junk 1970; Sioli 1984; Goulding 1980, 1985, 1990, 1993;
Smith 1981).

Open River   Figure 155

From a ship sailing upriver on the Amazon, the width of the vast river can
take on the appearance of a small inland sea. Strong currents and underwater
sediment bars make navigation more tricky than it might at first seem, and
large ships sometimes run aground. There is not much wildlife to be seen on
the open river, though large-billed (*Phaetusa simplex*) and yellow-billed (*Sterna
superciliaris*) terns are common, as are Neotropic cormorants (*Phalacrocorax
olivaceus*). An occasional fish might jump up momentarily, but little else will
be seen. The thermal currents above the river might support soaring vultures,
raptors, storks, and screamers (see below). In the central and lower Amazon,

Large-billed tern

skies are typically clear above the widest stretches of the river, while neighboring forests have clouds above, the result of forest transpiration (Salati and Vose 1984). But along the upper Amazon the humidity is often such that the river itself is cloud-covered and subject to intense cloudbursts.

Riverine and Stream Edge   *Figure 160*

For observing wildlife it is far more useful to navigate moderate or narrow tributaries than the main river itself. It is in these areas that species such as giant otter, capybara, various caiman, sunbittern, and a selection of herons and egrets are best seen. Black-collared hawks (*Busarellus nigricollis*) are common, perching by the riverside as they search for fish, as are several caracara species (yellow-headed [*Milvago chimachima*], black [*Daptrius ater*], red-throated [*Daptrius americanus*]). Many other bird species are partial to gallery forest, including Amazonian umbrellabird (*Cephalopterus ornatus*), bare-necked fruit crow (*Gymnoderus foetidus*), and gray-necked wood-rail (*Aramides cajanea*). While on the main river you might see only the large ringed kingfisher (*Megaceryle torquata*), but all five Neotropical kingfisher species may be encountered along a stream. At night streamside trees make ideal perches for various potoo species (*Nyctibius* spp.). Tucked cryptically along the bank of a stream, an anaconda may be digesting its most recent meal. Where streams or small tributaries meet the main river, both pink river and gray dolphins often congregate.

Plant life along riversides is typically successional, with many small, thin-boled trees, though occasional large kapoks as well as other giants may be evident on the floodplain. Cecropias and various leguminous trees are often abundant, and heliconias tend to dominate the understory. Other trees, such as rubber trees and Brazil nut trees, can be found on floodplain streamsides. Various arums are also common as are palms, especially moriche palm.

Beaches and Sandbars

The vast tonnage of sediment washed from the Andes and carried by the tributaries and Amazon itself is often deposited along the river edge or as bars in the river itself. Along the varzea forests of the upper Amazon the sediment is deep blackish-gray, the rich volcanic soil transported from the high Andes. Altered annually by the flood cycle, sediment deposit forms extensive beaches and sandbars that provide habitat for birds such as plovers, black skimmer (*Rhnchops niger*), and various herons and egrets, as well as good resting places for caiman. Swallows of various species, mostly seen in flight over the river, can often be seen perched on small snags and bushes along beaches and sandbars. Most common are southern rough-winged (*Stelgidopteryx ruficollis*), white-winged (*Tachycineta albiventer*), and white-banded (*Atticora fasciata*).

Sandbar Scrub

Andean sediment is rich in nutrients and does not go uninhabited for very long. Various plant species, including such familiar temperate genera as *Salix*, the willows, quickly invade and often become sufficiently dense to stabilize the soil. Sandbar scrub is typically dense, composed of a low diversity of fast-growing colonizing plant species. Once sediment is deposited from the action of riverine dynamics, the area is subject to colonization by pioneer plant species typical of early succession (page 64).

One comprehensive study of primary succession on western Amazonian floodplains recorded 125 species in the plant assemblage that initially colonized newly deposited sediments. Of these colonizing species, 81% were dispersed by wind or water. Colonizing species included many perennial herbs as well as river margin trees, grasses, and various climbers. These pioneer species are typically replaced by more competitive plants as succession proceeds, but they continue to thrive by colonizing newly deposited sediment (Kalliola et al. 1991).

River Islands

From the riverside walk at Iquitos you cannot see across to the far side of the Amazon River, but that's not because the river is too wide. It is because the river is interrupted by huge sediment islands that literally block one's view all the way to the other bank. Padre Island is the main island visible from Iquitos, though just to the west of the city one finds Timarca and Tarapoto islands. River islands can be of all sizes, but the big ones are stable, composed of years of sediment deposit stabilized by vegetation invasion. Many humans inhabit and farm the river islands (as well as varzea floodplain bordering rivers), planting rice, corn, peppers, beans, and bananas. Whole towns can be found on the larger river islands.

Forests that grow on river islands as well as mainland floodplain forests can be managed for sustained yield of various products. The riverine people, called *ribereños* in western Amazonia and *caboclos* in Brazil, have long inhabited riverine areas and actively alter the species composition of the forest in order to achieve economic gain. A study from the southeastern portion of the Amazon estuary in the vicinity of Belem, Brazil, and the Rio Tocantins demonstrated that the local people employ several management techniques, including the active removal of undesirable plant species (such as certain spiny palms), removal of firewood species, cultivation of such species as cacao, avocado, and mango, and maintenance of potentially useful species such as rubber trees (Anderson et al. 1995). The result of such management activities is that the forest is altered in species composition but is nonetheless maintained as forest in a sustainable manner. *Ribereños* also are active agriculturalists, planting maize, rice, manioc, bananas, and other crops. Fish are the major protein source for most river inhabitants (also see chapter 7).

Oxbows   *Figure 161*

Where the flow dynamics of the river become unstable (typically during the high-water period), the river may cut a new channel, effectively isolating a meander and creating what is called an oxbow lake, a habitat of essentially standing water. Oxbows are common in rivers subject to a variable flood cycle,

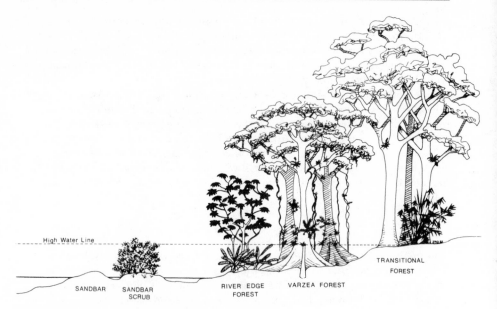

High Water Line

TRANSITIONAL
FOREST

SANDBAR   SANDBAR   RIVER EDGE   VARZEA FOREST
               SCRUB   FOREST

Schematic representation of five (presumably) successional stages in river-created habitats along large rivers in western Amazonia (drawing by John P. O'Neill). From Remsen and Parker (1983). Reproduced with permission.

and they provide yet another kind of riverine habitat, where water stagnates rather than flows rapidly. Such still water supports vast growth of water hyacinth, as well as the giant Victoria waterlilies (page 128). It is here that the peculiar hoatzin can be found (see below).

Floating Meadows

Entire islands of floating grasses can be encountered along the Amazon and within Amazonian lakes (Junk 1970). Along the main rivers, some of these grassy islands occasionally reach a size where they can be a hazard to navigation. Two grasses, *Paspalum repens* and *Echonochloa polystachya*, are abundant components of the floating meadows and together make up about 80–90% of all the floating grass species of Amazonia. *Paspalum* is adapted to float, forming dense, floating mats for the four- to five-month rainy season, when the river is high. The plants grow and spread asexually during this time, but they also flower and make seeds so that during dry season a multitude of seeds falls on the newly exposed ground, to germinate quickly. Thus *Paspalum* is adapted to be both a floating and a terrestrial plant. Terrestrial *Paspalum* has a distinctly different morphology from the aquatic form, even though they are the same species. Unlike *Paspalum*, *Echinochloa* has no floating morph but remains rooted throughout the flood cycle. This species is most common in lakes.

Swamps *Figure 29*

A swamp is generally an area of woody vegetation that is innundated by standing water for a significant part of the year. Swamps are typically lower in tree species richness than less wet sites. This is presumably due to the greater physiological stress imposed on the roots of trees that are forced to remain submerged, the result being that fewer species have succeeded in adapting to such protracted innundation. Many swamp tree species have stilted root systems, and buttressing is extremely common as well.

Throughout interior Amazonia the most characteristic swamp forests are composed of palms, notably the moriche palm, *Mauritia flexuosa*. Moriche palm is one of the most distinctive, abundant, and widespread Neotropical palms. Growing in swamps and along wet areas, often forming pure stands, a swamp forest of palms is termed an *aguajale*. Moriche palms are tall and slender, their fronds appearing as spike-tipped fans on elongate stalks that radiate from a common base atop the trunk. Stands of moriche palms are prime feeding areas for various large macaw species whose powerful bills are sufficiently strong to crack the hard palm nuts that occur in dense clusters below the fronds.

Other bird species have become moriche specialists, rarely if ever found away from moriche stands. These species provide examples of an important component of Neotropical species diversity, the tendency toward extreme habitat specialization. The moriche oriole (*Icterus chrysocephalus*), a small, striking black oriole with bright yellow crown, shoulders, rump, and thighs, feeds and nests within moriche fronds. The sulphury flycatcher (*Tyrannopsis sulphurea*) is a gray-headed, yellow-bellied bird, easily confused with the widespread tropical kingbird (*Tyrannus melancholicus*). Both species can be difficult to see well in the dense palm fronds. More obvious, the fork-tailed palm-swift (*Reinarda squamata*) is a common aerial feeder, streaking through the skies in the vicinity of moriche stands in search of insect prey. This pale swift with a deeply forked tail builds its nest on the underside of a dead moriche palm leaf. Finally, the point-tailed palmcreeper (*Berlepschia rikeri*) is a unique member of the Furnariinae, or ovenbird, family (page 283). It is entirely confined to moriche palm stands in Amazonia and is nowhere really common. It is difficult to spot among the dense, fanlike palm fronds, but its presence can be known from its song, a loud series of ringing notes (Ridgely and Tudor 1994). The bird itself is bright cinnamon on the wings and tail and streaked boldly with black and white on its head and breast. See it really, really well, and you'll see its bright red eyes.

Along the coast, where salt water incursion is normal, swamps are composed of various combinations of mangrove species (see chapter 11).

Flooded Forests

Floodplain forests within the Amazon Basin cover an area of approximately 150,000 km² (57,000 sq mi), which is roughly equivalent to an area the size of the state of Florida (Goulding 1985, 1993). Overall, floodplain forest comprises only about 4% of the total Amazon Basin, the remaining forest all being terra firme (Goulding 1993). Floodplain forests are so named, of course, because they are innundated by the annual flood cycle. Depending upon location, floodplain forests may be innundated anywhere from two to ten months out of the year. For example, the Amazon forest itself (from Manaus eastward) is flooded for about six months, whereas the upper Rio Madeira is in flood for only two to five months annually (Goulding 1985).

As mentioned previously (page 191), flooded forests may border whitewater, clearwater, or blackwater rivers. Because sediment load varies, forest productivity varies. Whitewater flooded forests are typically higher in stature and biomass (and probably species richness) than clearwater rivers. Flooded forests of blackwater rivers are typically low in stature, and species richness tends to be a bit lower and to vary less from site to site as compared with whitewater rivers (Goulding 1985, 1993).

During the flood cycle fish have direct access to the forest, and many species are important fruit and seed consumers as well as seed dispersers (see below). Certain monkey species, such as the pygmy marmoset and uakaris, are restricted to floodplain forest, as are many species of birds and arthropods. Some plant species are unique to flooded forests, though many have closely related species in dry forests, suggesting a recent speciation between dry and floodplain species. For example, two species of closely related palms in the genus *Astrocaryum* are distributed such that one is abundant in flooded forests, while the other is found only on terra firme (Goulding 1993).

It is important to understand that the term floodplain forest is not an absolute term. Some forested areas are located immediately adjacent to the river and flood frequently, while areas farther from the river may be in flood infrequently. For Tambopata Reserve in Amazonian Peru, Phillips et al. (1994) recognize seven types of flooded forest:

> *Permanently water-logged swamp forest*—former oxbow lakes still flooded but covered in forest
> *Seasonally water-logged swamp forest*—oxbow lakes in the process of filling in
> *Lower floodplain forest*—lowest floodplain locations with a recognizable forest
> *Middle floodplain forest*—tall forest, flooded occasionally
> *Upper floodplain forest*—tall forest, rarely flooded
> *Old floodplain forest*—subjected to flooding within the last 200 years
> *Previous floodplain*—now terra firme but historically ancient floodplain of Tambopata River

As discussed in chapter 7, floodplain forests are extensively utilized by the mestizo people of Amazonian Peru (Phillips et al. 1994)

Who Were Those Amazons?

The word Amazon has a romantic sound, a suitable name for the world's most massive river. But what does it mean? Could it be an Indian word for mighty river, or great waters, or some such thing? Well, no. Actually it means "without a breast." Amazon is derived from the Greek *a-mazos*, which translates as "no breast" (Smith 1990). The world's mightiest river apparently shares the name of a group of ancient, mythical women, the Amazons, who once lived in what was then Scythia, and who were alleged to be such committed warriors that each woman would have a breast removed so that the protruding mammary gland would not interfere with drawing a bowstring. But ancient Scythia, a land that is now part of Ukraine, is a long way in time and space from equatorial South America. Even the most committed female warrior, breastless or not, couldn't shoot an arrow that far. The connection between the legend of the ancient Amazons and the name for the world's biggest river is due entirely to a Spanish explorer named Francisco de Orellana.

Orellana, born in 1511, was the first European and quite possibly the first human being to sail the length of the Amazon River. Orellana gave the Rio Negro its name, and in all likelihood he unintentionally did the same for the Amazon.

The story of Orellana's arduous journey essentially begins with a broken promise. Orellana initially set out on a long trek heading east from Quito, Ecuador, into the vast tropical forest, accompanying expedition leader Gonzalo Pizarro, the younger brother of Francisco Pizarro, the conquistador who had adeptly conquered the Incas. The object of the trip was a search for two commodities that both explorers held very dear: gold and spices, particularly cinnamon. (The desire to search the forest for cinnamon was bound to be in vain. The plant in question is *Ocotea quixos*, a member of the Lauraceae, which contains the fragrant compound cinnamaldehyde. It grows only in the vicinity of Quito and is absent from Amazonia [Gottlieb 1985]. If they wanted cinnamon, they should have stayed in Quito.) Pizarro believed that somewhere in that vast green wilderness was an Indian king, El Dorado, who possessed so much gold as to cover his body routinely with gold dust. Both Pizarro and Orellana must have figured that such a king would have more than enough gold to share it with them. Or, more likely, they would simply find the king, conquer him, and take his gold. Orellana was an accomplished soldier, having already lost one of his eyes to the rigors of battle. The stunningly large expedition set out with 200 Spanish soldiers on horseback, another 150 Spanish soldiers afoot, 2,000 dogs, each trained to attack Indians, approximately 2,000 pigs to be used for food, plus 2,000 llamas to serve as pack animals, and about 4,000 Indians, also to be used as carriers, an astonishingly large host to assemble for a very long walk through a dense jungle (Gheerbrant 1988).

Things did not go so well. Many in the expedition sickened and died. Not only did they not find gold (nor, understandably, many cinnamon trees), they didn't even find enough food (thus forcing them to consume their animals), and they were subject to repeated demoralizing attacks by hostile Indians (apparently undiscouraged by the attack dogs). Crisis loomed and, on Christmas day 1541, almost a year after the start of the expedition, it was agreed that Orellana would take about fifty men and leave Pizarro and most of the others to proceed east, where a village was reputed to exist, and where they might find food. Orellana, who promised to procure provisions and return, took along a Dominican friar, Gaspar de Carvajal, who became the chronicler of Orellana's adventure. For those who enjoy history as told by its participants, Carvajal's account is readily available (Medina 1988).

Orellana broke his promise and never returned to help Pizarro. Instead he and his men continued east, following the strong current of the Napo River and eventually encountering the still unnamed Amazon River (just east of what is now Iquitos), which they followed all the way to the Atlantic Ocean. As Orellana sailed the big river, he and his party were frequently attacked by Indians. At one point is his account, Carvajal describes a vicious attack upon them by a mysterious group of tall white women with braided hair, who were all exceptionally talented in the use of bow and arrow. Orellana called them "Amazons," and the name has been used for the river ever since (which is too bad in a way for Orellana, since the river probably should be called the Rio Orellana, and even was so-named for a brief time).

Orellana, according to Carvajal's account, eventually captured an Indian woman and from her learned much about the Amazons—their riches, their warrior abilities, their curious sexual habits, and a number of other things. The one thing he apparently didn't learn was exactly where they lived, since no one has since relocated them, and not for lack of trying (see Shoumatoff 1978, 1986; Gheerbrant 1988). Alfred Russel Wallace, always the skeptic, suggested the obvious. Maybe Orellana made it all up. Maybe what he saw was not a tribe of ferocious women but one of ferocious men who vaguely resembled ferocious women. Wallace describes Indian men as looking rather feminine to European male eyes (or eye, in Orellana's case), especially when those eyes have perhaps not looked closely upon a woman for considerable time.

Now, to be fair, there is at least one hypothesis that argues that the word Amazon does indeed derive from an Indian phrase for tidal waters and has nothing whatsoever to do with the Orellana expedition. Most scholars, however, hold to the belief that the huge river is really named for a tribe of mythical women. It's a better story anyway.

Riverine Natural History

The Amazon, the Orinoco, and their main tributary rivers are wide, affording stunning panoramas. But the best way to experience the diversity of wildlife that populates the rivers and their neighboring habitats is to sail quietly near shore, exploring marshes, streams, and oxbows.

Where protected from human hunting pressure, groups of capybara, the world's largest rodent (see below), graze like small hippos along the water's edge, ever watchful to avoid falling prey to a caiman or anaconda. Riverine birds, such as the turkey-sized horned screamer (*Anhima cornuta*) and prehistoric-looking hoatzin (*Opisthocomus hoatzin*), nest along the banks. One study found that 15% of the *nonaquatic* Amazon bird species are directly dependent upon riverine habitats such as beaches, sandbar scrub, river edge forest, varzea, and transitional forest (Remsen and Parker 1983). Many species depend on the annual flooding cycle to preserve the particular habitats they require.

The Hordes of Fishes   *Figures 158, 159*

We all know fish come in schools. This being so, the Amazon could well be known as "Piscine University." There are more than 2,400 species of fish, an astounding variety, known to inhabit the waters of the Amazon and its tributaries, and up to 800 additional species may remain yet to be formally described (Lowe-McConnell 1987; Goulding 1980). Approximately 40% of the species thus far described are members of two groups, the characins and the catfish (Goulding 1985). These multitudes include the neon tetra (*Hyphessobrycon innesi*), cardinal tetra (*Cheirodon axelrodi*), pearl headstander (*Chilodus punctatus*), silver hatchetfish (*Gasteropelecus levis*), bronze corydoras (*Corydoras geneus*), and oscar (*Astronotus ocellatus*), all favorites of the aquarist (Lowe-McConnell 1987).

Any aquarist knows that one of the most attractive features of the popular Amazonian fish species is their small size. A large number of very small-sized species is a general characteristic of the Amazonian fish fauna. It has been hypothesized that small size evolved in response to a diet of tiny arthropods obtained during the flood cycle from within the flooded forest (Goulding 1993). Thus when the tiny neon tetras gather at the surface to grab up miniscule morsels of tropical fish food, they may be exhibiting a behavior originally evolved as they massed around flooded forest trees, gathering up the displaced insects and spiders.

The Amazon Basin is home for the infamous 35-cm (about 14-in) red piranha (*Sarrasalmus nattereri*) and its relatives, a group of fish whose reputation for collective ferocity is rarely merited. Though they are widespread and, to put it mildly, extremely abundant, piranha are only potentially dangerous when water levels are low and food supply is poor, concentrating the already hungry predatory fish and putting any potential protein source at risk of attack. When conditions are normal, piranha are of no danger to swimmers. At Yacumama Lodge, along the Yarapo, a blackwater river near Iquitos, Peru, I watched children gleefully swimming off one end of a dock while other children were happily catching small piranhas from the other side of the dock. Piranha do sometimes increase to an abundance where their collective predatory habits act to deplete more desirable species of fish. Along parts of the Brazilian Amazon, a plant known locally as timbo (*Lonchocarpus utilis*) is used to reduce numbers of piranha and piranha eggs. The plant contains rotenone, a toxin often used throughout Amazonia by indigenous people to paralyze fish temporarily and thus make them easy to catch. The rotenone, at the concentration used, is lethal to piranha and piranha eggs but does little if any harm to other fish (Balick 1985).

Besides piranha, electric eels (*Electrophorus electris*), which can attain lengths of 1.8 m (6 ft) and emit a jolt of 650 volts, are to be appreciated from a respectable distance. Electric eels are quite common and, because of their harmful potential (so to speak), are generally feared more than any other creature of the river by riverine people. There is also a group of bizarre catfish collectively called candiru, one of which, *Vandellia cirrhosa*, is particularly irksome. Slimmer than a pencil, this tiny fish is normally a parasite of other fish, attaching to gills. However, it allegedly has the disconcerting habit of (presumably) mistakenly swimming into the human urethral, vaginal, or anal opening, whereupon it lodges itself with an array of sharp fin spines. Not good. Once the fish is so located, its human host experiences significant discomfort (you probably guessed this much). Though candiru are widely reputed to have this unfortunate mistaken sense of direction, I have been unable to locate any

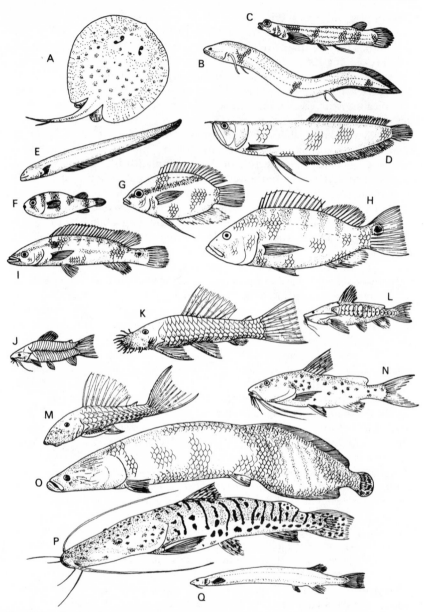

South American freshwater fishes I. (A) *Potamotrygon* stingray (30 cm, Elasmobranch);
(B) *Lepidosiren* lungfish (50 cm, Dipnoi); (C) *Anableps* foureyefish (15 cm, Anablepidae);
(D) *Osteoglossum* (50 cm, Osteoglossidae); (E) *Electrophorus* electric eel (60 cm, Electrophori-
dae); (F) *Colomesus* pufferfish (Tetraodontidae); (G) *Cichlasoma* (10 cm, Cichlidae); (H)
Cichla (40 cm, Cichlidae); (I) *Crenicichla* (25 cm, Cichlidae); (J) *Hoplosternum* (15 cm,
Callichthyidae; (K) *Ancistrus* (15 cm, Loricariidae); (L) *Megalodoras* (70 cm, Doradidae);
(M) *Hypostomus* (15 cm, Loricariidae); (N) *Pimelodus* (30 cm, Pimelodidae); (O) *Arapaima*
(150 cm, Osteoglossidae); (P) *Pseudoplatystoma* (120 cm, Pimelodidae); (Q) *Vandellia* (4 cm,
Trichomycteridae). From Lowe-McConnell (1987). Reproduced with permission.

South American freshwater fishes II, illustrating the adaptive radiations in characoid fishes. (A) *Gasteropelecus* (6 cm); (B) *Tetragonopterus* (12 cm); (C) *Brycon* (50 cm); (D) *Leporinus* (30 cm); (E) *Anostomus* (12 cm); (F) *Characidium* (4 cm); (G) *Poecilobrycon* (4 cm); (H) *Metynnis* (12 cm); (I) *Colossoma* (50 cm); (J) *Serrasalmus* (30 cm); (K) *Prochilodus* (40 cm); (L) *Boulengerella* (45 cm); (M) *Acestrorhynchus* (20 cm); (N) *Hoplias* (30 cm); (O) *Hoplerythrinus* (25 cm); (P) *Hydrolycus* (60 cm); (Q) *Salminus* (50 cm). From Lowe-McConnell (1987). Reproduced with permission.

documented accounts of actual invasion. Not that anyone would necessarily want to admit to such a mishap.

Most Amazonian fish are far less dangerous, though equally interesting. The huge pirarucu (*Arapaima gigas*) is an important protein source for people who live along the Amazon (Goulding et al. 1996). It reaches weights of 150 kg (about 300 lbs) and lengths of up to 3 m (10 ft), a colossal size now rarely seen due to intense fishing pressure from people. The pirarucu, which preys on many other fish species, occurs throughout Amazonia, especially in quiet lakes. A relative of the pirarucu, the smaller arawana (*Osteoglossum bicirrhosum*) is an elongate fish that always seems to be looking downward. It is one of many Amazonian fish that commonly show up in the tanks of the aquarium fancier, but it is also an important food fish for Amazonian people. Among the more unusual piscines is the South American lungfish, *Lepidosiren paradoxa*. This eel-like fish with large scales and thin, ribbonlike fins can gulp air in the manner of its ancestors that swam in stagnant lakes 350 million years ago. There is also a single lungfish species in Africa and one in Australia. The curious distribution of these three species of ancient freshwater fish on three widely separated continents is almost surely the result of plate tectonics and the breakup of Gondwanaland (page 111).

Some fish, like many terrestrial animals, have evolved to be fruit and seed consumers in Amazonia. Somewhere around 200 species of fish devour fruits and seeds in Amazonian waters, far more species than do so in tropical Africa or Asia (Goulding 1985, 1993). A frugivorous diet is facilitated by the flood cycle, which enables fish actually to swim well within the gallery forest, foraging for dropped fruits, many of which often float at the surface, making them easy for fish to find and consume. The tambaqui (*Colossoma macropomum*), an inhabitant of blackwater rivers and igapo forests, has been documented to be an important seed disperser, particularly for *Hevea spruceana*, a rubber tree, and *Astrocaryum jauari*, a palm species (Goulding 1980). Both of these tree species are widely distributed and relatively abundant, produce large seed crops, and have fruits that are laden with fat and protein, encased within hard nuts that many animals are unable to break. In the case of the rubber tree, seeds are contained in large capsules that eventually pop open and effectively toss their seeds as far as 20 meters (65 ft). The seeds float, and tambaquis gather around rubber trees where seeds are being released.

The tambaqui, a basslike, somewhat oval-shaped characin fish, weighs as much as 30 kilograms (66 lbs), and, with its specialized, rounded, molarlike teeth, it is capable of crushing and grinding very hard fruits, such as palm nuts and rubber tree seeds. Tambaqui feed almost exclusively on fruits for the first five months of the flooding season. The fruits consumed by the tambaqui contain sufficient protein and fat for the fish to survive during periods of low water from fat stored from its flood-cycle seed consumption. Seeds often contain toxins; thus, though fruit pulp is digested, seeds are not. Seed toxicity obviously enchances the probability of dispersal, as the seed tends simply to pass through the digestive system of the fish. Tambaqui also feed on fruits such as figs.

Juvenile tambaqui feed not in blackwaters but in whitewaters, along varzea floodplains and lakes. At the close of the flood cycle, when the waters drop, adult tambaqui migrate from nutrient-poor to nutrient-rich waters, where they spawn. Young tambaqui do not consume fruits but rather are zooplankton feeders, whose finely structured gillrakers aid in the capture of the tiny animals.

Even piranhas are known to consume seeds, removing the husk and masticating the soft seeds within (Goulding 1985). Characins are without a doubt important seed predators, possibly the most important seed predators in the flooded forests (Goulding 1985).

Catfish are not as destructive to seeds as characin species are because they gulp the fruit whole rather than macerate it (Goulding 1985). Thus catfish are somewhat the piscine equivalent of birds such as toucans, digesting the fruit pulp and passing the seed out of the alimentary canal unharmed.

Besides fruits and seeds, some Amazonian fish consume leaves and woody plant material, including detritus. One characin in the genus *Semaprochilodus* has evolved lips and mouth parts specialized for the removal of fine particulate detritus (Goulding 1993).

Because zooplankton are most abundant in varzea waters, this habitat forms an essential resource for many species of fish during the juvenile period of their life cycles. Zooplankton tend to be most abundant after the flood cycle, at times of low water, after the receding flood has carried them into the main river channels. Many fish, such as *Rhabdolichops zareti*, an electric fish in the Sternopygidae family, specialize on zooplankton feeding (Lundberg et al. 1987).

Two Dolphins and a Manatee

Aquatic mammals also live in the Amazon and Orinoco, including two species of freshwater dolphins and one manatee species.

The largest and most widespread of the two dolphin species is the pink river dolphin or boto (*Inia geoffrensis*). This animal is known by several common names depending upon region. In Spanish-speaking areas it is usually called *bufeo* or *tonina*, while in Portuguese Brazil it is known as *boto*. The common English spelling, "bouto," is apparently erroneous, as it is based on phonetic spelling of boto (Best and da Silva 1989). All dolphins, along with the whales, are mammals in the order Cetacea. Dolphins belong to the Odontoceti, the toothed whales. Pink river dolphins hunt for fish and other aquatic animals (turtles, crabs) in muddy waters of the Amazon and Orinoco basins, and they readily forage among the trees of flooded forests. Botos, which range in length from about 2 to 2.5 m (6.6–8.2 ft), have a long, slender snout, bulbous head, and modest dorsal fin. They tend to be pale in color, sometimes distinctly pinkish. Though common, they can be hard to see well as they typically reveal very little of themselves while surfacing to breathe (Emmons 1990). Botos are occasionally solitary but are usually in small groups of two to four animals, often at the mouth of a tributary or stream.

Unlike the majority of dolphins, botos have a flexible neck, which they probably put to good use as they search for food among the trees of the flooded forests. They are nearly blind but have effective sonar, permitting them to navigate among the dense trunks of the forest (Goulding 1993).

Boto

Amazonian animals are, indeed, the stuff of legends, and the pink river dolphin is no exception. The creature is rarely hunted, as many riverine people regard it as having most extraordinary powers. For instance, if a young woman is impregnated and the father is unknown, the pregnancy is often blamed on a nocturnal liaison with a boto, who is assumed to have lured the lady into the water for illicit porpoises (oops, sorry, I mean purposes). It has also reportedly been said that if a man adorns one of his wrists with the ear of a boto, he will enjoy large and lasting erections (Shoumatoff 1978). Uh huh.

The second dolphin species is *Sotalia fluviatilis*, the tucuxi (pronounced "too-coo-she") or gray dolphin. Tucuxis are smaller than botos, being no more than 1.5 m (4.9 ft) in length. They are also darker than botos and have a shorter snout, a much less bulbous head, and a more distinctive triangular dorsal fin. They tend to show much more of the head when they surface than is typical of the boto. In general, tucuxis are easier to see well as they often leap from the water and tend to be found in larger groups than botos, of from two to nine. Tucuxis are found along coastal South America as well as throughout the Amazon Basin (Emmons 1990). They are scarce or absent in most of the Orinoco Basin, probably due to seasonal changes in water depth. In the Amazon, tucuxis are typically found in deeper waters than botos. For instance, tucuxis do not swim into flooded forests as botos do (Goulding 1993). Tucuxis are so characteristic of deeper waters that river pilots rely on observing them for navigational aids. Because of the dramatic seasonal lowering of the Orinoco, it is doubtful that there is sufficient depth for tucuxis to survive. Hence, tucuxis have been reported only between the mouth of the Orinoco and Ciudad Bolivar, where a 10-m (33-ft)-deep navigation channel is maintained (Meade and Koehnken 1991).

The Amazonian manatee (*Trichechus inunguis*) is a close relative of the larger West Indian manatee, (*T. manatus*). Manatees are in the mammalian order Sirenia (the name refers to the fact that these odd, homely animals were allegedly once mistaken for mermaids, a tribute to the degree of loneliness or perhaps forgetfulness suffered by sailors long at sea). There are only four species on Earth. The Amazonian manatee reaches a length of about 2.8 m (9.2 ft) and can weigh up to 500 kg (1,100 lbs). It is hard to confuse a manatee with any other animal. Manatees have smallish, puffy-looking heads adorned with a wide, blunt snout and short whiskers. Their eyes are small but they will have a look around when they surface to breathe. The Amazonian species ranges throughout much of the Amazon and its major tributary rivers, but you will likely not see it, as it has been reduced throughout its range to the status of endangered species. Hunted for meat, oil, and hide, this once abundant creature, though now protected, still suffers persecution (Emmons 1990). Manatees are vegetarians, feeding in quiet waters that host an abundance of water lettuce or water hyacinth.

A Really Big Otter

Figure 175

The largest member of the weasel family (page 310) is the giant otter (*Pteronura brasiliensis*), found throughout Amazonia. Just as the Amazon has its giant snake (anaconda), giant turtle (arran), giant fish (pirarucy), giant rodent (capybara), and giant waterlily (Victoria), so has it a giant otter. This creature measures almost 1.5 m (5 ft) in length, not counting its meter-long tail! With a sleek, reddish brown coat, the giant otter is identified by its large size (a really good field mark!), fully webbed feet, and semiflattened tail, somewhat like an elongated beaver's tail. Giant otters are social, and groups forage diurnally in the quiet waters of the Amazonian tributaries, especially around oxbow lakes. Carnivorous, they feed on fish, mammals, birds, and other vertebrate prey, some sizeable. I was in a small canoe when I first came upon a giant otter. In a lapse of good judgment, we decided to play an audiotape of a giant otter's distress call. The beast reacted instantly and was soon swimming directly at us, teeth bared. Considering that giant otters are reputed to include anacondas in their diet, and as you may have already guessed, we quickly stopped playing the tape—and the otter quickly lost interest. Unfortunately, giant otters are much reduced by hunting pressure (for their skins and because they are perceived as competitors with humans for fish) and are listed by CITES as endangered.

Should you see an otter that is not really giant, you may well be looking at a southern river otter (*Lutra longicaudis*), a species that ranges throughout the Neotropics as far south as Uruguay, and to elevations as high as 3,000 m (9,842 ft) (Emmons 1990). The southern river otter is but half the size of the giant otter and can be identified by its all-white throat and belly and nonflattened tail. Though widely ranging, this species, like its larger cousin, has been sadly reduced by overhunting and is also on the CITES endangered list.

The Master of the Grasses

Figure 173

The capybara (*Hydrochaeris hydrochaeris*), at 1.2 m (4 ft) in length and 55 kg (120 lbs) in weight, is the world's largest rodent, a magnificent creature to behold. Ranging throughout most of lowland South America, from Panama to northeastern Argentina, capybara are fundamentally aquatic, ecologically similar to the African hippopotamus. A capybara is a cavimorph rodent (page 303), stocky, essentially tailless, with a light tan coat and short, thick legs. The toes are partly webbed, an obvious adaptation to the animal's aquatic habits. The head is squarish, with eyes, ears, and nostrils located on the upper part of the head, a probable adaptation to an aquatic lifestyle. Being a cavimorph, the capybara's vocalizations sound strikingly like those of guinea pigs, though you must be lucky enough to approach a contented one very closely to hear its

Young capybara

charming little squeeks, twitters, and grunts, sounds far more delicate than one might expect from a hundred pounds of rodent. Though usually found in small family groups, herds can grow to fifty or a hundred, especially in the Venezuelan Llanos or Brazilian Pantanal, where capybara remain abundant. Capybara groups typically spend most of their time in and along water courses feeding on water lilies, water hyacinth, leaves, and sedges that line Amazonian rivers, lakes, and swamps. The name capybara translates to "master of the grasses," a reference to their abundance on wet savannas. Their natural enemies, as might be expected, are caiman, jaguars, and anaconda. In many places, humans have hunted them and populations have been drastically reduced. In other areas, however, such as the vast ranches on the Venezuelan Llanos, capybara populations, when properly managed, provide a sustained yield of meat and leather (Ojasti 1991).

A Stupendous Serpent

The largest of all the New World snakes is the magnificent golden brown anaconda (*Eunectes murinus*), which ranges throughout Amazonia. Anacondas do not grow to be quite the length of some of the Old World python species, but they can reach nearly 10 m (approx. 30 ft), though such a length is rare. Anacondas are wider in body than pythons and are considered to be the bulkiest of the world's snakes. One 5.8-m (19-ft) specimen, photographed with seven men holding it, weighed 107 kg (236 lbs). Anacondas feed on agoutis, capybaras, peccaries, tapirs, large birds, and even crocodiles and caiman. They do not eat people and will avoid humans by taking shelter under water. Nonetheless, it's not a good idea to disturb a 6.1-m (20-ft)-long anaconda. I and a group I was with encountered a coiled anaconda along a riverbank, and we estimated the creature to be within this size range (it looked like a stack of colorfully patterned airplane tires). We left it very much undisturbed. Not particularly skilled swimmers, anacondas normally capture prey by lying in wait along quiet, muddy, marshy riverbanks, where even an immense snake can look remarkably camouflaged when coiled at the water's edge. Anacondas, like most snakes, are prolific breeders. One recorded birth included seventy-two baby snakes.

Crocodilians *Figures 176, 177*

Crocodilians are survivors of the reptilian dominance of the Mesozoic Era. Today's crocodiles and caimans look scarcely different from their ancestors, whose menacing eyes beheld such equally menacing creatures as *Tyrannosaurus rex*. There are only twenty-two extant species in the family Crocodylidae, of which nine occur in the Neotropics. Crocodilians eat fish and other water-dwelling animals, including capybaras, snakes, and birds. They typically are most active at night, spending the day basking. A night boat trip, flashlight in hand, should reveal the red eyeshine of caimans as they search for a meal. After an aquatic mating ritual, females build a nest mound and lay up to sixty eggs, depending on species. Parent animals, especially the female, aid the newly hatched young in moving from the nest to the water and remain with them for some weeks. Juveniles have many predators, including storks, egrets, raccoons, and anacondas. Adults have fewer predators but have been mercilessly overhunted by people. A comprehensive account of crocodilian biology can be found in Ross (1989) and Alderton (1991).

Two subfamilies of crocodilians inhabit the Neotropics, the true crocodiles and the alligators/caimans. They are similar in appearance and behavior, but crocodiles have more sharply pointed snouts than alligators and caimans, and the upper fourth tooth is visible on the outside when the jaws are closed. Alligators and caimans do not show the upper fourth tooth when the jaws are shut.

Crocodiles are in the subfamily Crocodylinae, which include the Old World Nile crocodile (*Crocodylus nilticus*) and the large and dangerous Indo-Pacific crocodile (*C. porosus*). There are four Neotropical species of crocodiles. The American crocodile (*C. acutus*) ranges from the Florida Keys and western Mexico southward to Ecuador, inhabiting coastal mangrove swamps. In eastern Central America, the Morelet's crocodile (*C. moreletii*) inhabits coastal mangroves and inland riverine habitats and is most common in Belize. The Orinoco crocodile (*C. intermedius*), as the name suggests, is found in Venezuela and eastern Colombia. Finally, the Cuban crocodile (*C. rhombifer*) is found only in a very limited range, the Zapata Swamp in Matanzas Province on Cuba.

Caimans and alligators, in the subfamily Alligatorinae, are generally much more abundant than crocodilians. They avoid salt water, occurring instead along riverine areas. The best-known species is subtropical, the American alligator (*Alligator mississippiensis*), which, with protection, has reestablished healthy populations throughout most of its range in North America. This success story is not generally repeated with South American caiman species. Once vastly abundant throughout the Orinoco and Amazon basins, caimans, like crocodiles, have suffered extreme population reduction in many areas because of human hunting pressure. Only in habitats such as the Venezuelan Llanos and the Brazilian Pantanal are caimans still numerous, particularly one species, the common caiman (*Caiman crocodilus*). The common caiman, which is also called the spectacled caiman, has a complicated taxonomy. Most authors recognize three subspecies, while some argue that five subspecies exist. Overall, common caimans can be found from southern Mexico all the way to parts of northern Argentina. They grow to lengths of 2.5 m (about 8 ft), occasionally larger, though such individuals are rare. The black caiman (*Melanosuchus niger*), which can grow to lengths of nearly 6 m (20 ft), is by far the largest of any Neotropical crocodilian. It inhabits the central Amazon Basin, from the mouth of the Amazon westward as far as northeastern Peru and Bolivia. The species has suffered serious population reduction throughout most of its range and is today probably most numerous in the eastern part of its range. Reductions in black caiman populations have allegedly led to increased numbers of capybara and piranha (Alderton 1991). The other caiman species are the Cuvier's dwarf caiman (*Paleosuchus palpebrosus*) and Schneider's dwarf caiman (*P. trigonatus*) both of which are found in both the Orinoco and Amazon basins, and the broad-snouted caiman (*Caiman latirostris*), found in southeastern South America.

Sadly, the CITES appendix of endangered species now includes the black caiman, one race of the common caiman, the broad-snouted caiman, the American crocodile, the Orinoco crocodile, the Morelet's crocodile, and the Cuban crocodile.

Turtles

Turtles are old creatures in the evolutionary sense, the turtle body a curious anatomical arrangement that has impressively stood the test of time. The largest known fossil turtle, appropriately named *Stupendemys*, with a carapace (upper shell) in excess of 2 m (6.6 ft), is known from the Pliocene in Venezuela (Carroll 1988). The turtles seen today in Amazonian waters belong to a relatively ancient group called side-necked turtles (Pelomedusidae), which date back to the Cretaceous Era. Side-necks do not pull their heads directly back under their shells, but rather tuck them sideways. Otherwise, side-necked turtles look similar to other freshwater turtles. The matamata (*Chelus fimbriata*) looks prehistoric, an imposing turtle that somewhat resembles snapping turtles. It is a bottom dweller that is frequently caught in fishing nets. You may well not see a turtle, however, when you visit Amazonia. There are surprisingly few, only about twenty species, considerably less than inhabit the Mississippi River and its tributaries (Goulding et al. 1996).

Riverbanks, at least in some areas, ought to be lined with basking giant arran turtles (*Podocnemis expansa*), which can weigh in excess of 45 kg (100 lbs). At one time the species was vastly abundant. When Henry Walter Bates (1863) explored Amazonia, he commented on the abundance of this big turtle and how good it tasted: "Roasted in the shell they form a most appetizing dish." Bates also said he was "astonished" at the skills of the Indians in shooting turtles as well as collecting eggs. Indians not only hunted the turtles, they kept many captured individuals penned for later use as food. Unfortunately, arran turtles have been so seriously reduced by hunting pressure, for both their meat and their eggs, that the species is now considered endangered, and the creature can be seen only in a few protected reserves (Goulding et al. 1996). However, the suggestion has been made that this species could be cultivated for food and thus brought back from the verge of extinction (Mittermeier 1978). The turtle meat is obviously (given their history) quite edible, and it has been argued that the yield from turtles would be 400 times that of cattle if rainforest were converted to equivalent pasture. But there is a problem. Unfortunately, the turtles don't appear to breed well in captivity, so thus far arran turtles have yet to become "hamburgers on the halfshell." And the species remains sadly endangered.

Birdwatching along Rivers and Streams   *Figure 171*

As rivers cut through rainforests they provide an excellent vantage point for observing bird activity. From a boat lazily moving downriver you can watch three and occasionally four species of vultures soaring overhead; observe hawks, caracaras, and falcons perched in riverside trees; and see parrots ranging from frantic flocks of small, screeching parakeets to the larger, more sedate macaws. Flycatchers are conspicuous, and various species of swifts and swallows skim above the water pursuing insects. Many remarkable Neotropical birds tenant streamsides and riverine habitats.

HOATZIN

Perhaps most unique among riverine bird species is the hoatzin (*Opisthocomus hoatzin*). This extraordinary bird is found along slow, meandering streams and oxbows within the Amazon and Orinoco basins. Hoatzins roughly resemble chickens in size and shape. However, their overall appearance suggests a primitive, almost prehistoric, bird. A hoatzin is somewhat gangly, its body chunky, its neck slender, its head small. The face is not fully feathered but rather consists of bright blue bare skin surrounding brilliant red eyes. A conspicuous, ragged crest of feathers adorns the bird's head. Adding to its antediluvian appearance is the bird's subdued plumage of soft browns with rich buff on breast and wings. Hoatzins are weak fliers, a feature that contributes to their primitive appearance. No one who sees a hoatzin ever forgets it. And usually you don't see just one. Hoatzins live in groups, noisy groups at that. Their nonmusical, gutteral vocalizations add to the auditory experience of Neotropical oxbow lakes.

Though originally believed to be related taxonomically to chickens, hoatzins are now considered most closely related to cuckoos, order Cuculiformes (Hilty and Brown 1986; Sibley and Ahlquist 1983; Sibley and Monroe 1990). The species is nonetheless sufficiently unique to be placed in its own family, Opisthocomidae. Hoatzins have an unusual diet, breeding system, and juvenile behavior.

Hoatzins are among the only avian foliovores, feeding mostly on leaves (over 80% of its diet), often from plants that are typically loaded with secondary compounds, such as plants of the arum family (e.g., philodendrons). Leaves are bitten off, swallowed, and ground into a large bolus in the bird's oversized crop. With the aid of an extensive microflora housed within the expanded crop and esophagus, the bolus slowly ferments and is digested. The birds benefit from some of the digestive products of their microflora, and the bacteria, which are as concentrated in hoatzins as they are in bovines, also help detoxify secondary compounds. The odd amalgamation of partially decomposed leaves gives the bird an unpleasant odor (rather like cow manure),

Hoatzin

a beneficial characteristic since it renders the flesh distasteful to human hunters. Though a few other bird species are known to eat leaves, hoatzins represent the only known case of a bird species that exhibits foregut fermentation, a unique adaptation (and coevolution with microorganisms) that enables the birds to survive on a diet of normally toxic plants (Grajal et al. 1989).

Hoatzins are communal breeders, and anywhere from two to seven birds cooperate in a single nesting. The pair responsible for the eggs is usually assisted by nonbreeding birds called "helpers." Studies have shown that nests with helpers are considerably more successful at fledging young than nests lacking helpers (Strahl 1985; Strahl and Schmitz 1990). The helpers aid in incubation and feeding young, enabling the juvenile birds to grow more quickly and thus reduce their vulnerability to predators. Their streamside nests are quite crude, consisting of a cluster of thin sticks so loosely constructed that the eggs are usually visible from beneath.

Baby hoatzins bear a superficial resemblance to *Archaeopteryx*, one of the first birds, whose fossilized remains established that birds evolved approximately 120 million years ago during the Mesozoic Era, when dinosaurs flourished. Young hoatzins possess claws on their first and second digits, enabling them to climb about in riverside vegetation. Juvenile hoatzins swim and dive efficiently. Should they be faced with danger, they escape by dropping from the vegetation into the water. When danger passes they use their wing-claws to help in climbing back on vegetation. Wing-claws were also present on *Archaeopteryx*, though no one suggests that the resemblance between the modern hoatzin and the first bird is other than coincidental. Young hoatzins lose their wing-claws as they attain adulthood.

SUNBITTERN

Figure 152

Stalking along quiet riverbanks, the heron-shaped sunbittern (*Eurypyga helias*) hunts fish, amphibians, crustaceans, and insects, which it captures by striking quickly, using its long neck and spearlike bill. With a sharply defined white line above and below the eye, and complexly patterned plumage, the sunbittern resembles the sunflecked forest interior. When displaying, it spreads its wings, revealing bright chestnut, yellow, black, and white linings that give the bird its name. Its legs are bright red. Like the hoatzin, the sunbittern is the only species in its family, Eurypigidae (order Gruiformes). Its low whistled call is commonly heard at dawn and dusk along streamsides.

SCREAMERS

Three species of screamers (family Anhimidae, order Anseriformes) are found along slow rivers, swamps, and marshes throughout South America. Screamers are most closely related to ducks and geese. The horned screamer (*Anhima cornuta*) occurs throughout northeastern South America and is common in the Venezuelan Llanos and throughout much of Amazonia. Nearly the size of a turkey (which it vaguely resembles), it is a shiny black bird with thick legs, large, unwebbed feet, and a smallish, chickenlike head. The horned screamer is named for both its long feather quill that tops off its head and its loud, piercing call. Screamers, though bulky birds, are excellent fliers and frequently perch in riverside trees. They are unique in their possession of a layer of air between their skin and muscle, and the buoyancy provided by this "inner tube" may aid them in soaring. Look closely at the vultures overhead: there may be a screamer or two soaring among them. DNA analysis suggests that screamers may be evolutionarily most closely related to the odd magpie goose (*Anseranas semipalmata*) of Australia and New Guinea (Monroe and Sibley 1993).

STORKS, HERONS, AND EGRETS

Figures 153, 168, 169, 170

Other tropical water birds are prone to soar high above rivers, marshes, and wet savannas. Anywhere from the southern Amazon through Central America, storks (family Ciconiidae, order Ciconiiformes), such as the wood stork (*Mycteria americana*), maguari stork (*Euxenura maguari*), and the huge jabirou (*Jabirou mycteria*), can be seen making lazy circles overhead during the heat of the tropical day. The jabirou is among the largest storks, topping 1 meter (about 3 ft) in height. It is all white on wings, tail, and body, with a bareskinned black neck at the base of which is a bright red patch. Its bill is long, thick, and slightly upturned.

Herons and egrets (family Ardeidae, order Ciconiiformes) are also common in marshland and riverside throughout the tropics. The long-necked herons and white-plumaged egrets are more slender than storks and fly with their necks held in an S-shaped curve. Storks fly with outstretched neck and head. The white-necked heron (*Ardea cocoi*) resembles the widespread great blue heron (*Ardea herodias*) of North America but has a white neck. The grayish-green striated heron (*Butorides striatus*) is the Neotropical equivalent of the little green heron (*Butorides virescens*), abundant throughout most of North America. Several Neotropical heron species have no equivalents in North America. The agami, or chestnut-bellied heron (*Agamia agami*), is a skulker, stalking prey along shaded streambanks. It has the longest bill relative to body size of any heron. The small and elegant capped heron (*Pilherodias pileatus*) is all white except for a black cap and blue skin around the face and bill. See it well and you'll see its long, white head plume. There are also several species of tiger-herons (*Tigrisoma* spp.) along Neotropical rivers and marshes. These birds are named for the fine barring that characterizes both juvenile and adult plumage.

Jabirou storks at nest

BOAT-BILLED HERON

The boat-billed heron (*Cochlearius cochlearius*) is an odd inhabitant of mangrove swamps and riverbanks, named for its extraordinarily wide, flattened bill. Taxonomically unique, it is the only member of the family Cochleariidae (order Ciconiiformes). Colonies of boat-billed herons leave their roosts at night and feed individually along rivers and marshes. The function of their seemingly oversized bill remains largely unknown, though the bill may be touch-sensitive, aiding the bird in searching for creatures inhabiting mud. They also feed on frogs, fish, and crustaceans. The species was believed to be most closely related to the night herons, and, indeed, the boat-billed heron somewhat resembles the widespread black-crowned night-heron (*Nycticorax*

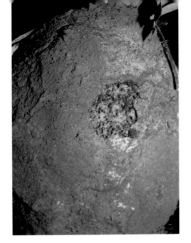

Figure 88. Termite nest, slightly damaged, revealing some of its occupants.

Figure 89. Bull's horn acacia with *Pseudomyrmex* ants patrolling the branch.

Figure 90–92. Leafcutter ants (*Atta*) gathering leaves to take to their huge subterranean nest, where the leaves will be used to culture fungi, which is the actual food of the ants. Note size variation among worker ants.

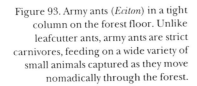

Figure 93. Army ants (*Eciton*) in a tight column on the forest floor. Unlike leafcutter ants, army ants are strict carnivores, feeding on a wide variety of small animals captured as they move nomadically through the forest.

Figure 94. *Doxocopa laure* (Nymphalidae) (left). Many other butterfly species have similar wing patterning. *Philaethria dido* (Heliconiinae) (right) is one of many heliconia species.

Figure 95. The lantern fly (*Fulgora laternaria*), also called "peanut head."

Figures 96–99. Katydids, showing diversity of cryptic patterns.

Figure 100. Small tree frog (coqui) on forest floor in Puerto Rico is very difficult to see because of its small size and cryptic coloration.

Figure 101. Boa constrictor on forest floor in Belize is a large animal but nonetheless easy to overlook as its cryptic coloration blends well with its background.

Figures 102 and 103. Mealy parrot and common iguana are each cryptic when in the canopy.

Figure 104. The cane toad is both large in size and not particularly cryptic in coloration, but it is protected by toxic secretions from its skin.

Figure 105. This highly toxic poison-dart frog (*Phyllobates*) from Costa Rica illustrates the obviousness associated with warning coloration.

Figure 106. Tamandua is cryptic and defends itself with slashing front claws, used here to tear apart termite nest.

Figure 107. Millipede, with its hard carapace, coils in a ball and emits a toxin.

Figure 108. White-fronted capuchin monkey.

Figure 109. Common woolly monkey demonstrating the utility of its prehensile tail.

Figure 110. Brown-throated three-toed sloth.

Figure 111. The face of a three-toed sloth.

Figure 112. Three-toed sloth in a cecropia.

Figure 113. Eyelash viper.

Figure 114. Basalisk lizard.

Figure 115. Owl butterfly.

Figure 116. Colorful spider (*Cupiennius*) from Costa Rica.

Figure 117. Jaguar.

Figure 118. Ocelot.

Figure 119. Kinkajou.

Figure 120. Silky (pygmy) anteater.

Figure 121. Woolly mouse opossum.

Figure 122. Large fruit-eating bat.

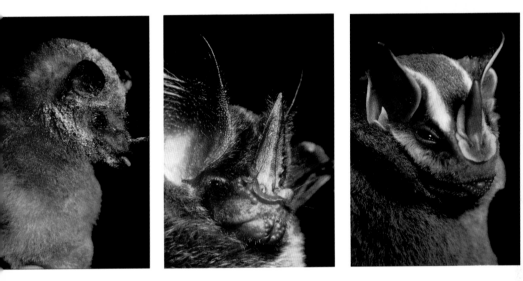

Figure 123. *Glossophaga longirostris* (Phyllostomidae: Glossophaginae), long-tongued bat, nectar feeder, from Grenada. Note pollen on face.

Figure 124. *Mimon crenulatum* (Phyllostomidae), hairy-nosed bat, from Ecuadorian rainforest. Feeds mostly on insects and spiders.

Figure 125. *Uroderma bilobatum* (Phyllostomidae), tent-making bat, from Costa Rica. Feeds on fruits, nectar, and insects.

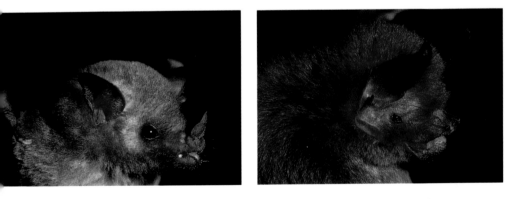

Figure 126. *Glossophaga soricina* (Phyllostomidae: Glossophaginae) from Costa Rica. Nectarivorous.

Figure 127. *Pteronotus parnelli* (Mormoopidae), mustached bat, from Costa Rica. Feeds on aerial insects.

Figure 128. *Saccopteryx letura* (Emballonuridae), wing-lined sac-winged bat, adult male, from Ecuadorian rainforest. Feeds on tiny insects.

Figure 129. *Hylonycteris underwoodi* (Phyllostomidae: Glossophaginae), long-tongued bat, nectar feeder, from Costa Rica.

Figure 130. *Noctilio albiventris* (Noctilionidae), lesser bulldog bat, a fish-eater, adult female, from Ecuador.

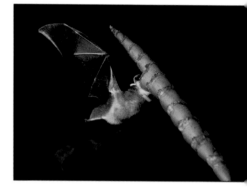

Figure 133. *Glossophaga soricina* (Phyllostomidae) hovering at *Vriesea gladioliflora*, a bat-pollinated bromeliad.

Figure 131. *Carollia castanea* (Phyllostomidae), short-tailed fruit bat, adult male, a frugivore from Costa Rica.

Figure 132. *Tonatia silvicola* (Phyllostomidae), round-eared bat, insectivore, from Ecuadorian rainforest.

Figure 134. *Glossophaga soricina* (Phyllostomidae: Glassophaginae), adult female with pup.

Boat-billed heron

nycticorax) in overall plumage as well as in nocturnal habits. However, examination of the DNA of the boat-billed heron has led to questioning of that assumption (Sibley and Ahlquist 1983), and thus its early evolutionary history remains elusive.

JACANAS

Eight species of jacana (family Jacanidae, order Charadriiformes) use their elongate, unwebbed toes to walk delicately atop lily pads searching for arthropod food throughout the world's tropical marshlands and riversides. Two species, the northern jacana (*Jacana spinosa*) and wattled jacana (*J. jacana*), are Neotropical. Both are chicken-sized, blackish birds with dark rufous wings that reveal bright yellow patches when the birds fly. The northern jacana is one of the few birds of which only males incubate the eggs, and any female will mate with several males.

KINGFISHERS

The 38-cm (15-in) ringed kingfisher (*Megaceryle torquata*) is abundant, noisy, and conspicuous along the Amazon as well as other Neotropical rivers and coastal areas. All kingfishers (family Alcidinidae, order Coraciiformes) have large heads and bills and make their livings plunging headfirst into streams and rivers in pursuit of fish. The ringed is by far the largest of the five Neotropical kingfisher species. Bluish gray above with a rufous breast and white neck,

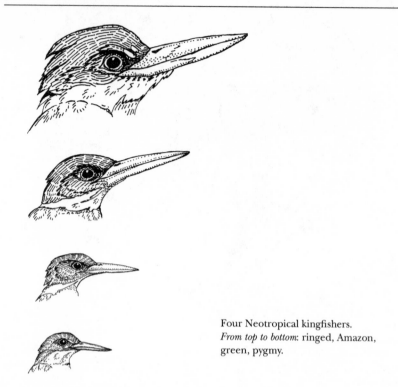

Four Neotropical kingfishers.
From top to bottom: ringed, Amazon,
green, pygmy.

its loud, rattling call, large size, and distinct color pattern make it easy to recognize. Like all kingfishers, the ringed is a swift and direct flier.

The other four kingfisher species, each in the genus *Chloroceryle*, are quite similar to each other in appearance but differ in size. From largest to smallest, there is the 11-inch Amazon kingfisher (*Chloroceryle amazona*), the 23-cm (9-in) green-and-rufous kingfisher (*C. inda*), the 20-cm (8-in) green kingfisher (*C. americana*), and the 14-cm (5.5-in) pygmy kingfisher (*C. aenea*). All of these, plus the ringed, may be spotted zooming along the same stretch of river (though the green-and-rufous fails to get into northern Central America). Each of the four is iridescent green above with various amounts of rufous on the upper and lower breast.

The kingfisher assemblage is yet another of the many examples of size gradation among related tropical bird species. As was discussed with flycatchers (page 102), the differing sizes may permit each species to specialize, taking a certain size range of fish. The degree of interspecific competition is, therefore, kept rather minimal, and the various species coexist within the same ecological community. Since captured fish are swallowed whole, it is hard to imagine that the diminutive pygmy kingfisher would ever be able to consume a fish that would comprise a square meal for a ringed kingfisher. Indeed, I have watched a pygmy kingfisher plunging into a shallow rainwater pool after tiny tadpoles and insects. Ringed kingfishers routinely eat fish larger in body size than a pygmy kingfisher! Kingfishers excavate tunnel nests in soft riverbanks.

Yellow-billed jacamar

JACAMARS

The sixteen jacamar species (family Gabulidae, order Piciformes) are exclusively Neotropical. Like kingfishers, they excavate nest tunnels in banks, hence their frequency along rivers. They are also commonly encountered both in forests and along forest clearings. Some of forest-dwelling species excavate nest holes in termite colonies. Jacamars are slender birds with rather long tails and long, sharp bills. They sit motionless on exposed limbs and sally forth to snap insects out of the air. Two basic color forms occur: one is metallic green above with rufous on the breast (a pattern also encountered among the kingfishers), and the other is dull blackish brown with white on the lower breast and belly.

Glossy ibis

9

Introduction to the Andes
and Tepuis

The Andes

Figure 4

YOU CAN'T miss them, especially as you gaze out from a comfortable cruising altitude of 10,000 m (33,000 ft) aboard an Aeroperu jetliner flying north from Lima, Peru, to Quito, Ecuador. Looking out from your window seat you see beneath you the Andes Mountains, at most a mere 20 million years old, their geologically young, ragged peaks densely clothed in a blanket of snow, an immense chain of granite stretching below as far as you can see in either direction. To the east, over the high peaks, lies the vastness of the Amazon Basin, while to the west is a narrow belt of coastline, much of it arid desert. In between, within the mountains themselves, are the high puna, paramo, and flat altiplano of the high Andes. The many puffy cumulous clouds that litter the sky, sometimes obscuring the view of the mountains below, were created in part by the effect of the massive chain. In this chapter I will introduce you briefly to the Andes and to Andean ecology. Good general accounts can be found in Morrison (1974, 1976) and Andrews (1982).

The Andes are the dominant topographic feature throughout all of western South America. Approaching Quito, which itself is at an elevation of 2,858 m (9,375 ft), you hope the clouds will lift sufficiently that you can see Cotopaxi, the currently quiescent volcano that looms above the city at an elevation of 5,897 m (19,344 ft), one of many potentially turbulent mountains along the almost 10,000-km-long (approx. 6,000-mi) Andes chain. The youthful mountain range below you stretches from Cape Horn, the southernmost tip of South America, all the way north to the Caribbean Sea, finally terminating in the gentle, densely forested Northern Range on the island of Trinidad.

The Andes Mountains, known in South America as *Cordilleras de los Andes*, run from south to north, the major ridges bending northeast when they reach Colombia and continuing into Venezuela, one ridge continuing northward into Panama. For most of their extent they are really a complex series of parallel chains divided by a flat tableland area, normally at about 4,000 m (approx. 13,000 ft) elevation. This tableland is called *altiplano* and is the site of several unique, isolated, high-elevation lakes, the largest being Lake Titicaca, which sits literally on the border between Peru and Bolivia.

As befits relatively young and geologically active mountains, the Andes are ruggedly tall, with peaks routinely ranging from 2,000 m (6,560 ft) to 6,000 m (19,680 ft). There are over a dozen peaks in excess of 6,100 m (20,000 ft), the tallest being Mt. Aconcagua in Argentina, at 6,962 m (22,834 ft). You must go trekking in the Himalayas or Pamirs of Asia to find mountains of similar stature (Everest is about a mile higher than Aconcagua).

The Andes cross the Equator, where snow falls at elevations between 4,500 and 5,000 m. Climb high enough and you can in theory stand directly on the Equator and toss a snowball off a very cold mountain. If you have exceptional throwing abilities, the snowball would melt in the hot, humid rainforest some 4,800 m (about 16,000 ft) below. As you move progressively north or south of the equator, climate becomes more severe and snowline occurs at increasingly lower elevations. Approaching the southernmost part of the Andes, snowline is at only 1,000 m (3,280 ft).

Geography of the Cordilleras: A Closer Look

The Andes Mountains extend well beyond the climatic zone of the Neotropics, beginning at the frigid southern tip of South America, just north of an area called Tierra del Fuego (Land of Fire). This land was once inhabited by the indigenous Yahgan (Yamana) Indians, a people living in an extremely harsh climate but who were physiologically capable of sleeping on snow, without benefit of blankets or down comforters (Bridges 1949). Ships rounding Cape Horn continually face strong westerlies, gale-force winds that create among the roughest seas known. It is here that Captain FitzRoy of the H.M.S. *Beagle* discovered the Beagle Channel. As he sailed through the channel, Fitzroy noted a 2,438-m (7,996-ft) mountain, which he named Mt. Darwin, after the *Beagle's* most distinguished passenger.

The Andes stretch northward, a relatively narrow ridge along the border between Chile and Argentina. Some of the tallest peaks occur east of the cities of Valpariso and Santiago, near the mountain city of Mendoza in Chile. Here, in close proximity, one finds Mt. Aconcagua (6,962 m or 22,834 ft), Mt. Tupungato (6,802 m or 22,310 ft), and Mt. Mercedario (6,772 m or 22,211 ft), and the Andes begin to widen into a series of ridges with extensive tracts of altiplano in between. The lower mountain slopes are temperate in climate, not yet tropical, and precipitation varies, depending on elevation, from 25 to 102 cm (about 10 to 40 in) annually. West of the mountain ridge, in northern Chile near the city of Còpaipo, the Atacama Desert begins, an arid coastal region extending northward almost 3,218 km (2,000 mi), finally becoming the Sechura Desert on the border between Peru and Ecuador.

Where the countries of Chile, Peru, and Boliva meet, the topography of the Andes becomes increasingly complex, as the mountain range diversifies into a series of ridges with vast area of intervening high-elevation altiplano. It is here that there was once an extensive inland sea, the legacy of which remains as unique salt flats (Salar de Uyuni and Salar de Coipasa) as well as Lake Titicaca and a few other scattered, small lakes. The main chain of the Andes, the Western Cordillera, continues west of the salt lakes and Lake Titicaca toward Machu Picchu (the great city of the Incas) and Cuzco, Peru. East of Titicaca, the Cordillera Real and Cordillera de Carabaya gradually descend on their eastern slopes through zones of humid montane forest, eventually terminating as tropical lowland rainforest in western Bolivia and eastern Peru.

In Ecuador and Colombia, the Andes diverge into three major ridges, the Western, Central, and Eastern cordilleras. The Western Cordillera extends north through Panama into Central America, eventually merging with the Mexican cordilleras, an orographic region that continues into the United States as the coastal ranges of California. The Central Cordillera extends roughly 800 km (approx. 500 mi) northeastward through Colombia. The Eastern Cordillera passes through Bogota toward the northeast, dividing into two ridges, the Cordillera de Perija and the Cordillera de Merida. The former continues northeast in Colombia and terminates on the Guajira Peninsula bordering the Caribbean Sea, while the latter bends further northeast, passing into Venezuela, terminating finally in the Northern Range in Trinidad (an island today but which was once part of Venezuela and only became isolated when sea level rose after melting of the glaciers several thousand years ago).

The effect of the complex topography of the Andes on the distributions of plants and animals cannot be understated. The immensity of the overall range, the divisions of ridges and intervening altiplano, and the elevational differences along the mountain slopes have provided ideal conditions for evolutionary divergence among many taxa. The fact that countries such as Colombia, Ecuador, and Peru have so many species is due in no small measure to the presence of the Andes.

Montane Cloud and Elfin Forests
Figures 5, 6

With increasing altitude up a mountainside, temperature decreases, and thus condensation and precipitation increase, supporting cool and humid *montane* forests. Typically, as moisture from the lowland forest rises and cools, montane forest becomes enshrouded in heavy mist for at least part of each day, giving rise to the term *cloud forest*. Clear morning skies soon yield to dense mist, clouds that persist from afternoon through dark. For obvious reasons, cloud forests look and feel overcast and damp.

Cloud forests are the dominant kind of ecosystem along a narrow altitudinal belt (from 2,000 to 3,500 m) along the east slope of the Andes. Usually, Andean cloud forests begin to replace lowland forests at about 1,200–1,500 m, a transition that may be very abrupt (Stotz et al. 1996). Cloud forests are also found in parts of southeastern Brazil (de Barcellos Falkenberg and Voltolini 1995), at higher elevations in the Greater and Lesser Antilles, and in parts of Nicaragua, Guatemala, Panama, and Costa Rica. In Central America, the transition from lowland to cloud forest occurs at about 600–700 m elevation. Cloud forests are also found in Hawaii, parts of central Africa, Madagascar, southern India, and much of tropical Asia. For a general though technical review of cloud forest ecology, see Hamilton et al. (1995).

Tropical cloud forests vary, of course, from one site to another, but all share general characteristics (Hamilton et al. 1995). Cloud forests typically are lush, with a high biomass and obvious abundance of epiphytic orchids, bromeliads, ferns, mosses, and lycopods, variously covering and draping branches and tree trunks. Vines are relatively uncommon, but shrubs are often abundant. Trees, most of which exhibit gnarled trunks and branches, are from 25 to 30 m in height, not as tall as in lowland rainforest, and normally densely spaced. There are often two tree strata. Buttressing is rare, and bark tends toward being dark and rough. Tree crowns are compact and do not spread, parasol-like, as is typical of lowland forest trees. Leaves are sclerophyllous, meaning that they are small, hard, waxy, and usually thick, similar to evergreen oaks. Though palms are less common and less species-rich than in lowland rainforest, there are montane palm species. Tree ferns (especially *Cyathea*), which are true ferns but grow to the size of small trees, are often common, adding a mysterious, almost prehistoric look to the landscape. Small ferns are also often abundant. Standing alone in such a mist-enshrouded, primeval-looking forest, one is sort of tempted to look for tyrannosaurs (and one should probably be glad not to be finding them). Bamboos thrive in humid montane forests, with one genus, *Chusquea*, often particularly abundant. Species richness of trees is less than in lowland forests but still generally high, especially species in the Melastomataceae. For a list of the most common genera in Andean cloud forests, see Stotz et al. (1996). Many endemic species occur in cloud forests (see below).

Cloud forests in southeastern Brazil are similar in structure to those in the Andes, though with some differences (de Barcellos Falkenberg and Voltolini 1995). Central American cloud forests have a somewhat different species composition from those in South America, with the addition of several oak species, conifers, and sweetgum (*Liquidambar*) (Stotz et al. 1996).

Because of the persistent cloud cover, sunlight is considerably reduced in cloud forests compared with lowland rainforests. Precipitation is abundant, in the form of both rainfall and interception of moisture directly from the clouds. Because the temperature is cool and moisture is abundant, evapotranspiration is generally low. Soil, not surprisingly, is wet, even waterlogged. Soils become very acidic (termed histosols) and often form peat, making the ground somewhat boglike.

Trees and shrubs at highest elevations (exact elevation varies with latitude) become noticeably shorter in stature, many extremely gnarled, heavily laden with epiphytes, especially mosses, lichens, and bromeliads. In addition, there is an abundance of microphyllous epiphytes such as lichens, green algae, and bryophytes. Such a stunted forest is called "elfinwood," characterized by an abundance of short twisted trees barely 2 meters tall. These diminutive forests can be so dense as to be impenetrable. In this climate of nearly perpetual mist, tree growth is significantly slowed by a shortage of sunlight as well as low temperatures. Not much energy is available for trees to invest in stems (nor is there much point to it since the light is so diffuse), hence stature is short (Grubb 1977a). Also, the atmosphere is so saturated with moisture that transpiration can be difficult, thus trees may be limited in their nutrient uptake from the soil. Elfin forests have a lower species richness of trees than lower-elevation forests, probably due to the greater rigor of the climate. Prominent genera include *Podocarpus, Clusia,* and *Gynoxys.*

At elevations between 3,500 and 4,500 m (11,400–14,800 ft) in the Andes, normally above timberline, islands of gnarled trees dominated by the genus *Polylepis* can be found among a landscape of puna or paramo (see below). *Polylepis* trees are confined to protected, rocky slopes. Though *Polylepis* can be found mixed among other species in lower-elevation cloud forests, the genus occurs in essentially pure stands at higher elevations. *Polylepis* is very drought resistant, with the largest trees reaching heights of about 18 m (59 ft). Though *Polylepis* woodlands occur all the way from Venezuela to northern Argentina and Chile, this ecosystem type is considered threatened throughout much of its range (Stotz et al. 1996).

Montane cloud forests support numerous species of plants and animals found in lowland rainforest, but they also harbor many endemic species (Gentry 1986b; Long 1995; Leo 1995; Stotz et al. 1996). Indeed, high levels of endemism may qualify as a defining characteristic for tropical montane cloud forests (Stadtmuller, cited in Leo 1995). For example, Stotz et al. (1996) point out that the tropical Andes and Amazon Basin each contain approximately the same number of bird species (791 and 788, respectively) but that the Andes have more than twice as many endemic species as the lowland area (318 compared with 152). In the tropical Andes, about 40% of the bird species are endemic, compared with 19% in Amazonia. Some cloud forests are so remote and difficult to reach that new species of birds, all obviously endemics, have been discovered only recently. Ornithologists have described a new species of cotinga (Robbins et al. 1994), a wren (Parker and O'Neill 1985), two antpittas (Schulenberg and Williams 1982; Graves et al. 1983), and an owl (O'Neill and Graves 1977), all from northern and central Andean cloud forests.

Though most Caribbean islands host at least some endemic species, they are often not confined only to cloud forest; thus, endemism of birds is less pronounced in Caribbean cloud forests than in those of the Andes (Long 1995). Nonetheless, it is interesting that the elfin-woods warbler (*Dendroica angelae*), an endemic strictly confined to Puerto Rican cloud forest, was discovered only as recently as 1971. In Central America, Long (1995) identifies sixteen bird species considered to have restricted ranges confined mostly to montane cloud forest.

Other vertebrate groups show considerable endemism in cloud forests, though additional studies need to be done to fully document patterns. For example, there are no published surveys of the fauna in the southeastern Brazilian cloud forest (de Barcellos Falkenberg and Voltolini 1995). Leo (1995) lists seventeen mammal species endemic to Peru, all of which occur mostly in cloud forest (some occur in other types of montane forest). These include one marsupial, one bat, two primates, one sloth, and twelve rodents. Leo (1995) also lists forty-two species of anurans (frogs, tree frogs, and toads) from five families that are endemic to Peruvian montane cloud forest.

Some Creatures of the Cloud Forests

The spectacled bear (*Tremarctos ornatus*) is named for its facial pattern of beige lines surrounding its eyes and cheeks. Otherwise, the creature is black. It is considered a medium-sized bear, weighing about 200 kg (approx. 900 lbs). The only species of bear found in South America, it inhabits low-elevation montane and cloud forests from Panama through Peru and Bolivia. The species is considered somewhat relict, as it once ranged from the southern United States (California to the eastern seaboard) and throughout Central America (Eisenberg 1989). Like most bear species, it is omnivorous, feeding on a wide variety of vegetation (including hearts of bromeliads) as well as small vertebrates and invertebrates. Mostly solitary, spectacled bears have been reduced in population in many areas by hunting and are among the most difficult Neotropical large mammals to see.

Cloud forests form the primary habitat for numerous bird species. One of the most spectacular Neotropical birds, the resplendent quetzal (*Pharomachrus mocinno*), inhabits Central American cloud forests, though it (along with other cloud forest bird species) migrates seasonally to lower elevations (see page 135). Among the more elegant Andean cloud forest birds are the four species of mountain-toucans. Mostly blue-gray with yellow rumps and long, variously colored bills, these birds are found only among the epiphyte-laden trees of cloud forests. Many of the most colorful tanagers and bush-tanagers are unique to cloud forests (Isler and Isler 1987). These include such spectacular species as the scarlet-bellied mountain-tanager (*Aniognathus igniventris*) and grass-green tanager (*Chlorornis riefferii*), which I saw in cloud forest near Cuzco, Peru. These tanagers were joined by an equally elegant male blackburnian warbler (*Dendroica fusca*), one of several wood-warbler species that nest in boreal forests of North America but winter in Andean cloud forests. Many hummingbird species are found in cloud forests, as well as high paramo and puna. Typical and ground antbirds also frequent cloud forests. One that is a particular challenge to birders is the giant antpitta (*Grallaria gigantea*), which, at 22.5 cm (about 9 in), is the largest and one of the most secretive of the ground antbirds. It inhabits stagnant, swampy areas inside cloud forest and is more often heard than seen. Much more gaudy is the Andean cock-of-the-rock (*Rupicola peruviana*), a close relative of the lowland Guianan cock-of-the-rock (*R. rupicola*), whose courtship behavior is detailed in chapter 12. The Andean cock-of-the-rock inhabits rocky ravines near streams in low- and mid-elevation montane forests. If you see a male well, you won't forget it. For a complete account of Andean birds, see Fjeldsa and Krabbe (1990).

Conservation of Cloud Forests

Globally, tropical montane cloud forests are at least as threatened as lowland rainforest, and, in some places, probably more so. Cloud forests are being cut for many of the same uses as lowland forests: commercial logging, expansion of subsistence agriculture, hunting for sport or commercial trade, exploitation of nonwood forest products, and clearing to make cropland for coca and other drug plants. With the increased human activity has come an increase in human-caused fires as well as threats from introduced plants and animals (Hamilton et al. 1995). Cloud forests are ecologically important both for their function as watersheds and for their unique biodiversity, and their value should be considered along with lowland rainforest in the development of comprehensive conservation and management plans for sustainable use (Doumenge et al. 1995).

Tropical Alpine Shrub/Grassland—Paramo   *Figures 7, 8*

Paramo is a shrubland ecosystem occurring at altitudes above cloud and elfin forest from Costa Rica south to Bolivia. Climate is wet and cool, often with nightly frosts throughout the year. Dominant vegetation consists of large, clumped grasses called tussock grasses with sharp, yellowish blades, along with a scattering of terrestrial bromeliads and ferns. Among the tussock grasses grow shrubs, of heights up to 4.6 m (15 ft), resembling small trees. Leaves grow from the base of the stem, surrounding it in a pattern termed a rosette. Most distinctive of these shrubs are the Espeletias, perhaps the oddest members of the immense composite family (to which daisies, asters, and goldenrods belong). Espeletias have short, thick, woolly trunks densely surrounded by withered dead leaves, topped by a rosette of thick, elongated green leaves, each covered by soft hairs that help minimize evaporative water and heat loss. Espeletias, scattered among the tussock grasses on the cold, windy Andean slopes, attract many hummingbirds and bees to feed on the nectar of the yellow flowers. Among them is the 23-cm (9-in) giant hummingbird (*Patagonia gigas*), the largest of over 300 hummingbird species (page 260).

Tropical Alpine Grassland—Puna

Figure 9

Puna, which may be dry or wet areas, is cold alpine grassland replacing paramo when conditions are more severe. Tussock grasses also occur abundantly on puna, as well as various succulents, such as cacti. But perhaps the most striking of the puna flora is *Puya raimondii*, the world's largest bromeliad, a huge and dense cluster of stilleto-like leaves with flowers on a stalk that may attain a height of 8–9 m (26–30 m). Puya flowers relatively infrequently. When it does, its hundreds of flowers cluster on a stalk that protrudes well beyond the dense basal cluster of bayonet-like, thick leaves. Puya, as well as other puna plants, is visited by numerous hummingbird species, including such spectacular species as purple-collared woodstar (*Myrtis fanny*), Peruvian sheartail (*Thaumastura coa*), and green-tailed trainbearer (*Lesbia nuna*).

Puna also is the favored habitat of several interesting mammals, including the vicuna (*Vicogna vicugna*), a South American member of the camel family. Dominant males group with up to ten females in herds that roam about the barren puna (Koford 1957). Another wild camel of the Andes, somewhat larger than the vicuna, the guanaco (*Lama guanacoe*) is perhaps ancestral to the llama (*Lama guanicoe glama*) and alpaca (*L.g. pacos*). No wild members of either of these species exist, and their origins date back to domestication by pre-Columbian peoples (Morrison 1974). Llamas are the beasts of burden for the mountain Indians, the modern Incas. The husky mountain viscacha (*Lagidium peruanum*), a member of the rodent family, is a close relative of the famed chinchilla (*Chinchilla sp.*), a species now quite local due to overtrapping.

High above the puna and paramo fly the black and white mountain caracara (*Phalcobaenus albogularis*) and the immense Andean condor (*Vultur gryphus*). The condor is surely the most majestic of the Andean birds as it soars effortlessly on thermal currents, flying from mountaintop to seacoast. Twice the size of a turkey vulture, the condor has a 2.9-m (9.5-ft) wingspread (surpassed, but only barely, by the largest albatrosses) and can weigh as much as 6.8 kg (15 lbs), making it one of the most massive of birds.

The high Andean salt lakes of the Altiplano are habitat for several flamingo species as well as other birds rarely encountered at lower elevations. The

James's flamingo (*Phoenicoparrus jamesi*) was actually considered extinct until rediscovered on a lake 4,400 m (14,450 ft) high in the Bolivian Andes in 1957 (Morrison 1974). More common are the Andean (P. *andinus*) and Chilean flamingos (P. *ruber*). All flamingos feed on brine shrimp and other small crustaceans skimmed from the water with their peculiar hatchet-shaped bills.

Also inhabiting high Andean lakes are the puna teal (*Anas versicolor*) and the Andean goose (*Chloephaga melanoptera*). The former is a small duck with a black upper head and white cheeks. The latter is a stocky white goose with black on the wings and tail. It has an extremely short bill. Black-headed Andean gulls (*Larus serranus*) are also common throughout Andean wetlands.

Along fast-flowing Andean rivers look for the torrent duck (*Merganetta armata*) and white-capped dipper (*Cinclus leucocephalus*). The sleek male torrent duck has a boldly patterned white head with black lines and sharply pointed tail. The female is rich brown. Both sexes have bright red bills. Six races of torrent duck occur from the northern Andes to the extreme southern Andes. Torrent ducks brave the most rapid rivers, swimming submerged with only their heads above water. The white-capped dipper is a chunky bird suggesting a large wren in shape. Like the torrent duck, it favors clear, cold mountain rivers, submerging itself in search of aquatic insects and crustaceans.

Remote mountain lakes, cut off for long periods from other lakes, often harbor unusual animals whose isolated populations have evolved in distinct ways. Lake Titicaca, which lies between Peru and Boliva, is home for the giant coot (*Fulica gigantea*), a ducklike bird with a chickenlike bill. On Lake Atitlan in southwestern Guatemala lives the endangered Atitlan grebe (*Podilymbus gigas*), a flightless grebe similar in appearance to, but larger than, the widespread pied-billed grebe, P. *podiceps* (Land 1970). Andean montane natural history is treated in greater detail by Morrison (1974, 1976).

Tepuis, Unique Examples of Ancient Erosion
and Modern Evolution
Figures 2, 3

The smooth terrain that prevails in much of southeastern Venezuela, part of the geologically ancient Guianan Shield, is broken by about 100 scattered, flat-topped mountains called *tepuis*, which together occupy an area of about 500,000 km^2 (200,000 sq mi^2). Tepuis are not part of the Andes Mountains, but are much older. Were they located in the United States, tepuis would be called mesas or tablemountains, in reference to their characteristic flattened summits. The word *tepui*, which is taken from the Penon Indian language, means mountain, and mountains they are. From the flattened lowlands of the Gran Sabana and its surrounding tropical rainforests, tepuis rise abruptly from their forest-enshrouded bases as steep, rocky escarpments to heights of over a mile. The tallest, Mt. Roraima, is 2,810 m (9,220 ft). The tepui region, located approximately 650 km (400 mi) south of the coastal city of Caracas, is home to the world's highest waterfall, Angel Falls, which plummets 979 m (3,212 ft) from atop Auyan-tepui. Angel Falls, named for the bush pilot Jimmy Angel, who accidently discovered it in 1935 (George 1989), is but one of hundreds of waterfalls spilling from various tepuis, continuing the ancient process of erosion of the tepuis. Flights to observe Angel Falls are a common tourist attraction from Ciudad Bolivar and Canaima National Park. The aircraft flies through dense banks of restless clouds that gather from moisture forced upward among the tepuis, passengers hoping (1) for an unimpeded view of the stunning waterfall and summit of Auyan-tepui and (2) that the pilot is well aware of the precise location of each of the mountains among which he is flying. The flight is both thrilling and spectacular.

Tepuis are of interest not only for their obvious stark beauty but for their intriguing geological history. They represent some of the most ancient geological formations in all of South America. Indeed, the sandstone and quartzite rock of which the tepuis are essentially all composed has been dated to be at least 1.8 billion years old (George 1989). This means that the rock (though not the tepuis—keep reading) was around when South America was still part of the huge continent of Gondwanaland (page 111), well in advance of the separation from which would be born South America, Africa, Antarctica, India, and Australia. Indeed, if one could project one's self back through time, to somewhere between 400 and 250 million years ago, the tepui region would be in close proximity to what would eventually become the division between South America and Africa, an area of lowlands in proximity to the sea. By somewhere between 180 and 70 million years ago, during the Mesozoic Era when dinosaurs were abundant, the future tepui region, known as the Roraima Plateau, became affected by the movement of Earth's plates, as tectonic activity was separating the continents. At this time the Andes were being uplifted to the west and the Roraima Plateau was being eroded by a combination of tectonic and meteorologic activity into what would become the tepuis. Evidence for the reality of continental drift can be seen in that the sandstone of the tepuis is virtually identical to that found in the mountains of the western Sahara (George 1989). Erosion continued throughout the Cenozoic Era, and the flattened tops of today's tepuis are all that remains of the once extensive Roraima Plateau. Most of the mass of sandstone that once comprised the plateau has long since found its way to the oceans through the continuous process of erosion. Today's tepuis represent but a fraction of that sandstone.

The flattened, eroded tops of the tepuis represent sky islands, an archipelago of isolated mountaintops. Receiving as much as 4,000 mm (about 157 in) of rain annually, much of it in the form of deluges from thunderstorms, the plants and animals that tenant the tepuis have evolved largely in isolation from populations in the surrounding lowlands and, for that matter, on other tepuis. Indeed, Sir Arthur Conan Doyle was so inspired by the splendid isolation of the cloud-enshrouded, wet tepuis that he chose the region as the setting for his 1912 science fiction novel *The Lost World*, a land where dinosaurs could still be found. Alas, no dinosaurs have as yet been located on any of the tepuis, nor are they likely to be. But the biota is nonetheless of great interest. At least half of the approximately 10,000 plant species are endemic, a clear example of the effect of evolution on isolated populations. Orchids abound, with sixty-one species found on Auyan-tepui alone (George 1989). Also common are various plants that consume insects, such as pitcher plants and sundews. The soil atop the tepuis is extremely poor, mostly just eroded rock. Insectivorous plants are advantaged in such a soil-impoverished habitat because they can supply their need for such nutrients as nitrogen and phosphorus by digesting insect bodies.

Animal populations have also evolved atop the tepuis. Among birds, there is a subspecies of paramo seedeater (*Catamenia homochroa*) and white-throated tyrannulet (*Mecocerculus leucophrys*) found only on the tepuis (Few 1994), and several warbler species in the widespread and diverse genus *Myioborus* (called "whitestarts") are found mostly on the tepuis.

Tree frog

10

Savannas and Dry Forests

ALTHOUGH most people conceive of the Neotropics as rainforest, those who have actually traveled in Central and South America know differently (figures 10, 11, 12). There are vast grassland ecosystems, some wet, some dry, some with scattered trees, some with scattered shrubs, located throughout tropical areas, collectively termed *savannas*. J. S. Beard (1953) was among the first researchers to define tropical savanna, calling it "a natural and stable ecosystem occurring under a tropical climate, having a relatively continuous layer of xeromorphic grasses and sedges, and often with a discontinuous layer of low trees and shrubs." In Beard's definition, "xeromorphic" refers to plants adapted to withstand periodic dryness, and the key element of the definition is that savannas are stable. They do not succeed to forest or any other ecosystem type. Where trees form a more dominant component of the vegetation, the ecosystem is a *dry forest* rather than savanna, but the two kinds of ecosystems intergrade and have much in common.

In South America, savannas and dry forest ecosystems are estimated to occupy about 250 million hectares (617 million acres), principally in Brazil (cerrado, caatinga, campo rupestre, and Pantanal), Colombia (Llanos), and Venezuela (Llanos), though large tracts can also be found in eastern Bolivia (Pantanal) and in northern Argentina and Paraguay (Chaco) (Fisher et al. 1994).

Many tree species populate savannas throughout Central America, the Caribbean islands, and equatorial South America, including acacias, palmettos, palms, cecropias, and others, depending upon location. Local plant species composition varies considerably. In general, savanna species diversity is higher in South America than in Central America. In much of Central America, the most abundant savanna tree species is Caribbean pine (*Pinus caribaea*), often adorned with bromeliads and orchids. Several species of oaks are common in Central American savannas, though no oaks are found in South American savannas. Fire-resistant tree species such as *Byrsonima crassifolia*, *Casearia sylvestris*, and *Curatella americana* are abundant on South American savannas as well as large swamps of moriche palm. Grasses and (in wetter areas) sedges form the ground vegetation. Soil, though it can vary widely, is either sandy or claylike, typically being described as "poor soil."

Many savanna and related ecosystems throughout the Neotropics are endangered from cutting and conversion to anthropogenic ecosystems. Indeed, savanna, cerrado, and dry forest are considered by many to be even more threatened than lowland ranforest.

Some Examples of Savannas

The Dry Pine Savanna of Belize
Figure 10

Throughout much of southern Belize east of the Maya Mountains, the dominant ecosystem type is savanna, abounding in Caribbean pine, but with many other species ranging from grasses, palms, and palmettos to cecropias and miconias. Compared with the nearby tropical moist forest nestled within the protective Maya Mountains, the pine savanna is an area of low species richness, a simpler, more arid and rugged-looking ecosystem. During the dry season, which extends from about February through most of May, the pine savanna is subject to occasional fires, the evidence of which can be seen as charred stumps and burned bark on many of the pines throughout the region. In this area, fire is an important ecological influent, a factor that provides the key ingredient in maintaining dominance of savanna.

Wildlife is less diverse in savanna than in interior lowland moist forest, but many animal species typical of forest occasionally range into savanna, including boa constrictors and jaguars. The pine savanna is a good place to look for gray fox, tayra, and white-tailed deer. It is also excellent for seeing birds of prey, including such species as roadside hawk, laughing falcon, and aplomado falcon.

Llanos—Seasonal Savannas of Venezuela

The wide floodplain of the Orinoco River extends over an area of grassy savanna, interrupted by gallery forest and hammocks of woodland. This habitat, which bears a strong physical resemblance to the Florida Everglades, is called the Llanos. Extending for approximately 100,000 km^2 (38,600 sq mi), the vast Llanos habitat can be found throughout southern Venezuela into parts of Colombia. Grasses and sedges, especially those in the genera *Panicum, Leersia, Eleocharis,* and *Luziola,* and *Hymenachne,* dominate much of the landscape, growing up to 0.9 m (3 ft) in height. Trees and shrubs are widely scattered, often occurring as island woodlots called *matas* (Blydenstein 1967; Walter 1973).

The Llanos is a highly seasonal wet savanna, with a pronounced dry season extending through most of the northern winter. Approximately 1,000 mm (39 in) of rain is received over the seven-month rainy season (Solbrig and Young 1992). But for nearly five months, rainfall is quite low. It is then when natural fires are common. During dry season, vast flocks of wading birds such as ibises, storks, herons, and egrets are concentrated in the relatively limited remaining wet areas. These birds, plus the presence of such spectacular species as capybara, giant anteater, anaconda, spectacled caiman, and jaguar, make the Llanos one of the best areas in the Neotropics for observing wildlife. Rainy season extends essentially from May through September and usually peaks in July. On average, rainfall exceeds 1,200 mm (47 in) per year and can be over 1,500 mm (59 in) annually. At peak rainy season, vast areas of Llanos are in full flood, though higher areas, hammocks (called *bancos*) of palms and other trees and shrubs, remain above water. Because of the strong degree of season-

Pinnated bittern, common in Llanos

ality, plant and animal species must be generally adapted to endure both drought and flood (Kushlan et al. 1985). Many immense cattle ranches can be found scattered throughout the Llanos, some of which also serve to host eco-tourist groups who wish to see the birds and other animals.

The Pantanal of Brazil and Bolivia
Figures 11, 169, 170, 172

The vast Pantanal, a name that means "swamp," is a more southern ecological equivalent of the Llanos, sharing many of the same species. Larger in area than the Llanos, the Pantanal covers approximately 200,000 km² (75,000 sq mi), of which about 70% is within the state of Mato Grosso do Sul in Brazil, with the remaining area in eastern Bolivia. It is a region of low elevation, only about 150 m (500 ft) above sea level, a vast, flattened basin created by deposited sediment eroded from the surrounding highlands. Eventually all of the many Pantanal rivers flow into the Rio Paraguay, the Pantanal equivalent of the Orinoco. Dry season ranges from May through October (essentially the opposite pattern from the Llanos). During rainy season, at its peak from late January through mid-February, water levels can rise as much as 3 meters, flooding much of the low-lying grasses and sedges.

In general, the human population is low in this hyperseasonal wet savanna, with but a few large cattle ranches and scattered small towns and villages. Consequently the wildlife diversity becomes a spectacle rivaling the African savanna. River banks are lined with myriad caiman kept well fed by the bountiful populations of piranha, tetras, catfish, and other fish. Giant otters and southern river otters make dens along riverine embankments. Marsh and red brocket deer can be seen among the tall Pantanal grasses, as well as giant anteater, Brazilian tapir, crab-eating fox, and crab-eating racoon. The Pantanal abounds with capybara and vast hosts of wading birds, including three stork species, four ibis species, and a dozen species of herons and egrets, most of which can be found on any given day. Overhead, snail kites search for apple snails and black-collared hawks hunt fish. The bulky southern screamer (*Chauna torquata*) can sometimes be seen by the dozens, foraging like cattle in the wide grasslands. Among the copses of palms, the huge hyacinth macaw can be found, the Pantanal being the final stronghold for this once abundant and majestic parrot.

White Ibis

The Brazilian Campos Cerrados   *Figure 12*

The largest area of savanna vegetation occurs in central Brazil, the *campos cerrados*, forming a wide belt across the country from northeast to southwest (Sarmiento 1983). Cerrados are unique, occurring on acidic, deep, sandy soil. Vegetation is not simply grassland but rather ranges from open woodlands with a 4-to-7-meter (13–23 ft) canopy to dense scrub thicket (Whittaker 1975). Cerrado soils are nutrient-poor, a factor probably partly responsible (see below) for their existence, rather than richer forest. Crop yields are dramatically increased when soil is fertilized with trace elements (Walter 1973).

The campos cerrados is the preferred habitat of the seriema, an odd bird that ecologically resembles the secretary bird from the East African savannas.

Pampas

In extreme southern Brazil continuing southward through Patagonia is an area termed *pampas*. Mostly grassland, the pampas are not considered to be true savanna, which is confined to the tropics and subtropics, but are part of the southern temperate zone. Biogeographers distinguish between the term *savanna*, for tropical and semitropical grassland, and *steppe*, for nontropical grassland. The pampas are steppes and consist of vast stands of tussock grasses

(*Stipa brachvchaeta, S. trichotoma*). In areas of sandy soil and decreased rainfall, dry woodland occurs, consisting mostly of a single species, *Prosopis caldenia*. Some unique animal species inhabit pampas, and I will briefly describe some of the most interesting later in the chapter.

Dry Forests
Figure 12

Depending upon the degree of dryness, which normally reflects a prolonged dry season, Neotropical woodlands may be partially or totally deciduous, dropping leaves periodically. Tropical broadleaf woodlands often merge with savannas, the latter usually being drier.

Dry forest is found throughout the Neotropics. It ranges from northwestern Mexico to northwestern Costa Rica, a belt of dry, deciduous forest that essentially extends with little interruption all along the western coast of Middle America.

In Brazil, a woodland type called *cerrado*, dominated by shrubs and small trees, occurs between semideciduous forest and savanna in which scattered trees are semideciduous and small, typically no taller than 7.6 m (25 ft) (see above). The stocky trees have dense, twisted branching patterns and thick bark. Cerrado areas are highly seasonal and experience frequent natural fires, and the soil is typically very sandy (Walter 1971).

A somewhat similar kind of ecosystem, *caatinga*, scattered throughout parts of Brazil, consists of highly seasonal (with prolonged dry season) deciduous forest dominated by spiny trees and shrubs with thick leaves and thick bark, their branches covered with an abundance of lichens and mosses. The ground may also be rich in mosses and other heathlike plants. Caatingas occur in climate that could support forest were it not for the nutrient-poor, sandy soil plus marked seasonality in precipitation. These ecosystems are not nearly as diverse as moist forests but nonetheless are characterized by a unique array of trees, grasses, and sedges (Pires and Prance 1985).

Thornwoods occur in semidesert areas from Mexico through Patagonia. Dominant trees are usually *Acacia* species and other leguminous trees, of short stature, spaced well apart, and often interspersed with succulents such as cacti and agave. In many areas of thornwood, large herds of goats can be seen wandering about. Thornwood is very common along the Pan-American Highway throughout Peru as well as in central Mexico and many West Indian islands.

What Causes a Savanna to Be a Savanna?

There is no simple or single environmental factor that determines that a given site will be savanna (Bouliere and Hadley 1970; Huber 1982, 1987). G. Sarmiento and M. Monasterio (1975) put the paradoxical nature of savannas well: "It is remarkable that savanna is the only ecosystem in the entire warm tropical region whose origin and permanence have been considered unanimously as a fundamental ecological problem." Neotropical savannas occur on a wide variety of soil types and experience all extremes of tropical climate. Rainfall may be seasonal or nonseasonal, and water drainage may be rapid or slow (Huber 1987). Fire is common, and savannas tolerate fire well, rebounding quickly after burning.

All savannas occur at low altitudes, less than 1,200 meters (3,937 ft) above sea level. Most savannas are located in environments where the amount of moisture is somewhere between what characterizes rainforest and dry deciduous or thorn forest. In other words, on a scale from very wet to very dry, savannas usually are in the midrange. This suggests that climate has a strong influence on savanna formation, but it cannot be the only influence because often savanna occurs in the midst of otherwise wet forest areas. For this reason, local soil type must also contribute to savanna formation. Soil and climate can interact in dramatic ways. In the central Llanos, heavy rains interact with soil to form a hardened crust of lateritic ferric hydroxide, usually at some depth in the soil but occasionally on the surface. This crust, termed *Arecife*, is sufficiently hard to impede the growth of tree roots, except where the woody species encounter channels through the crust. Tree groves or *matas* occur where roots have penetrated Arecife, resulting in the clustering of trees. Grass, with shallower root systems, usually thrives above the level of Arecife (Walter 1973).

Fire is believed to be an important influence on both savanna formation and propagation. Natural fires set by lightning are common, especially in areas with a pronounced dry season. Some savannas form on sites where rainforest has been repeatedly cut and burned, suggesting that human activity can change the site from forest to savanna.

Below, I review each of the major factors determined to be important in savanna formation.

Climate

Savannas typically experience a rather prolonged dry season. One theory behind savanna formation is that wet forest species are unable to withstand the dry season, and thus savanna, rather than rainforest, is favored on the site. Savannas experience an annual rainfall of between 1,000 and 2,000 mm (39–79 in), most of it falling in a five-to-eight-month wet season. Though plenty of rain may fall on a savanna during the year, for at least part of the year little does, creating the drought stress ultimately favoring grasses. Such conditions prevail in savannas throughout Venezuela, Colombia, Bolivia, Surinam, Brazil, and Cuba, but many savannas in Central America (Nicaragua, Honduras, Belize) as well as coastal areas of Brazil and the island of Trinidad do not fit this pattern. In these areas, rainfall per month exceeds that in the above definition. For only three months at the most is rainfall below 100 mm (3.9 in) per month. Other factors must contribute to savanna formation in these areas.

Soil Characteristics

In many characteristics, savanna soils are similar to those of some rainforest, though more extreme. For example, savanna soils, like many rainforest soils, are typically oxisols and ultisols (page 52) with a low pH (4–4.8) and notably low concentrations of such minerals as phosphorus, calcium, magnesium, and potassium, while aluminum levels are high. Some savannas occur on wet, waterlogged soils; others on dry, sandy, well-drained soils. This may seem contradictory, but it only means that extreme soil conditions, either too wet or too dry for forests, are satisfactory for savannas. More moderate soil conditions support moist forests.

Waterlogged soils occur in areas of flat topography or poor drainage. These soils usually contain large amounts of clay and easily become water-saturated. Air cannot penetrate between the soil particles, making the soil oxygen-poor. In extreme cases, hardened pans form, as in the case of Arecife cited above.

By contrast, dry soils are sandy and porous, their coarse textures permitting water to drain rapidly. Sandy soils are prone to the leaching of nutrients and minerals (page 49) and so tend to be nutritionally poor. Though most savannas are found on sites with poor soils (because of either moisture conditions or nutrient levels or both), poor soils can and do support lush rainforest. The white, sandy soils of the upper Amazon (page 56) support such forests, unless the forest is cut and burned (see below).

Fire

The vast savannas of Nicaragua, Honduras, and Belize are populated abundantly by Caribbean pine. Riding through miles of savanna along the Southern Highway in Belize, one notices that many of the pines have darkened fire scars on their trunks. Indeed, midway between the coastal towns of Dangriga and Punta Gorda is located a fire tower affording a magnificent panorama of savanna. For most of the year, a pair of barn owls (*Tyto alba*) are the only inhabitants of the tower. However, during the dry months of the spring, the tower is manned by someone on the lookout for fires. Lightning storms commonly set fires during dry season, and the effects of dryness and periodic fires combine to preserve savanna. Caribbean pines tolerate occasional mild fires better than other tree species in Belize. Grasses also thrive in an environment with periodic fire.

Most savannas probably experience mild fires frequently and major burns every two years or so. Many savanna and dry forest plant species are called *pyrophytes*, meaning they are adapted in various ways to withstand occasional burning. Frequent fire is a factor to which rainforest species seem unable to adapt, though ancient charcoal remains from Amazon forest soils dating prior to human invasion suggest that moist forests also occasionally burn. Experiments suggest that if fire did not occur in Neotropical savannas, species composition would change significantly. When burning occurs, it prevents competition among plant species from progressing to the point where some species exclude others, reducing the overall diversity of the ecosystem. But in experimental areas protected from fire, a few perennial grass species eventually come to dominate, outcompeting all others (Inchausti 1995). Evidence from other studies suggests that exclusion of fire results in markedly decreased plant species richness, often with an increase in tree density (Silva et al. 1991). There is generally little doubt that fire is a significant influent in maintaining savanna, certainly in most regions (Kauffman et al. 1994).

Human Influence

On certain sites, particularly in South America, savanna formation seems related to frequent cutting and burning of moist forests by humans. Increase in pastureland and subsequent overgrazing has resulted in an expansion of savanna. The thin upper layer of humus is destroyed by cutting and burning. Humus is necessary for rapid decomposition of leaves by bacteria and fungi and recycling by surface roots (page 54). Once the humus layer disappears, nutrients cannot be recycled and leach from the soil, converting soil from fertile to infertile and making it suitable only for savanna vegetation. Forests on white, sandy soil are most susceptible to permanent alteration.

In some areas deep-rooted grasses imported from Africa to furnish fodder for cattle have come to dominate savannas, replacing native species. These grasses, in particular *Andropogon gayanus* and *Brachiaria humidicola* (often called African elephant grasses), now are estimated to cover about 35 million hectares (86.5 million acres) of savanna (Fisher et al. 1994). Though these invading species have significantly reduced plant species richness, they do seem to sequester carbon dioxide effectively, perhaps offsetting somewhat the release of CO_2 from extensive burning of rainforests and savannas (see chapter 14).

Are Savannas "Natural"?

Some ecologists have suggested that virtually all Neotropical savannas have resulted historically from human activity. This claim is unsubstantiated by historical evidence, however. Analysis of preserved pollen suggests that savannas were present long before people arrived and thus are a naturally occurring ecosystem type in the region. Evidence exists that savanna vegetation grew in the Amazon Basin as recently as 13,000 to 30,000 years ago (Sarmiento and Monasterio 1975). Haffer (1969, 1974) has provided evidence of the importance of expansion and contraction of savannas during the Pleistocene in determining many present-day patterns of species diversity (page 115).

Though savannas in general contain far fewer species than rainforests, Neotropical savanas demonstrate the highest plant species richness of any savanna ecosystems on the planet (Huber 1987). In numbers of both herbaceous and woody species, Neotropical savannas rank first. Such a high species richness suggests that evolution of savanna species has been occurring for many millennia prior to human arrival and that savanna is as unique and intrinsic to the Neotropics as rainforest.

The origins of savanna clearly vary regionally. Sarmiento and Monasterio (1975) summarized savanna causation, concluding that three basic savanna types exist. The *nonseasonal savanna* is largely the result of poor soils. It is the savanna of white, sandy soils, where drainage is rapid and climate is wet most of the year. The *seasonal savanna* is the most widespread type in the tropics. It occurs on sites with a stressful dry season where soils are sandy and nutrient-poor. Fire is an important component of these savannas. The *hyperseasonal savanna* is characterized by an annual period of water deficiency plus a period of saturation. In other words, there is either too much or too little water in the soil, making soils nutrient-poor and tending to waterlog. These savannas are typically all grass, with very few trees and shrubs.

There is a dynamic, temporal interface among savannas, dry forests, and moist forests. One expands while the others contract in a climatically driven, edaphically influenced, long-term equilibrium that has produced and continues to produce far-reaching effects on evolutionary patterns of both plants and animals.

Conservation Issues

Savannas and dry forests have been overshadowed (no pun intended) by interest in preserving rainforests. Unfortunately, ecosystems such as the campo grasslands, cerrado, and caatinga of Brazil are among the most threatened ecosystems in the Neotropics (Stotz et al. 1996). Interest in the conservation of savannas and dry forest has lagged behind that of rainforests. These are, after all, far less glamorous ecosystems. Be that as it may, endemism among birds is far higher in dry forest than rainforest, and thus the likelihood of extinctions looms large should these ecosystems continue to be converted to such activities as soybean farming.

What Lives in Savannas and Dry Forests? *Figures 144, 145, 154*

Savannas have a rather monotonous look to them, as do dry forests (certainly in comparison with rainforests), but South American savannas are surprisingly rich in species. Beyond that, a high percentage of the species are endemic. In a survey of the *campo rupestre* savanna in Brazil, a 200-km^2 (77 sq mi) rocky area with an abundance of succulents and grasses, a total of 1,590 plant species have been found, of which about 30% are endemic to that particular savanna type (Burman 1991). Among the various endemics, many have fragmented, small populations.

Species richness of savannas, though reasonably high, is much less than rainforests. Expansion of savanna either by climate or by human alteration of the landscape will contribute to loss of rainforest species. Mammals such as monkeys, sloths, tapirs, and peccaries are either absent or highly uncommon in savannas. On the other hand, deer, gray fox, jaguarundi, tayra, and most armadillos inhabit savannas as well as forests. Anteaters venture into savannas in search of their formicine prey, as do the 1.5-cm (5-ft)-long giant armadillos (*Priodontes giganteus*). These creatures may attain weights of up to 59 kg (130 lbs). The giant anteater (*Myrmecophaga tridactyla*) is more common in savannas than rainforests. Many rainforest species venture into savannas that abut forest but will not extend their ranges far into savanna.

Savannas are ideal habitats in which to observe birds. The Llanos—vast, wet grasslands—support a diverse assemblage of waterbirds. Kushlan et al. (1985) studied the wading bird community of Venezuelan Llanos and found twenty-two species of large wading birds, including seven ibis species, one spoonbill, eleven herons and egrets, and three storks. They compared the Llanos with the Florida Everglades, where only fifteen species of large waders occur. Herons were the most diverse species in both ecosystems. Stork species, including the wood stork (*Mycteria americana*), maguari stork (*Euxenura maguari*), and the huge jabirou (*Jabiru mycteria*), were richest on the Llanos, probably because large fish, their principal prey items, were more abundant than in the Everglades. Only the wood stork occurs in the Everglades. The researchers noted differences among species relative to their foraging behaviors, prey selectivity, and habitat use. They hypothesized that the greater diversity of waders on the Llanos is due in part to greater habitat diversity, increasing the types of feeding areas available. They noted that several species, including the buff-necked ibis (*Theristicus caudatus*) and sharp-tailed ibis (*Cercibis oxycerca*), fed in very shallow habitats on high ground. Such areas were less available in the Everglades. They also noted that during the dry season many Llanos species fed together, using similar foraging behaviors and feeding sites. They speculated that prey availability is so high that competition among the wader species is minimal.

Vultures as well as many raptor species frequent savannas, searching for prey in the open habitat. The crested caracara (*Polyborus plancus*), with its ragged black crest, can often be observed walking about on the ground. Like the vulture, it is a carrion feeder. The rusty-colored savanna hawk (*Heterospizias meridionalis*) is also often on the ground, especially after fires, when it searches for small mammals, reptiles, and insects driven out by the flames.

Passerines provide interesting bird-watching in savannas. The gray and black fork-tailed flycatcher (*Muscivora tyrannus*) performs aerial acrobatics in quest of insects. Adult fork-tails have streamerlike tails twice the length of their bodies. The brilliant red and black vermilion flycatcher (*Pyrocephalus rubinus*) conspicuously dashes from a perch in pursuit of some luckless insect.

Central American savannas do not show bird species richness comparable to that found in nearby moist forests, nor do they contain the large number of endemic species that characterize South American savannas and dry forests. In a study of a lowland pine savanna in northeastern Nicaragua, there were only 56 bird species, compared with 116 species in nearby moist forest (Howell 1971). Of the savanna species, 26 were permanent residents, including such species as chipping sparrow (*Spizella passerina*), hepatic tanager (*Piranga flava*), eastern meadowlark (*Sturnella magna*), and red crossbill (*Loxia curvirostra*), all of which also occur in similar habitats in North America and are (in the Neotropics) savanna specialists, species that could not possibly colonize rainforest. There is little difference in bird species richness in savannas in southeastern North America compared with Nicaragua. A savanna in Georgia contained exactly the same number of nesting species (15) as that in Nicaragua, and far more individuals.

Snakes and lizards of various species occur commonly in savannas. I encountered my first boa constrictor as it crossed a savanna road in Belize.

Scorpions and tarantulas also frequent savannas, and certain insects, like tiny biting flies, can be an extreme nuisance.

A Quick Look at the Pampas

The wind-swept, barren pampas host a variety of animals not characteristic of rainforests. The pampas deer (*Ozotoceras bezoarticus*) is endemic to the pampas, as is the maned wolf (*Chrysocyon brachyurus*). Rodents are abundant, including the mara (*Dolichotis patagonum*), sometimes called the patagonian hare. Superficially resembling a hare (including long ears), this rodent, also called the patagonian cavy, can leap 1.8 m (6 ft).

Charles Darwin was fascinated by a group of burrowing rodents of the pampas collectively called the tucotucos (*Ctenomys* spp.). Writing in *The Voyage of the Beagle*, Darwin said, "This animal is universally known by a very peculiar noise which it makes when beneath the ground. A person, the first time he hears it, is much surprised; for it is not easy to tell whence it comes, nor is it possible to guess what kind of creature utters it. The noise consists in a short, but not rough, nasal grunt, which is monotonously repeated about four times in quick succession; the name Tucutuco is given in imitation of the sound."

The largest birds of the pampas are the flightless rheas, relatives of the ostrich. Two species, the common rhea (*Rhea americana*) and Darwin's rhea (*Pterocnemia pennata*), occur both on pampas and on more northern savannas. The common rhea, which can also be found in the campos cerrados and Pantanal, is the larger and more abundant of the two. Rheas have the unusual habit of laying eggs in a communal nest. Several females mate with one male, and each hen deposits two to three eggs in the same nest. Only the male incubates. Rheas run very swiftly but were successfully hunted by gauchos, the horsemen of the pampas, who brought them down using the bola, a twine tied to three balls, which was skillfully tossed from horseback to entangle the bird's neck or legs.

Ibis in flight

11

Coastal Ecosystems: Mangroves, Seagrass, and Coral Reefs

Mangroves and the Mangal Community

Figures 14, 15, 17, 18, 19

FORESTS of mangroves line tropical coasts, lagoons, and offshore islands (called cays [and pronounced "keys"] in the Neotropics). They range throughout the climatic tropical zone, from southern Florida (especially the Florida Keys) southward along both the Atlantic and Pacific coasts. The ecological community dominated by mangrove trees of various species (see below) is often called mangal. The West Indian term mangrove is not taxonomic, but rather refers to a series of characteristics that mangroves have in common. In other words, "mangrove" is based on physiological attributes. Tomlinson (1995) defines true mangroves as trees that are ecologically restricted to tropical tidal areas, where they tend to form pure or low species richness stands, and where they reproduce often by making new plants (viviparity) rather than by seeds. In addition, they often exhibit aerial roots of some sort and are strongly adapted to tolerate high levels of immersion in salt water. Finally, mangroves are taxonomically distinct from their nearest terrestrial relatives.

About thirty-four mangrove species occur globally, all restricted to tropical waters where the average water temperature is never less than 23°C (73°F) (Rutzler and Feller 1996). The Neotropics have the fewest species of mangroves of any global region, with eight species (some of which are relatively uncommon). Some mangroves are adapted to colonize shallow sandflats, trapping sediment and gradually building a dense, muddy, organic soil. As mangroves are far more salt tolerant than other tree species, they tend to line tropical coasts, and their abundance extends inland along tidal rivers. Though subject to disturbance (especially from hurricanes), they can eventually rebound, though recovery is often protracted and growth can be surprisingly slow (see below). Mangroves range in height from short and shrublike to anywhere from 10 to 20 meters (33–66 ft). Mangal provides important habitat for many kinds of insects and other arthropods, nesting sites for birds, as well as shelter and food for many fish and invertebrates. They are, to put it mildly, an essential component of coastal ecology. Mangrove ecology is summarized in Walsh (1974), Lugo and Snedaker (1974), Rodriguez (1987), and Tomlinson (1995).

Types of Mangroves

Unlike the names *oak* or *maple*, both of which refer to a specific genus of evolutionarily closely related species, mangrove is a general term describing a combination of physiological and anatomical attributes, and thus there is not total agreement among botanists as to exactly which species should be called mangroves (see Tomlinson's definition, above). However, in the Neotropics there are a handful of species that are considered to be mangroves under any classification. They are obvious for their abundance in suitable areas throughout the region.

All mangroves are woody plants that tolerate high internal concentrations of salt, a consequence of frequent immersion in high salinity tropical seas. These plants respond to the physiological challenges of high temperature and concentrated salt exposure with a variety of adaptations. For example, some have salt glands on their leaves that effectively remove excess sodium and chloride. Mangroves are also tolerant of soils low in oxygen. The thick, odorous, muddy substrate that anchors them is virtually anaerobic, devoid of any gaseous oxygen. But some mangroves have roots that extend above the soil, modified to obtain oxygen from air. Mangrove leaves are similar to leaves of many rainforest tree species in that they are simple, unlobed, and very thick, with a heavy waxy cuticle. This aids in storing water and preventing excess water loss through transpiration. Their major control of gas exchange is through stomatal openings on the leaves.

Red mangrove (*Rhizophora mangle*) is an abundant species that can grow as a bushy shrub, a stunted, "bonzailike" tree, or a full 10-meter (33-ft)-tall tree. It has reddish bark and numerous aerial prop roots, some of which are firmly anchored to the substrate and some of which grow downward toward it. Prop roots are important in providing a firm anchor for the plant since red mangrove is typically found on exposed flats. The broadly spreading roots help assure stability against winds, tides, and shifting sands. Prop roots also contain openings called lenticels, important in transporting air to the oxygen-starved deep roots. Leaves are oval and thick, dark green above, yellowish below. Flowers are pale yellow with four petals. Fruits are reddish brown and produce elongated green seeds that actually germinate while still attached to the parent plant. Seedlings resemble green pods and are about the length of a pencil. They drop from the plant and float horizontally in the sea, becoming flotsam in the tropical ocean. Dispersal is effective, and red mangrove has populated all tropical seas around the world. Seedlings eventually absorb sufficient water that they orient vertically as they float in the sea. Should the tide carry the vertical seedling to a shallow area, once it touches substrate it will anchor and begin to put out roots. Red mangrove ecology is discussed by Golley et al. (1962) and Onuf et al. (1977).

Studies conducted on Twin Cays in Belize have shown that red mangroves that do not exceed a meter in height may nonetheless be several decades old. The reason for the stunted growth is the shortage of but one nutrient, phosphorus (Rutzler and Feller 1987, 1996). This study demonstrates that mangrove growth can be exceedingly slow, an important point when considering how mangroves affect recolonization after disturbance.

Black mangrove (*Avicennia germinans*) is also abundant throughout the Neotropics, often forming pure stands in anoxic substate. Black mangroves tend to grow in less exposed areas than reds but are able to thrive in oxygen starved sediment. They are bushy-topped trees that can reach heights of up to 20 meters (66 ft). Leaves are oval, leathery, and downy white underneath. The flower is yellow and tubular, the fruit green and oval. Seedlings float and, like adult trees, are tolerant of low oxygen levels. The most notable feature of black mangrove is its root system. Shallow horizontal roots anchor it in thick, smelly, anaerobic mud, but these roots send up vertical shoots, above ground, called pneumatophores. Lenticels on the pneumatophores feed into wide air passages connecting with underground roots, providing a means for air transport to the oxygen starved root system.

Two other common mangroves are the white mangrove (*Laguncularia racemosa*) and buttonwood (*Conocarpus erecta*). White mangrove tends to grow at slightly higher elevations than red and black mangrove. It is less tolerant of prolonged immersion in the sea. White mangrove has scaly, reddish brown bark and greenish white flowers. It grows from 9 to 18 m (30–60 ft) tall. Buttonwood resembles white mangrove but occurs only well away from daily flooding by salt water. It is the least salt-tolerant of any of the four common mangroves. It was once not even considered to be a mangrove. As salinity declines, such as at higher elevations or with distance, pure stands of red or black mangrove give way to mixed stands of several species, not all of which will be mangroves. The transition from mangal community to upland community is generally gradual.

Exploring a Mangrove Cay

Man-o'-War Cay, 16 km (10 mi) east of the Belize mainland, is typical of most of the mangrove cayes throughout the American tropics. There is a bird colony located on it, mostly composed of magnificent frigatebird, or man-o'war birds (*Fregata maginificens*). It is the birds that give the cay its name. Our sailboat nears the island, passing over shallow beds of seagrass (see below) interspersed with scattered bare areas of white coral sand. A spotted eagle ray (*Aetobatus narinari*) glides along ahead of the boat, clearly visible in the transparent Caribbean Sea, and black longspined sea urchins (*Diadema antillarum*) dot the bottom. Brown pelicans (*Pelecanus occidentalis*) dive headfirst for fish in the shallow water. It is a hot, sunny January day, and most of the frigatebirds are airborne, gliding like stiff-winged avian kites against the deep blue sky.

Our attention is drawn to the mass of birds above us. Frigatebirds are slender, glossy black, with a forked tail and 2-m (7-ft) wing span. They use their sharply hooked bills to snatch fish skillfully from the water surface. They also routinely pursue and rob terns, pelicans, and boobies of their prey, a behavior that earned them the name "man-o'-war bird." Females have black heads and white breasts, and juveniles, which are equal to adults in size, have all-white heads and breasts. The breeding males are adorned with bright scarlet throat sacs, which, when inflated, dangle loosely like balloons from their necks. From a distance, the host of inflated red throats resemble so many Christmas balls dotting the mangroves. As the boat squeezes its way among the myriad red

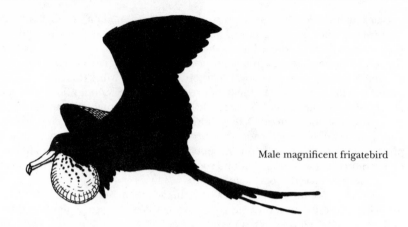

Male magnificent frigatebird

mangrove roots that line the outside of the cay, the birds become increasingly nervous, clattering their bills and flying from their flimsy stick nests. A group of about a dozen brown boobies (*Sula leucogaster*), smaller than the frigate-birds, with brown heads and bodies and white breasts, leave their nests and fly past us. One male frigatebird almost brushes the mast with his wing as he flies off, his huge crimson throat sac swinging awkwardly from side to side. The rank odor of the guano, the accumulated excrement on the leaves and mud, is a reminder of the role of these birds in concentrating nutrients. The mangroves are dense on this cay because the bird guano helps fertilize the soil.

The boat ties to some red mangrove prop roots that offer access to the interior of the cay. Still, we must acrobatically climb under and around the matrix of roots in order to set foot in the slippery, odorous mud that is the cay. White sneakers blacken as they become engulfed in sticky mangrove mud. Though the base of the cay is coral sand, the sand is thoroughly glued together by the sticky mud. Above us, birds still perched look down as we notice some scattered carcasses of dead chicks fallen from their nests, decomposing as they hang limply among the red mangrove branches. Here and there is an adult that suffered the same fate. Landing in mangroves is not all that easy for such large birds.

The mangrove roots are anything but barren of life. Among them are numerous mangrove tree crabs (*Aratus pisonii*) and land hermit crabs (*Coenobita clypeatus*), the latter sporting all manner of snail shells. Mostly active at night, some land hermit crabs burrow, and others tuck in for the day among the mangrove roots. They have one very large strong claw and do not hesitate to use it when suddenly disturbed. Also on board the trees are mangrove peri-winkles (*Littorina angulifera*), small snails that cluster about the branches. Here and there among the crabs and periwinkles, a tarantula goes about its business. The mud, like the tree branches, is also heavily populated by crabs, in this case mangrove fiddler crabs (*Macropipus puber*). As we try to walk through the morass of mud and roots, these tiny crabs scurry to their innumerable burrows. The mud seems alive with them. An odd thought occurs: one would not like to contemplate being staked out here at night.

Once inside the line of red mangroves we soon encounter an abundance of black mangroves. Small pools of stagnant water, where the intense tropical heat has concentrated salt, are interspersed among the engulfing mud, and the entire area is peppered with oxygen gathering pneumatophores from black mangrove roots, poking up from the mud. Walking over these stubby, pencil-like vertical roots requires some getting used to. An amazing array of flotsam has been trapped, including glass and plastic bottles, styrofoam, broken hunks of plywood, an old hat, somebody's cracked sunglasses. These are but a few of the many objects washed into the cay during storms, trapped by the mangrove roots and mud. Looking at the junk, its easy to see how these cays can grow by trapping sediment.

In the innermost part of the cay, where sediment is thickest, with maximum protection from the wind, we locate a clump of stately white mangroves. These trees are less bushy but distinctly taller than red mangroves, though a few black mangroves are about equal in size. Singing from the branches is a mangrove warbler (*Dendroica erithacorides*), now considered by some to be a subspecies of the widely spread yellow warbler (*D. petechia*). It is a small, slender bird with a dull red head and yellow body. Nearby is an iguana. How it got to the cay is anybody's guess, but iguanas are good swimmers.

How a Mangrove Cay Develops

Red mangrove lines the outer edge of the cay, and, in the sea just beyond the cay, red mangrove saplings grow. A careful look at the pattern of mangrove distribution in the cay suggests a zonation pattern among the various species. Outermost are "pioneer" red mangroves, followed by black mangroves, and innermost are white mangroves and an occasional buttonwood. This zonation pattern was once thought to correlate with the tolerance each species has for salt-water immersion, but, in fact, black mangrove is the most salt tolerant, so the pattern is not a simple case of response to salt exposure. The cay appears to be expanding outward. Red mangroves continuously colonize the outermost edges of the cay, but as sediment builds and the cay rises, black mangroves in turn expand their ranges outward, mixing with the reds. Whites and buttonwoods likewise expand their ranges as the sediment builds higher ground, as these species are most sensitive to immersion in salt water. Such a pattern has also been described for mainland coastal mangroves. Reds are outermost, blacks intermediate, whites and buttonwoods innermost.

Many authorities dispute the idea that mangroves are sharply zoned and represent a sort of successional sequence (Rodriguez 1987). Further, there is doubt about how routinely mangroves build up cays by trapping sediment. Experiments designed to test the efficacy of mangroves in cay reclamation following hurricane damage did not prove successful (Rutzler and Feller 1987). Geological evidence suggests that mangrove cays originate from coral deposition, not sediment accumulation by colonizing mangroves. Not all mangrove areas seem to accumulate sediment, and changing conditions caused by storms and tides certainly influence the pattern of mangrove distribution.

Some mangrove cays have remained stable for many years, without significantly expanding or contracting.

Experimental work from Belize examining the response of both red and black mangroves to a tidal gradient has demonstrated the complexity of factors influencing zonation between these two abundant species (Ellison and Farnsworth 1993). Only red mangrove propagules were capable of becoming established in the lowest low-water zone, where the innundation of water is greatest and available oxygen lowest. The species not only could survive but thrived, with high rates of root and stem growth, as it became firmly attached to the muddy substrate. Red mangrove propagules exposed to mean water depth grew less well, as a result of high rates of herbivory. Black mangrove could not become established in lowest low water, thus red mangrove enjoys both freedom from competition and lower predation pressure in this zone. Both red and black mangrove can succeed at mean water (on the tidal gradient), and thus they may engage in interspecific competition in this zone. Insect herbivory is also intensive in this zone and could affect competition between mangrove species. More research is needed to know to what degree herbivory is a factor. It has been suggested that black mangrove may be a gap-specialist species, dependent upon chance disturbance in order to become established (Ellison and Farnsworth 1993). Differential seed predation of mangrove species may also have a significant influence on zonation patterning (Smith 1987). Experimental work in Australian mangrove communities has suggested that seed predation by such creatures as crabs can result in a loss of about 75% of seeds. Most importantly, significant differences exist among the rates of seed consumption from one mangrove species to another. Should neotropical mangrove communities be similarly affected by seed predation differences, distribution patterning could be strongly influenced. Like temperate intertidal communities, mangal zonation at lowest elevation (maximum immersion) may be most affected by physiological tolerance differences among the various mangrove species, while at highest elevation biotic influences such as competition and herbivory are the predominant determinants.

Adjacent to Man-o'-War Cay, Cocoa Plum Cay has borne the full brunt of several hurricanes, being severed at least twice. What was one island became two, which, after another storm, became three. Cocoa Plum bears the scars of its storms. Uprooted mangroves, dead mangrove stumps, and wide-open sandy areas make for picturesque photographs. However, among the fallen and dead mangroves are numerous seedlings and saplings of red and black mangroves. The cay is gradually being recolonized and will eventually be rebuilt by the trees. Around the cay are shallow areas of submerged blackened peat, which were once part of the island. These areas illustrate the dynamic nature of a mangrove island. Storms change its boundaries, but the mangroves manage to persist.

In recent years Caribbean mangal and coral reef ecosystems have borne the brunt of hurricanes Allen (1980), Gilbert (1988), Hugo (1989), and Andrew (1992). Each has caused extensive damage to both reef and cays, but such periodic devastation seems a normal component of the Caribbean climate, and the organisms are adapted to recolonize eventually.

Seagrass

Closely associated with the mangrove cays are scattered, vast flats of seagrass that populate the clear, shallow sea inside coral reefs. Seagrasses are flowering plants, not algae, but they are adapted to a life underwater. They spread by means of horizontal underground stems called rhizomes. Once a seagrass seed sprouts, the plant can spread by rhizomes and colonize an entire sandflat. Seagrass supports interesting animal communities and aids, as does mangal, in providing concentrated energy that helps sustain the coral reef. The food webs of mangal, seagrass, and coral reef are all interlinked.

Types of Seagrass

The most common seagrass of the tropics is turtle grass (*Thalassia testudinum*), which has long, flattened green blades. Another species, less common than turtle grass, is manatee grass (*Syringodium filiforme*), which has rounded leaves resembling long, thin cylinders. Both turtle grass and manatee grass can occur singly or in mixed clumps. Both names are well chosen. Indeed, sea turtles and manatees (*Trichelchus manatus*) frequent seagrass flats and feed off the grasses. Among the grasses are clumps of various green algae, many of which are hardened by calcium deposits. *Penicillus* resembles a shaving brush protruding from the sand. *Udotea* is bright green, suggesting a small funnel anchored in the sand. *Halimeda* resembles thick clumps of green disks.

A Swim among Mangroves and Seagrass

After our strenuous walk through the mangrove cay, it is time for a refreshing swim. The Caribbean is irresistible as we don face mask, snorkel, and fins and plunge into the 27°C (80°F) water. The shallow, sandy bottom is a bit disturbed by the turbulence we create, and some coral sand momentarily clouds the water. Soon we swim into clarity and begin to inspect the inhabitants of the submerged red mangrove roots.

Attached to the roots are numerous oysters, sponges, barnacles, and tunicates, all feeding by filtering tiny plankton from the sea. Algae also abounds on the root surfaces. Hardly any bare area can be found on a submerged root. Many of the tiny animals contain mixtures of toxic substances in their bodies, protecting themselves from predators much as tropical rainforest plants utilize defense compounds (Rutzler and Feller 1996). The attached host of organisms, collectively called a fouling community, does no harm to the mangrove roots and may, in fact, protect them from the onslaughts of boring animals (Ellison and Farnsworth, cited in Rutzler and Feller 1996). Among the attached animals crawl the thin, brittle stars that randomly wave their five arms if picked off their chosen root. Tiny, colorful coral reef fish swim in and out among the stilted roots. Black-striped sargeant majors (*Abudefduf saxatilis*) poke at the roots as a school of juvenile French grunt (*Haemulon flavolineatum*) accompanied by a few juvenile schoolmasters (*Lutjanus apodus*) swim hastily past. Their speed may have something to do with the small great barracuda

(*Sphyraena barracuda*) lurking in the shade of the roots. The mangrove roots provide food and shelter for many of the juvenile fish that, as adults, will populate the coral reef.

As we swim away from the mangroves and over turtle grass, we feel a rhythm of gentle rolling, a mild swell waving blades of marine grass back and forth. There are shells of large queen conch (*Strombus gigas*) scattered about. Many of the shells are uninhahited by conchs but are instead the domains of tiny damselfish. These brilliant violet and yellow fish are so loyal to their conch shell that it is possible to catch them merely by lifting the shell from the water. They swim into it! Large, reddish orange sea stars (*Oreaster reticulatus*) are scattered here and there, as are the foot-long, orange-brown sea cucumbers (*Isostichopus badionotus*). These are two of the most conspicuous echinoderms of the seagrass beds. Schools of parrotfish of several species graze like herds of colorful piscine cows over the turtle grass. They make small puffs as they spit out sand that they ingested as they munched the grass. Listen closely and you'll hear a staticlike sound as the host of bony fish mouths grind coral, converting it into sand. More schools of grunt swim by, as well as a large school of silvery bonefish (*Albula vulpes*). As we continue over the turtle grass, we note scattered coral heads. Over there is boulder coral (*Montastrea annularis*). To our left is some elkhorn coral (*Acropora palmata*). Among the corals we see more of the various coral reef fish species. A small school of reef squid (*Sepioteuthis sepioidea*) darts by, disturbing some needlefish (*Strongylura notata*) at the water surface. As we swim back to the boat, we spot a small hawksbill turtle (*Eretmochelys imbricata*) swimming over beds of turtle grass. How appropriate.

The Food Web of Coastal Tropical Ecosystems

Both mangal and seagrass ecosystems exhibit high productivity. Red mangrove can produce 8 grams of dry organic matter per square meter per day, a rate of carbon fixation considerably higher than most other marine or terrestrial communities (Golley et al. 1962). Leaf and twig fall in a red mangrove forest was estimated to be over 1.2 tons per hectare (3 tons/acre) per year.

Such impressive productivity is not wasted. When a mangrove leaf or turtle grass blade drops off, it is almost immediately colonized by a wide range of bacteria, fungi, and protozoans. The colonization by microbes results in increasing the protein concentration of the fallen leaf (because the bodies of the microbes on it are rich in protein and because microbes consume carbohydrates, thus reducing the ratio of carbohydrate to protein), and the leaf will then be grazed by shrimp, worms, crabs, and various fish, including juvenile grunts and snappers. As the leaf is chewed, the microbes and part of the leaf are digested, but the remainder of the leaf either is spit out or passes through the alimentary canal. In either case it is recolonized by microbes, and the process begins yet again. In this manner the energy and minerals trapped by the mangroves and seagrasses is slowly and efficiently released to the animal inhabitants of the community. Mangal and seagrass export both energy and nutrients to neighboring estuaries and coral reefs by providing a continuous source of food for invertebrates, juvenile fish, and other creatures.

Coral Reefs

Figures 20, 21

In warm tropical seas around the globe, coral reef surrounds islands and parallels coastline. All reef-building coral species are confined to tropical waters where water temperature does not drop below 20°C (68°F). Coral reefs rival rainforests in gross productivity. Like rainforest, coral reef exhibits co-evolutionary associations among species, including the very animals that give the reef its name. Fortunately for the snorkeler or scuba diver, most coral reef inhabitants, in particular the colorful fishes, are relatively easy to observe. Kaplan (1982) provides an good introduction to the ecology of Caribbean reefs and identification of their inhabitants.

Coral is an animal, a member of the phylum Coelenterata (sometimes called Cnidaria), to which belong jellyfish, sea anemones, and hydras. Each coral animal is a tiny polyp housed in a calcium carbonate test (shell) of its own creation. The polyp is sessile and feeds by capturing tiny food particles (the reef plankton) in a ring of minute tentacles armed with poison cells. The cumulative efforts of many millions of coral animals build the reef. Coral animals are usually nocturnal, and the tiny coral heads embedded in the reef matrix are therefore often not evident to the snorkeler during the day.

Photosynthesis is done by plants. Coral reef accomplishes nearly as much productivity per unit area as rainforest (Whittaker 1975). However, if coral reef is such a productive ecosystem, where are the plants? What are the coral reef equivalents of the giant rainforest trees? Animals abound, fish seemingly everywhere, often in large schools. Plants, by contrast, seem conspicuous by their absence. Though it may seem paradoxical, plants on a coral reef are at the opposite end of the size spectrum from rainforest trees. The plants responsible for photosynthesis on a reef are mostly microscopic, one-celled algae that live among the matrix of reef material secreted by the coral polyps. These tiny organisms, most of which are protozoan dinoflagellates called *zooxanthellae*, live in intimate association with coral animals. Supplied with shelter and waste products from coral, zooxanthellae photosynthesize in the bright light that penetrates the clear waters over the reef. Coral animals use both the oxygen produced by the zooxanthellae as well as some of their products of photosynthesis. Zooxanthellae and coral represent an evolutionary mutualism: both groups of organisms are interdependent. Their interaction ultimately supports the bulk of the organisms that inhabit the reef.

There are approximately 60 coral species in the Caribbean, compared with over 700 species in the Indo-Pacific. This pattern of species richness is paralleled by fish. There are approximately 500 species in the Bahamas, but just over 2,000 species in the Philippines and 1,500 species on the Great Barrier Reef off eastern Australia. Although the richnesses of fish communities in the tropical Atlantic and Pacific differ, the pattern of species diversity is about the same for both areas (Gladfelter et al. 1980). This means that any small area of reef in the Caribbean will contain about the same number of fish species as an equal area, say, on the Great Barrier Reef. Fish diversity correlates with reef surface area, surface complexity, and height. However, the pool of species (no pun intended) is much greater in the Pacific, so the between-habitat diversity is greater there than in the tropical Atlantic. It is not clear why species richness is so much lower in the Caribbean, but one possibility is that because the Atlantic is a much younger ocean than the Pacific, there has been less time and less area for speciation to occur. Caribbean coral reefs are present from the Florida Keys through the Bahamas, Greater and Lesser Antilles, and along the Yucatan Peninsula. Indeed, the longest barrier reef (see below) in the Western Hemisphere is the 220-km (137-mi) reef that extends from the upper Yucatan Peninsula (around Cancun) southward all along Belize to the Gulf of Honduras.

Reefs grow as generation after generation of individual coral polyps add to the matrix. The living part of the reef rests atop the efforts of thousands of past generations. Three types of reef occur, and one type may, in time, develop into another. Fringing reefs surround most tropical islands. Atolls are also rings of coral, often incomplete, but no island is present. However, atolls are the remains of fringing reefs that surrounded an island that is now submerged. Only the ring of coral remains visible from the surface. From observations made during the *Beagle* voyage, Charles Darwin recognized that fringing reefs may become atolls (Darwin 1842). As an island subsides, the coral animals continue to build the reef. The island sinks, but the reef doesn't. Another type of reef, barrier reef, runs parallel to a coastline.

Coral species are adapted to different disturbance regimes. Some tolerate wave action better than others. Some do best in shallow waters, some in deeper waters. Because of wave action, differing tolerances and interactions among coral species, and relationships between coral and zooxanthellae (see below), reefs tend to be zoned.

The most windward part of a reef is often termed the *high energy zone*. It is here that the reef is most exposed to wave action. The reef front or fore reef is the area from the windward edge of the reef to the lower limits of coral growth, normally in excess of 70 meters (230 ft) deep. A swimmer moving toward the windward edge swims over shallow coral formations until coming to a point where the reef drops off abruptly, and only deep blue can be seen. If the swimmer parallels the beach, swimming over the reef, a structural pattern called *spur and groove* formation can be observed. Shallow walls of coral are interspersed with deeper canyons where coral sand and scattered brain corals and other species cover the bottom. The spurs are the dense coral walls that have built up fringing the mainland or island. The grooves are canyons created by erosion from strong wave action. Many coral species are common in this zone, including elkhorn and staghorns (*Acropora*), boulder coral (*Montastrea*), finger coral (*Porites*), brain coral (*Diploria*), starlet coral (*Siderastrea*), sheet coral (*Agaricia*), and stinging coral (*Millepora*; stinging coral is not a true coral, but rather is a hydrozoan, related to the potent stinging Portuguese man-o'-war).

The next zone is the windward reef flat. This is a shallow area on the reef crest, often dominated by stinging coral, various soft corals (*Palythoa* and *Zoanthus*), and encrusting red algae. This zone also typically contains scattered elkhorn and staghorn corals, boulder coral, and others. Coral sand covers the bottom, and scattered green algae (e.g., *Halimeda*, *Penicillus*, *Acetabularia*) are present.

Further leeward is the lagoon zone. Lagoons are protected areas, usually shallow (10–15 m, 33–49 ft), that support populations of turtle grass, scattered coral heads, and green algae. Mats of blue-green algae (cyanobacteria) may also occur. Sea stars, sea cucumbers, and conchs are common.

Coral species tend to exhibit vertical zonation. The deepest corals are encrusting species that occur at depths of up to 60 meters (197 ft). Hemispherical corals, such as brain corals, occur at depths of 5–35 m (16–115 ft), while branching corals, such as staghorn and elkhorn coral, are shallow, from surface to 10 m (33 ft). Shallow water corals are highly branched and have a high surface area/volume, a geometric growth pattern characteristic of multilayered trees. Their average polyp diameter is small compared with deeper-water species. Elkhorn and staghorn coral colonies have such high concentrations of zooxanthellae that they are functioning metabolically as autotrophs (Porter 1976). These corals, which indeed function essentially as plants, are a key reason for the high productivity of the reef. Hemispherical corals have large polyps, less zooxanthellae, and a smaller surface area/volume ratio. They are metabolically much more heterotrophic than autotrophic. Deep-water corals have large polyps and are total heterotrophs.

The diversity of coral species in any given zone is strongly influenced by disturbance history (Woodley et al. 1988) as well as interactions among the coral species (Porter 1972). Many examples of competition are known for corals. For example, stinging corals (*Millepora* spp.) often redirect their growth in the direction of gorgonian (sea fan) corals (*Gorgonia* spp.). The stinging coral contacts, abrades, encircles, and encrusts the target gorgonian (Wahle 1980).

A comparison between rainforest and coral reef reveals some striking similarities. Both ecosystems are highly productive, virtually the most productive of any natural ecosystems. Net primary productivities on reefs can approach 3,000 grams dry matter annually, sometimes higher. One difference, however, is that primary producers, the plants, are far more diverse and have vastly more biomass in rainforest. One-celled algae perform the same role, with the same success, as giant rainforest trees. There is a high species richness of animals in both coral reef and rainforest, including a high predator diversity. One difference, however, is that animals are generally easier to observe on reefs, though many nocturnal species occur. Territoriality is common among animals in both ecosystem types. Sexual selection is also evident. Fishes, such as wrasses and parrot fish, are highly sexually dimorphic and polygynous. Just as mixed flocks of birds forage in rainforests, mixed species schools of fish exploit coral reefs. There are impressive examples of coevolution in both ecosystems. Some coral reef fish are called "cleaners." They move into the mouths and around the skins of larger fish and remove attached parasites. There are also species that mimic cleaner fish but bite a chunk out of the host fish instead (Limbaugh 1961). Both coral reefs and rainforests contain many cryptic species and also some species exhibiting warning coloration. Toxic compounds and allelochemics are evident among species from both ecosystems. Rainforest species show both a high within- and high between-habitat species diversity. Coral reefs in the Caribbean lack a high between-habitat diversity, possibly because of historical factors.

But perhaps more than any other characteristic of similarity between coral reefs and rainforests is the profound importance of periodic disturbance. Strong hurricanes such as Andrew and Hugo have provided ecologists with excellent opportunities for studying the importance of these tropical monsoons in influencing diversity patterns. Particularly on exposed reef, high mortality rates typically follow in the wake of the passage of a hurricane that makes a direct hit (Fong and Lirman 1994; Hughes 1994; Witman 1992). Following a hurricane, recruitment of corals begins anew on the damaged areas. Further, different hurricanes can produce very different patterns of damage. By studying hurricane effects, it has become apparent that chance plays a strong role in the structuring of any given coral reef. Each hurricane is limited in the area it affects. Further, the local history of the reef will influence how that reef will be affected. The sum total of such a periodic disturbance pattern is that coral reef diversity differs from one location to another, and that species interactions within a coral reef ecosystem are eventually affected by meteorological disturbance. Connell's hypothesis (page 121) about how an intermediate level of natural disturbance is a key to maintaining high species richness appears equally applicable to coral reef and rainforest—just as Connell hypothesized.

12

Neotropical Birds

BIRDS are a magnet that helps draw visitors to the Neotropics. Some come merely to augment an already long life list of species, wanting to see more parrots, more tanagers, more hummingbirds, some toucans, and hoping for a chance encounter with the ever-elusive harpy eagle. Others, following in the footsteps of Darwin, Wallace, and their kindred, investigate birds in the hopes of adding knowledge about the mysteries of ecology and evolution in this the richest of ecosystems. Opportunities abound for research topics. There are many areas of Neotropical ornithology that are poorly studied, hardly a surprise given the abundance of potential research subjects. Like all other areas of tropical research, however, bird study is negatively affected by increasingly high rates of habitat loss. This chapter is an attempt to convey the uniqueness and diversity of a Neotropical avifauna whose richness faces a somewhat uncertain future.

Avian diversity is very high in the Neotropics. The most recent species count totals 3,751 species representing 90 families, 28 of which are endemic to the Neotropics, making this biogeographic realm the most species-rich on Earth (Stotz et al. 1996). But seeing all 3,751 species will present a bit of a challenge.

When Henry Walter Bates (1892) was exploring Amazonia, he was moved to comment on the difficulty of seeing Neotropical birds in dense rainforest:

> The first thing that would strike a new-comer in the forests of the Upper Amazons would be the general scarcity of birds: indeed, it often happened that I did not meet with a single bird during a whole day's ramble in the richest and most varied parts of the woods. Yet the country is tenanted by many hundred species, many of which are, in reality, abundant, and some of them conspicuous from their brilliant plumages.

The apparent scarcity of birds in Neotropical lowland forests seems surprising because more species of birds occur there than in any other kind of ecosystem. Entire families, including cotingas, manakins, toucans, ovenbirds and woodcreepers, typical antbirds, and ground antbirds, are essentially confined to the Neotropics, as are such unique species as screamers, trumpeters, sunbittern, hoatzin, and boat-billed heron. Bates put his finger on the irony of birdwatching in the tropics. Even birds with glamorous plumages can be remarkably silent, still, and difficult to spot in the dense, shaded foliage. Patience, persistence, and keen eyes are required of the tropical birder. Birds often seem to appear suddenly, because a dozen or more species may be moving together in a mixed species foraging flock (page 39), and thus the bird-watcher may face a feast-or-famine situation. One minute birds seem absent. Then suddenly they are everywhere. Bates described such an encounter:

There are scores, probably hundreds of birds, all moving about with the greatest activity—woodpeckers and Dendrocolaptidae (from species no larger than a sparrow to others the size of a crow) running up the tree-trunks; tanagers, ant-thrushes, humming-birds, flycatchers, and barbets flit-ting about the leaves and lower branches. The bustling crowd loses no time, and although moving in concert, each bird is occupied, on its own account, in searching bark or leaf or twig; the barbets visiting every clayey nest of termites on the trees which lie in the line of march. In a few minutes the host is gone, and the forest path remains deserted and silent as before.

In this chapter, I survey Neotropical birds, their adaptations, and basic ecol-ogy. Without question, the most notable characteristic of the Neotropical avifauna is its extreme species (and subspecies) diversity (Haffer 1985; Stotz et al. 1996). New species are still being discovered. For example, the chestnut-bellied cotinga (*Doliornis remseni*) was unknown to ornithology until discov-ered in 1989 in Ecuador and subsequently described as a new species (Robbins et al. 1994). Recent taxonomic work indicates that certain species, particularly among the typical antbirds and ovenbirds, should be split into several species rather than counted as one. Even large taxonomic divisions are being reorga-nized, such as the recent split that divides the antbirds (formerly all in the family Formicaridae) into two families, the typical antbirds (Thamnophilidae) and the ground antbirds (Formicaridae), as described in Ridgely and Tudor (1994). Taxonomy of Neotropical birds is also being affected by analyses of DNA similarities among species (Sibley and Ahlquist 1990; Sibley and Monroe 1990) and other molecular-based studies. In this account, I follow the classifi-cation given in Parker et al. (1996). For a most up-to-date description and summary of the orders of birds, see Gill (1995).

The finest reference on Neotropical birds is the four-volume series by Ridgely and Tudor, of which two volumes (1989, 1994, both on passerines) are now in print. The other two volumes will be published within a few years. In addition, the volume by Sick (1993), though confined to Brazil, contains a wealth of natural history information on Neotropical birds. The most up-to-date list of Neotropical bird species is found in Parker et al. (1996), though this list omits common names and may thus prove awkward for the birder or stu-dent unfamiliar with scientific names. For a reasonably complete, though out-dated, list of species, see Meyer de Schauensee (1966) or Howard and Moore (1980). Monroe and Sibley (1993) provide a checklist of the world's birds with a classification based on DNA analysis. Austin (1961) provides a dated but still useful survey of the natural history of the world's birds, including, of course, the Neotropics. Perrins and Middleton (1985) provide concise general natural history information on the world's birds. For detailed life histories of selected Neotropical species, see Skutch (1954, 1960, 1967, 1969, 1972, 1981, 1983). David Snow (1976, 1982) has written two books focusing on his studies of frugivorous birds. A comprehensive volume edited by Buckley et al. (1985) is invaluable for the serious student of Neotropical ornithology. For birders interested in identification, I include a list of regional field guides (page 316). The conservation of Neotropical birds is thoroughly treated by Stotz et al. (1996).

The dark, complex foliage of interior rainforest hosts the majority of tropical bird species, a diversity that increases markedly from Central America into equatorial Amazonia. From forest floor to canopy, hundreds of different species probe bark, twigs, and epiphytes for insects and spiders. Others swoop at aerial insects, follow army ants as they scare up prey, search for the sweet rewards of fruit and flowers, or capture and devour other birds, mammals, and reptiles. One bird, the harpy eagle (*Harpia harpyja*), stalks monkeys, sloths, and other large prey. A recent analysis suggests that there are currently 3,751 species of Neotropical birds (Brawn et al. 1996; Stotz et al. 1996), which represents about a third of all species of birds in the world. Even with such an abundance of diversity, patience and luck are needed to see birds well, especially when they may be 30 m (100 ft) or more above ground, or moving through dense vegetation.

Large Ground Dwellers

Tinamous

Though treetop species can be a challenge to see, even ground dwellers can be elusive. Forty-seven species of tinamous comprise the order Tinamiformes, a peculiar group of birds endemic to the Neotropics. A tinamou is somewhat chickenlike, a chunky bird with a short, slender neck, a small, dovelike head, and thin, gently downturned beak. Plumage ranges among species from buffy to deep brown, russet, or gray, often with heavy black barring. Some tinamous inhabit savannas, pampas, and mountainsides, but most live secretive lives on the rainforest floor, searching for fallen fruits, seeds, and an occasional arthropod. Forest tinamous are much more often heard than seen. One of the most moving sounds of the rainforest is that of the great tinamou (*Tinamus major*), a clear, ascending, flutelike whistle given at dusk, a haunting sound that heralds the end of the tropical day. One bird begins and soon others join in chorus. Evening twilight is the hour of the great tinamou senenade—they rarely sing during full daylight or dawn. Basically solitary, the tinamou may use their chorus to signal each other as to their various whereabouts.

The best way to see a tinamou is to walk a forest trail quietly (!) at dawn. You may suddenly come upon one foraging along the trail, and it will probably stare blankly at you for a moment before scurrying into the undergrowth. Tinamous are generally reluctant to fly but may abruptly flush in a burst of wings, landing but a short distance away. They cannot sustain flight for long distances because, even though their flight muscles are well developed, they are not well vascularized, and the limited blood flow greatly restricts their effectiveness (Sick 1993).

Though superficially resembling chickens, tinamou anatomy and DNA analysis show that they are closely related to ostriches, rheas, and other large, flightless birds. They are considered both an ancient and anatomically primitive group. Their rounded eggs are unusual for their highly glossed shells and range of colors, from turquoise blue and green, to purple, deep red, slate gray, or brown. Only the male incubates the eggs, another characteristic shared with ostriches and rheas.

Chachalacas, Guans, and Curassows   *Figure 151*

The fifty species of chachalacas, guans, and curassows (family Cracidae) are similar in appearance to chickens and turkeys and used to be classified with them but are now placed in their own order, Craciformes. They are found in dense jungle, mature forest, montane, and cloud forest. Though often observed on the forest floor, small flocks are also often seen perched in trees. All species nest in trees. Delacour and Amadon (1973) provide a detailed overview of the cracids plus individual life history accounts.

The nine chachalaca species are each slender, brownish olive in color, with long tails. Each species is about 51 cm (20 in) from beak to tail tip. A chachalaca has a chickenlike head with a bare red throat, usually visible only at close range. Most species form flocks of up to twenty or more birds. The plain chachalaca (*Ortalis vetula*) is among the noisiest of tropical birds. Dawn along a rainforest edge is often greeted by a host of chachalaca males, each enthusiastically calling its harsh and monotonous "cha-cha-lac! cha-cha-lac! cha-cha-lac!" The birds often remain in thick cover, even when vocalizing, but an individual may call from a bare limb, affording easy views.

Twenty-two species of guan and thirteen species of curassow occur in Neo-

Spix's guan

Female great curassow

tropical lowland and montane forests. Larger than chachalacas, most are the size of a small, slender turkey with glossy, black plumage set off by varying amounts of white or rufous. Some, like the horned guan (*Oreophasis derbianus*) and the helmeted curassow (*Pauxi pauxi*), have bright red "horns" or wattles on the head and/or beak. Others, like the blue-throated piping guan (*Pipile pipile*), have much white about the head and wings and a brilliant patch of bare blue skin on the throat.

Guans and curassows, though quite large, can be difficult to observe well. Small flocks move within the canopy, defying you to get a satisfactory binocular view of them. Like chachalacas, guans and curassows are often quite vocal, especially in the early morning hours.

Both New World turkey species occur in the Neotropics. The wild turkey (*Meleagris gallopavo*), which graces the Thanksgiving table with its domesticated cooked presence, once ranged south to Guatemala. Now only domesticated individuals are found throughout the tropical portion of its range. The spectacular ocellated turkey (*Agriocharis ocellata*) ranges, still wild, from the Yucatan south through Guatemala. Smaller than the common turkey, the ocellated has a bright blue, bare head with red tubercles. Its plumage is more colorful than its relative, particularly its tail feathers, which have bright blue and gold, eye-like markings that give the bird its name. Ocellated turkeys are easy to see at Tikal National Park in northeastern Guatemala and Chan Chich Lodge in Belize.

Trumpeters *Figure 150*

Nothing looks quite like a trumpeter except another trumpeter. These oddly shaped, rooster-sized birds of the rainforest floor are uniquely humpbacked, with long legs, slender necks, and a chickenlike head. There are only three trumpeter species in the world (family Psophiidae), and each is confined to a different region within Amazonia. Species are distinguished by the wing coloration, ranging from white to very dusky. Otherwise the birds are blackish but with iridescent violet and greenish colors when in direct sunlight.

Trumpeters amble along the forest floor in small flocks, feeding on such diverse items as large arthropods and fallen fruits. They are also reputed to chase snakes. The name trumpeter comes from their curious vocalization, a ventriloquial, muffled hoot, rather like the sound of air blowing over the opening of a bottle (Sick 1993). Trumpeters will occasionally run around in circles, strutting and prancing with wings outstretched, apparently a courtship or excitement display. They roost in trees and nest in tree cavities. They are generally considered to be weak fliers.

Doves and Pigeons

Doves and pigeons (order Columbiformes) are much alike in anatomy, but, in general, doves are more birds of edges and open areas (with some notable exceptions such as the quail-doves) while pigeons are found mostly in closed forest. Doves and pigeons feed heavily on seeds and fruits, and some species can be important seed dispersers. There are just over 300 species in the world, of which about 64 occur in the Neotropics (Parker et al. 1996). Some Old World doves are extraordinarily colorful, but Neotropical species tend toward a plumage of muted colors such as grays, tans, or rich brown. Some of the larger species make low, deep, cooing vocalizations that suggest the hooting of an owl. Doves and pigeons of various species are relatively common throughout Neotropical habitats. Skutch (1991) surveys the general natural history of the group.

Ruddy quail-dove

The Gaudy Ones

Several groups of Neotropical birds are known for their bright colors. Among them are the trogons, motmots, toucans, cotingas, manakins, parrots, and tanagers. The ecology of many of these birds is closely associated with their habit of eating much fruit.

Trogons

There are thirty-nine species of trogons (Trogoniformes) found in the world's tropics and subtropics, and twenty-five are Neotropical. The family is well represented in Middle America as well as South America, and two species are found in the Greater Antilles. A trogon is a chunky, squarish bird with a long, rectangular tail and short, wide bill. Brilliantly colorful, males have iridescent green and blue heads and backs, and bright red or yellow breasts. Females resemble males but are duller in color, often quite grayish. The pattern of black, gray, and white on the tail and the color of the eye-ring (a patch of colorful skin circling the eye) are important field marks to identify various species. They range in size from about 23 to 38 cm (9–15 in).

Trogons tend to sit upright with tail pointed vertically down. They remain still, and so they are often overlooked. The easiest way to spot one is to look for its swooping flight, flashing the bird's bright plumage, and note where it lands. Most trogons vocalize throughout the day, often a repetitive "cow, cow, cow," or "caow, caow, caow," varying, of course, from one species to another. Sometimes the note sounds harsh, but in some species it's softly whistled and melodious. A good way to see a trogon at close range is to try to imitate its call. If the imitation is good, trogons will "come in" to investigate. Some species are common along rainforest edges or successional areas. Look for their characteristic upright shape perched in cecropia trees.

Trogons are cavity nesters. Some species excavate nest holes in decaying trees, others dig into termite mounds. The violaceous trogon (*Trogon violaceous*), common from Mexico throughout Amazonia, utilizes large wasp nests. Alexander Skutch (1981) observed how a pair of violaceous trogons took over a wasp nest. The pair excavated their nest over several days in the cool early morning hours before the wasps became active. Skutch observed the trogons attack the wasps. Perching farther from the vespiary than they had done while watching each other work in the cool early morning, they made long, spectacular darts to catch the insects, sometimes seizing them in the air, sometimes plucking them from the surface of their home. A sharp tap rang out each time a trogon's bill struck the vespiary in picking off yet another wasp. The two trogons never eliminated all of the wasps but did successfully nest, snapping up fresh wasps daily. Oddly, considering that they are often very aggressive, few wasps attempted to sting the trogons or drive them away.

Trogons feed on fruits from palms, cecropias, and other species, which they take by hovering briefly at the tree, plucking the fruits. They also catch large insects and occasional lizards, swiftly swooping down on them or snatching them in flight. Trogon bills are finely serrated, permitting a tight grip on food items.

The most spectacular member of the trogon family is the resplendent quetzal (*Pharomachrus mocinno*), which appears on Guatemalan currency and is said to be the inspiration for the legendary phoenix (the Guatemalan currency is the *quetzal*). Quetzals inhabit the cloud forests of Middle America, migrating to lower elevations seasonally. Peterson and Chalif (1973) describe the quetzal as "the most spectacular bird in the New World," a debatable point (my favorite is the Guianan cock-of-the-rock), but you get the idea. The male is "intense emerald and golden green with red belly and under tail coverts" (Peterson and Chalif 1973). The male's head has a short, thick crest of green feathers (kind of an avian mohawk) and a stubby, bright yellow bill. Most striking are the male's uniquely elongated upper tail coverts, graceful plumes that stream down well below the actual tail, making the bird's total length fully 61 cm (24 in). Females are a duller green and lack the elaborate tail plumes.

Motmots

Motmots (family Momotidae) consist of nine species, all Neotropical. They are most closely related to certain kingfishers (page 215) and the todies (Todidae), a group of five species of small, brilliantly colored, kingfisher-like birds that inhabit various islands of the Greater Antilles. All of these birds share a similar foot structure, in which the outermost and middle toes are fused together for almost their entire lengths. Motmots are slender birds whose back and tail colors are mixtures of green, olive, and blue with various amounts of rufous on the breast. They have a wide, black band through the eye, and some species have metallic, blue feathers at the top of their heads. They range in size from the 18-cm (7-in) tody motmot (*Hylomanes momotula*) to the 46-cm (18-in) rufous motmot (*Baryphthengus martii*).

Two remarkable features of motmots are a long, raquet-shaped tail (present on most but not all species) and heavily serrated bills. The tail, which in some species accounts for more than half the bird's total length, develops two extraordinarily long central feathers. As the bird preens, sections of feather barbs drop off, leaving the vane exposed. The intact feather tip forms the "raquet head." One may first sight a motmot as it sits on a horizontal branch in the forest understory methodically swinging its tail back and forth like a feathered pendulum. Another distinctive motmot characteristic is its bill, which is long, heavy, and strong, with toothlike serrations. I have held motmots and can testify as to the strength of their bite. They feed on large arthropods such as cicadas, butterflies, and spiders and will often whack their prey against a branch before eating it. They also take small snakes and lizards and frequently accompany army ant swarms (see antbirds, below and page 40). Motmots also eat much fruit, especially palm nuts, which they skillfully snip off while hovering in a manner similar to trogons.

All motmots are burrow nesters, another characteristic they share with kingfishers and todies. They excavate a tunneled nest along watercourses or occasionally within a mammal burrow.

Motmots are vocal at dawn. The call of the common and widespread bluecrowned motmot (*Momotus momota*) may have given the family its name. The bird makes a soft, monotonous, and easily imitated "whoot whoot; whoot whoot." Often a pair will call back and forth to one another.

Toucans, Aracaris, Toucanets, and New World Barbets

Perhaps more than any other kind of bird, toucans symbolize the American tropics. With an outrageous boat-shaped, colorful bill almost equal in length to its body, the toucan silhouette is instantly recognizable. As it flies with neck outstretched, a toucan appears to follow its own oversized bill. *Toucan* comes from *tucano*, the name used by Topi Indians in Brazil. Altogether, there are fifty-nine species in the family Ramphastidae, including toucans, aracaris, toucanets, and New World barbets, all Neotropical. Their anatomy and DNA indicate a close alliance with woodpeckers (and thus they are in the same order, Piciformes), and both groups share certain characteristics of foot anatomy (two toes face foward, two face to the rear), as well as the habit of roosting and nesting in tree cavities. Rhamphastids occur in lowland moist forests and montane cloud forests. Toucans, aracaris, and toucanets range in body size from 31 to 61 cm (12–24 in). Barbets (see below) are smaller.

Toucans' seemingly oversized bills are actually very lightweight. The bill is supported by bony fibers beneath the outer horny surface (which is not very different from a fingernail). The upper mandible is slightly down-curved, terminating in a sharp tip. Highly colorful patterns adorn most ramphastid bills.

Toucans have rather slender bodies with relatively short tails. Like their bills, their bodies are colorful, including patches of green, yellow, red, and white. One major group has ebony body feathers offset by white or yellow throats and scarlet on the rump or under the tail. Most species have a colorful patch of bare skin around each eye.

The 51-cm (20-in) keel-billed toucan (*Ramphastos sulfuratus*) is one of the larger species, ranging from tropical Mexico through the upper Amazon. It is black with a bright yellow throat and breast, a white rump, and scarlet under the tail. The bill is green with an orange blaze on the side and blue on the lower mandible. The tip of the upper mandible is red, and there is bare, pale blue skin around the eye. Both male and female look alike, a characteristic of most rhamphastids. The keel-billed toucan has a call remarkably like a treefrog: "preep, preep, preep" (see below). Like most toucans, keel-bills associate in flocks of up to a dozen or more individuals. Typically, when one toucan flies, soon another follows, and then another. A loose string of toucans will move from one tree to another.

Toucans are primarily frugivores, taking a wide variety of' fruits from many genera, including *Cecropia* and *Ficus*. They show a preference for the ripest fruits, selecting black over maroon and maroon over red, the precise order of ripest to least ripe. Toucans are relatively large, heavy birds and prefer to perch on strong branches, reaching out to snip food with their elongate bills. Toucans are gulpers (page 139). A bird snips off a fruit and holds it near the bill tip. It then flips its head back, tossing the fruit into its throat. Though this may seem awkward, the birds seem to have little difficulty. The long bill may be adaptive in permitting the relatively heavy toucan to reach out and clip fruits from branch tips, which its weight would otherwise prohibit. In addition to fruits and berries, toucans eat insects, spiders, lizards, snakes, and nestling birds and eggs, all of which contain more protein than fruit.

Some sympatric species of toucans bear a close anatomical resemblance and are best identified by voice. For example, in northeastern Amazonia, the Cuvier's toucan (*Ramphastos cuvieri*) and the yellow-ridged toucan (*R. culminatus*) are both black with white throats and chest and a yellow rump. But the Cuvier's toucan, which is a bit larger, vocalizes with a loud series of whistled yelps, whereas the yellow-ridged toucan makes a sound much like the monosyllabic croaking of a frog (Hilty and Brown 1986). Many ornithologists and birders have noted that where two large and similar toucans co-occur, one typically is a yelper while the other is a croaker. The same vocalization pattern applies, for instance, in Panama, where the keel-billed toucan is a croaker and the similar chestnut-mandibled toucan (*R. swainsonii*) is a yelper (Ridgely and Gwynne 1989).

Aracaris (genus *Pteroglossus*) are 38–41 cm (15–16 in) long, mostly dark in color with banded breasts highlighted by bright yellow or orange red. Their bills are patterns of gray and black. They have longer pointed tails than typical toucans. Toucanets (genus *Aulachorhylnchus*) are about 33 cm (13 in) long and primarily greenish, with rufous tails. Their bills are dark below and yellowish above. Both aracaris and toucanets are gregarious fruit eaters.

Toucans bear an anatomical and ecological similarity to Old World hornbills (family Bucerotidae). Both families consist of colorful birds with huge, downcurved bills and slender bodies. Both nest in tree cavities, and both include fruit as a major part of an otherwise broad diet. Hornbills and toucans are not evolutionarily closely related and thus represent an example of convergent evolution.

Barbets are smaller than toucans but are colorful, frugivorous birds with prominent, wide bills. There are twenty-six barbet species in the Old World and eighteen in the Neotropics. Anatomically, except for plumage differences, they appear quite similar. However, analysis of DNA indicates that New World barbets are less closely related to Old World barbets than to New World rhamphastids (Sibley and Ahlquist 1983, 1990). Work on mitochondrial gene cytochrome c also supports this view (Barker and Lanyon 1996). Look for barbets in small flocks in fruiting trees. A barbet flock can be extremely territorial when defending a fruit tree, driving away larger birds such as pigeons.

Fruit and Nectar Feeders

Toucans are but one of several major bird families that concentrate on fruit for their diets. Throughout the tropical year there is at least some availability of both fruit and nectar. Though seasonality exerts important effects on animal communities (chapter 1), it is nonetheless generally true that some plants are fruiting or flowering every month of the year. In the temperate zone, fruits tend only to be abundant from midsummer through autumn. Many birds, including migrating species, switch over from predominantly insect to fruit diets at that time. In the tropics, however, no such dramatic switch need be made. The *constant* availability of at least some nectar and fruit has made it possible for several major bird groups to specialize and feed on one or the other (or both).

Hummingbirds

Nectar feeders consist almost exclusively of hummingbirds (family Trochilidae, order Trochiliformes). There are 322 species of these small, rapid fliers, all restricted to the New World. Most are tropical, but 16 species do migrate to breed in North America. The iridescent beauty of their plumage is reflected in their names: berylline, emerald-chinned, magnificent, garnet-throated, sparkling-tailed, ruby topaz, jewelfront, blossomcrown, and so on. Among the hummingbirds you'll meet trainbearers, sylphs, coronets, velvetbreasts, sapphires, hillstars, firecrowns, sabrewings, spatuletails, topazes, racket-tails, starthroats, fairies, and mangos. Skutch (1973) provides a general account of hummingbird natural history.

Hummingbirds are highly active and seemingly effortlessly fly forward and backward or hover. Some of the smaller species resemble large insects as they buzz by, and the smallest hummingbird is, indeed, named the bee hummingbird (*Calypte helenae*). They accomplish their remarkably controlled flight both by a unique rotation of their wings through an angle of 180° and by having an extemely high metabolism. Hummingbird heart rates reach 1,260 beats per minute, and some species beat their wings approximately 80 times per second! Hummingbird metabolisms require that the birds must eat many times per day to fuel their tiny bodies adequately. Some mountain and desert species undergo nightly torpor, an adaptation to the cold temperatures of the evening. One species, the bearded helmetcrest (*Oxypogon guerinii*), which lives in the cold, windswept paramo of the high Andes, feeds mostly by climbing about on *Espeletia* flowers, often actually walking on matted grass (Fjeldsa et al. 1990). By not hovering as much as other hummingbirds, the helmetcrest must save considerable energy, aiding its survival in such an inhospitable environment. Metabolic rate and body temperature drop during the nighttime hours, and the bird is thus able to sleep without consuming an inordinate amount of energy and literally starving itself.

Hummingbirds are both thrilling and frustrating to watch because they move so quickly. Suddenly appearing at a flower, its long bill and tongue reaching deep within the blossom to sip nectar, a bird will briefly hover, move to a different flower, hover, and zoom off. Others will come and go, and some will occasionally perch. The best way to see hummingbirds well is to observe at a flowering tree or shrub with the sun to your back so that the metallic, iridescent reds, greens, and blues will glow. In those hummingbird species that are sexually dimorphic, the male has a glittering red, green, or violet-blue throat patch called a gorget. The gorget is part of the male's display behavior when courting females. Depending on the sun's angle relative to the bird and observer, the gorget may appear dull, partially bright, or utterly brilliant and sparkling. When a male is courting, he positions himself such that the female is exposed to the gorget at its utter brightest.

All hummingbirds are small, the tiniest being the bee hummingbird found only in Cuba and the Isle of Pines. It weighs about as much as a dime. The largest is the 23-cm (9-in) giant hummingbird (*Patagona gigas*) of the Andean slopes. This bird is sometimes first mistaken for a swift as it zooms past.

The diversity of bill anatomy, plumage, and tail characteristics among hummingbird species represents a fine example of adaptive radiation (page 98). The Andean sword-billed hummingbird (*Ensifera ensifera*), which lives high among Andean dwarf forests, has a body length of 13.2 cm (5.2 in) plus an almost equally long bill! This extraordinary length is a probable case of co-evolution since the Andean swordbill feeds on *Passiflora mixta*, a flower with a very long, tubelike corolla. The booted racket-tail (*Ocreatus underwoodii*), also

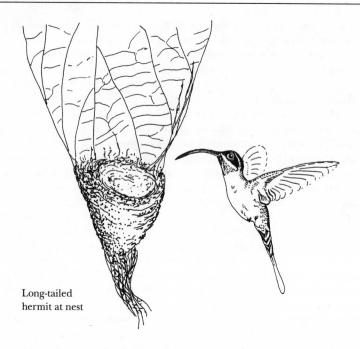

Long-tailed
hermit at nest

a cloud forest dweller, has two long, central tail feathers with bare shafts but feathered at the tips, somewhat like a motmot. The ruby topaz (*Chrysolampis mosquitus*), a lowland forest and open area generalist species, is certainly one of the most beautiful hummingbirds. Males have glowing, orange throats and bright, metallic, crimson heads.

Though most hummingbirds are brilliantly colored, not all are. The subfamily Phaethornithinae includes the twenty-nine hermit species, some of the commonest hummingbirds in lowland forests. Most are greenish brown with grayish or rufous breasts. Unlike most species where males are brighter than females, hermits have similar sexes. All hermits have a black line bordered by white through the eyes and a long, often downcurved bill. Hermits inhabit the forest understory and edge, and their more subdued plumage seems to fit well with the dark forest interior. Male long-tailed hermits (*P. superciliosus*) are both abundant and vocal throughout Central and South America. Males gather in courtship areas called *leks* (see manakins, this chapter) and twitter vociferously at each other as each attempts to entice a passing female.

Hummingbirds are attracted to red, orange, and yellow flowers, and a single flowering tree or shrub may be a food resource for several species. When a tree is abundantly covered by flowers, it is neither economical nor practical for a single hummingbird to try to defend it from others. Nonetheless, hummingbirds are generally pugnacious, and it is easy to observe both intra- and interspecific aggression among them as they jockey for a position at their favorite flower. This competition is exacerbated because, though a plant may have many flowers, very few may be nectar-rich (see below). Some hummingbirds are highly territorial, defending a favored feeding site, while others seem to circulate along a regular route visiting several flowers. The latter, which include some of the hermits, are called "trapliners."

Wolf (1975) reported a curious example of hummingbird territorial behavior for the purple-throated carib (*Eulampis jugularis*). Like many hummingbird species, males are pugnacious and territorial, defending favored flowers and dominating access to the nectar-rich food. Some females, however, employ a behavioral strategy that permits them to circumvent male dominance and gain access to desired flowers. Females court males, even during the non-breeding season (when they cannot become pregnant). Both during and after the courtship process, a normally aggressive male will permit the "cooperative" female to feed on "his" flowers. Males inseminate females, though to no avail. Wolf reasoned that the behavior is adaptive for females, since they gain access to food that otherwise they could not hope to acquire, but he was unable to identify any clear advantage to the male, since courtship and copulation use energy and no offspring result. Wolf titled his paper "'Prostitution' Behavior in a Tropical Hummingbird." Hmm.

Hummingbirds sometimes have a mutualistic relationship with plants, feeding on nectar but facilitating cross-pollination. Hermits, for instance, often feed on the nectar of heliconia flowers. Many heliconias produce relatively constant amounts of nectar per flower. However, one heliconia studied in Costa Rica, *Heliconia psittacorum*, exhibits a "bonanza-blank" pattern of nectar production (Feinsinger 1983). Some flowers contain abundant nectar (bonanzas), some essentially none (blanks). Many other tropical plants, especially those in open successional areas, also are bonanza-blank flowerers (see below). Hermits must visit many flowers in order to encounter one with high nectar content, thus the bonanza-blank pattern presumably aids *Heliconia psittacorum* in accomplishing cross-pollination

In a comprehensive study of ten successional plant species and fourteen hummingbird species at Monteverde cloud forest in Costa Rica, Feinsinger (1978) documented that flowering was staggered among plant species, resulting in a constant nectar supply to hummingbirds. In five plant species that were closely measured for nectar volume, the bonanza pattern was evident. Feinsinger speculated that plants may conserve energy by producing large numbers of "cheap" nectarless flowers and a mere few "expensive" bonanza flowers, forcing hummingbirds to visit many flowers to find satiation. By visiting many flowers, cross-pollination is promoted.

Hummingbirds display a range of foraging patterns. Feinsinger and Colwell (1978) identified six patterns evident in how hummingbirds exploit flowers: high-reward trapliners, which visit but do not defend nectar-rich flowers with long corollas; territorialists, which defend dense clumps of somewhat shorter flowers; low-reward trapliners, which forage among a variety of dispersed or nectar-poor flowers; territory-parasites of two types (large marauders and small filchers); and generalists, which follow shifting foraging patterns among various resources. Large marauders are species with large bodies that can intimidate normally territorial smaller species. They move in and take what they want. Small filchers are species with small bodies that "sneak" in to feed quickly, before being detected by a territorial bird. High-reward trapliners such as the hermits have a regular route that they visit and are most common in the forest understory. The other types of foraging are evident in the canopy and open, successional areas.

The complexity of interactions between hummingbirds and plants is further complicated by the fact that species of nectar-eating mites (*Rhinoseius*) depend entirely upon hummingbirds for their dispersal among flowers (Colwell (1973, 1985a, 1985b). The mites are transported in the nasal cavities of the birds! Mites are therefore dependent on the *mutualistic* interaction between birds and plants. This complex tapestry of ecological interdependence involves two mite species, three hummingbirds, one flowerpiercer, and four hummingbird-pollinated plants.

One other group of birds, besides hummingbirds, feeds principally on nectar. The eleven species of flowerpiercers, all members of the large tanager family (see below), do not probe into the center of the flower, as hummingbirds do, but instead snip a minute hole through the petal at the base. The bird pokes its bill in, sips nectar, but receives no pollen. Flowerpiercers are therefore nectar parasites. A few hummingbirds have occasionally been seen employing a similar behavior.

Tanagers

Tanagers are a group of unusually colorful, small, perching birds. *Tanager,* like the word *toucan,* comes from the Topi Indian language of Brazil. The diverse subfamily Thraupinae (part of the huge family Emberizidae, order Passeriformes) consists of 242 species of tanagers, euphonias, chlorophonias, honeycreepers, dacnis, conebills, and flowerpiercers. Most species are brilliantly colored and feed on fruit (mashers, page 135), nectar, and insects. All occur in the New World, and they are found abundantly from lowland forests to high montane and cloud forests. They are particularly common around forest-edge habitats and are easy to see at fruiting figs, palms, cecropias, and so on. Though four tanager species migrate to North America to breed, the remainder are all confined to the Neotropics. Storer (1969) reviews a general taxonomy of tanagers, and Isler and Isler (1987) provide a comprehensive field guide illustrating all species, with much discussion of natural history. Also see Ridgely and Tudor (1989). For a good general account of tanager natural history, see Skutch (1989).

In most Neotropical tanagers, males and females have similar plumage. The common names of tanager species reflect their multicolored, exotic feather patterns. On a trip to southern Central America, for instance, one may encounter the crimson-collared, scarlet-rumped, flame-colored, blue and gold, golden-hooded, silver-throated, and emerald tanagers, a list that is far from exhaustive. One of the most common and widely distributed birds of the tropics is the blue-gray tanager (*Thraupis episcopus*), which is well described by its name. Chlorophonias are bright green and yellow, highland tanagers. Among the most exotically colored tanagers is the paradise tanager (*Tangara chilensis*) of South America. Like a mosaic of neon colors, this incredible bird has a golden green head, purple throat, bright scarlet lower back and rump, black upper back, and turquoise breast. As if that were not enough, paradise tanagers travel in flocks, so you get to see more than one! Euphonias (genus *Euphonia*) are a group of small tanagers, also multicolored, that tend to feed heavily on mistletoe berries (family Loranthaceae). They are important seed dispersers of mistletoe, as their sticky droppings, deposited on branches, contain the seeds that begin life as epiphytes, before becoming parasitic. Euphonias often nest in bromeliads (Bromeliaceae).

Honeycreepers, which include dacnises and conebills, are nectarivorous, though they also include ample amounts of fruit and arthropods in their diets. Warbler-sized, they have fairly long, downcurved bills. One of the most common and widely distributed of this group is the bananaquit (*Coereba flaveola*). This small bird with a dark back, white eye stripe, and yellow breast is among the most ubiquitous of tropical birds. It is found in virtually all habitats from lowlands to cloud forests and often becomes quite tame around gardens. Like the flowerpiercers, bananaquits are prone to poke a small hole at the outside base of a flower and drink nectar without contacting pollen.

Some tanagers, such as the ant-tanagers (genus *Habia*), are army ant followers, and many tanagers, euphonias, and honeycreepers move with antbirds, woodcreepers, and other species in large, mixed foraging flocks.

Studies have revealed the high diversity and intriguing complexity of behavior within both canopy and understory mixed species flocks in the Peruvian Amazon (Munn 1984, 1985; Munn and Terborgh 1979). Each flock type consists of a core of five to ten different species, each represented by a single bird, a mated pair, or a family group. Up to eighty other species join flocks from time to time, including twenty-three tanager, euphonia, and honeycreeper species, a remarkably high diversity. Mixed foraging flocks occupy specific territories and, when another flock is encountered, the same species from each flock engage in "singing bouts" and displays as boundary lines are established. Adult birds tend to remain flock members for at least two years. Nesting occurs in the general territory of the flock, the nesting pair commuting back and forth from nest to flock.

Munn's work revealed an odd twist on interactions within mixed foraging flocks. One long-held hypothesis about mixed flocks is that being part of one serves to help protect against predation (Moynihan 1962). With so many eyes looking, predators have difficulty going undetected. Munn showed that actual sentinel species are part of every mixed flock. One species of shrike-tanager and one antshrike (an antbird, see below) aided the flock by giving general alarm calls when danger threatened. Both sentinel species, however, also gave "false alarm calls," a behavior Munn described as "deceitful." False alarm calls are exactly that: alarm calls uttered when no danger is present. The hypothesized reason for the false alarms is that the alarmist has a better chance of capturing food that is also being sought by another bird. When a white-winged shrike-tanager (*Lanio versicolor*) and another species were both chasing the same insect, the shrike-tanager's false alarm would cause momentary hesitation by the other bird, allowing the shrike-tanager to capture the insect.

Orioles, Oropendolas, and Caciques

The large avian family Icteridae (order Passeriformes) includes the blackbirds and their relatives. Among those occurring in the Neotropics are thirteen oropendola species, nine cacique species, and twenty-four oriole species. Oropendolas and caciques are colonial and make long, hanging, basketlike nests. An oropendola nest tree is easy to spot because it is out in the open and adorned with numerous pendulous nests. The isolation of the nest tree affords some protection against predation by monkeys, since the simians are usually loathe to leave the canopy and cross open ground. Oropendolas are large birds (some almost crow-sized), and caciques are robin-sized. In shape, both caciques and oropendolas are relatively slender, with long tails and sharply pointed bills. Oropendolas come in two color types. One group of species is mostly black and chestnut, with yellow on the bill and tail, and the other is quite greenish. Caciques are mostly sleek black but with bright red or yellow rumps and/or wing patches, and yellow bills.

Both caciques and oropendolas tend to locate their colonies near bee or wasp nests. Because these colonial insects can be very aggressive toward intruders, this behavior helps reduce the probability of predation by mammals. Robinson (1985a, 1986) learned that yellow-rumped caciques (*Cacicus cela*) employ other strategies that would also seem to protect the colony. Caciques often nest on islands in a river or lake, affording added security from both mammals and snakes, for would-be predators would have to cross a water body patrolled by otters and caimans. Caciques tend to mob potential avian predators, and unused, abandoned nests remain in the nest tree along with active nests. The presence of the unused nests may confuse a predator. Not surprisingly, each cacique attempts to locate its nest in the center (where protection is maximized), rather than the more risky periphery of the colony. For caciques, colonial nesting and group defense is a significant adaptation against nest predation.

Robinson documented, however, that yellow-rumped caciques were occasional victims of nest piracy by other bird species. One, appropriately named the piratic flycatcher (*Legatus leucophaius*), harassed caciques until they abandoned their nests to the flycatchers. Russet-backed oropendolas (*Psarocolius angustifrons*) destroyed cacique eggs and killed young, leaving empty nests. Finally, troupials (*Icterus icterus*), which are large, aggressive orioles, both took over cacique nests and destroyed eggs and young. Robinson hypothesized that the piracy is not related to competition for food because each of the nest pirate species has a diet different from that of caciques. Instead, the creation of many nearby empty cacique nests serves to confuse potential predators and confer protection on the nest pirate species (Robinson 1985b).

Orioles nest as territorial pairs and are not colonial like the oropendolas or caciques. They are colorful, with various combinations of orange, yellow, and black. Several oriole species migrate to nest in North America, but most remain in the tropics.

Orioles, oropendolas, and caciques feed on fruit and nectar, mixing various arthropods into an otherwise vegetarian diet. Like the tanagers, these birds move in mixed flocks, foraging in fruit trees and drinking nectar from blossoms.

Parrots   *Figures 102, 146, 147, 148, 149*

Like toucans, parrots are quintessentially tropical. Global in distribution, they occur mainly in tropical forests of the southern hemisphere. In the Neotropics, 136 species of the family Psittacidae (order Psittaciformes) can be

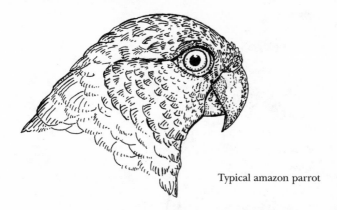

Typical amazon parrot

found, ranging from the spectacular large macaws (genera *Ara, Anodorhynchus, Cyanoliseus*) to the sparrow-sized parrotlets (genus *Forpus*). Among the most commonly encountered of the New World parrots are the chunky, short-tailed amazons (genus *Amazonia*). There are also moderate-sized, long-tailed parrots known collectively as parakeets (many in the genus *Aratinga*). Parrots are mostly green (though there are some dramatic exceptions) and can be remarkably invisible when perched in the leafy forest canopy, quietly and methodically devouring fruits. They reveal their presence by vocalizing, usually a harsh screech or squawk. Often a flock will burst from a tree like shrieking banshees, and it is amazing to see how many birds were actually in the tree when so few were readily apparent. Generally, there is little or no difference in plumage between the sexes. Forshaw (1973) has treated all of the world's parrots.

Parrots are gregarious frugivores. It is uncommon to find only one or two, though that may occur with large macaws. Flocks move about in forests and savannas searching out fruits, flowers, and occasionally roots and tubers. Parrots climb methodically around the tree branches, often hanging in awkward acrobatic positions as they attack their desired fruits. Their sharply hooked, hinged upper mandible is useful in climbing around in trees as well as in scraping and scooping out large fruits. Using their strong, nutcracker-like bills, they can crack many of the toughest nuts and seeds, which they eat with equal relish as the pulpy fruit itself. Their tongues are muscular, and they are adept at scooping out pulp from fruit and nectar from flowers. Because of their ability to crush and digest seeds, they are primarily seed predators rather than seed dispersers. The orange-chinned parakeet (*Brotogeris jugularis*) in Costa Rica is, for instance, a seed predator on the trees *Bombacopsis quinatum* and *Ficus ovalis*. Janzen (1983c) shot one bird, examined its crop, and found several thousand *Ficus* seeds, all of which were damaged in some way. Droppings from orange-chins contained no evidence of intact seeds.

The most spectacular Neotropical parrots are the nineteen macaw species, especially the larger species. These long-tailed parrots with bare skin on their faces range in plumage from the predominantly green chestnut-fronted (*Ara severa*), military (*A. militaris*), and great green (*A. ambigua*), to the bright red scarlet (*A. macao*) and red-and-green (*A. chloroptera*), to the brilliant blue-and-

Scarlet macaw

yellow (*A. ararauna*) and the deep indigo blue of the hyacinth (*Anodorhynchus hyacinthinus*). Macaws are most commonly seen flying to and from their roosting and feeding sites. Their slow wing beats and long tails make them distinctive in flight. Many macaws frequent gallery forests along watercourses or humid forests interrupted by open areas. They feed heavily on palm nuts, their huge bills being fully capable of crushing these dense fruits.

Unfortunately, forty-two species, or about 30% of the species of Neotropical parrots, are considered to be at some clear risk of extinction, principally from habitat loss and/or pet trade (Collar and Juniper 1992). In addition to threatened and endangered species, many, if not most, other species are also in decline for the same reasons. Nest trees are cut to procure the nestlings for the pet trade. Deforestation eliminates still more nest trees. Mortality rates among parrots shipped from Latin America for the pet trade are staggering. While it may prove possible to develop protocols to manage parrot populations such that there can be sustainable use for international trade (Thomsen and Brautigam 1991), doing so in countries where enforcement is at best problematic (and that would be most Latin American countries) suggests a dubious potential for success. Free-ranging large macaws may provide a unique opportunity for ecotourism, an example discussed in chapter 14.

Cotingas

Cotingas (family Cotingidae, order Passeriformes) are among the real glamour birds of the Neotropics. With names such as bellbirds, umbrellabirds, cocks-of-the-rock, pihas, fruiteaters, fruit-crows, and purpletufts, the sixty-six cotinga species are, indeed, a colorful and diverse family. They are birds of rainforests and, to a lesser extent, cloud forests, and they are described as "extreme fruit specialists" (Snow 1982). Large cotingids eat laurels (Lauraceae), incense (Burseraceae), and palms (Palmae), while smaller species eat smaller, sweeter fruits, sometimes plucking fruits while hovering. Cotingas typically have wide, flattened bills, shaped well for accommodating rounded fruits. Cotingas feed only on the flesh of the fruit and not the seeds and thus can be effective seed dispersal agents. Some species, such as the fruit-crows and pihas, mix insects among their fruits, but most cotingas feed exclusively on fruit.

Cotingas are diverse. Some, such as the umbrellabirds and cocks-of-the-rock, are large and colorful or have ornate plumage, while others, such as the fruiteaters and pihas, are smaller and relatively drab. Some are sexually monomorphic, the males and females looking alike, while others represent extreme cases of sexual dimorphism. Some form pairs and occupy territories, while others are highly polygynous, cocks mating with many hens. In a few species, such as the cocks-of-the-rock (genus *Rupicola*) and screaming piha (*Lipaugus vociferans*), males gather to court females in mating areas called leks. Bellbirds (genus *Procnias*) are known for their piercing, bell-like call notes, pihas for their loud scream, cotingas for their shiny metallic plumage, cocks-of-the-rock for their golden-orange or orange-red coloration and fan of head feathers, and umbrellabirds (genus *Cephalopterus*) for their extraordinary, umbrellalike head plumes and inflatable air sac on the breast. Cotingas generally make small, inconspicuous nests, incubate but a single egg, and have a prolonged incubation period. Bellbirds typically incubate for approximately thirty days, and cocks-of-the-rock for forty or more days. This long incubation period is probably related to feeding nestlings almost exclusively fruit, which is low in protein but high in fat and carbohydrate. Cotinga natural history is thoroughly reviewed by Snow (1982), and cotinga examples of sexual selection will be given below.

Manakins   *Figures 139, 140*

The fifty-three species of the family Pipridae (order Passeriformes), the manakins, are small, chunky, fruit-eating birds, most of which inhabit lowland forests. All are confined to the Neotropics. Manakins are extremely close evolutionary cousins of the cotingas and tyrant flycatchers, and recent taxonomic analysis has suggested that manakins of several generi (ex. *Schiffornis*) are not true manakins but intermediates between cotingas, tyrant flycatchers, and manakins (Prum and Lanyon 1989). Males of most species are quite colorful; females, drab olive green and yellowish. Manakins have short tails, rounded wings, and a short but wide bill with a small hooked tip. They pluck fruits on the wing, supplementing their largely frugivorous diets with occasional arthropods.

Manakins have among the most elaborate courtship displays of any birds. Like some of the cotingas, many manakin species are "arena birds" and court in concentrated leks, assemblages of males that display to transient females. Others court in dispersed leks, while still others have a unique cooperative courtship behavior in which several males display together in an extraodinarily coordinated manner. Manakin courtship is detailed in the next section. Only females build the nest, incubate, and feed young. Clutch sizes are typically small, one to two birds per nest. Manakin courtship is reviewed by Sick (1967), Lill (1974), Snow (1976), and Prum (1994) and is discussed further below. For a general review of courtship behavior among arena birds, see Johnsgard (1994).

Leks and Lovers—Sexual Selection among Cotingas and Manakins

Charles Darwin (1859, 1871) devised his theory of sexual selection in part to account for why certain bird species, among them many of the cotingas and most manakins, display extreme differences in plumage between the sexes. This sexual dimorphism almost always involves brightly and ornately plumaged males compared with subtle-plumaged, more cryptically colored females. Why females are cryptic seemed an easy question to Darwin. Females undergo natural selection for cryptic plumage because such coloration aids in reducing the risk of discovery by predators. But why are males so colorful? Adding to this mystery was the fact that elaborately colored males often augment their already gaudy selves by engaging in bizarre courtship displays.

The Amorous, Glamorous Cock-of-the-Rock   Figures 135, 136, 137, 138

The Guianan cock-of-the-rock (*Rupicola rupicola*), a large (grouse-sized) cotinga, provides an example of elaborate courtship and plumage. The courtship of this species has been studied by Snow (1982) and Trail (1985a, 1985b). Males are chunky, with short tails and bright, golden orange plumage with black on the tail and wings. In flight they resemble winged, day-glo orange footballs! Beaks, legs, eyes, and even the very skin are orange. And not just any orange: bright, vivid, magnificent orange. The first male Guianan cock-of-the-rock that I saw was some distance away, perched in the midlevel understory of thick Brazilian rainforest. I urgently asked Bob Ridgely, who had spotted the bird, to describe just where it was. He said something like "Just look for the orange beacon" and pointed in the general direction. He was right. It was not difficult to find the creature. The male's already striking plumage is further enhanced by delicate, elongated orange wing plumes and a crescent-like thick fan of feathers extending from the base of the bill to the back of the neck. Females are dull brown, with neither the wing plumes nor the head fan.

Males gather in the rainforest understory in confined courtship areas called concentrated leks. Each male clears an area of ground in which to display and defends perches in the vicinity of its display site. The lek can be a crowded place, with males as close to one another as 1.2–1.5 cm (4–5 ft) and several dozen males on the same lek. When a female approaches a lek, each male displays by landing on the ground and posturing to her. Each displaying cock strokes its wing plumes and turns its head fan sideways, so that the female sees it in profile, and stares at her with its intense orange eye set against flaming orange feathers. The object of the cock's bizarre display is to mate, presumably by suitably impressing the female. Females do not appear to be easily impressed. A hen will typically visit a lek several times before engaging in copulation. These visits, called mating bouts, always excite the males to display. Only one male on the lek will get to mate with a visiting female, who may return to mate with him a second time before laying eggs (Trail 1985b). No extended pair bond is formed, only a brief coupling. The cock returns to the lek, continuing to court passing hens while the newly fertilized hen attends to nest building, egg laying, incubation, and raising the young. The basis of her behavior in choosing a male from among many potential contenders is one facet of what Darwin called *sexual selection.*

Darwin reasoned that, in some species, female choice was the dominant factor in selecting males' appearances. Put very simply, males are pretty (or musical or noisy or perform complex dances) because females have tended through generations to mate mainly with males having these unique features. Since plumage color is heritable (as are behavioral rituals), gaudy coloration was selected for and continually enhanced. Recent work in sexual selection suggests that females may learn much about the evolutionary fitness of males by signals communicated both by plumage condition and male courtship behavior (Andersson 1994). In other words, females are not being frivolous in driving male evolution toward more elaborate, gaudy plumage and exotic behavior but rather are looking intently for the best vehicles in which to place their precious genes for their journey into the next generation.

The other facet of sexual selection recognized by Darwin is that males must compete among themselves for access to females. This may be accomplished by dominance behavior, guarding females, active interference with other males' attempts to mate (see below), injury to other males, or merely being sneaky and mating before other males can react. Gaudy plumage may contribute to a male's success by intimidating other males and thus make it easier to gain the attentions of a female. Male/male competition coupled with female selection of the winner is what Darwin defined as sexual selection.

Sexual selection has costs for both males and females. Though the hen exercises the most choice in the mating process, she is left solely responsible for the chores of nest building, incubation, and caring for the young. These are risky, energy-consuming tasks. Males may at first glance seem the luckiest, rewarded by a life of lust in nature's tropical "singles bar," the lek. The combination of male/male competition plus dependency on female choice makes life surprisingly difficult for most males, however. Though some cocks are quite successful, mating frequently, others, the losers, spend their entire lives displaying to no avail. After a lifetime of frustration, they die genetic losers, never selected even once by a hen. Pepper Trail (1985a, 1985b), who studied the Guianan cock-of-the-rock in Suriname, documented high variability in male mating success. He found that 67% of territorial males failed to mate at all during an entire year. The most successful male performed an average of 30% of the total number of annual matings, and the lek contained an average of fifty-five cock birds! One of these fifty-five mated 30% of the time. Many never mated. Such is the cost of sexual selection for males. In reproductive terms, females are the most fortunate sex. Most females do mate, though success in fledging young may certainly vary considerably among females.

Trail (1985a) also discovered another interesting twist in the mating process of the Guianan cock-of-the-rock. Some males were sore losers and habitually disrupted the mating of others. Trail found that aggressive males that disrupted copulations by other males fared better in subsequent mating attempts. He learned that males that were confrontational "were significantly more likely to mate with females that they disrupted than were non-confrontational males." He hypothesized that only the cost of confrontation in terms of energy expenditure, loss of time from the aggressive bird's own lek territory, plus risk of actual retaliation kept direct confrontational behavior from becoming even more manifest among the birds. On the other hand, Trail (1985b) found adult fully plumaged males remarkably tolerant of juvenile males that were still plumaged in drab colors, resembling females. Yearling males would actually attempt to mount adult males as well as females in a crude attempt at mating. Adult males did not respond aggressively to these misguided efforts, possibly because yearling plumage, being drab, does not stimulate an aggressive response.

Screaming Pihas and Clanging Bellbirds

Sexual selection has evolved in various ways, and thus courtship patterns differ among species. The screaming piha (*Lipaugus vociferans*), a common bird throughout much of Amazonia, courts on leks that are much larger in area than those of the Guianan cock-of-the-rock. These dispersed leks, as they are termed, are possible because it is not the plumage of the bird that matters, it is its voice, and its voice is mighty. The piha differs from the cock-of-the-rock in that it is downright nondescript, being a slender, robin-sized bird light gray on the face and breast and dark gray-brown on wings, back, and tail. Though certainly attractive (at least to human eyes), it is by no means the glamour bird that its cousin is.

Male and female screaming pihas look alike and thus would not seem to fit with Darwin's concept of sexual selection, focused as it is on sexual dimorphism. It is with voice, however, and not looks, that a male screaming piha attracts a female and tells other males that he is doing so. Piha leks are composed of up to thirty males, and David Snow (1982) reports that males' calls are "one of the most distinctive sounds of the forests where these birds occur, its ringing, somewhat ventriloquial quality seeming to lure the traveller ever onwards into the woods." Ringing indeed. To me it sounds like a strident, clanging "peeh-HE-hah!" reminding me vaguely of the cracking of a whip. And it really carries.

Barbara Snow studied the screaming piha in Guyana and found that one male spent 77% of his time calling on the lek, usually from a thin horizontal branch well below the canopy. (Screaming pihas, though understory birds, can be surprisingly hard to see, as they are so nondescript and their voice is so ventriloquial.) An excited male called at the rate of twelve times per minute. Calling seems to replace plumage and display behavior as the signal to the females. Sexual selection has occurred, but for characteristics of voice, not appearance.

Bellbirds (genus *Procnias*), like the screaming piha, rely heavily on voice as part of the courtship process. There are four species, each shaped generally like a starling though larger in size, ranging throughout lush montane forests of northern South and Central America. They tend to migrate vertically,

Female bearded bellbird

breeding in highland forests and moving downslope to lowland forests when not breeding. Unlike pihas, bellbirds are sexually dimorphic, the males having much white on the body along with ornate wattles about the head. In one species, the white bellbird (*P. alba*), the male is entirely white with a fleshy, wormlike wattle dangling from its face above the bill. The male bare-throated bellbird (*P. nudicollis*) is almost all white but has bare blue skin on the throat and around the eyes. The male bearded bellbird (*P. averano*) has black wings and a chestnut head with a heavy "beard" of black, fleshy wattles hanging from its throat, and the male three-wattled bellbird (*P.tricarunculata*) is chestnut on body, tail, and wings, but with a white head and neck, and three fleshy wattles hanging from the base of the bill. Females of all four species are similar greenish yellow, darkest on the head, with streaked breasts.

Male bellbirds establish calling and mating territories in the forest understory. Though not true lek birds, bellbirds have courtship territories that are closely spaced together. Each male spends most of his time on territory vocalizing to attract hens. Males take no part in nest building, incubation, or raising young. Two well-studied examples are the bearded bellbird in Trinidad and the three-wattled bellbird in Panama. Both court in a generally similar manner.

Male bearded bellbirds are among the first sounds one hears upon entering the Arima Valley in Trinidad. David Snow (1976, 1982) aptly describes their call as a loud "Bock!" The call carries amazingly well, and I thought the birds were nearby when, in reality, they were a quarter of a mile or more from me. The call definitely has a bell-like quality, though it is a muted clang, and the ventriloquial quality of the call note is evident. Even when very close to a calling male, it can be frustratingly difficult to locate him. Cock birds initially call from a perch above the canopy, often on a dead limb, but will drop down into the understory to complete the courtship. Females never call, and it is clear that male vocalizations are an essential part of sexual selection in bellbirds.

The object of calling is to attract a hen to the male's territory. Each cock bellbird has his own courtship site in the forest understory. The male "bocks" rather continuously, mixing the bocking with a series of "tock, tock, tock" notes. If successful in luring a female to his territory, the male initiates a series of courtship postures, performed from a horizontal branch upon which the female perches as his only audience. These postures include display of the beard wattles, a wing display, and a display in which a bare patch of skin on the male's thigh is revealed. All bellbird species include a "jump-display" as part of courtship. A cock bearded bellbird will leap from one perch to another, landing before the hen with his body crouched, tail spread, and eyes staring at her. You can guess what happens next, assuming the male has performed satisfactorily.

The three-wattled bellbird takes the jump-display one step further. The cock jumps over to the place occupied by the hen while at the same moment the hen skitters along the limb to occupy the place the male just vacated. Called a "changing-place" display, the male then slides across the branch to be right next to the hen, emitting a close-up call virtually in her ear. Following the successful execution of this maneuver, more bellbirds come into the world.

The Dancing Manakins   *Figures 139, 140*

Manakins carry the evolutionarily inspired art of courtship dancing to extremes. Male manakins are brightly colored, glossy black with bright yellow, orange-red, scarlet, or golden heads and/or throats, some with bright yellow or scarlet thigh feathers, and some with deep blue on their breasts and/or backs and long, streamerlike tails. A few species are sharply patterned in black and white. But fancy feathering notwithstanding, it is dancing in which these birds excel. Snow (1962a, 1976), Sick (1967), and Schwartz and Snow (1978) have made detailed studies of manakin courtship behavior and have provided much of the information outlined below, augmented by my own experience with these species.

The white-bearded manakin (*Manacus manacus*) courts on rainforest leks. I've observed its courtship in the Arima Valley in Trinidad. The male has a black head, back, tail, and wings but is white on the throat, neck, and breast. Its name comes from its throat feathers, which are puffed outward during courtship. Females are greenish yellow. Up to thirty or more males may occupy a single lek. Each male makes his "court" by clearing an oval-shaped area of forest floor about a meter across. Each court must contain two or more thin vertical saplings, as these are crucial in the manakin's courtship dance. The male begins courtship by jumping back and forth between the two saplings, making a loud "snap" with each jump. The snap comes from modified wing feathers snapped together when the wings are raised. When a female visits the lek, the snapping of many males is audible for quite a distance. In addition to the snap, the male's short wing feathers make a buzzing, insectlike sound when it flies, and thus active manakin leks can become a cacophony of buzzing, snapping birds. The intensity of the male's jumping between saplings increases until he suddenly jumps from sapling to ground, appearing to ricochet back to another sapling, from which he slides vertically downward like a firefighter on a pole. David Snow's film footage of the slide revealed that successful males slide down right to the female perched at the base of the sapling pole. Copulation is so quick that Snow only discovered the presence of the female in the film. He never saw her while he was witnessing the event!

Following copulation, the female leaves the lek and attends to nesting. The male starts to dance again. Male manakins spend most of their adult lives at the

Courtship dance of the white-bearded manakin. See text for details. From H. Sick (1967). Reproduced with permission.

lek. Some, as in the case of the cock-of-the-rock, are probably consistently successful and mate often. Others may never mate. Observations of banded males on Trinidad have revealed that life on a lek is usually fairly long for individual birds. Some live for a dozen years or more, a very long life span for such a small bird (Snow 1976). Males generally only leave the lek to feed on ripe fruits.

Another common Trinidad manakin that I have observed, the golden-headed (*Pipra erythrocephala*), is not a lek dancer, but rather each male displays in his own territory. As in the white-bearded, the dance begins when the male darts back and forth on selected twigs, calling "zlit" as he does so. Unlike the white-bearded, which dances close to the ground, the golden-headed usually displays about 3 m (10 ft) off the ground in an understory tree. The cock becomes increasingly vigorous in his dancing, crouching, his body at a 45-degree angle as he slides along a horizontal twig. His sparkling golden head and sleek black plumage are displayed very conspicuously, but more is yet to come. When a female arrives, the male skitters along the branch toward her, but *tail first!* As he advances, he bows, spreads his wings, and exposes bright red thigh feathers, all the while pivoting his body back and forth. The climax of the dance comes when the male suddenly flies from the dance branch and quickly returns, inscribing an "S-shaped" curve as he lands with wings upraised before the female. Various vocalizations accompany the performance.

If the white-bearded and golden-headed manakin performances amaze you, be warned that the blue-backed and swallow-tailed manakins (genus *Chiroxiphia*) seem to carry bird dancing to the point of incredulity. Blue-backed males

Blue-crowned manakin

have bright red on the top of their heads and shimmering turquoise blue backs on otherwise shiny black plumage. Swallow-tailed manakins, also called blue manakins, are similar but with blue on both back and breast and elongated central tail feathers. Even for male manakins they are extraordinarily beautiful, especially when seen in a burst of full sunlight as they dance in the forest understory. I say *they* because these manakins dance as a team. Two blue-backed males engage in a coordinated jump dance in which both birds occupy a thin horizontal branch, one jumping and hovering while the other crouches on the branch, the other jumping and hovering when the first lands. As they dance, they vocalize. The dance may occur in the presence or absence of a female, the males seeming to "practice" when a female is not present. The dance ends when one of the three cocks bows before the hen, head turned, exposing the bright red top, blue back upraised. In the case of the swallow-tailed or blue manakin, up to three males dance in perfect coordination before a single female. The three dancers align themselves horizontally on a thin branch, shoulder to shoulder before the female, each male facing in the same direction. The male farthest from the female jumps up, inscribes a 180-degree angle, and lands nearest the female, next to the other males. He immediately turns around, so once again all three dancers face the same direction. A second dancer, again the farthest from the female, repeats the first dancer's performance, and so on. The dance happens rapidly, and David Snow has described it as a spinning "Catherine wheel" of dancing males, jumping, displaying, and vocalizing in total coordination. No other case of such elaborate team dancing is known for birds. The termination of the performance occurs when one of the males vocalizes sharply, the effect of which is to "turn off" the other two males. The dominant male then erects his red head feathers as he perches before the female. She and he fly off into the underbrush.

One species, the wire-tailed manakin (*Pipra filicauda*), adds yet another element to the roster of manakin courtship techniques. Males, which are black with yellow breasts and a red cap, have stiff tail feathers that terminate in long, delicate filaments. Wire-tailed males dance in teams of two, rather like the blue-backed species. However, when the dominant male approaches a female, he performs a twist display in which he rotates his posterior side to side, gently touching the female on her chin with his tail filaments. Females apparently respond well to this maneuver, for a female will typically slide toward a male to receive the tail brushing. This is the only known example of tactile stimulation among manakins, and it appears that the unique tail is the product of sexual selection (Schwartz and Snow 1978).

Why do several male manakins cooperate in courting a single female? Only one will get to mate with her. Some evolutionary theorists believed that the males were perhaps brothers, sharing the majority of each other's genes. Co-operative behavior could result in reproduction of many of one's own genes, which happen to be shared with one's brother, a example of possible kin selection (page 100). However, Mercedes Foster (1977), who studied *Chiroxiphia linearis*, found little evidence for a close relationship among cooperating males. Foster found that one male was consistently dominant over the others. Though the assemblage remained together throughout the breeding season and even from one year to the next, cooperating males were not brothers and did not behave altruistically toward one another. Rather, subordinates were biding their time until they could replace the dominant male. Foster hypothesized that one male, acting alone, could never succeed in attracting a female. Only by being part of a pair or trio could a male hope eventually to succeed. When a dominant male dies, it is quickly replaced by a subordinate who "trained" under it.

Why Do Pihas Scream, Bellbirds Clang, and Manakins Dance?

The bizarre results of sexual selection in cocks-of-the-rock, pihas, bellbirds, and manakins are evident, but what sorts of selection pressures were responsible for their evolution? Both David Snow (1976) and Alan Lill (1974) have suggested possible scenarios for the "release" of males from postcopulatory reproductive chores, thus initiating the male/male competition and pattern of female choice that resulted in both the gaudy plumages and elaborate courtship behaviors.

Snow emphasizes the importance of a diet almost exclusively of fruit. He points out that both bellbirds and manakins feed so heavily on fruit that they are easily able to secure adequate daily calories with only a small percentage of their time devoted to feeding. Fruit is both relatively abundant and easily collected. It does not have to be stalked or captured and subdued. The male bellbird or manakin has lots of time in which to clang or dance.

Lill, who studied manakins, agrees that a fruit diet is significant in the evolution of sexual selection in these birds. He places his emphasis, however, a bit more on nest predation. A largely frugivorous diet has metabolic costs as well as benefits (Morton 1973). Incubation time is relatively long and nestling growth rates slow in highly frugivorous birds because fruit is nutritionally not well balanced for a baby bird (low in protein but high in fat and carbohydrate). Lill argues that because of the slow development time brought about by a diet of fruit (recall oilbird, page 140), nest secrecy is of paramount importance. Heavy egg and nestling predation are best minimized by having only one bird, the cryptically colored female, attend the nest. A male's presence at the nest could actually be detrimental to raising young, since one bird can easily find sufficient food for the small brood (usually two nestlings), and a second bird might inadvertently reveal the presence of the nest to potential predators. Lill argues that it is to the advantage of both female and male for the male to stay away because male absence actually increases the probability of egg and nestling survival. Males are dispensable, not needed for raising young. Lill concludes that this "male liberation" was followed by sexual selection and male "chauvinism" in the odd and varied forms described above.

Why Leks?

Given that a combination of factors have "released" males from attending nests, why have some species organized their courtship bouts in leks, especially the tightly clumped leks that are typical of manakins and cocks-of-the-rock? Several hypotheses have been suggested. One, called the "female preference model," argues that females prefer groups of males when making their selections of whom to mate with (Bradbury 1981). A male that stayed away from the lek would not attract any female, thus males have no choice but to join a lek. Another suggestion is that males might associate in leks because the lek area happens to be a place where females, for whatever reason, frequently occur. This idea, termed the *hotspot model*, presumes that leks form rather accidently, as males gather where they are most likely to encounter females (Emlen and Oring 1977; Bradbury and Gibson 1983). Both hypotheses place strong emphasis on female choice as causal to lek formation.

Beehler and Foster (1988) have critiqued both the female preference and hotspot models and have concluded that neither is sufficient to account for the evolution of lek mating systems. They offer yet another model, dubbed the *hotshot model*, that emphasizes the role of male-male dominance and interactions between dominant and subordinate males on a lek. Hotshots are individuals that control leks. Subordinates occasionally benefit from disrupting leks (recall Trail's observations of subordinate cocks-of-the-rock cited above), but mostly they bide their time while slowly advancing toward dominance. Beehler and Foster argue that novice males have little choice but to begin as subordinates, working their way up through the ranks to attain dominance status before they can reproduce. Subordinate birds congregate around the dominant cocks, since they have no hope for mating otherwise (recall Foster's observations on manakins cited above). The hotshot model places extreme emphasis on male-male interactions rather than male appearance and female choice. Dominance among cocks can be subtle, but it is real, and females will almost always select a dominant male with whom to mate. Beehler and Foster offer several predictions from their model. For instance, if all hotshot (dominant) cocks are removed from a lek, disruptions among the remaining males will increase (because none is dominant) and the lek may break up into several smaller leks, as new dominance rankings are established. Removal of the hotshots also predicts that females will visit the lek less and mate less until the lek restabilizes. These predictions are testable, and both the female preference and hotspot models predict different outcomes.

No model for lek evolution has as yet been shown to be conclusive. Indeed, evolutionary biologists routinely refer to the "paradox of the lek," an admission that leks are not easy to explain. The lek is by no means exclusively a tropical phenomenon. Leks occur among some shorebirds that nest in the arctic, among grouse that nest in grasslands, and among various other birds as well as some mammals.

Both the hotspot and hotshot models outlined above depend on proximate selection pressures operating now from within the environment. Prum (1994) has argued that evolutionary events dating back perhaps 14–35 million years ago, when frugivory may have permanently released males from parenting duties, may have set in motion an evolution of lek behavior such that lekking is now more readily explained by phylogenetic history than by any immediate selection pressures. Prum, perhaps a bit tongue in cheek, writes, "For manakins and a large majority of the lekking birds, the proximate answer to the 'paradox' of why they breed in leks is because their parents did; the ultimate answer lies in the ancient past when these behaviors initially evolved." For a general review of lekking, see Hoglund and Alatalo (1995).

Suboscines and Oscines

Of the more than 3,700 species of Neotropical birds, approximately 1,000 species are classified taxonomically as "suboscines." There are only about 50 suboscine species in all of the rest of the world, thus the Neotropics are unusual in harboring so many members of this group. The suboscines are part of the huge order of Passeriformes, or perching birds. Most passerines in the world are true oscines, which means that they have a complex musculature of the syrinx, the part of the trachea that produces elaborate sounds, such as the flutelike songs of various thrushes and solitares or the warbling of a canary. Suboscines, however, have a considerably less complex syringeal musculature and typically have far more limited singing abilities than true oscines.

Neotropical suboscines have undergone two major adaptive radiations, with the tyrant flycatchers, cotingas, and manakins representing one, and the woodcreepers, ovenbirds, true antbirds, ground antbirds, gnateaters, and tapaculos representing the other (Ridgely and Tudor 1994; Gill 1995). No one knows why suboscines have fared so well in the Neotropics, but the reason may simply be historical. At any rate, I have already discussed the tyrant flycatchers (page 102) as well as the cotingas and manakins (above). It is now time to turn our attention to the other major groups.

Insect-Arthropod Feeders

Several major groups of suboscine passerines utilize insects and other arthropods as their major food sources. These groups are among the most species-rich found anywhere. For instance, there are 218 species of ovenbirds (Furnariidae), 250 species of typical antbirds (Thamnophilidae) and ground antbirds (Formicariidae), 45 species of woodcreepers (Dendrocolaptidae), and an astonishing 393 species of tyrant flycatchers (Tyrannidae) (Parker et al. 1996). Of the above groups, only a few of the tyrannids venture to North America to nest. All others are entirely Neotropical.

Tyrannids, ovenbirds, and antbirds each represent a notable case of species diversification and adaptive radiation (chapter 4). Their primarily arthropod diets have probably provided major impetus in producing such diversity over evolutionary time. Eating insects per se does not cause species diversity nor speciation. It does, however, promote specialization, which produces divergence and can, therefore, be a factor in speciation. Insects require catching; they do not seek predators, but, on the contrary, are well adapted to avoid predation through either cryptic or warning coloration or escape behavior. Each insect-eating bird tends to develop a particular pattern of feeding, and its size, behavior, and bill shape become refined to focus on a particular size range and type of prey (Fitzpatrick 1980a, 1980b, 1985). Prey characteristics provide major selection pressures in shaping evolution among avian predators.

Second, species compete against each other. The presence of many other insect-eating species could generate continuous diffuse competition within a species assemblage, keeping each species ecologically focused on doing what it alone does best.

Insect eaters can roughly be categorized by overall feeding method. These are (1) flycatching (tyrant flycatchers, puffbirds, and nunbirds), (2) bark probing and drilling (woodcreepers and woodpeckers), (3) foliage gleaning (ovenbirds and many antbirds), and (4) ant following (some antbirds, others).

Flycatching

Tyrant flycatchers have been discussed (page 102). There is, however, another group, less diverse but deserving of mention here. The puffbirds and nunbirds (family Bucconidae, order Piciformes) consist of thirty-two species, all Neotropical, that feed on insects and spiders captured by darting from a perch and snatching them in midair. They are not passerines but are most closely related to jacamars (page 217), toucans, and woodpeckers. The black-fronted nunbird (*Monasa nigrifrons*) is typical of the group. Ranging throughout the Amazon Basin, this ubiquitous, robin-sized, forest-dwelling bird is easily recognized by its black upper plumage and tail, gray breast, and tapered, slightly drooping, bright red-orange bill. It perches in the understory, upright on a horizontal limb, and, typical of "sit and wait" predators, hardly moves a muscle until it spots potential prey, at which time it springs into the air in pursuit. Nunbirds often form noisy groups and typically join large, mixed foraging flocks and often follow army ant swarms.

Puffbirds are large-headed, heavy-bodied birds so named for the puffed appearance of their feathers. Though some species are boldly patterned in black and white, most species are brownish or tan. Their cryptic plumage plus their stationary behavior when perched in the shaded forest understory make them easy to overlook. The white-whiskered or brown puffbird (*Malacoptila panamensis*) is a common bird of the forest understory from southern Mexico through Ecuador. It is dark brown above and has a tan breast with brown streaking. Close examination reveals red eyes and white feathering around the bill. Higher in the canopy is the white-necked puffbird (*Notharctus macrorhynchos*), a larger bird with bold black and white plumage that ranges all the way from southern Mexico to Argentina. Both species have large rictal bristles, hairlike feathers around the base of the bill, which probably aid in capturing aerial insects. These two puffbird species are generally segregated vertically, the white-whiskered in the understory and the white-necked in the canopy. Such a distribution may reflect the outcome of both specialization for food capture

Brown puffbird

(canopy insects are not the same as those of the understory) and interspecific competition (since each species inhabits a different vertical area, they do not directly compete with each other).

Nunbirds and puffbirds excavate nests in termite mounds or in the ground. A puffbird pair seems undisturbed by the presence of an observer when the two birds excavate a termitary and tolerate termites crawling over them as they incubate. Those that burrow make very long nest tunnels (Skutch 1983).

Bark Drillers and Probers—Woodpeckers and Woodcreepers

Woodpeckers (family Picidae, order Piciformes) both probe and drill bark, extracting insects, mostly larval, by using their extremely long, extrusible, barbed tongues. They hitch vertically up tree trunks, their bodies supported by stiff tail feathers that act as a prop. The world's woodpeckers are treated by Short (1982), and Skutch (1985) provides a general natural history of woodpeckers.

Woodpeckers occur globally (except Australia) wherever there are trees, and thus many are temperate zone species. Neotropical woodpeckers vary in size from the 35.6-cm (14-in) ivory-billed types (genus *Campephilus*) to the diminutive 8.9-cm (3.5-in) piculets (genus *Picumnus*). The world's largest woodpecker is the 55.9-cm (22-in) imperial woodpecker (*Campephilus imperialis*), now possibly extinct from persecution and extreme habitat loss (Collar et al. 1992), but which formerly ranged through montane oak-pine forests of the Sierra Madre Occidental in western Mexico. Tropical woodpeckers range in color from bold black with red crest, to greenish olive, to soft browns and chestnut. Some species have horizontal black and white zebra stripes on their backs with varying amounts of red on the head. One species, the brilliant cream-colored woodpecker (*Celeus flavus*) of varzea forests in northern Amazonia, is bright yellow-buff with brown wings and a black tail. Another species, the boldly patterned yellow-tufted woodpecker (*Melanerpes cruemtatis*), named for its distinct facial stripe, is widespread and commonly seen along forest edge.

Neotropical woodpeckers excavate roosting and nesting cavities that are often usurped by other species. Skutch (1985) observed a group of collared aracaris (*Pteroglossus torquatus*) easily evict a pair of pale-billed woodpeckers (*Campephilus guatemalensis*) from their nest cavity. Skutch also reports that two tityra species steal cavities from several woodpecker species. Skutch portrays the 17.8-cm (7-in) tawny-winged woodcreeper (*Dendrocincla anabatina*), which attacks and forces several woodpecker species from their cavities, as "the most consistently aggressive bird that I have watched in tropical America."

Woodcreepers (family Dendrocolaptidae, order Passeriformes) look superficially like woodpeckers but bear no close evolutionary relationship to them. The anatomical similarity is a case of evolutionary convergence brought about by similar ecologies. Like woodpeckers, woodcreepers have stiff tail feathers that prop them vertically against a tree trunk. They tend to climb upward, often spiraling around the trunk. Woodcreepers evolved from the ovenbirds (Furnariidae), to which they are so closely related that some authorities consider them to be a subfamily (Dendrocolaptinae) within the Furnariidae (Raikow 1994). All woodcreepers have become bark-probing specialists that feed quite differently from woodpeckers. A woodcreeper moves methodically around the trunk, probing into crevices, poking its bill into epiphytes, and generally removing insects, spiders, and even an occasional tree frog. They rarely peck into the trunk, instead using their long bills as forceps to pick off prey. Woodcreepers may also join mixed flocks that follow army ant swarms (see below).

Like furnarids, woodcreepers are colored soft shades of brown and rufous. Many have various amounts of yellowish white streaking on breast, head, and back. The overall size of the bird, its bill size and shape, and its streaking pattern usually separate one species from another. The smallest is the 15.2-cm (6-in) wedge-billed woodcreeper (*Glyphorynchus spirurus*), which has a very short but sharply pointed bill. Several species reach about a foot in length, and the largest, the long-billed woodcreeper (*Nasica longirostris*), a sensational inhabitant of varzea forests, reaches just over 35.6 cm (14 in). Among the oddest of the group are the five species of scythebills (genus *Campylorhamphus*), whose extremely long, downward-curving bills are used to probe deeply into bromeliads and other epiphytes.

Many woodcreepers are ant followers, joining antbirds and other species to feed on insects and other animals disturbed by the oncoming army ants. With differing body sizes and bill shapes, several species of woodcreepers coexist and feed with little or no apparent competition. At one army ant swarm in

Ivory-billed woodcreeper

Wedge-billed woodcreeper

Belize, I observed six woodcreeper species. Two were large, three medium-sized, and one small.

Woodcreepers are common not only in rainforests but also along forest edges, disturbed jungle, and dry forests. Although suboscines, some species are highly vocal, their songs consisting of pleasant, melodious, whistled trills.

Foliage Gleaners—Ovenbirds

The ovenbirds (family Furnariidae, order Passeriformes) are "little brown birds" of the American tropics. All ovenbirds are generally nondescript, their plumage basically brown, tan, buffy, or grayish. Identification of individual species can be very difficult, since differences among species are often subtle and hard to see in the field. This highly diverse family occurs not only in lowland forests but in all types of habitat ranging through cloud forest, Patagonian pampas, Andean paramos and puna, and coastal deserts and seacoast. Many kinds, especially the spinetails, are common along forest edge and disturbed areas, and many are found in dry forests. The family takes its common name, ovenbird, from several species (most notably the horneros, genus *Furnarius*) that construct ovenlike, dome-shaped mud nests. Not all ovenbirds build such structures. Some species nest in natural cavities or in mud banks, and some make basketlike structures of twigs and grass. The thornbirds (genus *Phacellodomus*) construct large and conspicuous globular nests of sticks that are easy to see in dry forest. The evolutionary trends of furnarids are analyzed by Fedducia (1973). For a general natural history of the group, see Skutch (1996).

Ovenbird species have among the oddest common names of any birds. One may encounter a xenops, a recurvebill, a foliage-gleaner, and a leafscraper (not to be confused with leaftossers). There are also woodhaunters, tree-hunters, treerunners, palmcreepers, and earthcreepers (not to be confused with streamcreepers). There are barbtails, spinetails, tit-spinetails, softails, and thistletails (not to be confused with prickletails). Finally, there are thornbirds, miners, cinclodes, horneros, and canasteros. You will need patience and skill to sort out ovenbirds, as they are a challenging group.

All ovenbird species are basically insectivorous, but as a family they do not show the bill diversification that is so evident in woodcreepers. Rather, ovenbirds tend to be habitat- and range-specific and develop specialized feeding behaviors. Some, like the ground-feeding leafscrapers and leaftossers, systematically probe among the litter. Their bodies are chunky and almost thrushlike in shape. Others, like the slender foliage gleaners, search actively among the leaves, ranging throughout canopy and understory. Spinetails dart quickly from bush to bush while the small xenopses hang chickadee-like, searching the underside of a leaf. Ovenbirds of various species are common members of mixed foraging flocks.

Ant Followers—The Antbirds

Figures 141, 142

Antbirds (families Thamnophilidae and Formicariidae, order Passeriformes) include the antbirds, antshrikes, antwrens, antvireos, antthrushes, and antpittas. Until recently, all antbirds were placed in the family Formicariidae, but analysis of DNA patterns in the group resulted in splitting the family into two. About 75% of the antbirds are placed in Thamnophilidae, referred to as typical antbirds, while the other 25%, the antthrushes and antpittas, are placed in Formicariidae, the ground antbirds (Ridgely and Tudor 1994). Antbirds reach their peak species richness in Amazonia, where up to thirty or forty species may occur together. The name antbird, or formicarid, comes from the army ant–following behavior of some species. However, most antbirds do not follow army ant swarms. Some never do, some occasionally do, and some virtually always do. The latter group is often termed the "professional antbirds." For an introduction to the natural history of antbirds, see Skutch (1996).

Typical antbirds (Thamnophilidae) are more colorful than ovenbirds, with many sexually dimorphic species. Males are often boldly patterned in black and white. Some, like the widely distributed and common barred antshrike (*Thamnophilus doliatus*), are zebra-striped. Others are grayish black with varying amounts of white patterning on wings, breast, and flanks. Still others are chestnut or brown. Females tend to be rich brown, tan, or chestnut. Some antbirds have an area of bare blue or red skin around the eye, and in some species iris color is bright red.

Most antbirds are foliage gleaners, picking and snatching arthropods from

Great antshrike

foliage, and some capture insects on the wing. They forage at all levels from the canopy to the litter on the forest floor. They typically form mixed species flocks with other birds, and various antbird species tend to feed at specific heights above the forest floor. Mixed flocks of up to fifty bird species move through Amazonian lowland forests, of which twenty to thirty species may be antbirds. Certain species such as the flycatching antshrikes (genus *Thamnomanes*) occupy the role of "central" species in the flock (Willis and Oniki 1978). These antshrikes are highly vocal and act as sentinels, warning the others of impending danger should they spot a forest falcon or other potential predator (see above, tanagers). Willis and Oniki describe the relationship among the various mixed flock species as a "casual mutualism." They each benefit somewhat from the other's presence.

There are twenty-eight species of professional ant-following birds, each of which makes its livelihood by capturing arthropods scattered by advancing fronts of army ants. In addition, other bird species frequently, but not always, can be found accompanying the ants. There are even some butterflies that associate with army ants to feed on the bird droppings (Ray and Andrews 1980). In northern Amazonia, the white-plumed antbird (*Pithys albifrons*) is among the commonest professional antbirds. This bird is unmistakable, its face dominated by a tall white crest, its head black, its back and wings blue-gray, and its breast and tail rich chestnut in color. The spotted antbird (*Hylophylax naevioides*), the bicolored antbird (*Gymnopithys leucaspis*), and the black-faced antthrush (*Formicarius analis*) are among the most devoted ant followers in Central America. Where these three are found together, there are surely army ants about (Willis 1966, 1967).

Ant followers rarely feed directly on army ants. It is suspected that the high formic acid content of these insects deters birds from eating them. Instead, antbirds feed on anything from insects to small lizards scared up by the oncoming ant columns. Two army ant species, *Eciton burchelli* and *Labidus praedator*, are the ants most frequently followed. Birds such as woodcreepers, ovenbirds, motmots, certain tanagers, and other "less professional" antbirds come and go as part of the ant-following avian assemblage, but the professional antbirds always stay with the ants. Only when breeding do they become territorial and cease to follow ants for a time. Even then, they will quickly orient to army ant swarms within their territories.

Species such as the spotted and bicolored antbirds feed actively in trees and undergrowth, while the black-faced antthrush walks sedately on the forest floor. With the stature of a small rail, the black-faced antthrush, which can be found throughout lowland forests in Central and much of South America, walks with its short tail cocked upward and head held up and alert. It is easy to imitate its whistled, downscaled "chew, chew, chew, chew" call. In Trinidad, I called one almost to my feet as I whistled and it answered.

Antbirds tend to mate for life. Both male and female are active nest builders (Skutch 1969, 1996). One species, the ocellated antbird (*Phaenostictus mcleannani*), forms clans. Sons and grandsons of a pair return to the breeding territory with mates to form clans, and a clan will occasionally attack another intruding clan (Willis and Oniki 1978). Antbirds also sometimes intimidate migrant thrushes that attempt to gather at antswarms (Willis 1966).

Hairy-crested antbird, a typical
antbird, and black-faced
antthrush, a ground antbird

Birds of Prey

Birds of prey are diverse and abundant in the Neotropics. They range in size from the tiny bat falcon (*Falco rufigularis*) and pearl kite (*Gampsonyx swainsonii*) to the majestic harpy eagle (*Harpia harpyja*). Open areas, such as savannas, are excellent for searching out many of the larger species, since some habitually soar on thermal currents rising from the hot ground. Inside forests, birds of prey can be elusive. Many, such as the forest-falcons (genus *Micrastur*), sit motionless on a branch waiting for an opportunity to attack would-be prey.

New World Vultures   *Figures 144, 154*

Five vulture species, the black (*Coragyps atratus*), turkey (*Cathartes aura*), lesser yellow-headed (*Cathartes burrovianus*), greater yellow-headed (*Cathartes melambrotus*), and king (*Sarcoamphus papa*), can often be seen in various combinations in the sky together over rainforests and savannas. Not strictly raptors, these birds rarely if ever kill prey (black vultures are alleged to kill small animals) but rather devour animals that are carcasses, already deceased.

The black vulture is one of the most ubiquitous birds of the Neotropics, ranging from Argentina well into North America. Black vultures commonly congregate in vast numbers around garbage dumps and are thus common city dwellers, urbanite birds that work as sanitary engineers thoughout much of Latin America. Turkey vultures are named for their red heads, giving them a superficial resemblance to turkeys. Turkey vultures fly with wings distinctly upraised, in a dihedral pattern. One hawk, the zone-tailed (*Buteo albonotatus*), flies very much like a turkey vulture, a possible form of behavioral mimicry, since the plumage of the two species is generally similar. Perhaps a zone-tailed hawk can fly closer to potential prey if it's mistaken for a carrion-eating turkey vulture. Two other species, the greater yellow-headed and lesser yellow-headed, are closely related to the turkey vulture but have yellow-orange heads, not red heads. The greater yellow-headed is most common flying over forests and along watercourses. It is common throughout most of Amazonia. The lesser yellow-headed avoids closed forests, being common on savannas. The king vulture is the largest and most spectacularly plumaged Neotropical vulture. It is black and white, and its head is adorned with bright (but bizarre-looking) orange wattles.

Kites

Eleven species of kites gracefully skim Neotropical skies searching out small animals such as mice, birds, lizards, and arthropods. Kites have sharply hooked bills, a trait particularly evident in the snail kite (*Rostrhamus sociabilis*) and the hook-billed kite (*Chondrohierax uncinatus*). The snail kite specializes on one food source, large marsh snails, which it adeptly removes from the protective shell with its sharply hooked bill. Another common kite is the white-tailed (*Elanus leucurus*), often seen hovering over open fields and savannas seeking its small animal prey. The most graceful flier among the kites is the swallow-tailed (*Elanoides forficatus*), a slender black and white kite with a deeply forked tail. The 22.9-cm (9-in) pearl kite (*Gampsonyx swainsonii*) is one of the smallest tropical birds of prey. Mostly black with white underparts, it has a buffy forehead and face, and a white or rufous neck. Like most kites, it frequents savannas.

Hawks, Falcons, and Caracaras   *Figures 145, 154*

Forty species of hawks (family Accipitridae, order Falconiformes) can be found in the Neotropics. Included here is a mere sample.

The crane-hawk (*Geranospiza caerulescens*) is a slender, blackish gray inhabitant of wet savannas, mangroves, and swamps. The bird has exceptionally long, bright orange legs. It feeds both on the ground and in trees and has been reported to assume odd postures, such as hanging upside down, when probing epiphytes and branches for amphibians and reptiles. Another savanna species is the savanna hawk (*Heterospizas meridionalis*), which tends to be seen walking about on the open ground on its long yellow legs. It is largely rufous, with black tail and wing tips and dark barring across its breast.

The splendid white hawk (*Leucopternis albicollis*) is apt to be seen soaring on warm thermals over forests. As its name implies, it is virtually all white but for a black band across the tail and black on the wings and around the eyes. Other soaring hawks include the common black hawk (*Beautiogallus anthracinus*) and great black hawk (*B. urubitinga*). Both of these birds are almost all black but for white tail bands.

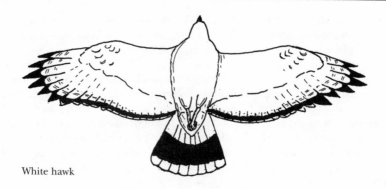

White hawk

The black-collared hawk (*Busarellus nigricollis*) is one of the most elegantly plumaged Neotropical birds of prey. Warm rufous in overall color, it has a white head and black throat, outer wings, and tail. This hawk has a distinct shape when seen overhead because its wings are quite wide and tail short. Found around marshes, it feeds mostly on fish.

The roadside hawk (*Buteo magnirostris*) is very well named. The hulking shape of this grayish rufous hawk can be seen perched on cecropias, palm trees, ceibas, and telephone poles all along tropical roads. This abundant and widely distributed species is also variable in plumage, and thirteen races have been recognized.

Falcons are small, speedy birds of prey known for their aerial agility. With long tail and sharply pointed wings, falcons are quick to pursue and capture rodents, small birds, and insects. One species, the diminutive bat falcon (*Falco rufigularis*), specializes in capturing bats at dawn and dusk. It is largely dark blue, with a white throat and orange on the thighs and lower belly.

The laughing falcon (*Herpetotheres cachinnans*) is often seen perched atop a snag along a forest edge, cleared field, or savanna. Very buffy on the head,

Bat falcon

neck, and breast, the laughing falcon has dark brown wings, back, and tail, with a black band through the eyes and around the back of the neck. Named for its penetrating loud call, these birds prey on snakes and other animals spotted by patiently sitting for long periods.

Forest falcons (genus *Micrastur*) are grayish falcons that skulk inside forests and are generally difficult to find. They are very inconspicuous, sitting motionless in the deep forest shade.

The yellow-headed caracara (*Milvago chimachima*) is common and conspicuous, often seen in groups along rivers and forest edges. Caracaras, like vultures, feed on carrion and, hence, are frequently encountered along roadsides. Yellow-headed caracaras are slender birds, buffy yellow on head, breast, and belly with blackish brown wings and tail. Several other caracara species also inhabit the Neotropics.

Crested caracara

Hawk-Eagles and Eagles

The largest Neotropical birds of prey are eagles and hawk-eagles. There are three species of hawk-eagles, each of which has a crest atop its head. The ornate hawk-eagle (*Spizaetus ornatus*) has a bright orange neck and a tall black crest. The black hawk-eagle (*S. tyrannus*) is uniformly dark, and the black-and-white hawk-eagle (*S. melanoleucus*) is black above and white below. Hawk-eagles are soaring hawks, usually seen above the canopy making circles high overhead.

The rare harpy eagle (*Harpia harpyja*) certainly ranks among the most splendid of Neotropical birds. In fact, I would argue that it is one of the most impressive avians on the planet. It ranges from southern Central America throughout Amazonia but is very difficult to find over most of that vast territory. This huge predator is just over a meter tall, with extraordinarily thick, powerful legs and feet. Mostly gray on face and belly, its wings, back, and upper breast are black. The head sports a tall blackish gray crest. No bird of prey equals the size of a harpy. Nonetheless, it is secretive, tending not to soar, and thus is difficult to

Harpy eagle

see well. Harpys feed on monkeys and sloths, including the largest species, captured by a swift pass, legs extended, grabbing the prey from its tree.

Strictly a forest dweller, now unfortunately quite rare over most of its range, the magnificent harpy eagle is a top prize for anyone seeking unique tropical birds. I visited a harpy nest at Imitaca Forest in Venezuela, seeing a nearly full-grown immature and an adult female. This forest was being heavily logged, and the long-term success of harpy eagle in this region is, I fear, dubious. Recent work done in the area with radiotelemetry, involving actually tracking the birds by satellite (Alvarez-Cordero et al. 1996), may aid in developing a management plan to ensure that harpy eagles have adequate protection and sufficient habitat. One promising result of this work is that some harpy eagles do not abandon an area even after years of selective logging.

Owls

Owls (order Strigiformes, families Tytonidae [barn owls] and Strigidae [typical owls]) are nocturnal birds of prey. Over two dozen species occur in the Neotropics. Below are several of the most wide-ranging.

The spectacled owl (*Pulsatrix perspicillata*) is the largest Neotropical owl, reaching 48.3 cm (19 in) in length. It is buffy yellow on the lower breast and belly, with dark brown back, wings, tail, and head. A dark brown band crosses its upper breast, and its bright yellow eyes are highlighted by white, giving the bird its name. Spectacled owls make a deep, hooting sound and can be easily attracted to tapes of their voice.

The black-and-white owl (*Ciccaba nigrolineata*) is just that. Its breast, belly, back, and face are barred black and white. This species, which feeds heavily on bats, is also responsive to tapes of its voice.

The mottled owl (*Ciccaba virgata*) is warm brown and tan with dark brown eyes. Unlike its black and white relative, it does not tend to be attracted to tapes.

During the daytime, it is not uncommon to see a small pygmy-owl (genus *Glaucidium*) with its bright eyes staring as it perches atop a snag. Several species of these 15.2-cm (6-in) owls occur in the tropics, but the most common is the ferruginous pygmy-owl (G. *barasilianum*), so named for its reddish brown plumage.

Amazonian pygmy-owl

North American/Neotropical Long-Distance Migrant Birds

Within the past two decades, research has burgeoned on Neotropical migrant birds, species that migrate north to nest in North America and then migrate south to winter in the Neotropics. With the realization that some species of Neotropic long-distance migrants are experiencing alarming declines in population (Terborgh 1989; Askins et al. 1990), the need to study these birds thoroughly on both their breeding grounds and their wintering grounds (as well as the migratory route in between) became a rallying cry for field ornithologists, me included. Below, I summarize some of what we have learned. For additional information, see Keast and Morton (1980), Hagan and Johnston (1992), Rappole (1995), and DeGraaf and Rappole (1995).

During autumn, 338 species, or about 52% of all North American migrant bird species, fly to wintering areas in the Neotropics (Rappole et al. 1983; Rappole 1995). This influx may total somewhere between five and ten billion birds. The majority winter in Central America, but many also winter in South America and the Caribbean Islands, especially the Greater Antilles. The density of North American migrants is high in the Neotropics from November through March. Not only are there many yearling birds in addition to adults, but the actual land area of Central America is less by about a factor of eight, compared with available nesting area in North America. Migrants ranging from Swainson's Hawks (*Buteo swainsoni*) to Least Flycatchers (*Empidonax minimus*) are packed into tropical America, and their abundance is evident.

Many North American migrants are from families that evolved in the Neotropics. Tyrant flycatchers, hummingbirds, tanagers, orioles, and wood warblers all originated in the Neotropics, though their speciation patterns may have been much affected by their breeding ranges in North America. Long-distance migrant species represent the relatively few that ventured northward into the temperate zone, extending their ranges, perhaps because the northern summer presents an abundance of proteinaceous insect resources for the rearing of young, longer days in which to feed, fewer predators, plus the availability of abundant nesting sites.

The host of Neotropic migrant landbird species, when on their tropical wintering grounds, use virtually all available habitat types. They can be found, often abundantly, in dry forest, mangrove forest, montane forest, and rainforest. Brushy successional areas are habitat for many species, such as gray catbird (*Dumetella carolinensis*), northern yellowthroat (*Geothlypis trichas*), and yellow-breasted chat (*Icteria virens*). Rainforests provide habitat for wood thrushes (*Hylocichla mustelina*), Kentucky warblers (*Oporornis formosus*), American redstarts (*Setophaga ruticilla*), and other wood warblers. But many of these species also utilize successional areas.

Some Neotropical migrants, like the black-and-white warbler (*Mniotilta varia*), range very widely in the Neotropics, occupying many kinds of wintering habitat. Black-and-white warblers, unique among wood warblers for their habit of foraging for arthropods on bark, somewhat like nuthatches, are extraordinary as they range widely and frequent virtually any terrestrial habitat. They can be found in oak-pine forests, in mangroves, in plantations, along any kind of forest edge and successional scrub, in dry forests, and in interior rainforests in western Mexico, Antilles and West Indies, all parts of Central America, and northern South America (Kricher 1995). In general, other migrants are more restricted (some much more so), putting them at risk should they suffer from habitat loss. An example is the cerulean warbler (*Dendroica cerulea*), which winters along a narrow altitudinal belt between 620 and 1,300 m (2,034–4,265 ft) in the eastern Andean foothills of Colombia, Ecuador, and Peru (Robbins et al. 1992). Unfortunately, this area has been and continues to be heavily deforested from conversion to agriculture, including cocaine fields, putting the future of this species at some risk.

Many North American migrants eat a diet high in fruit while in the tropics. Baltimore and orchard orioles (*Icterus galbula* and *I. spurius*) and scarlet and summer tanagers (*Piranga olivacea* and *P. rubra*) feed in cecropia and fig trees among mixed flocks of euphonias, Neotropic tanagers, and honeycreepers. Leck (1987) cites studies indicating that fruit availability is often high on disturbed habitats. Many researchers have noted that abundance of migrants is high in successional areas and young forests. Higher fruit availability may be one reason why migrants favor such areas. North American migrants are believed to be important fruit consumers and seed dispersers, especially for plants that typically grow in disturbed areas (Wheelwright 1988). Some Neotropical migrant species, such as the orchard oriole and Tennessee warbler (*Vermivora peregrina*), which feed on nectar, have been recognized as potential pollinators while on their wintering grounds.

Given that North American migrants spend perhaps the majority of their year in the Neotropics, it is not surprising to learn that some are relatively specialized when on their wintering grounds. For example, the worm-eating warbler (*Helmintheros vermivorus*) spends about 75% of its time foraging for arthropods that are hidden within dead leaf clusters, which abound in Neotropical forest and forest edge. *Cecropia* leaves alone are huge and curl when they drop, forming ideal habitat for arthropods. This wood warbler is specialized to reach deeply into the cavelike curled leaves. On its breeding grounds it spends about 75% of its time gleaning insects from live leaves (Greenberg 1987a, 1987b).

The degree to which North American migrants interact with Neotropical resident birds is by no means certain and probably varies considerably depending on the ecology of the birds. Willis (1966), working on Barro Colorado Island, noted that North American thrushes are prevented by resident antbirds from access to swarms of *Eciton*. Many other researchers, however, argue that North American migrants may compete relatively little, if at all, with resident species. Much research remains to be done on this question.

My colleague William E. Davis, Jr., and I (Kricher and Davis 1986, 1987, 1992) have investigated winter site fidelity among migrant species in southern Belize. We, along with many others working throughout the Neotropics, have shown that such species as the wood thrush, ovenbird (*Seiurus aurocapillus*), Kentucky warbler, and gray catbird occupy exactly the same locations from one winter to the next. Although these birds migrate north to nest, they return in the fall to precisely the same local wintering area used the previous year, a behavior called winter site fidelity. We placed fine mesh nets (called mist nets) in selected locations in both successional areas and rainforest. We captured some of the same banded individuals in the same nets, at the same locations, over three succeeding winters.

As one might expect, winter site fidelity often means that the birds are territorial, defending winter sites. Work by John Rappole and others has shown this to be true, at least for some species. Wood thrushes, for example, establish and defend winter territories, using subtle vocalizations and body posturing. Each wood thrush has its own turf within the rainforest. Survivorship among "floater" wood thrushes, birds that have not succeeded in acquiring and holding a winter territory, is diminished (Rappole et al. 1989; Rappole et al. 1992; Winkler et al. 1990). In another example, observations suggest that male hooded warblers (*Wilsonia citrina*) obtain territories inside interior forest and females are territorial in disturbed, successional habitats (Rappole and Warner 1980). Other researchers have also reported high site loyalties and as well as winter territoriality for other migrant species (Hagan and Johnston 1992; Rappole 1995).

Though many North American migrant species inhabit disturbed areas and montane forests, many also inhabit mature rainforest, and the degree to which deforestation (see chapter 14) may be eliminating wintering sites is an increasing concern. Some migrant species are now rare in North America, possibly due to loss of their wintering areas. The Kirtland's warbler (*Dendroica kirtlandii*) nests only in successional jack pine forests in Michigan. The bird was once probably more widely spread and could possibly occupy a larger nesting area today, but winter habitat loss in the Bahamas gives it no place where large populations could winter. The Bachman's warbler (*Vermivora bachmanii*) is an extremely rare nester in southern hardwood swamps. Loss of cane habitat in Cuba, where the bird winters, is thought to have been responsible for its populational demise. On the other hand, increase in second-growth habitat could actually favor such species as chestnut-sided warbler (*Dendroica pensylvanica*) and indigo bunting (*Passerina cyanea*). The effect of tropical deforestation on migrants eludes a simple answer. Like virtually all of tropical ecology, simple answers to complicated questions are just not there.

Male bare-crowned antbird

13

A Rainforest Bestiary

ALFRED Russel Wallace (1876) was deeply impressed by the animals he observed in the tropics:

> Animal life is, on the whole, far more abundant and more varied within the tropics than in any other part of the globe, and a great number of peculiar groups are found there which never extend into temperate regions. Endless eccentricities of form and extreme richness of colour are its most prominent features, and these are manifested in the highest degree in those equatorial lands where the vegetation acquires its greatest beauty and its fullest development.

In this chapter I will try to convey some of the wonder Wallace felt when he met the "eccentric" creatures that dwell within rainforests. I present here an array of Neotropical mammals, reptiles, amphibians, and invertebrates, by no means intended to be all-inclusive (what a foolish idea that would be), but rather with the focus of the discussion on those that the rainforest visitor is most likely to see or at least want to see.

Mammals

As a group, rainforest mammals tend to be secretive and mostly nocturnal, making it a challenge to see them well. Unlike the game herds of the African plains, rainforest mammals do not stand out in the open for easy viewing but rather scurry through the canopy or over the forest floor, usually well ahead of the naturalist. Still, by careful stalking or quiet sitting, especially in preserves where they enjoy protection from hunting, it is often possible to obtain excellent views of mammals. Many mammals are primarily nocturnal, and a walk at night with a good flashlight can be rewarding. Emmons (1990) provides an excellent field guide to Neotropical mammals, which I highly recommend. Eisenberg (1989) and Redford and Eisenberg (1992) provide a well-illustrated, comprehensive, two-volume review of Neotropical mammalian taxonomy, biogeography, and ecology.

Monkeys   *Figures 108, 109, 174*

Neotropical monkeys all belong to a group named the platyrrhines, referring to the position of the nostrils, which open at the sides in contrast with Old

World monkeys, the catarrhines, whose nostrils are more closely spaced and point downward (humans, having evolved from apes in Africa, are in this category). New World monkeys are less known for their nostrils, however, than for their tails. Some common and widely distributed platyrrhines, such as the capuchins, spiders, woollys, and howler monkeys, all have prehensile tails, which they skillfully use as a one-fingered fifth limb. It is with no little admiration that I have watched woolly and spider monkeys nonchalantly hanging by their tails high in the canopy, freeing both hands to feed calmly on fruits or aggressively throw branches at me, depending on their mood at the time. It's not possible to see such a sight anywhere else but in a Neotropical forest.

At present there is considerable question about exactly how many species of platyrrhines exist. Emmons describes forty-nine but emphasizes that there are many taxonomic uncertainties (Emmons 1990). The taxonomy of the group is complex, with some species having well-defined geographic races, some of which may, in future studies, be recognized as full species. What is clear is that platyrrhines have adaptively radiated to occupy many ecological niches in various kinds of Neotropical forest. There are large, apelike monkeys (spider, woolly, and howler monkeys), medium-sized, "typical" monkeys (capuchins and squirrel monkeys), monkeys with bald faces (uakaris), monkeys with long, shaggy fur (sakis), nocturnal monkeys (owl monkeys or douroucouli), small, lemurlike monkeys (marmosets), and squirrel-like monkeys (tamarins).

New World monkeys are inevitably forest animals, avoiding savannas. There are no Neotropical equivalents of the baboons that roam the African plains. Monkeys occupy interior forest, disturbed forest edge, gallery forests, and dry forests. They are most diverse and abundant in lowlands, but some occur in humid montane forests. Arboreal, most species rarely come to the ground, though some, like the capuchins and uakaris, occasionally do so to feed. Taxonomically, although all are platyrrhines, the "typical" monkeys are placed in the family Cebidae, and the tamarins and marmosets, sometimes called the squirrel-like monkeys, are in the family Callitricidae. Marmosets and tamarins are small, some of which, like the 15-cm (6-in), 185-gram (3-ounce) pygmy marmoset, are among the world's smallest primates. In addition, there is one species, Goeldi's monkey (*Callimico goeldii*), that is placed in its own family, Callimiconidae. The largest Neotropical monkeys are cebids, and the largest of the group are the 9-kg (20-lb) howlers. There are no Neotropical equivalents of the great apes, the orangutan, gorilla, and chimpanzees, though spider monkeys are ecologically quite similar to gibbons. Moynihan (1976) and Coimbra-Filho and Mittermeier (1981) provide a general review of the natural history and behavior of Neotropical monkeys.

Capuchins (genus *Cebus*) range from Amazonia through southern Central America and tend to be among the most commonly seen monkeys. There are four species, each 0.3–0.6 m (1–2 ft) in length (excluding the 45.7-cm [18-in] prehensile tail) and weighing 0.9–4.1 kg (2–9 lbs). They vary in color from pale brown to black, and each species has a pale face surrounded by whitish hair. Troops, typically numbering from five to thirty or more (depending on species), move quickly through forests foraging for fruits, leaves, and arthropods. Some also take birds' eggs, baby birds, and even small mammals. Capuchins are found in a wide variety of forest habitats, and one is apt to encounter a capuchin group anywhere from low tangle along the forest edge to the canopy of interior forest. They also frequent gallery forests, dry forests, mangrove forests, as well as disturbed and mature rainforest. Capuchins feed heavily on palm nuts and thus are frequently seen among the fronds in palm stands. Research on Barro Colorado Island in Panama on the white-throated capuchin (*Cebus capuchinus*) has shown that capuchins are important seed dispersers of several tree species (Foster, cited in Oppenheimer [1982]). Their habit of eating certain insect grubs, such as bruchid beetles, which they remove from fruits, makes them beneficial to trees parasitized by these insects. Capuchin troops are noisy, and their sounds attract agoutis and collared peccaries, which feed on fruits dropped by the simians (Oppenheimer 1982).

Titi monkeys (*Callicebus* spp.) are a bit smaller than capuchins. They have small faces and are thickly furred, body color ranging from gray to black, with nonprehensile, hairy tails. Titis are found in small groups of two to six and are diurnal and highly active, very skilled treetop jumpers. They feed on a wide variety of fruits, buds, and various arthropods (including spiders and millipedes), and one species, the widespread dusky titi (*C. moloch*), feeds especially on leaves. Titis seem to seek out thick jungle growth in which to feed and rest, often frequenting bamboo thickets. Titis are known for their dawn chorus, a loud, ringing duet performed by pairs. They are confined to Amazonia, and at present three species are recognized. However, the dusky titi has eight subspecies spread throughout central, southern, and western Amazonia (Emmons 1990), so the possibility exists that some of these may eventually attain full species status. At any rate, this racially polymorphic species has clearly been evolutionarily active.

Squirrel monkeys (genus *Saimiri*) are widely spread through Amazonia into Central America. There are three species, the common squirrel monkey (*S. sciureus*), found throughout much of upper Amazonia, the bare-eared squirrel monkey (*S. ustus*), found only south of the Amazon in central Amazonia, and the Central American squirrel monkey (*S. oerstedii*), an endangered species whose range is restricted to a small area along the Pacific Coast of Costa Rica and Panama. Like the situation with some other New World monkeys, taxonomists are not in total agreement about the systematics of the group. At any rate, a squirrel monkey, whatever its species, does not resemble a squirrel. It is a bit smaller than a capuchin, but its tail is equally long, giving it the appearance of a slender little animal with an immensely long, thin, black-tipped tail. The eyes are dark, appearing quite large, surrounded by white "spectacles" and a black nose and mouth. Ears are white. Body hair is grayish with rich rufous on the back, arms, and tail. The Central American species is more rufous than those found in South America. Squirrel monkeys favor gallery forests, lowland rainforest, and successional areas. They are usually obvious, as their troops number anywhere from 20 to over 100 animals, and they tend to be very active among the outer branches. They sometimes come around villages to feed on bananas, plantains, and citrus. They eat all manner of fruits as well as many insects. Squirrel monkeys can be abundant. One team of researchers estimated between 50 and 80 per square mile (Moynihan 1976).

Night monkeys or douroucouli (*Aotes* spp.) are well named, as these smallish, 0.9-kg (2-lb) monkeys are the only genuine nocturnal monkeys. Night monkeys range from northern Argentina and Paraguay throughout the Amazon Basin and north into Panama. One look at the owl-like, rounded head with immense dark eyes surrounded by white fur and soft grayish brown pelage is enough to discern the creature's nocturnal way of life. Groups of two to five night monkeys while away the daytime hours cuddled together in a hollow tree or among dense vines and other vegetation. At night they forage anywhere from almost ground level to the top of the canopy. They search for fruits, buds, insects, and, occasionally, nestling birds. Their densely furred, black-tipped tails are nonprehensile and thus droop straight down. Their loud calls, heard at night, keep the foraging troop together. They communicate by "body language," positioning the whole body, rather than by facial expressions, a probable behavioral adaptation to life in the dark (Moynihan 1976). Taxonomists are not agreed as to how many species there are, but up to nine species have been recognized, all very similar in appearance (Emmons 1990).

Uakaris (*Cacajao* spp.) (pronounced "WOK-a-ree") are medium-sized to large monkeys of gallery forests along the upper Amazon. Two species are currently recognized, one with long, thick, reddish or silvery white body hair, the other with entirely black body hair. In the red/white species, found in varzea forests, the head is both utterly bald and bright red, making the animal quite unmistakable. In the black species, found in igapo forests, the face (but not the entire head) is bald and black. Uakaris have short, hairy, nonprehensile tails. Groups of ten to thirty or more feed on fruits, leaves, flowers, and arthropods, especially caterpillars. They sometimes descend to the ground to forage on seedlings and fruits. Unfortunately, both species of uakaris are considered endangered due to hunting pressure and loss of forest to logging.

The sakis (*Pithecia* spp.) and bearded sakis (*Chiropotes* spp.) resemble uakaris in general body shape and long hair, but they have hair on their heads and faces, and their bushy tails are much longer than tails of uakaris (and also not prehensile). Most are brownish gray or black, and one species, the Guianan saki (*P. pithecia*), is strongly sexually dimorphic, the male being black with a white face, the female gray-brown. There are debatably four saki species and two bearded saki species, found in various places throughout Amazonia. Sakis are reported to be very skilled at leaping from tree to tree, giving them the Spanish name *volador*, or "flier." Small groups, rarely exceeding five animals, feed primarily on fruits and are found mostly in well-developed rainforest. Sakis are not nearly as uncommon as uakaris. Bearded sakis also feed primarily on fruits but are found in larger troops numbering up to thirty individuals.

Spider monkeys (*Ateles* spp.) occur from Mexico through the lower Amazon. There are five species, only three of which are common, and they are generally geographically separated, though there are many areas where the ranges of two species are parapatric (share a common border but do not significantly overlap). Because there is much variation in color within some of the species, the species are at present best identified by range (see Emmons 1990). The three common and wide-ranging species are the Central American spider monkey (*A. geoffroyi*), which ranges throughout Central America into southern Mexico, the black spider monkey (*A. paniscus*), which has two population centers, one east of the Rio Negro and north of the Amazon and one in western Amazonia west of the Rio Madeira, and the white-bellied spider monkey (*A.*

belzebuth), found widely southeast of the Amazon as well as northwest of it in parts of Peru, Colombia, and Venezuela.

Spiders are rather large but distinctly slender (hence the name), generally weighing about 6.3 kg (14 lbs). Their prehensile tail ranges from 50.8 to 88.9 cm (20–35 in) in length. They vary in color from black to pale brown and reddish. Troops of spiders typically consist of about eight adult males, fifteen adult females, and ten babies and juveniles. At any given time, four females will be either pregnant or in estrus. Bachelor male troops also occur. Often fewer animals are seen together, because troops frequently fractionate in a given area during the day, reassembling at their sleeping tree at night. Spiders forage together in the treetops, often quite actively. Their slender bodies adapt them well for graceful movement through the canopy, and they seem to prefer mature forest. Spiders often move by brachiation, swinging arm over arm from branch to branch in a manner similar to Old World gibbons. Studies on BCI have shown that 80% of their diet is fruits and 20% leaves (Moynihan 1976). Like many other monkeys, they tend to strongly prefer young leaves (page 153).

It is interesting to contrast primate behavior in areas where the animals are hunted regularly compared with those in which they enjoy complete protection. Near Manaus, Brazil, capuchins have apparently learned to descend to the ground so as to run through underbrush and avoid hunters (Andrew Whittaker, pers. comm. 1994). In unprotected forests where monkeys are regularly shot for food, I have seen troops of spider monkeys frantically dash through the canopy, apparently attempting to avoid the humans below. However, in a protected forest near Alta Floresta, Brazil, south of the Amazon, the presence of my group actually attracted a troop of white-bellied spider monkeys that behaved belligerently, approaching us quite closely in the forest, climbing into the understory and lower canopy, vigorously shaking branches, loudly vocalizing, and eventually urinating and defecating on us, the interlopers. It was a fulfilling experience to see these animals so closely, though prudent to wear a hat.

Monkey species are hunted to varying degrees throughout Amazonia, where several of the larger species are regarded as important food sources. In general, where human populations are low, monkey hunting is usually not so severe as to markedly deplete the primate populations. Where human populations are dense, local monkey populations may be reduced significantly (Mittermeier 1991). The most important resource in maintaining primate populations is large tracts of natural forest.

One of the most endangered Neotropical monkeys is the wooly spider monkey (*A. arachnoides*) or muriqui, an inhabitant of the threatened coastal forest of southeastern Brazil. It is estimated that only 300–400 of these creatures remain, scattered widely in fragments of what was at one time continuous forest (Emmons 1990). As is the case with some other endangered species, this animal was initially hunted for food and is now mostly threatened by habitat loss. Yet another species, the brown-headed or Colombian black spider monkey (*A. fusciceps*), is also experiencing severe habitat reduction throughout its narrow range in northwestern South America and could very well become endangered.

Woolly monkeys (*Lagothrix* spp.) are named for their thick, woolly fur, which may be black, brown, grayish, or reddish, depending upon both species and geographic location. There is much variation. By far the most widespread of the two species is the common woolly monkey (*L. lagothricha*), which ranges throughout western Amazonia. The rare and endangered yellow-tailed woolly monkey (*L. flavicauda*) occupies a very restricted range in northeastern Peru. Woollies are slightly larger than spiders, weighing up to 10 kg (22 lbs), and in body shape they more closely resemble howlers (see below). Like howlers and spiders, woolly monkeys have prehensile tails and are highly skilled arboreal acrobats. They prefer mature forest, where groups of two to sixty feed on fruits, palm nuts, seeds, foliage, and some arthropods. They feed at varying heights, not being confined to the canopy, and can be found in both seasonally flooded and terra firme forests.

Woollies are perhaps the most intensively hunted Neotropical monkeys (Emmons 1990). The meat is tasty and the animals themselves are large, so a kill provides lots of meat. Woollies have a generally low reproduction rate and cannot maintain their populations against strong hunting pressure.

Howler monkeys (*Alouatta* spp.) are as large or larger than woolly monkeys and are scientifically the best-studied Neotropical monkeys. Many primatologists have focused on howlers, including studies of their various behaviors, troop sizes, communications, territoriality, vocalizations, and feeding habits. Howlers are large and robust monkeys with prehensile tails and bearded faces. As a superspecies group they are widespread, distributed from the Amazon Basin south to Argentina and Paraguay and north to Trinidad, Central America, and the Yucatan Peninsula. There are six species, two of which are considered endangered. The most widespread is the red howler (*A. seniculus*), recognized by its bright reddish fur, which occurs throughout northern South America north of the Amazon and east of the Andes, including Trinidad. It is replaced south of the Amazon and throughout central and eastern Brazil by the red-handed howler (*A. belzebul*), which has all-black fur except on the hands, feet, tail, and (on males) scrotum, where it is rusty red. The black howler (*A. caraya*) is found in extreme southern Amazonia into Paraguay and Argentina. It is sexually dimorphic in coat color, the males being all black, the females tan. The brown howler (*A. fusca*), whose coat color is uniformly warm brown, is found in southeastern Brazilian coastal forests and in northeastern Argentina, and, though not yet considered endangered, it is experiencing rapid loss of habitat. Both species of endangered howlers occur largely (one entirely) in Middle America. The Mexican black howler (*A. pigra*), an all-black species quite similar in appearance to the black howler, is found only on the Yucatan Peninsula, in Belize, Guatemala, and Mexico. The mantled howler (*A. palliata*), essentially black but tan-brown on its sides and back, is found west of the Andes fron northern Peru to Colombia and throughout Central America to southern Mexico. Though this range is rather extensive (and does not overlap with the Mexican black howler), the animal has suffered from extreme fragmentation of its habitat throughout most of its range, as well as hunting pressure, and is thus, unfortunately, endangered.

Howlers are named for their ferocious voice, which echoes through the rainforest at sunrise and sunset, reminding one of the roar of an enraged jaguar. Males have an enlarged throat sac and tracheal cartilages that act as a resonator, dramatically amplifying their calls. Their infamous howling serves to mark the troop territory, and, because the howling carries for nearly a mile through rainforest, two troops will come to a mutual agreement about the real estate without necessarily ever meeting one another. Males are approximately 30% larger in body size than females, and this difference, plus the unique male vocal apparatus, suggests strong sexual selection in this species complex. An average howler clan consists of three adult males, seven to eight females, and varying numbers of juveniles. Clans vary in size, however, ranging from four to thirty-five. Males are dominant over females, and young animals tend to be dominant over older animals.

Howlers live in many forest types (mature and disturbed forests, gallery forests, semideciduous forests, lowland and lower montane forests) and, wherever they occur, specialize on a diet of leaves, with fruit and flowers making up only 30.7% of their diet. Because they rely so heavily on leaves, they tend to have fairly small territories. Glander (1977) found that the mantled howler consumed 19.4% mature leaves, 44.2% new leaves, 12.5% fruit, 18.2% flowers, and 5.7% leaf petioles (see page 150). This diet takes in substantial protein (mostly from mature leaves) and minimizes the amount of undigestable fiber and potentially toxic defense compounds ingested (Milton 1979, 1981). Howlers are more often heard than seen as they tend to remain well up in the canopy of mature rainforest. However, though endangered or threatened in many places, there are some areas where they remain common and relatively easy to find.

Howlers on Barro Colorado Island suffer a high natural mortality rate. Seasonal changes in food availability, plus periodic unpredictable shortages of high quality foods, were believed to place major stresses on the population. Some monkeys were heavily parasitized by botflies (page 379), though most monkeys seemed to survive their botfly wounds (Milton 1982). Still, such parasitic afflictions can't be pleasant for the simians.

Marmosets and tamarins, family Callitrichidae, are diminutive monkeys, the gnomes of the rainforest, scurrying through the branches, peeking out from behind leaves often larger than they are. They resemble hyperactive squirrels as they scatter about in the low tangles of branches, constantly scanning in all directions. They have the characteristic habit of leaping from one branch to another, landing vertically. Their tails, while used for balance, are nonprehensile. These small monkeys may be found in interior forest but also favor gaps and disturbed areas where insects and small fruits are abundant. Many feed in the lower story of forest as well as forest edge.

Marmosets (*Callithrix* spp. and *Cebuella* spp.) include four species, all of which are confined to South America. One, the tufted-ear marmoset (*Callithrix jacchus*), has four subspecies, which, taken together, range from central through southeastern Amazonia and coastal Brazil. Another, the pygmy marmoset (*Cebuella pygmaea*), ranges throughout much of western Amazonia. The other two species are found along the central and eastern Amazon, but only south of the river. There is substantial uncertainty about the exact ranges of some of the species and subspecies. Marmosets in the genus *Callithrix* have somewhat shaggy fur and long tails. Their faces are small but accentuated by prominent white fur about the ears. They range in color from nearly white to grizzled gray and black. The pygmy marmoset is yellowish-brown, with thick fur surrounding its tiny face.

The first pygmy marmoset I saw was clinging to the bars inside a bird cage in a Lima garden. Unfortunately, this and other species are frequently kept as pets. However, like other callitrichids, pygmy marmosets are too small to be regularly hunted, thus they often occur in good numbers near human habitations. Pygmy marmosets, which are barely 15.2 cm (6 in) long, feed on berries, buds, fruits such as bananas, and various arthropods. They also have the odd habit of "sap-sucking," which involves gnawing holes into a favorite tree trunk and drinking the oozing sap. Their lower incisors are unusually long, an aid in chewing holes to harvest sap (Goldizen 1988).

There are eleven species of tamarins, all in the genus *Saguinus*, plus one species in the genus *Leontopithecus*. As a group they are fairly similar in general appearance to marmosets. They have long tails and shaggy fur, often with striking coloration patterns, their small faces accentuated by topknots, moustaches, or ruffs. Like marmosets, tamarins search mostly for insects and other arthropods as well as fruits, but they do not drill holes in search of sap. Tamarins are essentially confined to South America, though one species, Geoffrey's tamarin (*S. geoffroyi*), ranges into Central America as far north as southern Costa Rica. The most widely distributed species are the midas tamarin (*S. midas*) of northeastern Amazonia and the saddleback tamarin (*S. fuscicollis*), found in western Amazonia. Other species have much narrower ranges.

The golden lion tamarin (*Leontopithecus rosalia*) is perhaps the most beautiful of the group, with uniformly glistening reddish gold fur, especially thick as it forms a ruff surrounding its diminutive black face. This species is the most endangered of any Neotropical monkey and is considered to be in extreme danger of extinction. Three subspecies each occupy limited ranges in highly fragmented forests in southeastern Brazil. Because this species is much less adaptable to forest disturbance than tamarins in general, it has suffered severely from deforestation.

Tamarins and marmosets are unusual among primates for their flexible breeding systems. In some ways they seem to exhibit the reverse of the normal primate pattern. They tend strongly toward monogamy, though most primates are polygynous (Hrdy 1981). However, studies have shown that one female tamarin may mate with several males (polyandry) without creating aggression among the males (Abrahamson 1985). Tamarin groups normally consist of four to six adults, typically two or more females and several males. Females are aggressive toward one another, and one female does all of the breeding in the group. Several males mate with the alpha female, and males devote much energy, more so than females, to parental care.

The saddleback tamarin (*S. fuscicollis*) displays both polyandry and cooperative breeding (Terborgh and Goldizen 1985; Goldizen 1988). A dominant female will mate with several males. Juveniles are normally raised with the help of older, nonreproductive individuals (brothers and sisters) as well as reproductively active males. Why should some individuals help others raise young? One possible explanation, at least for this species, may be that the cost of parental care is so high that helpers are needed. Tamarins normally birth twins, and the combined weights of the infants can equal 20% of the mother's body weight. Infants must be carried as the troop moves on its foraging route. Males help carry infants, reducing the burden on the mother, who still must feed her twins with milk. But, why should helpers help? Tamarin troops tend to consist mostly of closely related individuals. Those individuals that help raise a close relative such as a brother or sister are promoting their own genetic fitness, since each sibling shares 50% of their genes. By helping raise twins, a brother tamarin is promoting one full set of his genes, a clear case of kin selection.

Rodents: Agouti, Paca, and Others

Most species of mammals in the world have a large pair of continuously growing, chisel-like incisors, with which they gnaw all manner of food. That is because, of the approximately 4,000 species of mammals in the world, about 1,750 (44%) are in the order Rodentia, the rodents (Emmons 1990). By far and away the most diverse order of mammals, rodents are rivaled in diversity only by bats, but there are less than half as many bat as rodent species. The Neotropics harbor some of the most ecologically interesting as well as the largest of the vast assemblage. Several are aquatic, others arboreal, and still others are burrowers. Some are familiar, similar to those found away from the tropics. These include such groups as tree squirrels, pocket mice, rice rats, and the all too familiar house mouse (*Mus musculus*), each of which has ancestors that colonized the Neotropics from elsewhere. Others, like the porcupines, spiny rats, agoutis, paca, and capybara, are evolutionarily unique to the Neotropics, members of a suborder called the cavimorphs, which originated in the Neotropics. Cavimorph rodents include the familiar domestic guinea pig (*Cavia aperea*), chinchillas (*Chinchilla* spp.), and the North American porcupine (*Erethizon dosatum*).

Agoutis (*Dasyprocta* spp.) are one of the most common of the larger rodents, with six species that range from tropical Mexico to northern Argentina and Paraguay. Each of the six species has a similar ecology. Primarily diurnal, agoutis are apt to be encountered anywhere inside forests. One of my students described an agouti (as well as its close relative, the paca) rather well by asking me, "What are those little piggy things running across the trail?" Agoutis do not look like "little piggy things" when seen closely, but, from a distance, their chunky 63–76-cm (25–30-in) bodies, long legs, and delicate, prancing gait give an impression of a small, tail-less (though they have very short tails), hoofed animal with the head of a mouse. Depending mostly upon species, they range in coat color from buffy reddish-brown to grizzled gray-black. They eat by sitting upright on their hind legs, holding their food (usually a fruit or seed) with their front paws in a manner suggestive of mice and squirrels. Agoutis are often important seed dispersers, collecting more seeds than they can consume at once, burying the remainder in a widely scattered pattern. The wide pattern, termed "scatterhoarding," is possibly adaptive in protecting the agouti's cache from discovery by other seed consumers such as peccaries. During times of shortage, agoutis dig up their buried seeds. Agoutis are vocal, and, when frightened by a predator, they often emit a high-pitched alarm bark.

There are two species, both called acouchys (*Myoprocta* spp.), that look very much like smaller versions of agoutis but have short but obvious tails. Both are found in northern Amazonia and are ecologically similar to agoutis.

When agoutis retire for the evening, pacas (*Agouti paca*) come out. There is but one species of paca, and it ranges widely, all the way from tropical Mexico to northern Paraguay. Pacas are common but are less frequently seen than agoutis because of their nocturnal habits. Pacas seem to prefer to be near water, and daytime hours are typically spent resting in a burrow along a stream bank. Monogamous, males and females share a single territory but nonetheless forage alone and occupy separate dens. A paca resembles an agouti in shape but has larger eyes and longitudinal white stripes and spots along the sides of its reddish-brown coat. Their bodies are larger and legs proportionally shorter than agoutis, and pacas weigh more, up to 10 kg (22 lbs). When threatened by a predator, a paca will retreat to water and remain immersed until out of danger (or air). Pacas, like agoutis, feed on fruit but also take leaves and other plant materials including certain tubers. Unfortunately, paca meat is considered highly tasty by humans, and this species has been much over-hunted in various regions.

The capybara (*Hydrochaeris hydrochaeris*), the world's largest rodent, is discussed in chapter 8.

There are fifteen species of porcupines and dwarf porcupines to be found in the Neotropics. All of them bear stiff hairs that often form sharp quills, but they differ from the familiar and related North American species in that they all have prehensile tails. All porcupines have short faces with large, bulbous noses and rather small eyes and ears. One of the most wide ranging and frequently encountered species is the Brazilian porcupine (*Coendou prehensilis*), which ranges throughout Amazonia, from extreme northern South America to southeastern Brazil. It is nocturnal and almost entirely arboreal (aided by a strong prehensile tail), climbing through the trees like either a slow monkey or a fast sloth, depending upon your frame of reference. Neotropical porcupines feed on a combination of fruits (including palms), bark, and leaves, though the diets of some species are not well studied.

Spiny rats and tree rats (family Echimyidae) are the most diverse group of cavimorph rodents, with somewhere between 60 and 100 species ranging throughout the Neotropics. It is believed that spiny rats (*Proechimys* spp.) are the most abundant mammals in the rainforest (Emmons 1990), though they are solitary and nocturnal and thus not commonly seen unless one searches at night. The fur has spines, particularly in the region of the lower back and hind legs, but the spines lack barbs. Spiny rats, like some lizards, have tails that can break off with a minor tug, an adaptation that permits escape from a predator, though with the cost of a tail. Spiny rats are found on the forest floor, but closely related tree rats (*Echimys* spp.) are arboreal.

Neotropical trees harbor just short of twenty species of Neotropical tree squirrels (mostly in the genus *Sciurus*) that look quite similar in appearance to North American tree squirrels (also *Sciurus*). The ancestors of these animals apparently ventured southward during the great faunal interchange of the Pleistocene (page 113). Some Neotropical sciurids are reddish, including the Amazonian red-tailed squirrel (*S. granatensis*). One particularly striking species is the Central American variegated squirrel (*S. variegatoides*), which ranges in color from white with a black back, to reddish black, to blackish gray, depending upon range. To my eyes, the most wonderful of the lot is the diminutive Amazon dwarf squirrel (*Microsciurus flaviventer*), an adorable little brownish, hyperactive rodent, quite undersized for a squirrel, with a thin, straight tail and big, curious eyes. This small, largely solitary creature lives most of its life in the canopy.

Finally, there is a group of mice collectively called spiny pocket mice (genera *Heteromys* and *Liomys*). These little mice look very much like North American deer mice and white-footed mice (genus *Peromyscus*). They feed heavily on seeds.

Collared and White-lipped Peccaries

Peccaries are members of the huge even-toed ungulate order, the Artiodactyls. They closely resemble wild pigs but are in their own family, the Tayassuidae. They differ from pigs in that their upper canine teeth are extremely sharp and point straight downward, whereas in pigs these teeth curve outward as tusks. Peccaries also have a dorsal scent gland, located toward the posterior of their backs. This gland exits through a large and conspicuous opening that was once mistaken to be the animal's navel! The secretion of the gland is quite musky, and the animals rub their faces vigorously against one another's scent gland as a means of recognition and solidification of the herd. Peccaries are highly social animals and are rarely seen singly.

The most common and also the smallest of the two peccary species is the 23–29-kg (50–65-lb) collared peccary (*Tayassu tajacu*). This species is abundant in forests and savannas throughout tropical America as far south as Argentina and ranges northward through the Mexican deserts into the American Southwest. Collared peccaries form herds of anywhere from three to thirty or more. Their bristly hair is a mixture of black and gray as adults, but they are quite brownish as juveniles. The name *collared* is derived from a band of whitish hair that separates neck from shoulder. Their faces are unmistakably piglike, and their snouts have a hard, fingernail-like rhinarium that acts as a trowel in rooting up vegetation. Herds of peccaries forage like pigs for roots, bulbs, and underground stems as well as leaves and fruits. They also eat arthropods and small vertebrates, if they can catch them. Loose soil from their rooting efforts is a common sight in rainforests along with the prints of their small cloven hooves. During dry season they often congregate at favored watering places. As they forage, they communicate with soft, continuous grunts, but, should danger threaten, they emit a loud, deep "Woof!" reminiscent of a large dog's bark. When cornered, they erect their bristles, chatter their teeth, and display their large canines. They put on quite an impressive show of threat but will (usually) not charge unless no escape is possible. Collared peccaries have an undeserved reputation for aggression. They are fundamentally peaceful, highly social animals (Sowls 1984). Should you encounter a band of them, give them a wide berth and they will go about their business, leaving you totally intact.

The larger white-lipped peccary (*Tayassu pecari*), identified by white hair around its mouth, congregates in herds of 50 to 300 (Emmons 1990) and is essentially confined to rainforest. Its range is limited to the lowland forests of South America and southern Central America. Herds of white-lipped peccaries have been alleged to charge people who stumble upon them, but Emmons suggests that their reputation for aggression is considerably exaggerated. Probably so. White-lipped peccary herds range widely in search of fruits. Their odor is quite distinct from that of the collared peccary.

A study of the feeding behavior of both peccary species in the Peruvian Amazon concluded that white-lips can exert a bite force 1.3 times that of collareds (Kiltie 1982). In other words, white-lips can crack tougher fruits and seeds than collareds and thus can make use of food items unavailable to their smaller cousins. Kiltie speculated that the large herd size of white-lips may be related to diet. Some of the hardest fruits, the palms, drop many fruits at once, representing a temporarily abundant but highly patchy resource. In large herds, white-lips can find and exploit such a resource effectively. White-lips have a narrower rhinarium than collareds and do not dig as deeply for roots. They probably depend more heavily on hard fruits.

Both collared and white-lipped peccaries are extensively hunted, comprising a major source of protein and skins for the scattered human settlements (Redford and Robinson 1987).

Tapirs

Tapirs (genus *Tapirus*) are Perissodactyls, or odd-toed ungulates, evolutionary relatives of rhinoceroses and horses. Only four species of tapirs occur in the world, and three of them are in the American tropics (the other is in Indochina). Tapirs are stocky, almost hairless animals, brownish to black depending upon species, with a short elephantine proboscis and a dense but short mane of stiff hair on their upper neck. The mane probably aids the animal in making its way through dense undergrowth. Tapirs have an acute sense of smell and select food plants at least in part on the basis of odor. They eat only vegetable matter, including leaves and fruits of various species. Re-

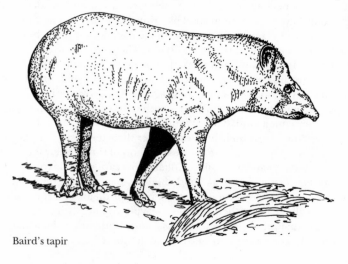

Baird's tapir

search on a captive tapir in Costa Rica indicated that most of the common rainforest plant species were unacceptable as food plants (Janzen 1983d). If true of free-living animals, tapirs could experience food shortages easily.

The three Neotropical tapir species are separated by range and habitat (page 108). The most widespread is the Brazilian tapir (*Tapirus terrestris*), which can be found east of the Andes from northern South America throughout Amazonia as far south as Paraguay. The Baird's tapir (*T. bairdii*) ranges throughout Central America and northern South America west of the Andes. The mountain tapir (*T. pinchaque*) has the most restricted range and is, as the name implies, essentially confined to higher elevations. It inhabits the paramos of the central and eastern cordilleras of the Andes, from Colombia to Ecuador (Eisenberg 1989).

Both tapirs and peccaries are widely hunted and thus tend to be wary. A study conducted in Belize showed clearly how local hunting pressure can significantly deplete tapir populations (Fragoso 1991). Tapirs are most active at night, and you will be very lucky to see one well. Look for tapir along water courses.

Sloths, Anteaters, and Armadillos   *Figures 106, 110, 111, 120*

Sloths and anteaters are among the most characteristic animals of tropical rainforests. Along with the armadillos, one species of which ranges into North America, they comprise the order Xenarthra (also called Edentata), creatures with peglike teeth plus a number of other anatomical characteristics that unite them. Anteaters, due to their skull modification for ingesting ants, are toothless, but sloths and armadillos have peglike teeth on the sides of their mouths. Front teeth are lacking. For a comprehensive account of these creatures, see Montgomery (1985).

The most commonly seen of the anteaters (Myrmecophagidae) is the 0.6-m (2-ft)-long tamandua (genus *Tamandua*). Two species occur, one (northern tamandua, *T. mexicana*) in Central America and in northern South America west of the Andes, and the other (southern or collared tamandua, *T. tetradactyla*) in Amazonia as far south as northern Argentina. Both species look similar, with long, pointed snouts, formidable curved claws on the forelegs, prominent ears, and a long, prehensile tail. The coat color is variable depending upon range. Some animals may be pure blond, others "vested" with black (Emmons 1990). Tamanduas are equally at home digging up ant nests on the ground or sampling the delicacies of termitaries in the trees. They excavate with their sharp front claws and extract the insects using their extensible sticky tongues. Tamanduas eat many kinds of ants as well as termites and bees. They tend to shy away from army ants and ponerine ants, both of which give nasty stings. When threatened, a tamandua may sit up on its hind legs and brandish its sharp curved claws. Be aware that such behavior is potentially dangerous to anything that bothers the anteater. The forearms are strong and the claws are sharp. The animal may look innocuous, but it isn't. Tamanduas are largely solitary and are active day or night.

The giant anteater (*Myrmecophaga tridactyla*) is much larger than the tamandua and is totally ground-dwelling. Its body measures about 1.2 m (4 ft) in length, and its huge, bushy tail adds almost another yard. Its head is shaped like a long funnel with eyes and ears placed well back of the small mouth, from which can protrude a 51-cm (20-in) sticky tongue. The grayish black coat color is punctuated by a broad black side stripe lined with white, the entire animal terminating in an immensely thick, ragged tail. Like the tamandua, the front claws are curved and sharp, an adaptation to digging into the hardened ant and termite nests that contain dinner for the anteater. It ranges through Amazonia and southern Central America, though it is now rare throughout much of its range. The animal remains common along forest edge and savanna areas such as the Llanos of Venezuela and the Brazilian Pantanal. Like the tamandua, the giant anteater will rear up and brandish its sickle-shaped, sharpened front claws if danger threatens. Some people who have been foolish enough to challenge such a display have allegedly been eviscerated.

A third anteater species is the silky or pygmy anteater (*Cyclopes didactylus*). Smallest of the three, it measures only 0.5 m (1.5 ft) in length. Don't count on finding one of these little creatures, for they are nocturnal and arboreal, climbing about in thick lianas. They have soft, golden buffy fur, short snouts and claws, and a prehensile tail. Their large black eyes testify to their nocturnal habits. They seem to eat only ants.

The term *sloth*, when applied to a person, has come to mean sluggish, lethargic, dull, perhaps even dim-witted. Real sloths (Bradypodidae, Megalonychidae) are probably all of these things. They do, however, have the tremendous advantage of staying around once you find one. If you care to watch a sloth move from one tree to another, a distance that would take a few seconds for a monkey, plan to spend about a day or so. Sloths lead slow-motion lives.

The most common sloth is the three-toed sloth (genus *Bradypus*), the favorite of Charles Waterton (see page 79). Three species occur, ranging throughout Neotropic forests, the most common of which is the brown-throated, *B. variegatus*. The three-toed sloth looks somewhat like a deformed monkey. It has shaggy, tan-colored fur, long forearms and hind legs (but no tail), and a rounded face with very appealing eyes. Its sadly vacuous expression gives the impression that the gleam in its eyes is but the reflection off the back of its skull. The common name derives from the three sharp, curved claws on each of its four feet, which serve as hooks as the animal hangs upside down from a branch, like an odd mammalian Christmas tree ornament. The easiest way to find a three-toed sloth is to scan the cecropia trees (page 40). Sloths rest motionless in cecropias, often in the centermost part of the tree.

For years it was assumed that three-toed sloths eat only cecropia leaves, but a study in BCI showed differently (Montgomery and Sunquist 1975). Sloths fed on leaves of ninety-six species of trees other than cecropia, though they did enjoy plenty of cecropia leaves. Interestingly, when sloths were in the crowns of these other tree species they were next to impossible to spot. Only by attaching tiny radios to the sloths were researchers able to locate them as they foraged in noncecropias. This is probably the reason why it was thought their diet was exclusively cecropias. They were just never seen in their other food trees! Further studies by the same researchers showed that sloths move to a different tree about every day and a half. They come to the ground only to urinate and defecate about once a week, which they do just at the base of a tree. They dig

Three-toed sloth

a small depression and cover their excrement, an operation requiring about thirty minutes, an odd behavior that may facilitate uptake of nutrients by the tree and thus enhance the growth of new leaves for the sloth. Another possible explanation is that because sloths defecate on the ground, predators cannot cue in on the odor of the sloth's fecal matter to find the sloth. Sloths have been estimated to have a population of from five to eight per hectare in Panama, a high population density attributable to their ability to ingest many kinds of leaves and tolerate the defense compounds contained therein. Because they have an extraordinarily low metabolic rate, they do not eat as much as their numbers and body size (0.6 m [2 ft] long, up to 4.5 kg [10 lbs]) might suggest. They are relatively efficient digesters, having a complex and long digestive tube. Three-toed sloths have few predators, among them the rare harpy eagle.

The other kind of sloth is the two-toed sloth (genus *Choloepus*), of which there are two species. The two-toed is similar in habits to its better known relative but is identified by its larger size and two, rather than three, prominent claws on the front feet and four, rather than three, on the hind feet. They weigh up to 9 kg (20 lbs). Two-toed sloths have habits similar to three-toeds but are more confined to primary forest than three-toed species. Although they look very similar, the two kinds of sloths, three-toed and two-toed, are only distantly related (Emmons 1990). The two-toeds are most closely related to the extinct giant ground sloths (page 113).

The other members of the Xenarthra are the ubiquitous armadillos (Dasypodidae), of which there are several species. One (*Dasypus novemcinctus*) ranges into southeastern North America and is expanding its range northward. Armadillos are slow-moving ground dwellers whose hard, bony skin protects them from most predators. When attacked, they curl up in a tight ball with their vulnerable soft parts tucked in. Mostly nocturnal, they are quite common especially in savannas, and they feed on a variety of insects and other

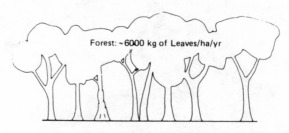

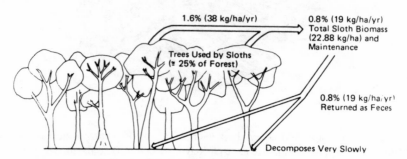

The role of three-toed sloths in mineral cycling. See text for details. From Montgomery and Sunquist (1975). Reproduced with permission of Springer-Verlag, Berlin, Heidelberg, New York, Tokyo.

arthropods. The largest species is the giant armadillo (*Priodontes maximus*), a huge creature that ranges throughout South America as far south as Argentina but is listed by CITES as rare and endangered over much of its range and extinct from some parts of it, all due to overhunting.

Tropical Raccoons and Weasels   *Figure 119*

The familiar northern raccoon (*Procyon lotor*), raider of garbage cans throughout much of North America, ranges southward as far as Panama. Should you be driving along a Central American road at night or camping inside a rainforest, do not be shocked if this black-masked, ring-tailed beast makes an appearance. Raccoons are members of the family Procyonidae, which also includes the coatimundi, kinkajou, and olingo (see below). In addition to the northern raccoon, the Neotropics boasts a species called the crab-eating raccoon (*P. cancrivorus*), which ranges throughout Central and South America as far south as northeastern Argentina. This animal is similar in appearance to the northern raccoon, but its legs and feet are darker in pelage. Crab-eaters frequent swamps and other aquatic areas and are generally nocturnal. Northern raccoons get as far south as Panama and overlap in range with crab-eating raccoons in Central America. However, where these

Figure 136. Lek of Guianan cock-of-the-rock, *Rupicola rupicola*.

Figure 135. Guianan cock-of-the-rock (male).

Figure 137. Male Guianan cocks-of-the-rock in courtship display.

Figure 138. Male Guianan cock-of-the-rock wing display on forest floor.

Figures 139 and 140. Male (left) and female (right) white-collared manakins.

Figures 141 and 142. Male (left) and female (right) barred antshrikes.

Figure 143. The great kiskadee is a tyrannid flycatcher found throughout the Neotropics.

Figure 144. King vulture, the largest and most colorful of the Neotropical vultures.

Figure 145. The savanna hawk, a common predatory bird of savannas, widespread throughout suitable habitat.

Figure 146. Scarlet macaws at clay lick in Amazonian Peru.

Figure 147. Two scarlet (left) and one red-and-green (right) macaw ingesting clay.

Figure 148. Blue-and-yellow macaw.

Figure 149. Hyacinth macaw.

Figure 150. Gray-winged trumpeter.

Figure 151. Razor-billed curassow (male).

Figure 152. Sunbittern.

Figure 153. White-necked heron.

Figure 154. Crested caracara and black vultures.

Figure 155. The Amazon River near Iquitos, Peru.

Figure 156. Cloud cover over rainforest, clear skies over the wide Amazon east of Manaus, Brazil.

Figure 157. Amazon, a white water river (below), mixes with Rio Negro, and blackwater river (above), where the confluence of the rivers occurs near Manaus, Brazil.

Figure 158. Piranha, abundant in Amazon.

Figure 159. Catfish are the second most diverse of the various groups of fish inhabiting the Amazon basin.

Figure 160. Quiet stream in southern Amazonia.

Figure 161. The giant Victoria water-lily, common in quiet waters along the Amazon and its tributaries.

Figure 162. Settlement along the Amazon, near Peru, on varzea floodplain. Note the accumulation of rich Andean sediment deposited during the annual flood cycle of the river.

Figure 165. Manaus opera house, recently restored, is a historical reminder of the brief rubber boom era in Amazonia.

Figure 163. Characteristic settlement of rivereños, people who farm the varzea floodplain.

Figure 166. Typical Amazonian transport.

Figure 164. House along the Amazon.

Figure 167. The "floating city," a section of Iquitos, Peru, that abuts the Amazon River and is inundated by the river at peak flood. The river was very low when this picture was taken in June.

Figure 168. Roseate spoonbill, capped heron, and great egret (gray-necked wood-rail and rufescent tiger-heron behind egret along the bank).

Figure 169. Myriad egrets, storks, and ibis concentrate during the dry season around marshy areas of the Brazilian Pantanal.

Figure 170. A pair of jabirou storks feed in a wet depression in the Brazilian Pantanal.

Figure 171. The widespread (and wingspread) anhinga can be found throughout the Neotropics and extends its range as far north as Georgia. Sometimes called "snake-bird" for its long, flexible neck, it feeds on fish.

Figure 172. Pond in the Brazilian Pantanal.

Figure 173. Capybara
with young.

Figure 174. White uakari.

Figure 175. Giant otter.

Figure 176. Spectacled or common
caiman remains fairly abundant and
widespread throughout much of
lowland South America, especially
in the Venezuelan Llanos and
Brazilian Pantanal.

Figure 177. Morelet's crocodile,
once widespread and abundant
throughout the Yucatan in Central
America, is now endangered, and
remains common only in a few
places mostly in Belize.

two similar species overlap, northern raccoons frequent coastal mangrove swamps while crab-eating raccoons are more partial to interior riverine areas (Emmons 1990).

Coatimundis (*Nasua* spp.) are pointy-nosed, familiar diurnal denizens of forests throughout the Neotropics. There are two species: the white-nosed coati (*N. narica*), which ranges from southern Arizona and New Mexico through Central America and into South America on the west side of the Andes, and the South American coati (*N. nasua*), which is found throughout Amazonia west of the Andes as far south as Argentina. Coatis have broad habitat tolerances and can be found in the Andes, in deserts, and in savannas as well as in rainforest. Coatis usually travel in small bands, mostly composed of females and young. The males tend to be solitary except during breeding time. Coatis shuffle along, resembling streamlined raccoons, with highly pointed snouts, black and white face masks, slender, 63.5-cm (25-in)-long grayish brown bodies, and a 33–71-cm (13–28-in) slim tail, usually with faint rings (you must be close to see this) and held upright, like a cat holds its tail. Though their tail is not prehensile, they are adept at tree climbing and are as apt to be seen in trees as on the ground. They feed on all manner of things, including fruits, ground-dwelling invertebrates, lizards, and mice. When they are not hunted, they become easy to observe.

The kinkajou (*Potos flavus*) is smaller than a coati and uniformly grayish tan, with an extremely long, prehensile tail. The species ranges throughout forests of Central and South America. Kinkajous are as nocturnal as coatis are diurnal. They scurry about the tree branches at night, often making loud, squeaking vocalizations, the banshees of the rainforest canopy. They can often be seen if you search with a flashlight with a strong beam that will penetrate the canopy at night. Kinkajous have forward-placed, large eyes and wide, rounded ears and thus look a bit like monkeys. They feed mostly on fruits (look for them in fig trees), supplemented by small animals.

The olingo (*Bassaricyon gabbii*) in face and body shape resembles a kinkajou, but it has a more gray-brown coat color and a faintly ringed, nonprehensile tail. Olingos range less widely than kinkajous, occurring mostly in humid forests of western Amazonia and much of Central America. They are not well studied and are rarer than kinkajous, with which they share similar habits. They are nocturnal, rarely leave the trees, and feed on fruits and small animals. Some taxonomists have suggested that there are as many as six olingo species, all geographically separated (Emmons 1990).

The weasel family, Mustelidae, is represented in the tropics by several noteworthy animals. The most common is the 0.6-m (2-ft)-long tayra (*Eira barbara*). Resembling a large mink or marten, the tayra is a sleek, blackish brown animal with a buffy face and a 45.7-cm (18-in), black bushy tail. It occurs throughout the Neotropics and, because it is diurnal as well as nocturnal, is observed frequently. Tayras lack prehensile tails but are good tree climbers. They are omnivorous, feeding on rodents, nestling birds, lizards, as well as eggs, fruits, and honey. They occupy many different habitats as well as rainforests, including savannas and coastal areas. They raise young in a den usually located in a hollow tree trunk.

The grison (*Galictis vittata*) resembles a sleek badger. Smaller than the tayra, grison are short-legged, slinky-looking animals with grizzled gray backs, black legs and throats, and a prominent white stripe behind the eye. Their gray tails are short and thick. Grisons range throughout lowland areas of the Neotropics. Like the tayra, they are not confined to forests but can be spotted in savannas and other open areas, and they are commonly encountered near human habitations where chickens and other tempting morsels are present. Strictly carnivorous, they feed on rodents and other small vertebrates. There is a second, similar species (*G. cuja*) found at higher elevations.

If you think you see or smell a skunk in the tropics, chances are you are right. The spotted skunk (*Spilogale putorius*) occurs as far south as Costa Rica, the hooded skunk (*Mephitis macroura*) gets into Nicaragua, and the hog-nosed skunk (*genus Conepatus*), of which there are several species, ranges as far south as Patagonia and up into the Andes.

The Neotropical Felines
Figures 117, 118

Most people who travel to the American tropics for the thrills of seeing wildlife, when asked which animal they would most like to see, would probably name *El Tigre*, the jaguar. The lion may reign supreme on the African savanna, but in the Neotropical rainforest, *Felis onca* is the top cat. Though this 1.8-m (6-ft) cat ranges from northern Mexico through Patagonia, it is now quite rare over much of its range and is listed by CITES as endangered. Only in the interior Amazon, remote montane forests, and other areas out of the immediate reach of hunters' gunfire does the 136-kg (300-lb) leopard of the New World remain unmolested. In the Cockscomb area of central Belize, a reserve, the first of its kind, has been created specifically to preserve jaguars. Jaguars closely resemble leopards in spotting pattern but are generally heavier. Size varies considerably among individuals, and some jaguars are known to top 181 kg (400 lbs). An adult jaguar has no nonhuman predators, with perhaps the rare exception of the anaconda, the serpentine giant of the Amazon (page 208).

Jaguars are ecological generalists. These cats are found in lowland and montane forests, along rivers, in jungles, savannas, and coastal mangroves. They feed on many kinds of animals, including deer, tapirs, peccaries, sloths, capybara, giant otter, fish, birds, reptiles, and caiman. Largely solitary and basically nocturnal, jaguars are seen far less frequently than are their footprints. A jaguar attacks its prey with a vigorous leap, quickly attempting to sever the neck vertebrae. The name *jaguar* derives from the Indian word *yaguar*, meaning "he who kills with one leap."

Much smaller and considerably more common than the jaguar are the ocelot (*Felis pardalis*) and its close relative the margay (*Felis wiedii*), both of which range throughout the Neotropics. Both ocelot and margay are placed in the genus *Felis*, along with the common domestic house cat, which they resemble but for their fancier pelts. The big-eared, bright-eyed ocelot is about 1.1 m (3.5 ft) in length with a thick, 40.6-cm (16-in) tail. A large individual weighs about 11kg (25 lbs). The margay is similar in appearance but slightly smaller in size. It is difficult to separate these species on the basis of a quick look because their spotting patterns are similar. However, ocelots tend to look more striped, whereas margay spots do not tend to run together as with the ocelot. Both small cats are essentially nocturnal, though ocelots can sometimes be seen during the day, usually in dense cover. Margays are believed to be more nocturnal than ocelots. Ocelots are terrestrial but margays are skilled tree climbers. Both animals are carnivores, feeding on anything from monkeys to insects. Like their domestic brethren, they spray to mark their territories. Jaguars, ocelots, and margays are the unfortunate victims of pelt seekers who make coats and then profits from killing these magnificent cats.

There is another Neotropical *Felis*, the oncilla (*F. tigrina*), a small cat only about half the size of a margay, which it otherwise resembles. Oncillas are poorly studied and much about their ecology and even their range (from southern Central America through Amazonia) is uncertain.

Of the various Neotropical cats, the jaguarundi (*Felis yagouroundi*) is probably the most frequently seen. It is common and diurnal, often found in savannas as well as forest. It is also the most frequently misidentified cat; many who are unfamiliar with it assume it to be a large weasel and not a cat at all. A jaguarundi looks superficially like a weasel because it is long and sleek, with proportionately short legs and a long tail. Its coat color varies from dark brown to gray to black, depending on the individual. It's a bit over 0.9 m (3 ft) in length (not counting the 0.6 m [2-ft] tail) and is totally terrestrial. Like the other cats, jaguarundis are essentially solitary.

The other large cat of the Neotropics is the puma (also called mountain lion, or cougar; *Puma concolor*), which is widespread in Central and South America as well as North America. It is an animal of the open savannas, forests, and mountains. It would be very much more common were it not for persecution by people. This large, light tan cat with a long tail is solitary, secretive, and wide ranging.

The Neotropical Marsupials

Figure 121

Most people associate marsupials—the kangaroos, wallabies, wombats, and bandicoots—with Australia. Marsupials are mammals that give birth to premature young that migrate to a pouch on the mother's abdomen where they attach to a teat and complete their development. Almost everyone has seen pictures of a mother kangaroo with her "joey" in her pouch.

Though Australia is the world's undisputed marsupial capital, until relatively recently South America boasted a diverse marsupial community. Today, though there are forty-one species of opossums living in the Neotropics, most of the original South American marsupial fauna is extinct. During the Pleistocene faunal exchange (page 113) and mixing of the North and South American faunal assemblages, a few species, like the durable Virginia opossum (*Dedelphis virginiana*), migrated northward where they continue to expand their population today (Carroll 1988). However, many placental mammals migrated into South America, and their arrival coincides to a degree with a decline in richness of marsupials. For example, North American placental sabre-toothed tigers entered South America via the land bridge, and relatively soon thereafter marsupial sabre-toothed tigers became extinct (Carroll 1988). South America's marsupial fauna today is but a remnant of what it once was.

The common opossum (*Didelphis marsupialis*), found throughout Central America and Amazonia, has hardly changed in appearance from its ancestors who roamed the planet 65 million years. The fossil record indicates that the common opossum dates back about 35 million years, and its family dates to the Upper Cretaceous period, contemporary with the last of the dinosaurs (Carroll 1988). Superficially ratlike, with pointed snout and scaly naked tail, the common opossum weighs between 2.3 and 4.5 kg (5–10 lbs) and is largely gray with some black. It inhabits almost any kind of terrestrial habitat other than desert and high mountains. Opossums are good tree climbers and often hang upside down, clinging by their prehensile tail. Totally omnivorous, the opossum will try eating almost anything. Its most unique behavior, "playing possum," is an act that feigns death when the animal is threatened.

In addition to the common opossum, the Neotropics host numerous other opossums. Most are nocturnal, but many are common and seen relatively often in daytime. There are furry little woolly opossums (*Caluromys*), tiny mouse opossums (*Marmosops, Marmosa, Micoureus, Gracilinanus*), bushy-tailed opossums (*Glironia*), four-eyed opossums (*Philander, Metachirus;* four-eyed in name only!), short bare-tailed opossums (*Monodelphis*), and the yapok, or water opossum (*Chironectes minimus*). Though these species are different sizes, they are all basically similar in anatomy, and the many species range throughout the various habitats of Central and South America. What this indicates is that opossums have undergone a successful adaptive radiation throughout the American tropics and are the survivors of what is, for other marsupials, a bygone era.

Reptiles and Amphibians

The tropics seem always to be associated with snakes. The fear of snake-infested trails and trees taints many people's views about the allure of rainforest. How can you admire the scenery when you always have to be looking out for poisonous snakes? In reality, venomous snakes are not frequently encountered. They tend to be secretive and nocturnal, and it's actually not easy to find them, even when you search diligently. There are, however, many species of snakes in the tropics, both poisonous and (most) nonpoisonous, and as a group they are fascinating. Once people conquer their initial fear of serpents, I have seen them develop intense curiosity about them, followed by admiration of their beauty. This section will address that curiosity and will focus on other reptiles and amphibians as well. But let's start with snakes. Venomous snakes at that.

The definitive reference on venomous reptiles of Latin America is Campbell and Lamar (1989). Ranging variously from northern Mexico to southern Patagonia, there are 145 venomous snakes to be found. These include 54 species of coral snakes, one sea snake, the cantil and the copperhead, 7 species of palm-pitvipers, 8 species of forest-pitvipers, 31 species of lancehead pitvipers, 14 species of hognosed and montane pitvipers, 26 species of rattlesnake and pygmy rattlesnake, the Mexican horned pitviper, and, largest of the lot, the bushmaster. Maybe it is a good idea to keep an eye on your feet as you walk rainforest trails. For discussion of risk and snakebite, see the appendix.

Snakes—Pitvipers

Figure 113

Pitvipers are all quite venomous. There are many species found throughout North America, as pitvipers (subfamily Crotalinae) range throughout both tropics and temperate zone. North American rattlesnakes, plus the well-known copperhead (*Agkistrodon contortrix*) and cottonmouth (water moccasin) (*A. piscivorous*), are all pitvipers. The "pits" referred to in the name are sensory depressions located between the nostrils and eyes. They sense heat and aid the snake in locating warm-blooded prey. Pitvipers have long hypodermic fangs in which a poison duct from modified salivary glands can deliver a lethal dose of biochemically complex toxin that attacks blood cells and vessels, surrounding tissue, and sometimes nerve tissue. Pitvipers tend to rest in a coiled position, which they also assume when danger threatens. Any pitviper may be aggressive in display, raising its head high and vibrating its tail. Rattlesnakes enhance this habit by the presence of the noisy rattles. Pitvipers normally have large, tri-angular-shaped heads and slitted, catlike eyes (with elliptical pupils), helpful features in recognizing them.

The most infamous Neotropic pitviper is the fer-de-lance (*Bothrops asper*), one of thirty-one species of lanceheads, all in the genus *Bothrops*, many of which look alike. Lanceheads are mostly lowland species found at elevations below 1,500 m (4,900 ft). They tend to have a close resemblance to one an-other, and it is often not possible to get a certain identification without a careful in-hand examination of the animal (NOT RECOMMENDED). The fer-de-lance will serve to introduce you to the group. Known variously as yellow-tail, yellow-jaw, tommygoff, as well as many Spanish and Portuguese names, its English name is terciopelo, though it is best known by its Trinidad name, fer-de-lance. The name refers to the lancelike shape suggested by the long, serpentine body and conspicuously large, triangular head. In Brazil a very similar species is called *jararaca*, meaning "arrowhead." The fer-de-lance, like most *Bothrops*, is a tan snake with dark brown diamond patterning along its sides (intensity of body color and blotching patterns are quite variable). The creature averages a length of 1,200–1,800 mm (approximately 4–6 ft), with some individuals reaching 2,500 mm (about 8 ft) (Campbell and Lamar 1989).

Regardless of size, and as with all *Bothrops*, a fer-de-lance is a potentially lethal snake: even the juveniles are highly venomous, and up to fifty young are born at a time (so take care to look around a bit if you should encounter a juvenile—it may have siblings lurking about). Campbell and Lamar (1989) note that *Bothrops* "are responsible for more human morbidity in the New World than any other group of venomous snakes." Venom is fast acting and painful. It rapidly destroys blood cells and vessels and produces extensive ne-crosis (decomposition) of tissue around the bite site. Infection can follow and can be massive. Mortality without treatment is about 7% but is reduced to between 0.5% and 3% if properly treated (Campbell and Lamar 1989). A person bitten by a *Bothrops* must receive antivenin quickly. Occasionally a *Bothrops* will bite in self-defense but not envenomate. It may have exhausted its venom on a recent catch or simply not discharge it. Note that most *Bothrops* are inactive during the day as they tend to hunt at night. However, they can be quick and aggressive if disturbed. Heed the words of Campbell and Lamar (1989): "Specimens can (and often do) move very rapidly, reversing directions abruptly, and defending themselves vigorously. An adult *B. asper*, if cornered and fully aroused, is a redoubtable adversary and must be regarded as extremely dangerous." Remember that.

Though most common in lowland forests, the fer-de-lance is also regularly encountered in overgrown tangle, and care should be taken whenever walking through dense jungle. The fer-de-lance feeds on various mammals and some birds. The species ranges mostly throughout Central America to northern South America, and throughout the Orinoco Basin. To the south, throughout Amazonia, it is largely replaced by the similar *B. atrox, B. brazili,* and *B. jararaca* as well as other *Bothrops* species.

Similar to *Bothrops* are the forest-pitvipers, of which there are eight species in the genus *Bothriopsis*. Most have restricted ranges in northern South America, but two species, *B. bilineata* (two-striped) and *B. taeniata* (speckled), are widely distributed throughout Amazonia. Forest-pitvipers tend to be slender and may exceed lengths of 1,500 mm (almost 5 ft). They all have prehensile tails and all are found in interior forest. Their habits appear to be similar to other pitvipers.

The genus *Bothriechis* contains seven species called palm-pitvipers. These snakes tend to grow to shorter lengths (600–800 mm, or 2–2.5 ft) than the lanceheads but are potentially dangerous nevertheless since they are arboreal and usually cryptic (often greenish), coiled among palm fronds. Most species are Central American and found in montane areas. However, one species, *B. schlegelii*, ranges throughout Central America and as far south as central Ecuador, being common in lowland rainforest. This species, called eyelash palm-pitviper because of enlarged scales that grow outward over each eye, is generally considered abundant throughout its range. Highly variable in color, most specimens are some combination of green, finely suffused with black, though some are bright yellowish-gold. A good climber, with a prehensile tail, the eyelash palm-pitviper feeds on small rodents as well as tree frogs and anolis lizards.

There are fourteen species of montane or hognosed pitvipers, all in the genus *Porthidium*. One, the jumping pitviper (*P. nummifer*), which ranges from southern Mexico through Panama, is alleged to sometimes hurl itself at its perceived attacker, a claim that is substantially exaggerated (Campbell and Lamar 1989). A short snake with a very thick body, gray-brown with black diamonds, its venom is apparently not as potent as that of other pitvipers.

Rattlesnakes are mostly North American, with the majority of species found in northern and central Mexico. However, there is one species, the Neotropical rattlesnake (*Crotalus durissus*), that occurs from Central America to Brazil and Paraguay, usually in dry forests and uplands. It is a regionally variable species, but it's basically the only rattler in South America, so if a scary-looking, coiled serpent rattles at you, it's this one. And be careful. The snake has an elaborate threat display and it is not bluffing. Campbell and Lamar (1989) note that a description of its "clinical venom effects" emphasized "progressive neuromuscular paralysis and hemolysis of red cells with resultant renal failure," resulting in a mortality rate of 72% without treatment!

Always save the best for last. And that would be the bushmaster (*Lachesis muta*), giant of the pitvipers. Not only is it the largest pitviper in the Neotropics, it is the world's largest, reaching lengths of 2.7–3.0 m (6.5–11.7 ft). The largest reported specimen was 4.2 m (13.8 ft), but that specimen has not been substantiated (Campbell and Lamar 1989). I encountered a bushmaster as it was slowly crossing a road at night in Trinidad and estimated the snake to be approximately 9–10 feet in length (it seemed prudent not to try for a more precise measurement). As its head was descending into the gully on the right side of the road, its tail had yet to emerge from the gully on the left side of the road. It was a truly magnificent creature to behold, seen to advantage in the headlights of our van, as my group remained at a respectful, safe distance. It is generally yellowish tan, reddish brown, or pinkish tan, with dark brown, diamondlike splotches, its thickest splotches on its back, not its sides, as in a typical *Bothrops*. The curious Latin name translates to "silent fate," so named for the snake's ability to strike without audible warning (though it usually does vibrate its tail and is also known to threaten with its neck inflated). Because of its length, a bushmaster can strike over a long distance. It has large fangs and can deliver a high dose of venom. There are not a great many reports of bushmaster envenomation, but those that exist are peppered with the word "fatality." It is an inhabitant of lowland rainforest throughout lower Central America, Amazonia, and the southeastern coastal forest of Brazil, and it is reportedly often seen coiled within buttresses of large trees. It's crepuscular and nocturnal and feeds primarily on mammals and birds. Careful now.

Pitviper Wannabes

On a recent trip to Peru, my wife and I along with a friend encountered a small, slender, gorgeous little emerald green snake resting in the middle of the trail, looking innocuous, even for a serpent. I could see no pits, its eyes were not vertically slitted, and its head was slender, not triangular. I decided that it was nonvenomous, as, of course, are most Neotropical snake species. Nonetheless, always erring on the side of caution, I fetched a small stick and (I thought) gently prodded the creature's tail end so as to encourage it to vacate the trail. I assumed it would hasten along. It didn't. It apparently took offense, coiled, lifted its head high, opened its mouth very wide, and vigorously struck at me (though it came nowhere near me—I am no stranger to evasive action). As it continued this aggressive behavior, its head seemed actually to flatten and become increasingly triangular. Good act. Many Neotropical nonvenomous snake species have evolved a behavior that suggests that the best defense is a strong offense, and thus their behavior when threatened is similar to that of venomous species, especially pitvipers. In effect, they are pitviper behavioral mimics (Campbell and Lamar 1989). Given that nonvenomous snakes may become Neotropical thespians, how do you know for sure when you are dealing with the real thing, an actual venomous snake? Just assume you are and you'll be fine.

Coral Snakes

"Red and yellow, kill a fellow; red and black, friend of Jack." This is a little rhyme to help remember the distinction between potentially lethal North American coral snakes and nonpoisonous, harmless kingsnakes (if red bands touch yellow, it's a coral snake, but if red bands touch black, it's not). Also remember that this lyrical distinction DOES NOT WORK in the Neotropics. There are fifty-four species of coral snakes in the Neotropics, and many (probably most) do not have red bands touching yellow. It's safest just to avoid colorful snakes with any combination of red, black, and/or yellow rings. They may be coral snakes, and you don't want to be holding a coral snake.

Coral snakes (fifty-three species in the genus *Micrurus*, one species in the genus *Micruoides*) are members of the global family Elapidae, to which the deadly cobras and mambas belong. Their powerful venom mostly affects the nervous system and acts quickly, producing paralysis and death by suffocation. Coral snake lengths vary among species but typically reach between 600 and 1,200 mm (about 2–4 ft), though some species can be more than 1,600 mm (just over 5 ft). Unlike pitvipers, coral snakes have rather short fangs and bite with force in order to inject their lethal venom. Coral snakes can be active both day and night but are most often found beneath logs or rocks in habitats ranging from deserts to lowland rainforest. They eat mostly lizards and other snakes.

The bright patterning that typifies all coral snake species is considered by most authorities to be warning coloration (page 82). Coral snakes exhibit various escape behaviors, such as raising the tail and moving the tail in tandem with the head, possibly making the animal appear as two snakes rather than one (Campbell and Lamar 1989). When threatened, a coral snake will actively thrash its body and wave its tail, and may try to bite. Handling one is therefore dangerous because its narrow head can easily slip through fingers, giving the snake an opportunity to strike. Many species of nonpoisonous snakes, often collectively called "false coral snakes," converge in color pattern with coral snakes wherever their ranges overlap (Greene and McDiarmid 1981; Campbell and Lamar 1989). For example, coral snake mimicry is shown by the nonpoisonous *Lampropeltis triangulum hondurensis*, which, in Honduras, looks stunningly similar to the coral snake *Micrurus nigrocinctus divaricatus*: wide, bright-red bands alternating with thinner black bands and a black snout. Campbell and Lamar (1989) illustrate this and several other examples of the concordance in body patterning between coral snakes and false coral snakes where their ranges overlap. If you should encounter a coral snake, or any serpent that looks like it could be one, promise me you won't touch it.

Snakes—Constrictors   *Figure 101*

The world's largest snakes are constrictors (family Boidae). In the Old World, these snakes are called pythons, but in the Neotropics, they are the boas. One species is called anaconda (page 208). A few boas are found in Madagascar and the Indo-Pacific islands, but most are in the Americas. Boas are nonvenomous, though their teeth are needle sharp, and their bite can be nasty. With wide heads, wide bodies, and elaborate patterning, they are often confused with pitvipers. They capture and kill prey through constriction, a process whereby the serpent coils around its victim tightly enough to prevent it from breathing, killing by suffocation. Following death of the victim, constrictors, like all snakes, swallow their prey whole, opening their mouths widely with jaws attached only by elastic ligaments.

The best known of the boas is the boa constrictor (*Boa constrictor*), a broadly ranging snake common throughout the Neotropics. Variable in color pattern, South American boa constrictors are generally lighter and more sharply marked than their Central American counterparts. Boa constrictors are warm tan with dark brown, diamond-shaped patterning. Their heads, like all boidae, are rather long with pointed snout. Boas average about 1,500 mm to 1,800 mm (5–6 ft) in length, but they do not commonly attain the huge lengths sometimes reported. The largest boa constrictor on record was 5,640 mm (18.5 ft), most extraordinary for this species. Boas can be aggressive and will coil, hiss, and bite if attacked. In captivity they can become docile (but they don't belong in captivity, regardless of disposition). They are mostly nocturnal, feeding on all manner of mammals, including small cats. They also take birds and lizards. Boa constrictors inhabit a wide range of habitats, from wet lowland forests to dry savanna.

The emerald tree boa (*Boa canina*) is probably the most beautiful of Neotropical boids. Deep green above, yellow green below, with a dorsal white line and scattered white spots, it has burning yellow eyes with catlike, slitted pupils. Confined to South America, this 1,800-mm (6-ft) boa is a tree climber, coiling itself tightly in such a way as to be most cryptic. It has been said to resemble a bunch of bananas when coiled in a tree, and, indeed, small individuals have occasionally been found on banana boats among the bunches. The tail is prehensile, and these snakes are skilled at moving about in the trees, preying on squirrels, opossums, birds, and lizards. Both the emerald tree boa and boa constrictor tame when handled frequently by humans.

The rainbow boa (*Epicrates cenchris*) is one of the smaller boas, growing normally to be about a meter long. To be appreciated, this little constrictor must be observed in full sunlight. Its dull blackish brown scales sparkle with iridescent colors, giving the snake its popular name. Rainbow boas are also skilled tree climbers and often prey on bats. They range from Central America all the way south to Patagonia.

The largest of the Neotropical constrictors is the anaconda, most common along rivers and in marshes, discussed on page 208.

Other Nonpoisonous Snakes

There are many species of nonpoisonous snakes in the Neotropics. They include the various vine snakes (genus *Oxybelis*), thin brown, gray, or green snakes that climb about the foliage capturing and feeding on lizards. The beautiful indigo snake (*Drymarchon corais*) can reach lengths of 3,000 mm (10 ft), eating virtually any kind of animal from fish to birds. The odd large-eyed, chunk-headed snake (*Imantodes cenchoa*, page 74) is extraordinarily thin, coiling in outer branches where it preys on small tree frogs and lizards. There are also rear-fanged snakes, which are mildly venomous.

Iguanas and Other Lizards   *Figures 103, 114*

The prehistoric-looking common iguana (*Iguana iguana*) is a ubiquitous inhabitant of rainforests. These lizards are among the largest of their clan to occur in the Neotropics. They are greenish when small but become dark brownish black when full-sized. A large iguana can exceed 1,800 mm (6 ft) in length, but much of it is tail, which tapers into a slender, whiplike tip. The face, with large mouth and wide, staring eyes, plus the antediluvian-like body of the reptile, provide a dinosaur-like countenance. Two short spines adorn the nose above the nostrils, and a loose membrane of skin called a dewlap hangs below the throat. The head is flat and covered by heavy, tubercle-like scaling, and the neck and back are lined with short, flexible spines. The legs sprawl alligator-like to the sides, and the feet have long toes with sharp claws. Iguanas do not often hurry, but they are capable of moving quickly if necessary. They spend most of their time in trees, usually along a stream or river, into which they jump should danger threaten. Excellent swimmers, they can remain underwater for considerable time. When small they concentrate on insect food but feed more heavily on fruits and leaves when full-sized. Should you encounter an iguana, even a large one, you have nothing to fear. They usually do not bite unless thoroughly harassed, they are not particularly effective scratchers, and they are nonvenomous. The most aggressive iguana I ever encountered was directing its hostility at a rat. Both the mammal and the reptile were contesting access to garbage dumped alongside the Amazon River in Iquitos, Peru. The iguana lost.

Adult iguanas feed heavily on fruits and leaves, and thus, because they are near the base of the food chain, can be abundant in rainforests. They provide a potentially important protein source for humans. Studies are under way to determine how to manage natural iguana populations for maximum sustainable yield (Werner 1991).

Iguanas are members of the large family Iguanidae, which includes the many anolis lizards, basilisk lizards, and ctenosaurs. Anolis lizards (genera *Anolis* and *Norops*) are generally abundant. Some are bright green, some are brown, and some are mixtures of both. Some can change color, rather like chameleons (an erroneous name frequently given them when in captivity). During the heat of the day the sounds of these and other small lizards scurrying over dry leaves precedes you as you walk along. Anolis have sharply pointed noses and large, conspicuous dewlaps, which males use during courtship. They are skilled tree climbers and are often seen facing downward on a tree trunk with necks stretched out horizontally. They are also common on foliage. Small arthropods make up their diet.

Basilisk lizards (*Basiliscus* spp.) are also called "Jesus Christ lizards" because of their ability to scurry at high speed (not walk!) across water. With long toes on the hind feet, lined with skin flaps, these odd lizards run full tilt on their hind legs across small streams and puddles. They look even more like dinosaurs than iguanas do because they are adorned with elongated spiny fins on the back and tail. They feed on invertebrates, vertebrates, and various fruits and flowers. Studies in Costa Rica indicate that basilisks are abundant lowland animals. Estimates indicated that there are between 200 and 400 per acre (Van Devender 1983). Basilisks are found primarily in Central America.

Ctenosaurs (genus *Ctenosaura*), or black iguanas, are among the larger iguanid lizards. Ctenosaurs closely resemble iguanas but have a more banded pattern (though they can change pigmentation easily and pattern varies widely from animal to animal; some are dark, some light), and they tend to frequent drier areas such as open fields, farmyards, savannas, roadside edges, and coastal areas. They are adept burrowers and skilled tree climbers. Like iguanas, the larger ctenosaurs concentrate on vegetable food, though they are not adverse to sampling such delicacies as bats, baby birds, and each other's eggs.

Iguanas and ctenosaurs shift their diets from primarily arthropods to primarily vegetation as they increase in body size. Pough (1973) hypothesized that this diet shift may be related to reptilian energetics and constraints of large size. Small lizards can be active and successful in catching small but scurrying prey such as beetles and spiders. Large lizards require more energy (simply because they are larger) but actually need less energy per gram of body weight (because large animals have slower metabolisms). Perhaps they are not as well served by spending time and energy trying to capture fast-moving insects, none of which individually contain much energy. Plants don't move, and they require little energy to "capture." Though harder to digest, more can be swallowed at a single sitting. Thus a vegetarian diet is more optimal for a large lizard, presuming digestion of the plant matter is not problematic to the creature. Even some of the medium-sized and large carnivorous lizards do not pursue prey, perhaps because such pursuit would expend more energy than would be contained in the prey item. These animals sit and wait and capture any slow-moving arthropod (such as a caterpillar or grub) that happens to blunder past.

Tropidarus, a typical genus
of rainforest lizards

The lizard family Teiidae includes the largest of the South American lizards, the tegus (genus *Tupinambis*). Three species, the common tegu (*T. teguixin*), the northern tegu (*T. nigropunctatus*), and the red tegu (*T. rufescens*), each have large bodies, and they range from Central America through Argentina. The common tegu reaches a length of 1,400 mm (55 in), making it the largest lizard species of the Americas. Tegus live in forested areas and forest borders, eating small animals, including chickens and chicken eggs, and local farmers hunt them as a protein source. Some members of the Teiidae, the water teiids, are partly aquatic, with flattened tails that function well as paddles. The caiman lizard (*Dracaena guianensis*) of northeastern South America bears a resemblance to its namesake. This 1.5-m (4-ft) lizard lives in floodplain forests.

Geckos (Gekkonidae) are a major lizard family of the world's tropics. They are usually light with dark splotches. With suctionlike scales on their feet, geckos cling comfortably to smooth walls. They feed exclusively on arthropods and are nocturnal, inhabiting dwellings and living in harmony with humans. Geckos are considered valuable for their abilities to keep numbers of cockroaches and other vermin within tolerable limits. The name "gecko" comes from their loud calls, given only at night, often while hanging on a wall rather near where you're sleeping. It takes some getting used to.

Tree Frogs, True Frogs, and Toads   *Figures 100, 104, 105*

Amphibians are animals that do not lay eggs in protective shells. Instead, they require water to reproduce. Typically, gelatinous eggs are laid in ponds or streams either as floating masses or attached to rocks or debris. Larval animals hatch and pass through a developmental stage in water (referred to as a tadpole in the case of frogs and toads). During the aquatic phase the animal breathes by external gill tufts, but these are resorbed when the larva passes through metamorphosis to adulthood. Adult amphibians usually require moisture for their skins, though toads are able to survive with dry skins.

Salamanders are not very diverse in the tropics. They represent exceptions to the general tendency for taxonomic groups to show high species richness in the tropics. By far the most abundant, diverse, and interesting amphibians are the anurans, the frogs, tree frogs, and toads.

There are approximately 4,000 species of anurans in the world, and some 1,600 occur in the Neotropics. In some areas censused in western Amazonia, there can be as many as 80 species of anurans at single sites within lowland rainforest (Duellman 1992; Rodriguez 1994).

Although many species of anurans reproduce in the manner described above, many also show dramatic departures. In Costa Rica, where 119 anuran species are found, Scott and Limerick (1983) have described a dozen different modes of amphibian reproduction. These include live birth, with fully formed miniature adults ("skipping the egg and larval phases"), eggs laid on plants where larvae hatch and drop into water, and eggs laid on land in foam nests, in bromeliads, or in tree cavities. For example, the species-rich genus *Eleutherodactylus* reproduces by "direct development," with eggs that hatch into tiny but fully formed frogs. Courtship patterns are also sophisticated. Frogs and toads vocalize, the males emitting a specific call that serves to attract the females. Many species are territorial. Anuran reproductive behavior has undergone an impressive adaptive radiation in the tropics. This is largely possible because the tropical rainforest maintains a constantly high humidity that facilitates keeping the anuran's skin constantly moist, a prerequisite to its survival (Duellman 1992).

One of the most intriguing examples of the complexity of frog life cycles is provided by the dart-poison frog *Dendrobates pumilio*. In this species, whose life history has been painstakingly documented, both male and female parents are involved in caring for the egg and tadpole. The male guards the egg for ten to twelve days. Once the eggs hatch into tadpoles, the female takes over, transporting each tadpole to a bromeliad containing water. Each tadpole is placed in a different bromeliad, and the female returns regularly to deposit an unfertilized egg that serves as food for the developing frog (Duellman 1992).

Tree frogs are arboreal, aided in attaching to leaves and stems by tiny suction disks on their feet. Most are quite small and well camouflaged, though some are very brightly colored and exhibit warning coloration. One of the most common (and most frequently photographed) is the gaudy leaf frog (*Agalychinis callidryas*). This Central American species has blazing, bulging red eyes, bright green on the upper body along with a scattering of white spots, bluish on the sides, white on the belly, and orange on hands and feet! Males and females have a rather prolonged mating ritual in which the female, with the male clinging to her, attaches eggs to a leaf as the male fertilizes them. Hatching occurs in approximately five days, and the larvae drop off the leaf into water.

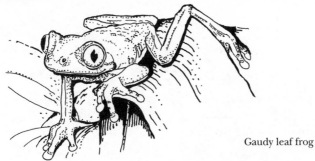

Gaudy leaf frog

Glass frogs (genus *Centrolenella*) are small green tree frogs with somewhat transparent bellies, revealing the beating heart and intestinal system. They attach eggs to leaves over streams, and larvae hatch and drop into the water. Eggs tend to hatch in heavy rain, facilitating the release of tadpoles into the water below. Tadpoles become bright red and burrow in stagnant litter in slow pools. Their color is the result of a concentrated blood supply, an adaptation to low oxygen levels in the mud.

The most colorful and dangerous of the tree frogs are the Dendrobatidae, the dart-poison frogs. Shiny black, some have bright red or orange markings, some bright green. The Colombian Choco Indians utilize the poisonous alkaloids from the frogs' skins in making potent darts for hunting (Maxson and Myers 1985). The poison affects nerves and muscles, producing paralysis and respiratory failure. These colorful frogs hunt by day, feeding on termites and ants, and it has been suggested that their warning coloration evolved in response to their long feeding periods when they would otherwise be vulnerable to predators. Males make their insectlike vocalizations during the day and attract females. Eggs are laid on land, after the male deposits sperm. The female remains with the fertilized eggs and carries the tadpole to water after it hatches. The female also feeds the developing tadpole highly nutritious unfertilized eggs that she produces. These eggs are deposited following signaling behavior by the tadpole, similar to solicitations of nestling birds when their parent brings food. Also see page 81.

Other anurans also contain skin toxins. The giant cane toad (*Bufo marinus*), largest of the New World anurans, secretes irritating fluid from its skin and is quite toxic if eaten. The large size of the animal (easily the size of a softball) would make it a tempting target for predators, but its toxic integument is so dangerous that dogs and cats have reportedly died just from picking up the toad in their mouths. Cane toad tadpoles also have toxic skin. One very curious human application of toad toxin has been studied in Haiti. Along with extract from puffer fish and two plant species, toad toxin is used to induce the death-like trance observed in victims of voodoo rituals (Davis 1983). The smoky frog (*Leptodactylus pentadactylus*) is another species possessing highly irritating skin secretions. These frogs are aggressive when threatened, inflating their bodies, rising on their hind legs, and hissing.

Giant cane toad

Besides warning coloration and toxic irritating chemicals, frogs have one other defense trick. Some species, which are normally cryptic, display *flash colors* when threatened. A flash color is revealed when the frog raises a foot or other body part, revealing a bright patch of color, usually red, orange, yellow, or blue. The sudden bright coloration may momentarily confuse the predator, interrupting its search image long enough for the frog to use its most obvious defense tactic, its jumping ability.

Invertebrates

Bugs. In the Neotropics they come in petite, small, medium, large, extra large, double extra large, and outrageous. They abound from treetops to litter, in closed forests and successional gaps, in field stations and tents, and, occasionally, in shoes. They are out by day, dusk, night, and dawn, and some of them even glow in the dark. And they are not just insects. Myriad other animals without backbones—centipedes, millipedes, various arachnids, and worms—are at home in the Neotropics. Seemingly innumerable invertebrate creatures crawl, hop, slither, climb, burrow, and fly through rainforests. In no way can this modest volume catalog even a representative sample of all the invertebrate life forms that can be found by diligently searching rainforests and other Neotropic habitats. Indeed, new species are found on virtually all collecting trips. Included here is a mere sample of some of the most interesting and frequently seen members of this vast host. These are some of the invertebrates you are most apt to encounter by casual searching. See the appendix for an introduction to the more noxious invertebrates (mosquitos, tarantulas, wolf spiders, scorpions, centipedes, bees, and wasps). Also see Penny and Arias (1982) and Hogue (1993).

The Social Insects

Two great insect orders, Hymenoptera and Isoptera, have evolved species that display complex social systems where societies of close relatives (genetic sisters) support a fertile queen, their mother. Hymenopterans are bees, wasps, and ants. Not all species are social, but many are. Isopterans are termites, and all are social insects. Termites are evolutionarily closely related to cockroaches. Both orders, Hymenoptera and Isoptera, are abundantly represented in the world's tropics. Workers are usually separated into differing morphological castes, consisting of various-sized workers and soldiers. Sterile animals are usually females, all sisters, but male workers occur in termites. Otherwise, males are produced only during mating of the queen and are otherwise superfluous. Wilson (1971) provides a comprehensive account of the social insects.

TERMITES   *Figure 88*

No one can visit the Neotropics without seeing basketball-sized termite nests attached to tree trunks and branches. They occur in all habitats, from rainforest to savanna and mangroves. In addition, especially in pastures and savannas, cone-shaped termite nests erupt directly from the ground. From the rounded, blackish brown nests radiate termite-constructed tunnels in which the workers pass to and from the colony. Termites also nest underground in vast subterranean colonies.

The most abundant of the Neotropical termites are the many species of *Nasutitermes*, which range throughout the region (Lubin 1983). These come in three castes: worker, soldier, and queen. If you wish to see workers and soldiers, locate a nest and cut into it. Nests are made of paperlike material called "carton," which is a combination of digested wood and termite fecal material that forms a glue. When a nest is cut into, scores of workers and soldiers swarm out to investigate the disturbance. Workers are pale whitish and lack the con-stricted abdomens that characterize ants. They have dark mahogany-colored heads. Soldiers are similarly colored but larger than workers, with prominent heads and long snouts. Soldiers can eject a sticky substance with the odor of turpentine that apparently irritates would-be predators, including the anteater *Tamandua*. Termites are fast workers and will quickly repair an injured nest. They swarm about the surface laying down new material to replace that which was damaged. Another way to see termites is to break open their tunnels. Workers are blind and follow chemical trails laid down by other workers. They will continue to pass en masse along an opened tunnel, though some will eventually repair it. The queen termite (or queens—some species have multi-ple queens) is located deep within the nest and cannot be seen unless the entire nest is dissected. Queens are immense compared with workers and sol-diers. Virtually immobile, weighed down by a monstrous, gourdlike yellow abdomen, their needs are attended to by workers as they pass their lives pro-ducing the colony's eggs.

Termites ingest wood, but they need lots of help to digest it. The hard cellu-lose is digested only with the aid of a complex intestinal fauna made prin-cipally of flagellate protozoans. It is actually the protozoans that digest the cellulose. Removal of these one-celled organisms will prevent a termite from digesting cellulose and other large molecules of wood. It and the protozoans are obligate symbionts, another example of a coevolutionary mutualism (chap-ter 5). The flagellate protozoans benefit by being in the termite in that it provides them with continuous food, shelter, and a means of dispersal, since they go where it does. The very hard woods of many tropical trees are probably in part an evolutionary response to selection pressures posed by termite her-bivory. Some woods are apparently toxic to termites.

Termites are so abundant in the world's tropical areas that they may contrib-ute to global climatic warming by enhancing greenhouse effect. Their com-bined digestive abilities can produce significant quantities of atmospheric methane, carbon dioxide, and molecular hydrogen. Forest clearance and the conversion of forests to agricultural ecosystems tend to increase termite abun-dance, thus accelerating the production of the atmospheric gases (Zimmer-man et al. 1982).

Termitaria, the termite nests, form patches of high nutrient concentrations in otherwise nutrient-poor tropical soils. A study conducted along the Rio Negro in Venezuela showed that termites consumed between 3 and 5% of the annual litter production, taking it to their termitaria. Termitaria contained more nutrients than litter, and litter was more nutrient-rich than the soil. When termites abandon termitaria, which they do rather regularly (nest rate abandonment averaged 165 nests per hectare per year), these sites form patches of high nutrient level ideal for many tree species. Termite activity has a potentially important influence on nutrient cycling and tree establishment in this area of very poor soils (Salick et al. 1983).

ARMY ANTS   *Figure 93*

The 1954 Hollywood film *The Naked Jungle* starred Charlton Heston as a besieged South American plantation owner under attack by billions of army ants. The film, taken from the 1938 Carl Stephenson short story "Leiningen Versus the Ants," portrays the six-legged marauders (called "fiends from hell" in the short story) as moving in a column 16 km (10 mi) long and 3 km (2 mi) wide. Nothing could turn the ants from their chosen path, right through Heston's plantation, and the ants ate everything in their path, including attempting to eat Heston's horse, with him on it. Their persistence and ingenuity knew no bounds as they forded streams by making little ant boats out of leaves. As they approached ever closer to the plantation, the landscape behind them was utterly denuded of all plant and animal life, a virtual moonscape. Heston survived, but only by flooding his plantation, washing away the formicine horde with righteous diluvian retribution. The portrayal of the ants in the film was, well, overstated by an order of magnitude or two or perhaps three. Sorry to disappoint, but real army ants are far less fearsome than the *mobutu*, as they were called (in hushed tones) throughout the film.

South American army ants are nearly sightless (African ones are totally blind), eat no plants whatsoever, and eat no animals larger than baby birds or small lizards. Should you encounter a column of army ants, simply step over or around them. The only way to be bothered by them is to sleep on the ground in their path (this did happen to someone on one of my trips) or inadvertently to overlook them and step in them. They won't eat you (or your horse), but they will bite, and their bite hurts. Gotwald (1982, 1995) provides a comprehensive overview of the world's army ants. Also see Holldobler and Wilson (1990).

Neotropical army ants are all members of the subfamily Ecitoninae, of which there are 5 genera and about 150 species. One of the largest and most frequently encountered species is *Eciton burchelli*. Eciton varies both in size and color. The largest individuals are soldiers, with extremely large mandibles, sharply hooked inward. The smallest workers are only about one-fifth as large as the soldiers. Color varies from orange and yellowish to dark red, brown, or black depending on subspecies. The easiest way to identify *Eciton* is by seeing soldiers with the huge pincer jaws. Some Amerindians use these formidable mandibles to suture wounds.

Eciton armies are large, often containing in excess of a million recruits. Armies are nomadic, moving through the forest, stopping at temporary bivouacs during their reproductive cycles. They may bivouac only for a night or for several weeks in the same area. When the entire mass moves, it is usually a nocturnal migration. Bivouacs are either underground or in hollow logs or trees and consist of massive clusters of the ants themselves. There is but a single queen, who remains in the bivouac except when the entire army is on the move. Then, like any matriarch or well-known show business personality, she is attended by an entourage of workers and soldiers.

Each morning the raiders stream from the bivouac, making a dense column in search of prey. Soon the column begins to fan out, raiding parties moving in different directions. Though virtually blind, the ants communicate well by chemical signals, and, once prey is discovered, the ants, being very fast runners, converge on it quickly. This behavior is termed *group predation* and is somewhat the insect equivalent to a wolf pack (Rettenmeyer 1983). The raiders do not restrict their plundering to the ground. They routinely climb trees, even ascending into the canopy. They will enter human dwellings and poke around the corners and under the boards searching for cockroaches and other small vermin. Smart humans leave the ants to their work. They won't remain long, and the house will be freer of insects for their visit. Captured food is stored temporarily along the marauding columns and later taken back to the bivouac.

Prey consists of anything alive and small enough to subdue, most commonly insects (especially caterpillars), spiders, millipedes, and other denizens of the litter and leaves. Small vertebrates such as tree frogs, salamanders, lizards, and snakes are routinely attacked, and baby birds in the nest are frequent army ant victims. Raiding parties attract much attention. Potential prey attempts to flee, and antbirds and other species gather at the edges of the ant swarm to feed on escaping animals (page 285).

THE MEAN AND NASTY GIANT TROPICAL ANT

Beware of the giant tropical ant (*Paraponera* spp.), also called the "bullet ant." This 2.5-cm (1-m)-long black ant can both bite and give a very painful sting. The normal mode of attack is for the creature to bite hard, then, once attached by its jaws, to twist its abdomen around and deliver a wasplike sting of substantial potency. Bullet ants occur throughout Central and South America and are both terrestrial and arboreal. Unfortunately, more than one hiker has been stung by a bullet ant that dropped from branches above. Janzen reports excavating the nest of *Paraponera*, and the ants received him aggressively. In addition to being stung, he noted a musky odor released as he disturbed the nest, which he hypothesized to have been an alarm chemical. He also heard the worker ants emit a loud squeaking noise (Janzen and Carroll 1983). Not surprisingly, such a formidable ant has its mimics. A beetle species resembles the giant ant when at rest, but the deceptive coleopteran looks like a wasp when in flight. Bullet ants tend to be rather solitary, but one is quite enough.

Cockroaches

It is hard to imagine a trip to the Neotropics without seeing *La Cucaracha.* Most of the world's 4,000-plus cockroach species live in the tropics, and the Neotropics can certainly lay claim to its share. Most people can identify cockroaches easily, as they routinely cohabit human dwellings, much to the dismay of the primate occupants. But such an attitude is wrongheaded. You ought to be willing to share with them. Cockroaches have resided on the planet for over 300 million years, a wonderful evolutionary success story. They do not bite or sting, nor do they carry vile diseases. So relax and enjoy them. You might as well. In the Neotropics, at least, they are always close by.

Cockroaches are oval in shape, the dorsal side covered by a pair of large wings, the head sporting a pair of very long antennae. Excellent fliers and basically nocturnal, they are often seen fluttering around lights at night. They are remarkably fast runners and have an uncanny ability to squeeze between floorboards and other tight places.

One of the most striking tropical species is the giant cockroach (*Blaberus giganteus*). At full size, this insect behemoth easily fills the palm of your hand

and then some. It lives in hollow trees and other reclusive places during the day and scurries about in search of food at night. *Blaberus* is a good mother cockroach. I came upon one on the forest floor, near an outhouse (they like outhouses) in Amazonian Peru. I noticed lots of odd little white things around the animal. Closer inspection revealed that the little white things were unpigmented baby cockroaches. The adult was brooding and was very protective about getting the pale little animals under the protection of her wings. With such devoted parental care, no wonder they've been around for so many millions of years.

One favorite food of this and other cockroach species is bat guano. Floors of bat caves literally oscillate, a sea of insects and excrement, as thousands of cockroaches devour the products of bat digestion. In Guatemala I took my students on a nocturnal field trip. With large, powerful flashlights we trekked to the outhouses and, in the dead of night, shined the beams down the holes. The cockroach show, mostly *Blaberus*, was something the students discussed long after returning home. Some claimed to have even dreamed about it.

Harlequin Beetle

There are many impressive tropical Coleopterans (beetles), and *Acrocinus longimanus* is one of the finest. It is both large, up to 7.5 cm (about 3 in), and colorful, being complexly patterned red, black, and yellow, with very long antennae. Males have extremely long and thick front legs, used during mating. Larvae live inside bark, forming galleries inside the wood. Adults inhabit a variety of trees, including figs, and are strongly attracted to sap. If you are fortunate enough to find one of these large insects, look carefully under its thick outer wings (elytra). Don't worry, it can't bite you, at least not to the degree that you ought to care about it. Beneath the wings you can usually find a tiny pseudoscorpion (not a real scorpion, so don't worry about being stung). The pseudoscorpion uses the harlequin beetle as a host, an example of commensalism. The host gains no benefit from the pseudoscorpion, but the little hitchhiker does no harm.

Rhinoceros Beetle

The rhinoceros beetle (*Megasoma elephas*) is named for the long, upcurved, hornlike projection possessed by the males. A huge and bulky insect, it can reach lengths of over 8 cm (3 in). It is a member of the Lamellicornia, a beetle group that includes the scarab and stag beetles. *Megasoma* is a scarab, mostly brownish, its elytra covered by tiny hairs. The combination of large size, long horns, and hairy body makes this insect extremely distinctive. If you see one, you'll recognize it. Females are similar to males in size but lack the long horns, a characteristic true of all scarabs.

Darwin (1871) hypothesized that males evolved the horns by sexual selection, somewhat like the male plumages of the cotingas and manakins (page 270). The horns, thought Darwin, would aid the males in combat for lovely lady scarabs. Males do use the horns in combat. Scarabs and stag beetles all over the world, from Costa Rica to Africa to the Solomon Islands, jostle like wrestlers, locking horns, the victor lifting the loser and tossing him out of the tree. What is not abundantly clear is the degree to which females care about all this pugilism. Males seem more oriented to fighting for favored feeding sites in the trees than for females (Otte 1980). Females may care about the battles insofar as victorious males have access to good sources of nutrition. From a female's viewpoint, these would be the males to get to know. Nonetheless, in many species, males have short horns (Brown and Kockwood 1986), and the adaptive significance of beetle horns remains to be solved.

There are many scarab species in the Neotropics, most very beautiful. This, of course, has led to their downfall since beetle collectors relish them. In addition, species such as *Megasoma* require mature lowland rainforest because the larvae must live in large decaying logs. The cutting of the rainforests and conversion from forest to brushy areas may significantly reduce the beetle's reproductive success (Howden 1983). Already, *Megasoma* is considered to be a rare species.

Lantern Fly

Figure 95

This remarkable insect (*Fulgora laternaria*) is really worth seeing. It is large, with a 12.5-cm (about 5-in) wingspan, and, when oriented vertically along a tree trunk, it vaguely resembles a lizard. The reason for the resemblance is its long head, with markings reminiscent of a cross between the heads of a lizard and an alligator. Indeed, one Spanish name for it is *mariposa caiman*, meaning "alligator-butterfly." A member of the Homoptera, or sucking insects, the lantern fly is unusual in almost every way, including its contribution to Neotropical folklore. Its name is derived from the mistaken belief that its huge head is bioluminescent, glowing in the dark. It isn't and it doesn't. Another even more startling piece of folklore is that if a young girl is stung by a lantern fly (or *machaca*, as the locals call it), she must have sex with her boyfriend within twenty-four hours or she will die from the bite (Janzen and Hogue 1983). This legend was probably authored by a creative but anonymous young Indian man from centuries past who kept company with a coy and perhaps gullible young Indian woman.

The lantern fly comes equipped with several survival strategies (Janzen and Hogue 1983). When oriented on a tree trunk, it is highly cryptic, its soft, mottled grays making it appear as part of the bark. If disturbed, it will climb away or drum its head against the tree, making a rapping sound. If the disturbance persists, the insect discharges a skunklike odor and flies to another tree. When it takes flight, it reveals bright yellow eye spots on its hind wings, rather like those of the owl butterfly (see below). These spots, which are also revealed by a quick flash of the wings, may act temporarily to confuse a would-be predator.

A Lepidopteran Sampler

Figures 94, 115

There are some butterflies that most every visitor to the lowland rainforests notices. I've previously discussed *Heliconius*, its relationship with *Passiflora*, and its mimicry complexes (chapter 6). Now I'll briefly introduce you to a few more butterflies and one moth. DeVries (1987) provides a guide to Costa Rican

Owl butterfly

butterflies (with much information on natural history—it is more than a mere guide to identification), and D'Abrera (1984) has a brief but useful guide to butterflies of South America.

Put out a few overripe bananas at night, and soon both a large, brownish butterfly and a moth will appear. The owl butterfly (*Caligo memnon*) and the black witch (*Ascalapha odorata*) are both among the largest of the tropical lepidopterans, and they tend to be strongly crepuscular, active at dusk and dawn, though adults fly during the day in deep rainforest shade. They feed on rotting fruits, hence their orientation to bananas. Black witches are occasionally mistaken for bats in the evening twilight as they flutter about a banana bunch, and this species is frequently flushed from shady sites such as hollow trees during the day, again giving rise to the mistaken notion that it is a bat.

Like the lantern fly, the owl butterfly has a prominent eye spot on the base of each wing. The spot is on the underside of the wing and is commonly flashed when the otherwise cryptic animal is disturbed on its tree trunk resting place. Only one spot is flashed, not two, although both wings have spots. Many individuals show damage from bird beaks in the vicinity of the eye spot, giving rise to the notion that the spot "redirects" the bird's attack, allowing the butterfly to escape with little damage. Another suggestion is that the eye spot serves to mimic a distasteful tree frog (Strandling, cited in DeVries 1983). The upper side of the wings is soft bluish, somewhat similar to the blue morpho.

The blue morphos (genus *Morpho*) are perhaps the most spectacular of Neotropical butterflies. Large, with brilliant, deep blue upper wings that seem to glow in sunlight, morphos are deceptively swift fliers able to elude the most persistent wielder of an insect net. Particularly common around streams and other sunlit areas, they feed on a wide variety of plant species. When on the ground or on a tree, morphos are cryptic, but they are (to put it mildly)

Satyrinid butterfly

obvious in the air. Their striking patterning and color, visible only in flight, has been termed the "flash and dazzle" strategy of capture avoidance (Young 1971). There are about eighty species of blue morphos, and all are members of the family Morphidae.

There is a group of butterfly species that accompanies army ant swarms, especially *Eciton burchelli* (see above). These so-called antbutterflies are all members of the large family Nymphalidae and subfamily Ithomiinae, and their collective common name is "army ant butterfly." They generally resemble *Heliconius* and are part of the large "tiger-stripe" mimicry complex (page 161) that includes heliconiines, danaines, satyrids, other ithomiines, and a moth or two. Only female army ant butterflies actually orient to the army ant swarms. Anywhere from eight to twelve will be flying about the swarm, and researchers captured up to thirty within a few hours. The butterflies were feeding on the droppings of ant-following birds. Females were probably using the droppings as a nitrogen source, necessary in making eggs (Ray and Andrews 1980).

One common sight throughout the Neotropics is the massing of butterflies along exposed riverbanks or where cattle or some other creature has recently urinated. Most of these are yellow, gold, or white butterflies, often in the genus *Phoebis* in the family Pieridae. Males are bright yellow with orange on the fore-part of the wing, and females are more uniformly orange with black lining the outer part of the wings. Sulphurs range from the United States through the Neotropics and are inhabitants of open areas and forest edges. They feed on many of the flowers, such as *Lantana*, which are the common tropical roadside weeds. They aggregate at pools containing urine from cattle or humans. The urine supplies them with sodium and nitrogen, just as bird droppings do for army ant butterflies. Other butterflies, especially swallowtails (*Eurytides*), may be present as well.

Forest-floor Millipede   *Figure 107*

This species, *Nyssodesmus python*, is a common denizen of the leaf litter. Females attain a length of about 10 cm (approximately 4 in), and males are a bit smaller. They look well armored, with a flattened, shiny carapace protecting their delicate undersides. The little beast ranges in color from yellowish orange to grayish.

Millipedes are harmless ambling herbivores and should not be confused with swift-moving carnivorous centipedes (see appendix). When threatened, millipedes roll up in a little ball and tough it out. They also have an impressive array of chemicals at their command. The hindgut can squirt a volley of noxious liquid, containing both hydrogen cyanide and benzaldehyde (Heisler 1983). You'll notice this trait if you handle one of these little creatures. Your hands will smell distinctly like almonds, from the cyanide. Wash your hands.

Peripatus

A *Peripatus* is a most unusual, wormlike animal with characteristics of both annelids (segmented worms) and arthropods (joint-legged animals). For this reason, *Peripatus* is usually classified in its own phylum, Onychophora. About thirty species of Onychophorans crawl about in the world's tropics, including Australia, New Zealand, New Guinea, Malay Peninsula, Africa, and the Neotropics. Though segmented like an annelid, *Peripatus* has the beginnings of an exoskeleton, a basic arthropod characteristic. Its stubby, unjointed legs terminate in claws, another arthropod characteristic. Eyes are simple, not compound.

Found among the denizens of forest floor litter, *Peripatus* ambles along on short unjointed legs, resembling a slender caterpillar with stubby, segmented antennae. It captures insects by shooting them with a gluelike substance that engulfs prey in a network of sticky threads. This odd adaptation also helps defend *Peripatus* from would-be predators (Ghiselin 1985).

Terrestrial crab

14

Deforestation and Conservation
of Biodiversity

In many ways, Brazil today is reminiscent of the United States of generations ago. Vast areas of the country are unoccupied. Were the potentials of the country realized, it could probably sustain more people at a higher standard of living than could the United States.
—Philip H. Abelson, 1975

If we ever allow tropical forests to disappear from Earth we shall have to tell ourselves not only that we have lost something of value. We shall have to admit that in certain significant senses, we have not gained much since we left our caves. For all that we are now firmly established on our hind legs, our response to tropical forests serves as a measure of how far we are ready to stand tall as citizens of the Earth, as joint members of our planetary home.
—Norman Myers, 1984

THESE divergent views reflect a debate that has become widely known in recent years. The popular press has abounded with editorials and stories about the ecology and economics of utilizing global tropical forests. Periodicals from *National Geographic* to *The New Yorker* have published feature stories on tropical forests and the struggle over their fate. And the issue of most concern is deforestation.

Deforestation refers to the cutting, clearing, and removal of rainforest or related ecosystems, and subsequent conversion into other, less biodiverse, anthropogenic ecosystems such as pasture, cropland, or plantation. It is well known that rates of deforestation have been high throughout the global tropics (Myers 1986, 1988). Because of its magnitude, its rate of increase over the past two decades, and its potential effects on loss of biodiversity and perhaps even alteration of world climate, deforestation is arguably the most important global conservation issue today.

Though deforestation of rainforests has engendered much attention and concern, deforestation is occurring in many other ecosystems in the Neotropics, and their loss is receiving far less attention. Coastal mangrove forests, for example, have been cleared in many areas. Deforestation has been accelerating in the Brazilian cerrado, a region of dry forests and arid scrub, much of which is being converted to large-scale agriculture, particularly soybeans (Stotz et al. 1996). These ecosystems are, in many cases, even more threatened than lowland rainforest or montane cloud forest.

Deforestation is often done for short-term profit at the expense of long-term sound economic and ecological policy. Economic pressures in countries where population growth is high have resulted in a labor force that typically increases more rapidly than the economy, thus promoting deforestation as a means of reducing unemployment and enhancing government revenues (Repetto 1990). Though here I will consider deforestation in the Neotropics only, you should note that tropical deforestation is a serious issue in tropical forests of Africa and Asia as well (Repetto 1990). At present, the estimated annual loss of the world's rainforests by cutting and conversion remains at approximately 2% (Lovejoy 1995).

Deforestation in Amazonia: The Burning Season

Figures 77, 78, 79, 80

September is the height of the dry season in southern Brazil. It is the time for burning the rainforest. Vast tracts of rainforest, cut in order to convert the landscape from forest to farm or ranch, are set afire; fires that are so extensive that they are visible from space; fires that cloud the atmosphere such that aircraft must sometimes be diverted from landing because airport runways are occluded from view by the haze from the smoke; fires that make the noon sun look more like an orange-colored setting sun; fires whose potential global impact made the cover story of *Time* magazine on September 18, 1989. Smoke lingers over the burned land. An acrid odor permeates the air. It is at ground level that deforestation is most stunning, for along much of the Transamazon highway, as far as one can see in any direction, there is nothing but burned forest. Scattered, blackened snags of trees are all that remain of what had been lush, diverse rainforest, cerrado, or grassy savanna.

The Extent of Deforestation in Amazonia

In September 1994 I traveled overland between Alta Floresta and the city of Cuiaba in the state of Mato Grosso, Brazil, a distance of approximately 804 km (500 mi). Alta Floresta is a rainforest area, known for its high number of endemic species, near the Teles Pires River, which drains into the Tapajos, flowing north to the Amazon about 1,100 km (683 mi) away. The highway from Alta Floresta heads east and south, eventually intersecting with br-163, one of the main north/south roads of the ever increasing Transamazon highway network. For the approximately 322-km (200-mi) distance from Alta Floresta to br-163, I passed nothing but cleared, burned forest, most of which had been only recently cut; there were not many functioning ranches, and certainly no settlements of people, only hectare after hectare of charred stumps. It was obvious that human population density was low. These forests had not been cleared to make room for masses of migrant humans trying to escape urban poverty.

From BR-163 to Cuiaba, a drive that took me from lowland humid forest into the vast Brazilian cerrado, the landscape became one of extensive agricultural fields that are machine cultivated, mostly for soybean production. Brazil is now the world's leading exporter of soybeans, its soybean production having risen sixfold in the decade from 1970 to 1980 (Skole et al. 1994). Several greater rheas, ostrichlike birds that were once much more abundant throughout the cerrado-dominated ecosystems near Cuiaba, were taking shade beneath giant powerline towers that bisected the dry, flattened, dust-covered soybean fields. A crowned eagle (*Harpyhaliaetus coronatus*), a rare cerrado-dwelling relative of the majestic harpy eagle, sat atop a utility pole as trucks loaded with Amazonian lumber sped by. My brief but insightful overland journey left me sobered at the reality of deforestation in Amazonia.

The deforestation that I witnessed was not confined to a limited region. I could have seen similar levels of deforestation in many areas of Brazil, especially in the western state of Rondonia, or in eastern Amazonia. During the period from 1980 to 1985, the World Resources Institute estimated an annual average loss of nearly 1.5 million hectares (3.7 million acres) of forest in Brazil (Hartshorn 1992). Taken as a whole, however, though certainly extensive, Amazonian deforestation is still confined to a relatively small percentage of the total area of the Amazon Basin, roughly 10–12%. I have also flown over Amazonia, and from the air it becomes obvious that deforestation is essentially restricted to areas adjacent to roads and rivers, especially the former. But it is important to keep in mind that road building continues; thus deforestation continues. For example, in 1996 the Brazilian government, partially funded by the Andean Development Corporation (which represents several different countries with interests in South America), made the decision to pave the entire length of BR-174, a road running 997 km (620 mi) from Manaus to the Venezuelan border. This road now exists but is far from fully paved and thus virtually impassable during rainy season. The paved road will pass through one major Indian reservation (Waimiri-Atroari) as well as several tracts of ecological reserves. Once paved, the road will make all of the adjacent area readily accessible to whomever wants to develop it or settle there.

At this writing, most of the deforestation in western Amazonia has occurred in the state of Rondonia, principally along BR-364. Deforestation is also fairly extensive in the Brazilian state of Acre, home of the late rubber tapper Chico Mendes, who was murdered by cattle ranchers for his role in helping persuade the Brazilian government to designate a large tract of rainforest as an extractive reserve, thus preventing it from being cut and used for ranching (see below). Other areas of significant deforestation within Brazilian Amazonia are found around the cities of Manaus, Obidos, Santarem, and Macapa (all of which are along the lower Amazon), and in the eastern states of Para (especially around Belem) and Maranhao, where the Transamazon highway system is very extensive. In general, there have been two major areas of increasing deforestation in the Brazilian Amazon: the main north-south highway between Belem and Brasilia and the major east-west roads between Mato Grosso and Rondonia (Skole et al. 1994). Along the upper Amazon, in the countries of Peru and Ecuador, there is limited deforestation around Iquitos, Peru, as well as along the river itself, including its major tributaries, the Napo, the Maranon, and the Ucayali (Collins 1990).

Deforestation estimates change annually, vary from survey to survey, and are subject to differences in interpretation. Satellite data are increasingly used to determine rates of deforestation (Booth 1989). Techniques of satellite imagery (Landsat, radar, thermal scanning, imaging spectrophotometry) have been continually refined and now can be accurately correlated with ground vegetation, as discussed below (Skole and Tucker 1993; Skole et al. 1994; Moran et al. 1994). For examples of satellite imagry as utilized to study land patterns in Belize, see O'Neill (1993).

Brazil is certainly not the only country in South America to have experienced accelerated levels of deforestation. From figures obtained by World Resources Institute, Hartshorn (1992) noted that four South American countries—Brazil, Colombia, Ecuador, and Peru—were among the top ten tropical countries ranked by annual loss of total forest area from 1980 to 1985. Brazil ranked number one, but because of its vast area its annual loss was only 0.4%, compared with Colombia, which ranked number two but had an annual loss of 1.7%, and Ecuador, which ranked fifth but had an annual loss of 2.3%, a deforestation rate fully 5.75 times that of Brazil.

Current estimates of deforestation in Amazonia range from 1.5 to 2 million hectares 3.7–4.9 million acres per year (Skole et al. 1994). Figures taken from the Food and Agricultural Organization (FAO) suggest a cumulative loss of approximately 100,000 km^2 of Brazilian rainforest from 1981 through 1985. This compares with FAO estimates of 6,000 km^2 (2,300 sq mi) from Colombia, 3,400 km^2 (1,300 sq mi) from Ecuador, and 2,600 km^2 (996 sq mi) from Peru (FAO, cited in Collins 1990). From estimates that were made in 1989, taken as a whole, approximately 90% of the Amazonian region remained at that time in natural forest (Brown and Brown 1992). The rate of deforestation from 1978 to 1989 was approximately 21,218 km^2/yr (8,129 sq mi), peaking in 1987 and dropping since that time (Fearnside et al. 1990, cited in Brown and Brown 1992; Skole et al. 1994). What this means is that of the total Brazilian Amazon, a region of nearly 5 million km^2 (1.9 million sq mi) (essentially the area of the lower 48 United States), where forest originally occupied about 3.96 km^2 (1.5 million sq mi) somewhere around 400,000 km^2 (153,240 sq mi) have thus far been deforested (Brown and Brown 1992; Molion 1991). This estimate has been challenged and modified in a study of satellite data indicating that the deforested area is about 230,000 km^2 (88,113 sq mi), thus only a little over half the estimate cited above, suggesting an Amazon-wide average of 0.4% deforestation annually (Skole and Tucker 1993). However, this same study suggested that a total of 588,000 km^2 (225,263 sq mi) of forest have sustained a negative impact from deforestation when one considers edge effects of fragmented forested areas (see below). In other words, when disturbance is considered along with outright deforestation, the satellite data indicate a greater impact compared with previous estimates. Landsat Thematic Mapper images now suggest that the rate of deforestation in Amazonia is considerably lower than the rates occurring in the previous decade (Skole et al. 1994).

The Brazilian states of Maranhao and Tocantins have suffered the most deforestation, 46% and 44%, respectively (Brown and Brown 1992). Again, figures vary among sources. A more recent study agrees that Maranhao has experienced the most deforestation but estimates its total to be only 27%, not 46% (Skole et al. 1994). Regardless of which figure to use, Maranhao and Tocantins, taken together, represent only about 11% of the total area of the Brazilian Amazon and only about 5% of the original forest. The huge state of Amazonas, with a total area of about 1.5 million km² (0.57 sq mi) (about 32% of Brazilian Amazonia), still had 99% of its original forest in 1989 estimates. Roraima had 98%, Amapa 99%, and even Rondonia and Acre, where deforestation has been extensive along the Transamazon highway, still had 85% and 94%, respectively, of their original forests. In addition, 143,390 km² (54,933 sq mi) was in protected areas as of 1990 (Brown and Brown 1992).

The process of ecological succession (page 63) has been active following deforestation throughout Amazonia. Many farmers who clear land are surprised by how quickly the successional process reclaims the land, and many people unfamiliar with the ecology of the region are surprised that once deforestation has occurred, the land is not ruined, a doomsday scenario that has been frequently overstated (see below). In fact, succession normally occurs rapidly throughout Amazonia. The increase in area covered by successional vegetation following initial deforestation is well documented by Landsat analysis using Landsat Thematic Mapper, providing a 30-meter resolution with pixels of approximately 0.09 hectares,(Moran et al. 1994). In this study, Landsat data were standardized against ground data, eventually providing an accuracy (in interpreting satellite spectral imagery) of about 92%. Using two study areas, one in eastern Altamira, one in western Altamira, and comparing estimates from 1985 and 1991, the Landsat data document the rapidity of succession following forest clearance. For example, in 1985, about 13% of the land was bare in the eastern Altamira site and 7.5% bare in the western Altamira site. In 1991 only 1.4% was bare in the eastern site and 0.6% in the western site. Areas of initial secondary succession, on the other hand, had increased from 1985 to 1991 by 15% in the eastern site and by 5.5% in the western site. Intermediate secondary succession changed much more, increasing 125% in the eastern site and 156% in the western site. The Landsat study indicated that forest had declined by a little over 1% annually in the eastern portion and about half as much in the western portion (Moran et al. 1994).

A similar study, confined to the state of Rondonia in western Brazil, also documented the extent of secondary successional vegetation (Skole et al. 1994). As Rondonia was colonized extensively by small landowner farmers from the mid to the late 1980s, a significant area of agricultural land came from clearing secondary forest, not just from clearing primary forest. For example, between 1988 and 1989, 58% of the land cleared was primary forest, while the remainder cleared was secondary vegetation. In 1989 approximately 33% of the land was in secondary vegetation, and farmers routinely had to reclear the land approximately every five years.

Deforestation in the Brazilian Atlantic Coastal Forest

In contrast to Amazonia, the Brazilian Atlantic Coastal Forest, a region of high endemism, has experienced an astonishing 88% reduction in forest area

(Brown and Brown 1992), with some estimates even higher (Wilson 1992). Of a total forested area of 1,205,780 km^2, the total deforested area is 1,059,756 km^2, with only 30,950 km^2 (2.5%) in totally protected areas. The Fundacao SOS Mata Atlantica (SOS Atlantic Forest Foundation) was created by Brazilians in 1986 in the hopes of preserving what remains of this unique forest, as well as creating increasing environmental awareness of the forest's conservation significance (Collins 1990).

Deforestation in Central America

The percentage of deforestation in Central America (which includes the countries of Panama, Costa Rica, Nicaragua, El Salvador, Guatemala, Honduras, and Belize) vastly exceeds that of Amazonia. Of course, the overall area is far smaller, only about the size of the combined areas of Utah and Nevada (Hartshorn 1992). Most deforestation in Central America has occurred since 1950, and by 1980 almost 60% of the lowland forests in Central America had been felled, converted primarily to pasture and agriculture. Of course, that figure has continued to accelerate. Four Central American countries—Costa Rica, El Salvador, Nicaragua, and Honduras—were among the top ten countries ranked by percentage loss of forests during the period 1981–85 (Hartshorn 1992). For example, during that period, Costa Rica averaged an annual loss of 3.9% of its forests, largely due to the vast expansion of cattle ranching. By 1983 about 83% of Costa Rican forests had been felled, mostly for beef production, and much of that was shipped to the United States for use as hamburger.

Today, large tracts of lowland forest can still be found in the Darien region of Panama, in the Peten region of eastern Guatemala, in the Maya Mountain region of western and southern Belize, and scattered throughout parts of southern Costa Rica, Honduras, and Nicaragua. El Salvador is almost entirely deforested, and Costa Rica, a country where immense amounts of research is done on tropical ecology, is also largely deforested north of San Jose, except for scattered preserves and parks. It should be noted, however, that Costa Rica has in recent years developed an enlightened attitude that favors conservation goals. Ultimately, such a view is in the country's best interest, as ecotourism is thriving in Costa Rica (see below).

Given the extent of deforestation in the lowlands of Central America, it is somewhat heartening that several large conservation units have been established in recent years. These include the Cockscomb Basin Jaguar Reserve and Bladen Reserve in Belize, the Tikal World Heritage Site in Guatemala, the Rio Platano Biosphere Reserve in Honduras, and the Barra del Colorado-Tortuguero complex in Costa Rica (Hartshorn 1992).

Deforestation in the Caribbean Islands

Deforestation is extremely extensive throughout the Caribbean Islands. Considering first the Greater Antilles, virtually all of the lowland forests in Cuba have long since been felled, replaced largely by sugar cane fields. A few isolated tracts remain, plus scattered montane forests, but Cuba, the largest of the Caribbean islands, is essentially totally deforested. Haiti, which suffers from abject poverty, is also totally deforested. There is some forest remaining in the Dominican Republic, though not much. Jamaica still has fairly extensive forest, proportionately more than any of the other Greater Antilles. Puerto Rico is largely deforested, but the U.S. Forest Service oversees management of the Caribbean National Forest, a 6,300-ha (15,567-acre) primary montane rainforest in the Luquillo Mountains. This tract, plus a few scattered areas of primary old-growth forest, represents less than 1% of the land area in Puerto Rico but supports by far the largest biodiversity on the island (Lugo 1994).

Among the Lesser Antilles (all of which are small islands), some forest (often montane) remains on most of the islands: Guadeloupe, Dominica, Martinique, St. Lucia, St. Vincent, Grenada, Trinidad, Tobago. Barbados has been entirely converted to sugar cane production and has no remaining natural forest. The finest rainforest among the Lesser Antilles is to be found in the Arima Valley of Trinidad and in central Tobago. (Note that Trinidad and Tobago are geologically part of the South American continent and thus not technically part of the Lesser Antilles, at least with regard to geologic origin.)

A Closer Look at Amazonian Deforestation

Cattle Ranching

Most North Americans probably harbor the belief that deforestation in Amazonia is the inevitable result of ever-increasing population pressures. Aware of the fact that human population growth rates are generally high in most Latin American countries, particularly Brazil, and further that there are millions of people living in what is often extreme poverty, it seems logical enough to presume that deforestation is the price that must be paid to open space for human colonization and farming. If Brazil is taken as the example, however, deforestation has not resulted in highly significant influxes of humans into land that was formerly rainforest (though this might be changing with the expansion of the Transamazon highway system). Quite the contrary, it can be argued that governmental policies promoting deforestation were initially designed to increase cattle raising at the expense of human populations (Moran et al. 1994). Beginning as early as 1966, owners of large cattle ranches were provided with generous tax breaks and subsidies, while small-property farmers were awarded only modest short-term loans (Fearnside 1987; Moran et al. 1994). Thus a small number of wealthy individuals were handsomely rewarded for locating vast cattle ranches within Amazonia. Indeed, more than 600 cattle ranches, each in excess of 20,000 hectares (49,420 acres), were government subsidized in the form of long-term loans, tax credits to cover most of the investment costs, tax holidays, and write-offs. These ranches failed to generate any profits, losing in excess of half of their invested capital within fifteen years (Repetto 1990). The overall meat output from these subsidized ranches averaged only 9% of what was projected, and most ranches essentially were used as tax shelters rather than productive components of the Brazilian economy. The Brazilian government, though no longer supporting new cattle ranches, continues to subsidize existing ranches, costing the government's treasury more than $2.5 billion in lost income (Repetto 1990).

Deforestation was, of course, not possible in most of Amazonia before the construction of the Transamazon highway system, which began in 1958 with a dirt road between Belem and Brazilia. After about two decades, approximately two million people had colonized the area opened by the road, which was finally paved in 1973. In that same time period, however, five million cattle were moved into the area, a trend that essentially was repeated as other roads were added linking northern and southern Brazil and eastern and western Amazonia. Though part of the road-building objective was to move 100,000 families into the Amazon Basin, economic setbacks and political changes within Brazil resulted in a net migration of only 6,000 families (see below). The focus shifted from moving humans to increasing cattle production, and today cattle ranches occupy somewhere around 8.4 million hectares (20.7 million acres), averaging 24,000 hectares (59,300 acres) each, with some as large as 560,000 hectares (1,383,760 acres). As noted above, these ranches operate at a profit only because of continuing generous governmental subsidies. A ranch of about 20,000 hectares (49,420 acres) typically receives a shocking 75% subsidy. These ranches do not employ large numbers of humans, typically only one cowboy for every 300 acres. There can be no doubt that cattle ranches have been a principal causal factor in Amazonian deforestation (Moran et al. 1994).

Three other sources of deforestation continue to affect Amazonia: resettlement of small-landowner farmers, increase in mining activity, and timber harvesting.

Resettlement of People

Following the initial construction of the Transamazon highway, the Brazilian government made an attempt to resettle families from the impoverished northeastern region of the country to the newly opened Amazonian frontier. The effort was fraught with difficulties, ranging from unforeseen ecological effects to spread of diseases by the colonists themselves (Smith 1981).

The Transamazon highway cuts through interior forest, almost entirely through terra firme, well off of floodplain. Soils are nutrient-poor, with high concentrations of aluminum. A few fertile soils occur near the highway, but overall only 3% of the soils in this region are considered to be fertile. As a result, to farm effectively, fertilizer is necessary, and fertilizer costs money. After cutting and burning, high rainfall in the region subjects the exposed land to erosion. Not only do farmers' fields erode, but so does the highway itself (which, at this time, was unpaved). Side roads may be cut off from main roads during the rainy season, and the main roads may be damaged, cutting off suppliers from the cities.

The intent of the Brazilian government was for the newly arrived farmers to plant upland rice as their basic crop. This was an unfortunate decision on the planners' part since upland rice requires fertilizer, herbicide, and insecticide treatment to remain healthy. The impoverished farmers were not in any position to buy such expensive agricultural aids. To make matters worse, poor strains of rice were selected. The commonest strain, IAC 101, was developed barely within a tropical climate and was subject to stem breakage and flattening during heavy rains in the wet Amazon. Fungi invaded, rotting the flattened plants, resulting in dramatically reduced yields. The average farmer, though planting 8 hectares (19.8 acres) annually, grossed only $1,900 for the crop. Rats, ground doves, fungi, and erosion claimed a large share of each crop.

Manioc (page 184) would have been a much better choice, as it is adapted to the wet climate and poor soils. An estimated annual yield of 20 tons per hectare could permit a family that can harvest 3 hectares to earn $3,000 from the sale of flour made from manioc. The harvestable part of manioc, the tuberous root, can be collected anywhere from six months to three years following planting. Farmers can cooperate and aid each other's harvest since harvest time is not critical, as it is with rice. Roots can remain unharvested during the worst of the rainy season when roads are flooded. Manioc flour is easily stored, and the food itself actually yields more calories than rice does.

The Brazilian government was too slow to embrace manioc as a staple crop. Instead it attempted to introduce new crops, not only rice, but also pepper, banana, and cacao, better choices but still not ideal. It also continued to promote the growth of cattle ranching along the highway (as described above). Though the region did not initially harbor agricultural pests, they soon began to appear. The highway, lined with newly planted crops, served as a corridor for the invasion of pest species into the interior Amazon. Pepper plantations were injured by a fungus that appeared in 1975 and was resistant to fungicides. Fungi also attacked the numerous banana plantations along the highway.

People brought their own diseases. In 1973 reports indicated that regional public hospitals had treated 1,285 cases of malaria. The rainforest itself did not initially harbor the malarial parasite, a minute protozoan called *Plasmodium* (see appendix). After settlement, three species of malaria were reported from the region, and all three were brought from the north by the people themselves. The mosquito vector of the parasite *Anopheles darlingi* thrived in the water culverts bordering the highway, and the disease became distressingly common. Unfortunately, malaria has two annual peaks, one of which occurs when the farmers are sowing their crops, and the other approximately at harvest time. Such timing obviously couldn't be worse from the standpoint of productivity. In addition to malaria, gastrointestinal diseases and parasite infections derived from poor sanitation also plague the people. Diarrheal diseases remain today the most serious causes of infant mortality in the region.

This attempt at colonization of the interior Amazon cannot be called a success. The overpopulated northeastern region of Brazil has grown by over 6 million since the Transamazon highway was built, and relocation of the people has accounted for less than 1% of the northeastern population growth. The cost to Brazil has been approximately $500 million, a very steep price for a project with so few benefits. As Smith (1981) wrote in his paper, "Along the Transamazon, many settlers soon fell victim to a biased and inefficient credit system, a poor selection of crops, infertile soils, and isolation from large markets."

In an ambitious attempt, Fearnside (1986) developed a complex computer model to predict sustainable human carrying capacity of Brazilian forests under different scenarios. His rather pessimistic conclusion was that potential carrying capacity for people under both simple agriculture and larger enterprises, such as cattle ranching, is disappointingly low. Nonetheless, recent studies have suggested that a wiser approach to achieving sustainable agriculture in Amazonia is possible (see below).

Mining Activity

A variety of minerals (copper, tin, nickel, bauxite [aluminum ore], manganese, iron ore, gold) exists, often abundantly, within scattered areas throughout Amazonia, and these potential economic resources have been increasingly sought. As with cattle ranching, government policies, most principally tax incentives, have led to large-scale investment in mining, which soon results in deforestation and/or forest pollution. For example, wood from surrounding forest was cut for charcoal to fuel pig iron plants in the $3.5 billion Grande Carajas region in the eastern Brazilian state of Para, resulting in an annual deforestation of 610,000 hectares (1,507,310 acres) (Moran et al. 1994).

Gold mining has increased dramatically in Amazonia. Many people are drawn to the region in an attempt to get rich from gold prospecting. I visited the small town of Las Claritas in southeastern Venezuela, locally called "Dante's Inferno," a nickname that makes reference to the fact that gold-mining towns often do not attract the most genteel clientele. Gold miners, many of them carrying sidearms reminiscent of the style of the North American Old West (understandably, they are protective of their gold), brought their gold to various buyers who manned a row of small trailors lined up in the town, where the precious mineral would be weighed and purchased.

There has been so much gold mining in Amazonia since the early 1980s that it is fair to describe the activity as a gold rush (Hacon et al. 1995). In Alta Floresta, Brazil, where I witnessed extensive deforestation activities, the 1980s gold rush slowed rural development in favor of increased gold-mining activities. The area now has twenty-five gold-dealing shops where 25 tons of gold are purified annually (Hacon et al. 1995). One study, cited in Barbosa et al. (1995), suggests a staggering total of a half million gold prospectors (*garimpeiros*) working throughout Amazonia, most of them mining small-scale operations called *garimpos*. While most gold-mining operations are small, some can be vast, employing many thousands of workers. For example, the Serra Pelada mine in Brazil yielded 29,000 tons of gold from 1980 to 1986 and employed 50,000 workers (Collins 1990).

Gold mining has had some strongly negative side effects, largely because the ore is processed on the site, resulting in ecosystem damage. Though mining activities do not result in much deforestation, there is accompanying regional pollution by mercury (used in gold extraction, as explained below). Methyl mercury has been found in high concentrations in Amazonian fish, has been shown to concentrate up the food chain, and thus enters indigenous Indian populations dependent on fish. Gold miners and others living near mining activities are subject to mercury pollution. Even without considering mercury pollution, gold mining frequently encroaches on lands legally belonging to indigenous tribes, creating conflict and sometimes violence between miners and Indians. For example, in a conflict between Yanomamo Indians and gold prospectors in the Brazilian state of Roraima, near the Venezuelan border, the Brazilian government supported the Indians and, with military-style tactics, evicted the gold miners from the Indian lands (Brooke 1993). This act created a large measure of political tension and disenchantment among non-Indian people throughout the state.

Mercury Pollution

Mercury is a potentially dangerous pollutant. It is a heavy metal, toxic to humans when exceeding normal concentration, and, when in the form of methylmercury, particularly dangerous to *in utero* brain development. Toxic levels of mercury can produce cerebral palsy syndromes in infants (Barbosa et al. 1995) and damage to the central nervous system of adults.

When gold is mined in Amazonia (see above), metallic mercury (Hg) is used to agglutinate the fine gold particles through amalgamation (Barbosa et al. 1995). This process results in the release of shockingly large amounts of mercury into rivers and soils. When the mercury-amalgamated gold is burned to liberate the gold, additional mercury is released into the atmosphere. In one study, it was estimated that of the 1,200 tons of Hg discharged into Amazonia during the past decade, between 800 and 1,000 tons may have entered the atmosphere as vapor (Barbosa et al. 1995). Mercury in the atmosphere is active: it oxidizes, becoming divalent, and can return to aquatic and terrestrial ecosystems in precipitation. It will then interact with organic material, becoming methylmercury, a form in which it can bioconcentrate up the food chain.

Mercury is also normally present in wood and other plant tissue and is released as vapor when wood is burned. Because of the extensive deforestation throughout Amazonia, mercury released from combustion of wood becomes a highly significant component of mercury pollution in the region. One study suggests that the annual Hg emissions from above-ground wood ranges between 103 and 515 tons annually, amounts that exceed estimates for mercury release due to gold mining (Velga et al. 1994).

The severity of mercury pollution has been well documented in various places throughout Amazonia, particularly around gold-mining areas. The normal concentration of mercury present in human urine should measure less than 10 parts per billion (ppb), but concentrations as high as 840 ppb have been recorded from gold-mining areas (Velga et al. 1994). Mercury contamination can also be documented in hair and blood samples. People who live in gold-mining areas usually take in mercury as vapor from the atmosphere, whereas indigenous people in the rainforest ingest methylmercury that has bioconcentrated in fish (Akagi et al. 1995; Barbosa et al. 1995). For example, at urban Alta Floresta, mercury in the form of coarse aerosol particles accounted for about 70% of the inhalable particulate nitrogen (Hacon et al. 1995). In contrast, 90% of the fish caught by rural villagers south of gold-mining areas on the Tapajos River were contaminated with methylmercury

(Velga et al. 1994). Because of the addition of methylmercury to the riverine food web, and the high prevalence of mercury vapor in the atmosphere, mercury is now considered a general environmental pollutant throughout much of Amazonia.

Timber Harvesting
Figures 81, 82

A ride essentially anywhere along the Transamazon highway system will demonstrate the degree to which timber harvesting significantly contributes to deforestation. Lumber mills are common along roadsides, and lumber-bearing trucks are among the most frequently seen vehicles. In 1990 in the Brazilian state of Para there were 238 sawmills near the town of Paragominas, a heavily deforested area directly south of Belem in eastern Brazil (Holloway 1993). In the Brazilian states of Rondonia and Roraima, it is estimated that wood products provide about 60% of the economic output, and the number of lumber mills has increased more than eightfold since 1965 (Moran et al. 1994). In Brazil the number of wood processors increased from about 100 in 1952 to in excess of 3,000 in 1992 (Holloway 1993). Part of the reason for the striking increase in timber harvesting is the failure of cattle ranches. As succession replaced cattle pastures with young forest, ranchers sold logging rights. About 100 species of trees, many which grow in disturbed areas, are considered to be satisfactory for harvest (Holloway 1993).

The increase in wood production from Amazonia is correlated with the declining output of wood from Asia, much of which is already deforested. In Amazonia wood is usually taken by selective logging, rather than by clear-cutting, because only a few species of trees are desirable for their lumber. Because of the heavy equipment necessary to extract single large trees, selective logging has a strongly negative impact on the surrounding area, in some instances resulting in as high as 40% mortality of nearby trees (Uhl and Vieira 1989). A research team from Amazon Institute of Man and the Environment (IMAZON) documented that for every tree removed by selective logging, 27 other trees that are 10 cm (3.9 in) or more in diameter are injured, 40 meters (131 ft) of road is created, and 600 square meters (6,458 sq ft) of canopy are opened (Holloway 1993). Selective logging is sufficiently destructive that the rate of succession is effectively slowed, and it takes decades longer for forest to become reestablished. The extent of selective logging is difficult to ascertain due to the similarity with large- and medium-scale natural gaps. But at least in most areas thus far, selective logging is as potentially degrading to habitat as cattle ranching.

One approach to logging that has shown promise is called strip logging (Jordan 1982; Hartshorn 1990). This process is designed to mimic natural succession (page 179). As such, it maximizes the potential for regeneration of forest. Basically, strip logging involves the clear-cutting of a relatively thin strip of forest that parallels a river along a slope. Gallery forest (bordering the river) is left intact, but a strip is cut immediately upslope and the desirable timber removed via a road that is also designed to parallel the river. Following this, another strip is cut several years later immediately upslope to the first strip and road. Nutrients eroded from the newly cut strip wash downslope and aid in speeding the recovery of the first strip. Finally, a third or fourth strip may be cut, each progressively upslope from the previous strip. As new strips are cut, previous strips are regenerating through normal succession. The strips are sufficiently narrow that there is no difficulty in dispersal of species to repopulate the clear-cut areas (as there can be when forest fragments are isolated— see below). Also, many trees regrow from stump sprouts. In essence, strip logging is really the creation of a narrow gap, and the recovery process is fundamentally the same as gap phase dynamic succession. And that's the whole idea.

Strip logging is well suited to sloping terrain, which is a common feature within Amazonia and, of course, along montane areas. About 27% of the soils in the Amazon Basin are on slopes exceeding 8%. As such, these soils are subject to high erosion potential. Elsewhere, even where slopes are more gradual, heavy rainfall can cause high erosion, such that in total, 40–50% of the soils in the Amazon Basin are regarded as having at least a moderate erosion risk (Nicholaides et al. 1985). Strip logging is specifically designed to minimize erosion by allowing recapture of eroded nutrients in regenerating areas. Thus far, however, strip logging is not intensively practiced.

Economics of Deforestation

Why is it (or was it) in the interest of governments such as Brazil to encourage deforestation? The answer to this question is a bit complex and demands some understanding of global economics. During the decades of the 1970s and 1980s, when Amazonian deforestation greatly accelerated, part of the reason was due to changes in the price of oil on the global market. Oil prices greatly increased, resulting in immense amounts of money being transferred from oil-dependent countries to the various oil-producing countries (OPEC). The price of a gallon of gasoline rose dramatically. Where did most of the money paid to OPEC go? It essentially went back to U.S. and European banks, a process termed "petrodollar recycling." While the banks increased their holdings, aware that they obviously must pay interest to their investors, they were, of course, anxious to encourage borrowers. That's where tropical nations like Brazil enter the picture.

Developing countries were encouraged to borrow, and, with the price of oil so high, it made sense to borrow to develop domestic energy sources such as hydroelectric power to lessen dependency on foreign oil. Borrowing was also a way of financing agricultural modernization, with the goal of exporting crops to reduce debt load (Skole et al. 1994). Consequently, in the past two decades, total farmland in Brazil has increased more than 60% and the land in crops increased 176%, much of it financed by government and much of it large-scale, using modern machinery (Skole et al. 1994). Obviously, this increase in farmland came about with the loss of lowland forest and other tropical ecosystems.

The result of agricultural modernization has been, among other things, that Brazil became a leading exporter of soybeans, wheat, coffee, and oranges (Skole et al. 1994). Though cattle ranching largely has failed, other intensive agriculture programs have so far succeeded (see below). This success, however, is not uniformly shared throughout the population. For example, coffee is a labor-intensive crop, employing many people, whereas soybean production is facilitated principally by machinery and can result in loss of small farms and increased unemployment (Skole et al. 1994). As soybean production expands at the expense of coffee production, laborers lose jobs. In recent years Brazil has attempted to develop a more pro-environmental stance and as such has curtailed many of its former policies that encouraged such rampant deforestation; thus the rate of deforestation has declined in many areas (Bonalume 1991; Skole et al. 1994).

Effects of Deforestation: Forest Fragmentation

Brazilian law requires that each landowner maintain at least 50% of the land as forest (Holloway 1993). Given that most activities result in deforestation, the result of the Brazilian policy has been that landscapes have become highly fragmented, with varying-sized islands of forest remaining amid a landscape of pastures, croplands, and other anthropogenically created ecosystems. This pattern of landscape fragmentation should come as little surprise to inhabitants of North America, where a similar pattern prevails in many areas. Land development has often resulted in chopping forests into many increasingly small and isolated units. Forest fragmentation on their North American breeding grounds is thought to be contributing to current population declines of some species of Neotropical migrant songbirds (Askins 1995; Pimm and Askins 1995; and see page 291). Forest fragmentation, though common throughout the Neotropics, has been generally poorly studied, with one outstanding exception.

Approximately 70–90 km (44–56 mi) north of Manaus, Brazil, on *terra firme* forest, there is an ongoing study that was established specifically to evaluate the numerous ecological effects of forest fragmentation, with the objective of learning how best to structure biological preserves. The project was formerly known as the Minimum Critical Size of Ecosystems Project but has since been renamed the Biological Dynamics of Forest Fragments Project (BDFFP). The project was initiated in 1979 with the support of INPA, the Brazil National Institute for Research in Amazonia, and the World Wildlife Fund. In 1989 the administration of the project was assumed by the National Museum of Natural History at the Smithsonian Institution, working in cooperation with INPA. The study has involved twenty-five principal investigators working with such taxa as woody plants, birds, primates, bats, small nonflying mammals, ants, butterflies, euglossine bees, and scarab beetles. As would be expected, there are numerous publications from such a comprehensive study, but excellent general reviews can be found in Lovejoy et al. (1986), Lovejoy and Bierregaard (1990), and Bierregaard et al. (1992).

The essence of the BDFFP is to evaluate the ecological effects created by varying area size and degree of isolation of forest fragments. The impetus for the project developed from the reality that deforestation is a threat to Amazonian biodiversity coupled with an ambitious attempt to test a model called the theory of island biogeography (MacArthur and Wilson 1963, 1967). Island biogeography theory describes a hypothesized mathematical relationship between the biodiversity of an island and the island's size (total area) and distance from the colonization source (mainland). The model predicts that all islands eventually reach an equilibrium with regard to species richness (where immigration rate equals extinction rate) and that large islands close to the mainland will have a higher species richness at equilibrium than small islands more distant from the mainland. Further, and much more important, the difference in species richness among various islands is *predictable* from the model. The power of the model is that it is quantitative and therefore testable. Since fragmented forests are, in a sense, islands, it was soon recognized that the model had potential conservation implications with regard to the structuring of nature preserves and parks (Diamond 1976; Simberloff and Abele 1976). The Minimal Critical Size of Ecosystems project was structured in essence as both a test and an application of the theory of island biogeography. The objective was to learn how best to design preserves (analogous to islands) amid an area of deforestation such that biodiversity was maintained at as maximum an equilibrium level as possible. The original name of the project reflects the objective of learning the minimum area that must be preserved to maintain reasonably normal biodiversity.

The researchers worked with cattle ranchers in designing the project (who, you will recall, were required by Brazilian law to leave 50% of their lands forested). The ranchers were persuaded to clear forest in such a way as to create forest fragments of different size and distance from an undisturbed, protected 1,000-ha forest area that served as a control (analogous to the "mainland" or source). Fragments varied in area as follows: 1 ha (5 fragments), 10 ha (2 fragments), 100 ha (2 fragments), 200 ha (1 fragment). Fragments were separated by varying distances (100 m, 250 m, 500–700 m) from the 1,000-ha control forest. What, for instance, would be the differences in biodiversity between two 10-ha plots, one of which was 500 m from the source and one of which was 100 m from it? What are the differences between a 1-ha and a 10-ha plot, both of which are 200 m from the source forest? Are understory bird species more sensitive to area effects than canopy species? Which species of monkeys are most sensitive to area effects and isolation? These are the kinds of questions that were posed.

The results of the BDFFP have been insightful, and the study is still ongoing. Various researchers have shown that virtually all taxa studied are sensitive in varying ways to both area and distance effects. For example, immediately after isolation, isolated fragments experienced an influx of understory birds (presumably immigrating from the surrounding deforested area), but after about 200 days, the total number of birds dropped to well below what it was before the forest fragment was isolated. In other words, biodiversity plummeted, what Bierregaard and Lovejoy describe as a "faunal collapse." Army ant–following birds (page 39) were particularly negatively affected. As another, very different example, euglossine bees, which are important long-distance pollinators (page 128), were apparently reluctant to fly into fragments isolated by 80 m or more from other forest. Primate diversity was also strongly area dependent. The number of primate species in four 10-ha fragments combined was less than the number in a single 100-ha fragment. Not all effects of fragmentation were manifested in population reductions. For example, small nonflying mammals actually increased in species richness, biomass, and abundance in 1-ha fragments compared with 10-ha fragments and continuous forest. In areas where forest was cut but burning was either not done at all or done very lightly, rate of succession was faster, and the dense vegetation supported a larger community of butterflies than either isolated forest fragments or continuous forest.

Some fragmentation effects were subtle but will come as little surprise to those acquainted with the complexity of ecological interdependencies that characterize virtually all ecosystems. Species of carrion and dung beetles declined in small fragments, presumably from the lack of excrement and carcasses, which disappeared as larger animals migrated out (Klein 1989). Three species of *Phyllomedusa* frogs were lost from small fragments, probably because peccaries had abandoned these areas and peccary wallows formed the breeding pools for these small frogs. With the peccaries gone, the frogs could not breed (Wilson 1992). Fragmentation thus produces a kind of ecological domino effect, with first-, second-, and third-order effects.

Edge effects were also (as expected) pronounced in isolated fragments. Many injured and dead trees were found along edges, and the overall turnover rate of trees was highest on edges. Rates of litter accumulation accelerated near edges. Seedling recruitment patterns varied as well.

There are obvious lessons to be learned from the BDFFP with regard to conservation of biodiversity. Isolation is problematic. Small, isolated forest fragments are ecologically depauperate and function differently from normal continuous forest. This should come as little surprise given that so many species of Neotropical trees, for instance, are dependent on long-distance pollinators or seed dispersers or both. A tree isolated in a small fragment well over 100 m from other forest could, in effect, be made sterile for want of pollinators or seed dispersers. But isolation, the creation of forest fragments, is exactly what is happening throughout Brazil and other Neotropical countries. In general, fragmentation will exert its most severe impact on those species that require a large area but are reluctant to cross small gaps. There are many such species in the Neotropics. The small frog *Chiasmocleis shudikarensis* requires only 0.0001 ha (.00025 acre) of forest but, because the species is very patchy in distribution, it is estimated that it requires a total of 500 ha (1,236 acres) as the minimum area to sustain a viable population (Dale et al. 1994). Any forest fragment of less than 500 ha will not sustain this species.

One possible way to mitigate the effects of fragmentation is to connect isolated fragments by corridors of uncut forest. Instead of islands of forest amid a sea of pasture, create islands of pasture amid a sea of forest—or at least a grid of forest "canals," so that forest-sensitive species can maintain their normal foraging ranges as much as possible.

Even an uncut forest of 1,000 ha (2,471 acres) may not be sufficient to meet the ecological requirements of certain species. Terborgh (1974) argues that national parks must have between 100 thousand and a million hectares to assure that maximum biodiversity is maintained. There is an ongoing debate among some conservation biologists, nicknamed the SLOSS debate (Quinn and Hastings 1987). Translated, SLOSS refers to "single large area or several small areas" (which would be of equal total area to one large reserve). Some ecologists strongly favor protection for huge tracts of land, arguing that such vast areas are necessary to ensure maximum species richness at all trophic levels. Other ecologists favor positioning smaller reserves in crucial areas, where certain species or certain unique ecosystems are most at risk, or where corridors connect small reserves. There is, as yet, no consensus on how best to preserve biodiversity. What is a consensus is that biodiversity is in decline. I will revisit the question of how to prioritize areas for conservation later in the chapter.

Effects of Deforestation: Loss of Biodiversity

No one knows exactly how many species of plants, animals, and microbes inhabit Amazonia, but there is little disagreement that the total Amazonian biodiversity probably exceeds that of any other ecosystem on the planet. And realize that biodiversity encompasses more than lowland rainforest habitat. It also includes such ecosystems as savannas and dry forests. Although all tropical rainforests exhibit high species richness, Amazonia, probably because of its vastness, achieves a total species richness that is collectively thought to be higher than that of virtually all other tropical forest regions. South America, including the Atlantic Coastal Forest and areas west of the Andes as well as central Amazonia, contains twenty-six areas of uniquely high diversity and endemism (Prance 1982a). Though endemism is noteworthy in rainforest, even higher levels of endemism, at least for birds, are found in dry forest, savanna, and grasslands (Stotz et al. 1996). Amazonian patterns of species richness, in general higher than in Central America, also show substantially more rarity and more localized populations than those of ecosystems in Central America or other tropical regions.

As discussed in chapter 4, biodiversity is generated by speciation and reduced by extinction. If speciation exceeds extinction, biodiversity increases. If extinction exceeds speciation, biodiversity declines. At the present time, biodiversity is presumably declining because habitat loss from deforestation is causing numerous local extinctions. The extent of these extinctions is unknown, though Wilson (1988) suggests that globally the extinction rate is anywhere from 1,000 to 10,000% above the natural rate of extinction. But no one knows for sure, and estimates certainly vary (Lugo 1988, and below). Myers (1988) suggests that if deforestation in Amazonia were to reach the point that only those areas now designated as protected reserves of some kind were to remain, about 66% of plant species and 69% of bird species known from the region would become extinct. Myers (1986) asserts that extinction rates (due to deforestation) are higher in tropical rainforests now than in all other ecosystems combined, resulting in what he calls a "mega-extinction spasm." Raven (1988), supporting Myers' view, argues that deforestation has resulted in a current global extinction rate (mostly caused by tropical deforestation) that rivals what occurred at the end of the Mesozoic Era, 65 million years ago, when the dinosaurs (and many other animals) became extinct.

It is very difficult, however, to estimate rates of species loss. As stated above, published estimates are certainly varied and often rather vague as to their database and methodology (Lugo 1988). On the more conservative side, one estimate suggested that if current rates of deforestation continue, within a century (from 1984, when the study was published), 12% of Amazonian bird species and 15% of Neotropical plant species would become extinct (Simberloff 1984). This represents a loss of about 84 bird species and nearly 14,000 plant species. Other estimates are more extreme. Estimates for rates of species loss from all tropical forests, cited by Lugo (1988), range from loss of a half-million species to several million by the year 2000, or, in percentages, loss of from 20% to 50% of all species throughout the global tropics. But these are only estimates. No one really knows, nor is there any reliable way to predict. Granted, the guesses may be "educated guesses," but thus far prediction of extinction patterns is, indeed, a guessing game. Bear in mind that there are likely many more species (for such groups as insects, for instance) in tropical rainforests than have thus far been discovered and described (Erwin 1988; May 1988, 1992). This, of course, makes it even more difficult to predict extinction patterns accurately.

Many Amazonian species may be particularly susceptible to extinction because they have relatively limited ranges, and, even within those ranges, many species tend to be rare, present in low population density. Rare species with limited ranges are obviously at risk of extinction from habitat loss. For example, at the Rio Palenque Biological Station in Ecuador, an area of only 0.8 km^2 (0.3 sq mi), there are 1,033 species of plants, and about one in every four is only known from coastal Ecuador, a relatively limited area (Gentry 1986; Wilson 1988). Thus the loss of any particular tract of rainforest such as Rio Palenque may result in the elimination of some of these range-limited species.

Neotropical forests have a much higher proportion of endemic species than are typically found in temperate forests. It is estimated that about 20% of the plant species in the Choco Department of Colombia are endemic, and within certain genera that figure rises to as high as 70% (Gentry 1986). Endemic species are also commonly encountered on Andean slopes, areas where deforestation has been very active. For example, Gentry estimated that almost every cloud forest in Panama has at least some endemic plant species. Related to endemism is the tendency for extreme habitat specialization in Neotropical forests. Gentry noted that four very similar species of *Passiflora* (page 155) occurred in the Iquitos area, each restricted to a different substrate: one on terra firme, one on seasonally inundated floodplain forest, one on white sand forest, and one on noninundated rich alluvial soil. The combination of high endemism and extreme specialization makes many species obviously susceptible to extinction by loss of habitat.

Patterns of endemism in birds have been well documented, and, perhaps surprisingly, humid forests have proportionally fewer endemic species than ecosystems such as dry forest and grassland. For example, 90% of all bird species restricted to dry forests are local endemics, and about 80% of grassland bird species are endemics. Endemism level among bird species of humid forests is 45% (Stolz et al. 1996).

Gentry (1986) recognized two kinds of endemism, both of which are significant in the Neotropics. One kind, paleoendemism, describes a species with a restricted range today but that had a much wider range in the past. For example, in California the giant sequoia tree (*Sequoiadendron gigantea*), now restricted to about seventy-five small, scattered groves on the western side of the Sierra Nevada, ranged widely in North America during the Cenozoic. Today it is a California paleoendemic. In the Neotropics there are many examples of paleoendemic species, especially in such places as the tepuis (page 225) of the Guayana Highlands. Another form of endemism, pseudoendemism or anthropogenic endemism, occurs when human activities have so reduced suitable habitat that a once widespread species is severely fragmented and reduced in range to being extremely local, sometimes only a single individual. Anthropogenic endemism is well documented for the region around Rio Pelenque in Ecuador. As an extreme example, the tree *Persea theobromifolia*, once a widespread species in western Ecuador, was overharvested for timber as well as suffered from loss of habitat to the extent that its total population is now less than a dozen trees (Gentry 1986).

In trying to ascertain the probable patterns of extinction resulting from deforestation, it is instructive to examine the Atlantic Coastal Forest of Brazil. This area has been deforested by at least 88%, and estimates are that only 3% of the original vegetation will remain by the year 2000 (Brown and Brown 1992). The Atlantic Coastal Forest is a region of high endemism among many taxonomic groups and thus would seem a region where extinction from habitat loss ought to be distressingly common. Indeed, some of South America's most endangered species occur here: the golden lion tamarin (*Leontopithecus chrysomelas*) is one of thirteen endangered primates; the maned sloth (*Bradypus torquatus*) is confined to the same forests as the tamarin; the black-throated piping-guan (*Pipile jacutinga*) is faced with extreme habitat loss and undue hunting pressure. But these species, while clearly endangered, are not yet extinct. A study by the Brazilian Society of Zoology listed 171 animal species from the Atlantic Coastal Forest considered to be vulnerable, endangered, or extinct. Only 6 of the 171 (2 birds and 4 butterflies) were considered to be actually extinct (Brown and Brown 1992). Given the immense regional loss from deforestation, it seems surprising that extinction has, as yet, been relatively low (though species that are vulnerable or endangered are certainly not assured an optimistic future).

However, Brown and Brown (1992) suggest three compelling reasons why extinction rates have been less than might be expected. First, the topography of the region is quite complex, making the region physically very heterogeneous, a patchwork of microhabitats. The suggestion is that most species have historically lived in small, isolated populations and are thus adapted to remain so. Second, the area is subject to a high degree of natural disturbance (from heavy rains, cold spells, varying seasonality), again resulting in possible species-specific adaptive responses to such unpredictable and potentially catastrophic events. Third, there may be high levels of adaptive plasticity among endemic plants and animals, allowing rapid adaptive responses to sudden perturbations. In other words, organisms that live in variable environments adapt to such realities, thus buffering them against extinction.

Not everyone agrees that species loss from the Atlantic Coastal Forest has, as yet, been as minimal as was suggested above. Myers (1986) suggests that the actual number of extinctions from that region is unknown but that "the total must surely be many thousands." Myers cites no data to support that contention, however.

Susceptibility to extinction varies among species, and there are certainly "extinction-prone species" (Terborgh 1974). It has been well documented that forty-five bird species have become extinct from Barro Colorado Island in Panama since the island was initially created by flooding when the Panama Canal was built (Willis 1974). Some of these species were typically found only in early successional ecosystems, and, with regrowth of forest at BCI, their habitats literally disappeared. However, eighteen interior forest species (whose habitat has remained intact) also have been lost from the island, and others continue to decline. Most of the now extinct species are either large in size, ground nesters, or ground foragers. It is not clear why some of these species have disappeared from the island, but Terborgh (1974) has suggested some characteristics that may typify species most prone to extinction. These include species that are large in body size and on the top of the food web, such as harpy eagle and jaguar. These animals require large areas in order to survive, and their populations cannot be sustained in fragmented areas such as BCI. But Terborgh also notes that widespread species with poor dispersal and colonizing abilities are also likely to be extinction prone, as are endemics. The Neotropics, whether it be lowland rainforest, campo cerrado, or savanna grassland, abounds with such species. As noted earlier (page 109), there are many bird species that apparently will not routinely cross rivers, even when the other side is clearly in view! Species with colonial nesting habits are also sensitive to habitat loss and thus potentially extinction prone. An obvious example is the oilbird (page 140), which nests in large colonies in widely scattered caves, but many bat species could fit this description as well. Finally, migratory species are potentially extinction prone because of their dependency on several habitat types that are widely separated. Evidence is accumulating that some species of Neotropical migrant birds are in serious decline (page 291).

Ecologists recognize that in most ecosystems there are certain species that are extraordinarily important resources for many others, so-called keystone species. Terborgh (1986) suggests that the abundance of fruit-dependent large animals in rainforests makes certain kinds of fruiting plants keystone species. For example, about three-fourths of the biomass of birds and mammals in Cocha Cashu rainforest in Peru is from animals that are in large part frugivorous. Most of these creatures are dependent on palm nuts, figs, and various miscellaneous nonfig fruits. Indeed, Terborgh suggests that without figs in the ecosystem, "one could expect to see it collapse." Combining nectarivorous species with frugivores, Terborgh concludes that at Cocha Cashu, only 12 plant species (out of a total of 2,000) "sustain nearly the entire frugivore community for three months of the year." The lesson here is obviously that the loss of certain plant species, such as figs, would have much greater impact than the loss of species upon which most animals are less dependent. And bear in mind that the argument works both ways. The loss of a keystone pollinator or fruit disperser could bring about the loss of a keystone plant species.

The study of biodiversity and its importance ecologically and pragmatically is still emerging as a branch of conservation biology and ecology. Ehrlich and

Wilson (1991) offer three broad reasons why there should be concern for loss of biodiversity (and efforts to prevent it):

1. Biodiversity provides aesthetic satisfaction, and it can be argued that humans should be ethically bound to provide protection and stewardship for the rest of nature.

2. It should be obvious that the richness of biodiversity provides humans with many goods of direct economic benefit (foods, medicines, and industrial products), and there is future potential for even more pragmatic usage.

3. Natural ecosystems, and thus biodiversity, provide humans with many essential services having to do with regulation of the atmosphere and climate, recycling of materials, and maintenance of soil fertility, to name but a few.

It is probably global deforestation and the plight of tropical rainforests (as well as other tropical ecosystems) throughout the world (and perhaps most particularly in Amazonia) that has brought the issue of biodiversity to the attention of the general public and has also made it much more sharply focused as a topic of scientific investigation. In 1992 the United Nations Conference on Environment and Development was held in Brazil, in the shadow of the world's largest rainforest area, and protection of biodiversity in light of the need for increased sustainable development was widely discussed and debated. And it will continue to be and should be. For more on biodiversity, see Wilson (1988, 1992). For a thoughtful introduction into ethical and other issues of biodiversity, see Norton (1987).

Effects of Deforestation: Potential Alteration of Regional and Global Climate

Approximately 336 million tons of carbon emissions enter the atmosphere each year from deforestation in Brazil, making Brazil the fourth major contributor of carbon to the atmosphere (Moran et al. 1994). For each hectare of forest that is cleared and burned, about 220 tons of carbon are released into the atmosphere (Holloway 1993). This is not an inconsequential addition. Carbon is released primarily in the form of carbon dioxide with the burning of fallen vegetation from cleared forest. Thus deforestation contributes directly to raising CO_2 levels in the atmosphere. (It also acts indirectly, because termites heavily colonize pastures, and their metabolic activities, as well as those of bovines, add CO_2 as well as methane in significant amounts.) In addition, the cleared forest is no longer there to act as a sink for CO_2, to take it up in the process of photosynthesis. Ecosystems such as crops and pastures that replace rainforest are far less productive and thus take up far less CO_2 than rainforest. Thus much of the CO_2 liberated from fallen rainforests tends to accumulate, raising its global concentration in the atmosphere.

Carbon dioxide is, of course, the principal compound implicated in greenhouse effect (water vapor, methane, and carbon monoxide also contribute, but to lesser degrees). Greenhouse effect is so called from the analogy with the action of glass in a greenhouse. Light passes through the glass, some of it becomes longer wavelength heat, and the glass then blocks the heat from escaping, thus warming and keeping warm the greenhouse. Carbon dioxide molecules collectively act somewhat like glass on a greenhouse to block heat waves from readily escaping the atmosphere, thus CO_2 helps to insulate the planet, greatly moderating temperature fluctuations. Greenhouse effect, in moderation, is an essential component of Earth's conduciveness to life. However, greenhouse effect can potentially result in dramatic warming of the planet, far beyond its current levels. Runaway greenhouse effect, such as is apparently the case on the planet Venus, can result in surface temperatures far beyond what living systems could endure. Because heat is a critical component of climate, greenhouse effect is intimately related to patterns of global climate.

Since 1800, the early onset of the industrial revolution, carbon dioxide from fossil fuels has been added to the atmosphere, steadily raising CO_2 levels from around 280 parts per million (ppm) to the present level of about 355 ppm (Vitousek 1994). In recent years, the extensive deforestation occurring not only in the Neotropics but in other areas as well has added yet another significant source of CO_2 to the atmosphere (Woodwell et al. 1983; Detwiler and Hall 1988; Houghton and Woodwell 1989; Post et al. 1990). Of course, not all CO_2 released into the atmosphere remains there. Some studies suggest that nearly half ends up in surface waters, shallow-water sediments, and the deep oceans (Stuiver 1978). But no one suggests that carbon dioxide has not, in fact, been accumulating at nontrivial amounts in Earth's atmosphere and does not continue to do so.

As carbon dioxide accumulates, many climatologists predict that global heat patterns and climate will change (Intergovernmental Panel on Climate Change 1990). One study suggests that the average global temperature has increased from .5 to .7°C (.9–1.3°F) since 1860, with the increase accelerating. It should be sobering to realize that the six warmest years on record (up to 1988) were 1988, 1987, 1983, 1981, 1980, and 1986. Predictions about changing patterns of precipitation, temperature, soil moisture, and the distribution of snow and sea ice vary among studies, and no one really knows with any degree of certainty what increasing greenhouse effect portends. But consider, for example, that as temperature increases, it will speed respiratory rates for both plants and microbes. Plants will thus expire more carbon dioxide, as will microbes, thus further adding carbon dioxide to the atmosphere (Houghton and Woodwell 1989). For good general reviews of the global carbon cycle, see Houghton and Woodwell (1989), Post et al. (1990), and Vitousek (1994).

The immensity of the Amazon Basin, largely dominated as it is by rainforest, assures that the area has a profound effect on regional climate. It has been estimated that the Amazonian flora is responsible for the generation of 50% of the total regional rainfall (Salati and Vose 1984; Holloway 1993). The dark, green rainforest absorbs a great deal of solar radiation, returning most of it as heat by evapotranspiration. There is relatively little reflectivity. Transpiration activities of the vegetation continually recycle water vapor to the atmosphere, making the vegetation itself largely responsible for the patterns of regional precipitation. And tropical vegetation is obviously dependent on maintenance of high rates of precipitation. What would be the climatic effect, both regionally and globally, if the rainforest were to be entirely replaced by grassland pasture? Complex computer models have been used to investigate this question.

One study (Shukla et al. 1990) examined a worst-case situation in which all Amazonian rainforest was replaced by grassy pasture. Though such a large-scale, total loss of Amazonian rainforest is unlikely, the authors of the study point out that if deforestation were to continue at rates typical of the 1980s, all Amazonian rainforest would, indeed, be gone within a century at most. The computer model coupled atmospheric and biospheric parameters and demonstrated that loss of rainforest would result in significant increase in surface temperature and a decrease in evapotranspiration and precipitation over Amazonia. These results, in and of themselves, were unsurprising. It stands to reason that pasture surface and soil temperatures should be higher than rainforest, and, indeed, such measurements have been taken and such proves to be the case. In general, deforested areas are from 1° to 3°C (1.8–5.4°F) higher in surface temperature than forests. In addition, absorbed solar radiation is less in pastures than in forests due to the higher reflectivity, or albedo, of pastures, though long wave heat radiation reflected from the ground is increased, which results in reduced storage capacity for soil moisture, reduced transpiration, less interception of precipitation, thus less evapotranspiration. The computer model predicted a reduction in calculated annual precipitation of 642 mm and in evapotranspiration of 496 mm, significant reductions. The implications of this magnitude of reduction are sobering. The dry season would increase in length, and, most important, the longer dry season may prevent any reestablishment of rainforest. Thus, the model suggests that irrespective of the biological problems inherent in reestablishing rainforest after such large-scale deforestation, the loss of rainforest would be essentially irreversible due to climatic changes alone.

Another computer-model study simulating the replacement of Amazon rainforest and savanna by pasture gave similar results, indicating an increase in temperature and decrease in precipitation by up to 20% (Lean and Warrilow 1989). However, the authors noted that prediction of regional climate change from effects of deforestation is highly dependent on the assumptions and formulations of the computer models.

It is very difficult to ascertain just how anthropogenic changes will alter such complex variables as climate. For example, deforestation has also been blamed for an apparent increase in the discharge rate of the Amazon River. Gentry and Lopez-Parodi (1980) reported that the height of the annual flood crest at Iquitos, Peru, has increased dramatically over the decade from 1970 to 1980. The increase correlated with increased deforestation in the region; thus the authors suggested that deforestation was responsible for the added discharge of the river. But correlations are nothing more than that. They do not prove cause and effect. Nordin and Meade (1982) and Meade (pers. comm. 1990) strongly disputed the claim that deforestation was causing the increased flood crest, arguing in part that the period of time was much too short to exclude natural cyclic flux, which would have nothing to do with deforestation. Richey et al. (1989) support Nordin and Meade, showing that an eighty-three-year record from 1903 to 1985 indicated no statistically significant change in discharge rate. Further, a two-to-three-year cycle was evident in the data set, creating oscillations in discharge that occur much earlier in the century than deforestation but appear identical to those more recent changes attributed to deforestation.

Alternatives to Deforestation

Brazil tends to dominate the literature about Neotropical deforestation and loss of biodiversity because Brazil both is an immense country (with a total area greater than half of South America) and has a vast rainforest (57% of Brazil is in the Amazon Basin). Indeed, the Brazilian Amazon Basin is estimated to contain about one-third of the remaining tropical moist forest on Earth (Myers 1991). Brazil is also a country in which economic concerns loom particularly large, resulting in annual deforestation rates of 1–2 million hectares (Skole and Tucker 1993). Brazil's huge international debt is reported to be 121 billion dollars (Holloway 1993), and its population continues to grow at about 2.5% annually. By the year 2025, it is estimated that Brazil will have a human population of 246 million (Holloway 1993). Today, more than two of every three Brazilians are urban, and fully half of the population is less than twenty years of age (which, of course, portends the predicted high rates of population growth in the future). Distribution of wealth within the vast country is anything but even. Large enclaves of extreme poverty can be found in both urban and rural areas (there are about 12 million *abandonados*, children without parents or homes). About 4.5% of the landowners control 81% of the farmland (Holloway 1993). But Brazil is not unique. Other Latin American countries face similar demographic realities. If rainforest conservation is to happen as the future unfolds, it will be necessary to confront the challenges imposed by population increase and economic needs. When so confronted, conservation goals often seem threatened. Need that be the case?

The answer must be and is no. As concerns over deforestation and subsequent loss of biodiversity have magnified in recent years, and as activities such as large-scale cattle ranching have proven generally unsuccessful, various alternatives have been suggested, some of which are promising (Anderson 1990). In combination, these innovations may succeed in significantly attenuating the trend toward loss of biodiversity.

Extractive Reserves

Given the high biodiversity of Neotropical rainforests and the fact that indigenous people are generally fluent in utilizing an impressive array of forest resources for pragmatic purposes (chapters 6 and 7), it seems possible to view the rainforest as a renewable, sustainable resource from which various products can be extracted on essentially a continuous basis. If this is the case, then preservation of large areas of rainforest would make economic sense as well as serve the interests of conservation and preservation of biodiversity. Thus, the establishment of extractive reserves has been suggested as an alternative to deforestation.

The term "extractive," as implied above, describes any renewable component of rainforest biodiversity that has economic value and can be used relatively continuously without significantly altering the ecosystem. In the most general sense, extractives refer to such usage categories as food, spices, fodder, palm oils, fiber, floriculture, fuel, medicinals, and biochemicals such as antioxidants, enzymes, vitamins, sweeteners, even chlorophyll (Duke 1992). Brack (1992) has suggested an analogy between biodiversity and the legend of El Dorado, the mythical land of gold and riches (page 199). Summarizing the many and diverse uses of rainforest flora and fauna, Brack claims that "both the conquistadors and the proponents of development failed to recognize that the true wealth of the area [Amazonian rainforest], the true El Dorado, is in the forests and the waters." True wealth? Perhaps. The fundamental point regarding extractivism centers on determining the economic value of bio-diversity (Gentry 1992). And the fundamental question is this: is a hectare of rainforest more valuable if left uncut but used for its extractable (and presumably sustainable) biological resources, or is it more economically sound (in the long as well as short run) to convert the rainforest to agriculture, pasture, or some other use? The answer, like the ecology of the rainforest itself, is complicated.

One way to approach the economics of extractive reserves is to select an area of rainforest, preferably one of the most species-rich in Amazonia, and survey that area for extractive potential, applying an economic market analysis to calculate the net present value (NPV) of a hectare of rainforest. Such an area is typified by the lowland rainforest near Iquitos, Peru, in which Gentry (1988) documented 289 species among a total of 858 trees and lianas of diameters greater than 10 cm (3.9 in). Gentry and two colleagues (Peters et al. 1989; Gentry 1992) calculated the economic yield from 72 tree species and 350 individual trees, in terms of products that could be readily sold in the Iquitos markets. Marketable products included rubber (1 species), timber (60 species), and edible fruits (11 species). Factoring in the need to permit regeneration of species whose products were extracted, the NPV of one hectare was calculated at $6,330. This figure is impressive given that the study also indicated that the same land would have an NPV of only $2,960 if given over to cattle ranching, and an NPV of $3,184 if devoted entirely to managed plantation. It would seem that extractive use would, indeed, be the wisest use, but the study cited above has been criticized as unrealistic in several respects, with the probable economic returns exaggerated (Browder 1992). We need to look more closely.

The concept of extractive reserves was initiated in Brazil by the National Council of Rubber Tappers and the Rural Workers' Union (Fearnside 1989, Allegretti 1990). Worldwide attention was focused on the Brazilian state of Acre when, on December 22, 1988, Francisco Alves (Chico) Mendes Filho was murdered by local cattle ranchers. Chico Mendes had by that time become somewhat of an international celebrity as a spokesperson for Brazilian rubber tappers. Conservationists, anxious to find economic and social arguments that favored rainforest preservation, were quick to embrace his cause. Mendes' murder gained a great deal of international sympathy for the plight of the rubber tappers, families whose livelihood is dependent on rainforest, and who were being forced out by cattle ranchers cutting vast tracts of rainforest. Mendes had helped facilitate a political process resulting in the establishment of the first two extractive reserves in Acre in February 1988. Conservationists applauded the creation of extractive reserves, hailing them as a useful means of rainforest preservation. Today Brazil has fourteen extractive reserves that cover a total of 3 million hectares (Holloway 1993). A census conducted in 1980 indicated, however, that there are only about 68,000 rubber-tapper families in the Brazilian Amazon. Because rubber tapping requires a large tract of forest (rubber trees are widely scattered), human population density must be low (1.0–1.7 persons/km^2), only one family per 300–500 ha (Fearnside 1989). It is therefore obvious that rubber tapping will support but a small population, which would seem to limit significantly its potential for large-scale forest preservation in a country that faces significant pressures from population growth. And there are other limitations, both economic and ecological.

Rubber is produced most economically in Asia, not South America (page 187). For rubber tapping to be economical in Brazil, it was necessary for the government to establish a tariff policy that fixed the domestic price of rubber at approximately three times the average world price (Fearnside 1989; Browder 1992). This tariff policy has been abandoned, causing a significant drop in rubber prices in Brazil, which has obvious negative impact on rubber tappers. The result is that many rubber tappers are cutting forest in order to establish farms (Browder 1992). In an analysis of three detailed studies of rubber-tapper communities, two from Brazil, one from Bolivia, Browder (1992) suggests that rubber tapping is at best a marginal way of making a living, with the practitioners among the poorest and most illiterate in the general population. He further suggests that, because of the economics of rubber prices, rubber-tapper households are often unstable, requiring frequent nomadlike moves to a different area; that rubber-tapper families frequently deplete local game and cut forest to augment their income through agriculture; and that rubber tappers are typically in heavy debt to landowners, bosses, and middlemen. Some rubber tappers actually work part of the year for logging companies. Finally, many have serious chronic health problems related to the lack of overall medical care. These sobering considerations make rubber tapping seem rather far removed from anything resembling El Dorado.

But rainforest can be used for more than rubber extraction. Recall that the initial definition of extractive reserve is an area where perhaps *many* useful products can be harvested, not merely one. Fearnside (1989) lists twenty-seven plant species that are currently collected by extractivists throughout Amazonia. Recall from chapters 6 and 7 that there is considerable potential for the discovery of new medicines (Farnsworth 1988) as well as for new agricultural and industrial products (Plotkin 1988).

Some areas are likely to be more conducive to extractivism than others. In a study comparing nontimber forest product extraction in Peten, Guatemala, with West Kalimantan, Indonesia, it was clear that the probability of successful, sustainable extractivism in Peten was considerably higher than in West Kalimantan (Salafsky et al. 1993). In Peten, three products—chicle (latex from the chicle tree, page 25), xate (fronds from a common palm, *Chamaedorea* spp.), and allspice (fruits from *Pimenta dioica* trees)—are each effectively extracted on a sustainable and profitable basis. One important reason why extractivism is successful in the Peten forest is that the species richness of trees in Central America is less than that of Amazonia or West Kalimantan. This means that there are proportionally more of each of the three species available for harvest, resulting in less search time, less travel time, and less carrying time (thus more economical harvesting). In addition, the annual time of harvesting is staggered among the three species: chicle is collected only during rainy season from August to January, allspice fruit is fully ripened only in July and August, and xate, though it can be harvested anytime during the year, is best collected during the time of peak demand, March through June. Thus extractivists have products available for convenient harvest essentially throughout the year, an ideal system, and one that was well understood by the ancient Mayans. In West Kalimantan, however, it was concluded that "prevailing ecological and socio-economic conditions make it unlikely that extractive reserves will play a major part in saving the rain forest" (Salafsky et al. 1993).

Though the creation of extractive reserves has value as conservation strategy for sustaining some rainforest, it should be clear that the concept is not a universal solution to deforestation. Browder (1992) identifies several important limitations. For extractivism to be successful, the product must have a market, be profitable to extract, and the work of extraction must be done at a level that assures the species will remain viable in that tract of forest. While it is encouraging to search for new medicines and other products, these commodities obviously must be found and shown to have utility before they can be extracted. Much surveying and research remains to be done. More to the point, however, extractors are not necessarily defenders of rainforests. Many in Amazonia combine extraction with logging, gold mining, raising livestock, and agriculture, to say nothing of depleting local populations of large animals through continuous hunting. Unfortunately, if the prices of extracted products begin to rise sharply, extractors are likely to increase their efforts only to benefit profitwise, an action that could result in overharvesting. On the other hand, if prices drop, the same result can occur because the extractor will need to take more to make up the profit loss. What this means is that extractivism depends on a very delicate economic balance, easily upset. And, to reiterate and broaden what was said earlier, all studies of extractive reserves thus far, whether rubber, Brazil nuts, or chicle, have concluded that large areas of forest are required and that such areas at best will only support low-density human populations.

Reclamation of Degraded Ecosystems

The science of restoration ecology is relatively new, a branch of applied ecology. The journal *Restoration Ecology*, established in 1993, publishes studies that focus entirely on some aspect of directed recoveries of ecologically damaged lands, using applied principles of general ecology. In this journal Brown and Lugo (1994) have presented a comprehensive overview of how ecologically damaged tropical lands could be rehabilitated and restored to sustainable use, as well as reviewed several examples of how such efforts have already met with success. One of their examples is a discussion of Janzen's efforts in the restoration of the dry forest in Guanacaste in northwestern Costa Rica and the establishment of Guanacaste National Park. Another of their examples is the rehabilitation of degenerated Amazonian pastures. Let's have a brief look at both examples.

Guanacaste forest is a tropical seasonal dry forest. As such it differs ecologically in many significant ways from moist forests. Janzen (1988a) points out that tropical dry forests have been so extensively cut and converted to anthropogenic use that they are among the most endangered Neotropical ecosystems, a point also strongly supported by Stotz et al. (1996). Janzen (1988b) took it largely upon himself to organize a comprehensive restoration effort directed toward the eventual establishment of Guanacaste National Park (McLarney 1988). When Janzen began his efforts, most of the area was deforested and had, for four centuries, been subjected to essentially continuous intensive human usage: burning, overcutting, overgrazing, and farming. Making use of scattered, remnant patches of dry forest, Janzen directed a well-conceived plan to restore the forest to as close to its original state as possible. These efforts involved exclusion of human-created fires, a full ban on hunting, and a labor-intensive program to recolonize the forest by hand-seeding and planting operations. The ultimate success of Janzen's plan rested on the need to convince the surrounding community of both the economic and the cultural benefits of the Guanacaste project, and, in a pragmatic way, to include members of the community in the employment opportunities that accompanied the project (Janzen 1988b; Brown and Lugo 1994).

Given that up to 12% of Amazonian forest has already been cleared, and given that numerous cattle and agricultural ventures that followed deforestation have been abandoned, the potential exists to reclaim the abandoned pastures for sustainable economic use. Recall from the discussion above (and see chapter 3) that the rate of succession has been rapid on both abandoned agricultural and pasture lands. Though soil is relatively poor throughout the region (chapter 3), it is still possible that cleared lands may be reclaimed and restored to productive use, not only in Amazonia, but throughout much of the Neotropics. If such activities were to succeed economically as well as ecologically, the rate of subsequent forest clearance could be slowed (Holloway 1993).

However, before considering if deforested land can be successfully reclaimed, the question looms as to whether or not such lands could support *sustained* agriculture. There are numerous potential difficulties awaiting anyone who attempts to farm the tropics using methods more appropriate for higher-fertility, more temperate soils (Janzen 1972, 1973, and above). This is the obvious reason why the traditional slash-and-burn, rotating method of farming has prevailed. Nonetheless, sustainable, high-yield farming is potentially possible in much of Amazonia. Farming will, of course, support a much greater human population than extractivism.

Nicholaides et al. (1985) review a study conducted on terra firme near Yurimaguas, Peru, in western Amazonia, which was designed to test the potential for continuous cropping. Up to twenty-five consecutive crops were planted and harvested, rotating among rice-peanuts-soybeans or corn-peanuts-corn. The results showed that without the careful application of fertilizer and lime, crop yields deteriorated rapidly. Continuous cropping was unsustainable. But with proper amounts of fertilizer and lime, sustainable, rotating crop production was achieved and was comparable to that on similar soils in more temperate climates. The key here is intensive and careful management. For this system to succeed, regional farmers would have to have the means of affording it, have the necessary knowledge to utilize it, and, of course, accept substitution of the new methodology in place of traditional shifting agriculture. Recall that all of the above proved quite difficult in the Brazilian attempt to farm the Amazon described earlier in the chapter. But in the Yurimaguas study, demonstrations of the success of the intensive management system were provided to local farmers, and economic analysis indicated that profits to the farmers would exceed their additional expenses in utilizing the technology.

Rondonia, in southwestern Brazil, has been subjected to significant deforestation caused mostly by an influx of immigrants, many of whom cleared forest on terra firme to create farms. The colonists employ the usual pattern of slash-and-burn farming (chapter 7), and soil quality continually decreases until the land must be abandoned, usually within six to eight years. However, this need not necessarily be the case. In a study that involved interviews with eighty-seven farmers, three of the farmers demonstrated that the land could be utilized much more efficiently than was typical (Dale et al. 1994). These individuals had been on their land an average of eight years, had enjoyed high productivity, and had thus far deforested no more than 20% of their total land area. This was done by using what the authors of the study called "innovative farming practices," which included growing a higher diversity of crops, making use of knowledge from agricultural extension agents, employing agroforestry, establishing beehives for honey collection, and digging ponds for fish farming. Computer simulations were developed to model the effects of extremes of land use practices projected over a forty-year period, and the models correlated well with actual observations. Under both typical and worst-case scenarios, the model predicted continued and rapid deforestation as the region becomes essentially all agricultural (and degraded, since that form of agriculture is not sustainable). Only under the innovative best-case scenario, as described above, was the coexistence of forest and farm achievable. Though it was clear that efficient farming not only is possible but is being practiced by a few Rondonia farmers, the authors recognized the significant socioeconomic problems that plague the region and combine to retard the spread of innovative usage of the land.

It is also possible to consider restoration of pasture to forest, which it once was. Uhl (1988) and Nepstad et al. (1990) have aptly summarized the difficulties of restoring forest when the land has been significantly abused. Major problems include dispersal of seeds into degraded lands, postdisperal seed predation, and the physical harshness of degraded pastures. Recall from the work of Uhl and colleagues described in chapter 3 (page 66) that regeneration pathways vary, with large gaps being the most difficult to regenerate to normal biodiversity. Also recall that Silman's work, described in chapter 4 (page 123), demonstrated that seed dispersal is generally low for most tree species. Finally recall from the discussion of the Biological Dynamics of Forest Fragments Project (described above) that the farther a forested site is from a source of colonizing forest species, the more problematic it becomes for that site to retain normal biodiversity. Uhl notes, however, that humans could be employed literally to act as seed dispersers and that such techniques as use of shade cloth and irrigation could mitigate the severe environmental conditions that prevail on degraded lands. Repellant chemicals could be used to help discourage seed predators. Uhl cites a study where topsoil was stockpiled and later restored following bauxite mining. The soil was replaced and sliced to increase aeration and facilitate root penetration, and the area was planted with native tree species. Uhl noted, however, that soil nutrient levels were low and that tree growth was slow, though survivorship was good.

With perhaps some touch of irony, it has recently been suggested that degraded pastures could be restored for, of all things, profitable cattle ranching (Holloway 1993). Doing so would involve use of chemically balanced fertilizers and specific plantings on cattle range to maintain the health of the pastures. But numerous questions remain.

The effect of pasture creation on biogeochemical cycling is complex and not yet fully understood. While it is obvious from slash-and-burn agricultural practices that most Amazonian terra firme soils soon lose their ability to support crops, the kinds of specific changes imposed on mineral cycles by conversion of forest to pasture are just being fully elucidated. Researchers from the Marine Biological Laboratory (MBL) at Woods Hole, Massachusetts, are collaborating with colleagues in Brazil to investigate how forest to pasture conversion alters biogeochemical cycles in Rondonia, where deforestation has been particularly prevalent (Bonalume 1991). Soil biogeochemical cycles are disrupted and altered by conversion of forest into pasture. For example, soil pH increases, and there is generally a lower rate of nitrogen cycling and plant nitrogen availability in pasture soils compared with those of forests (Neill et al. 1995a, 1995b). Nonetheless, some data suggest that pastures may be sustainable if managed properly. For example, established pastures may lose nitrogen much more slowly than forests, where annual net nitrification rates are high, with consequent loss of nitrate nitrogen through leaching and gaseous emission (Neill et al. 1995a, 1995b). Pasture soils show a generally higher concentration of carbon (in soil, not biomass!) than forest soils. Certain species of pasture grasses are of critical importance in the development of carbon stocks, as well as enhancing microbial respiration (Neill et al. 1996). Promoting growth of these grasses will maintain pasture health. In some cases, conversion of forest to pasture can actually increase carbon and nitrogen stocks in the soil and enhance the overall soil quality (Feigl et al. 1995). Further, soil carbon and soil nitrogen do not seem to decline in older pastures (C. Neill, pers. comm. 1996). The results of these ongoing studies may lead to the development of sustainable pastures, thus slowing the rate of forest clearance.

Increasing Ecotourism   *Figures 86, 146, 147, 165*

In 1983, 12,000 tourists visited the city of Manaus, Brazil, near the confluence of the Amazon River and Rio Negro, in the very heart of Amazonia. Five years later, in 1988, the number of tourists stopping in Manaus had increased to 70,000, an impressive jump over just five years. These people didn't come merely to see the unique opera house, a vestige of the old days of the rubber boom era, nor the facinating fish market, where dozens of species of freshly caught Amazonian piscines can be seen and bought for consumption on any given day. The tourists came for the rainforest. In 1988–89 alone, twenty-one cruise ships, luxury liners, docked at Manaus so that their clientele could, for a brief time, experience something of the Neotropical rainforest. When I visited Manaus in 1991 the main dock was dominated by the huge cruise ship *Pacific Princess* (yes, even the "Loveboat" made it to the rainforest). I also observed much construction of hotels and other tourist facilities. And I noticed the T-shirts people were wearing and selling, adorned with images of macaws, jaguars, river dolphins, and other wildlife, the signature species of the Amazonian rainforest. The economic gains to be had from ecotourism were obvious almost wherever I looked. Estimates now suggest that ecotourism in Amazonian Brazil is as lucrative as cattle ranching ever was. Such an interest in seeing the rainforest, if it can be sustained (no pun intended), is certainly economic incentive for preservation.

There are lodges located throughout the Neotropics that cater to ecotourists (Castner 1990), and more are being built every year. Most such lodges are reasonably accessible, are comfortably apportioned, serve safe food, have reliable radio or telephone service, and employ knowledgeable local guides. The best of these lodges are located on lands where all hunting is prohibited, making it possible to see exciting species ranging from peccaries to ocelots (see below).

Economic studies of the impact of ecotourism in various areas within the Neotropics are not abundant, but those that have been done suggest that ecotourism not only is lucrative but can be structured such that it is compatible with conservation interests and serves the local economy as well. For example, Yacumama Lodge near Iquitos, Peru, employs only local people, offering good wages and a health-care plan, in addition to providing numerous other community services. The lodge is profitable not only to its owners, but to the community as well. As a second example, the International Zoological Expeditions' Field Station at Blue Creek, Belize, is managed by local Kekchi Indians from Blue Creek Village. Many members of the community are employed in various ways to service the lodge and its clients. Other Neotropical lodges follow similar policies, making for a harmonious relationship between lodge and local people. This is as it should be.

The growth and impact of tourism has been studied in Madre de Dios in southeastern Peru, near the town of Puerto Maldonado (Groom et al. 1991; Munn 1992). Located in one of the most species-rich areas within Amazonia (a biogeographic center of endemism), the region was initially protected in 1973 with the establishment of Manu National Park and was further enlarged in 1977 with the addition of a reserve zone and a cultural zone. The entire park is now designated a UNESCO Man and the Biosphere Reserve and includes an immense total area of 1.88 million ha (4.6 million acres), with habitats ranging from montane grassy pampas to lowland rainforest (Groom et al. 1991; Foster et al. 1994). Nearby the reserve are several tourist lodges, including the well-known Explorer's Inn, plus the Tambo Lodge, Cusco-Amazonico Lodge, and Manu Lodge, all of which are accessible by jetliner (Lima to Cuzco to Puerto Maldonado). In addition, the Cocha Cashu Biological Station (mentioned numerous times in this volume) is located within Manu National Park but some distance away from Manu Lodge.

Tourism began in the region only around 1975, with the establishment of Explorer's Inn. It has grown such that, by 1987, annual revenues of the various tourist companies and independent tour guides grossed more than a million U.S. dollars (Gross et al. 1991). Looking only at Manu Nature Tours and Lodge, a facility of twenty-five beds, the gross income in 1985 was $7,000, with 37.5 employees and dependent family members supported. However, by 1989, gross income had climbed to an impressive $240,000, and the facility supported 270 employees and dependents (Munn 1992). Clearly, a well-run lodge in an optimal location for seeing wildlife can operate at an impressive profit. The Puerto Maldonado region hosts only about one-tenth the number of tourists that visit Iquitos (Gross et al. 1991), which is situated directly on the Amazon. However, Puerto Maldonado offers much better opportunities for seeing wildlife, and ecotourism will probably continue to increase there.

Having led tours to various sites in the Neotropics as well as been a participant in many, I realize only too well how difficult and challenging it can be to see rainforest animals. The rainforest is not like the African savanna, where the creatures generally stand above the grass for convenient observation. There are exceptions, of course. The vast Llanos of Venezuela and the extensive Pantanal region of southern Brazil are both extraordinary for the panoramas they offer for viewing wildlife. But many has been the tourist who traveled briefly to the rainforest and never got a really satisfying view of a monkey or a wild, free-flying macaw. The reasons are varied. First, rainforest is structurally complex and enclosing, making for many potential obstructions in locating and obtaining a satisfactory view of animals. Second, and more important, some tourist lodges are located either on or nearby land where local hunting is tolerated. Hunting and wildlife viewing are incompatible. If hunting is banned entirely, as it is, for instance, at Chan Chich Lodge in Belize, a walk through the rainforest will usually yield looks at such creatures as agoutis, coatis, spider monkeys, guans, and curassows, and perhaps even an ocelot (my wife saw one). And this is why people go to the Neotropical rainforest. Charles Munn (1992) conducted interviews with 300 ecotourists in Peru and learned (unsurprisingly) that 60–70% wanted mostly to see wildlife, in comparison with but 10–20% who were primarily interested in seeing Indians, and 10–15% whose interest focused principally on trees or the whole rainforest ecosystem. In all the experience I've had, I can certainly attest to the fact that visitors to the rainforest want to see animals more than anything else.

Munn (1992), acknowledging the reality that wildlife viewing can be difficult in rainforests, suggested that the unique macaw clay licks scattered along riverbanks of the Manu River are of great importance as ecotourist attractions. For reasons still poorly understood, large macaws and other parrots habitually gather on clay embankments where they ingest the clay. The parrots may be obtaining essential minerals, or they may be utilizing the clay to help digest defense compounds that they ingested from their diets of seeds and unripe fruits (clay, after all, is just a natural form of Kaopectate). No one knows. But dozens of macaws of several species can be seen regularly on clay licks, including the scarlet (*Ara macao*), the red-and-green (*A. chloroptera*), and the blue-and-yellow, each among the most sought-after species by ecotourists. Munn determined that approximately 356 red-and-green macaws were using one clay lick near Manu, each bird visiting the lick every other or every third day. (He was able to identify individual birds by noting the unique "signature" of red facial line feathers, different on every bird.) A trip to a clay lick tenanted by dozens of highly photogenic macaws of three or more species can be extremely satisfying to any ecotourist, and Munn estimated that at least twenty-six clay licks exist just in southeastern Peru. Indeed, he estimated that, taken as a whole, 20–50% of the Peruvian Amazon had enough wildlife (including species other than just macaws) to be viable for ecotourism. But Munn carried his analysis further. Using similar logic and calculations to those made to determine the worth (in tourist dollars) of a single African lion, Munn determined, assuming that each major macaw lick would be used by 150–400 large macaws, that each free-flying, wild macaw could potentially generate between $750 and $4,700 annually in tourist revenue. Because macaws are long-lived animals, each large macaw might therefore generate 22,500 to 165,000 tourist dollars during its lifetime. And even where macaws are protected from hunting, Munn and his colleagues determined that large macaw populations are limited by availability of nest cavities. Adding artificial cavities (nest boxes) could further enhance local macaw populations. Recently Charles Munn has discovered licks that attract tapirs, yet another potential attraction to ecotourists who crave the sight of these often hard to find mammals (Munn, pers. comm.).

National Parks and Preserves—
The Essential Need for Outright Protection   *Figures 84, 85, 87*

As noted above, large macaws are a major attraction for ecotourists, especially birders. But macaws are often not easily seen, even in areas where they should be common. Their scarcity is usually due to hunting pressure, for either meat (subsistence and market), feathers, or the pet trade. Munn (1992) notes that 17 counts of macaws taken along the Manu River, where macaws enjoy total protection, produced a mean of 6.2 sightings per hour, with a high of 18.9 macaws per hour. In contrast, in areas where macaws are routinely hunted, similar counts yielded virtually no large macaws. Munn argues that even moderate hunting "is sufficient to extirpate large macaws from large regions of the Amazon where the forest cover is still mostly intact." I completely agree. Everywhere I've been in the Neotropics where hunting is permitted has been a disappointment with regard to seeing wildlife, especially large birds and mammals. The local people, whether Amerindians or colonists, are skilled hunters, and their efforts have great impact.

Regardless of the conservation merits of reclaimed land and extractive reserves, there must be land (and lots of it) set aside and protected from *all forms of hunting* if there are to be rainforest ecosystems that accurately reflect the diversity and abundance of large animals. In an article poignantly titled "The Empty Forest," Kent Redford (1992) summarizes how both subsistence and commercial hunting have reduced populations of large animals ranging from caiman to curassows. Amerindians, as noted in chapter 7, typically take a range of prey but tend to concentrate on certain species, especially certain monkeys, as well as squirrels, peccaries, guans, curassows, large toucans, macaws, pacas, and agoutis. Nonindigenous hunters, such as caboclos, focus more narrowly, thus having a disproportionate influence in reducing local populations of such animals as peccaries, pacas, agoutis, capybara, giant otter, guans, and armadillos (Redford 1992). Commercial hunting for edible products such as eggs and meat, leather, mammalian skins to be used in the fur trade, and bird feathers have all resulted in significant declines of various large animals in many regions. As mentioned earlier, Redford has estimated that as many as 60 million animals are killed annually by all forms of hunting in the Brazilian Amazonian states. In addition, Redford notes that other activities apart from hunting can reduce large animal populations, including some of the practices of extractive users. For example, removal of nuts and fruits reduces important food sources for animals ranging from agoutis to various parrots. Logging can eliminate both fruiting trees and trees with potential nest cavities. Large animal populations then decline.

The result of such depredations is that you may walk in a rainforest that looks normal enough but isn't, because the large animals simply are not there. Don't be fooled into thinking these animals are merely elusive. They are gone, extirpated, or they are very significantly reduced from what would be their normal populations. If some remain, they will be extremely wary and difficult to observe. The absence of large animals is not merely regrettable because of lost opportunities to observe these creatures. Of far greater significance is the impact their absence creates on the ecological functioning of the rainforest itself. Large animals have direct and important influences with regard to herbivory and seed predation, seed dispersal, and predation (Redford 1992). The forest functions differently without them.

It is clear, therefore, that large and totally protected reserves must be maintained throughout Amazonia and other Neotropical areas if the full spectrum of rainforest biodiversity is to persist. It should be obvious that top carnivores such as jaguars and harpy eagles require large tracts of forest. What is less obvious is the fact that even pacas and agoutis, animals with relatively high reproductive rates and small home ranges, require a respite from human hunting pressure in order to attain normal population densities.

Unfortunately, protection of reserves in Amazonia is problematic at best. One study found that 40–100% of the area in all existing nature reserves in Brazilian Amazonia is directly accessible to would-be poachers by either river or road (Peres and Terborgh 1995). Further, the reserves are poorly protected at best, with a single guard given the responsibility, in some cases, for patrolling an area the size of the state of Delaware. As discouraging as this might seem, the authors suggest that it is possible cost-effectively to design reserves in such a way as to minimize the probability of illegal hunting. Minimizing illegal use thus becomes yet another important criterion for design of nature reserves in the Neotropics.

The best ecotourist facilities are located within or in proximity to national parks and reserves where all wildlife is protected, affording visitors the most chance to really experience the creatures of the rainforest. Redford (1962) notes, however, that the trend has been not to set aside fully protected reserves but rather to allocate lands for multiple uses, particularly to Amerindian tribes or as extractive reserves. In Brazil, for example, there are about 6 hectares of Indian-controlled lands for every hectare of protected conservation lands. In Colombia, the ratio is about 7 to 1.

Priorities for the Future

The natural ecosystems of the Neotropics are not assured an optimistic future. Areas such as the Atlantic Coastal Forest of Brazil and much of the northern Andes are under critical threat of loss from deforestation. Such areas as the Greater Antilles, central South America, and the central Andes are under severe threat. Tropical dry forests and cerrado ecosystems are even more threatened than rainforest (Stotz et al. 1996). On the other hand, much of Amazonia is not yet deforested, and rates of deforestation have in recent years been reduced in many areas. What steps should be taken, what policies should prevail to help ensure that conservation goals will be influential in future decisions about the fates of Neotropical ecosystems?

Basic Research

I hope you have read this book carefully, have learned from it, and have enjoyed it. But there is one mistake you might be prone to make after having read it. You might mistakenly believe that researchers really understand the ecology of the Neotropics and that there is relatively little else to be elucidated. That is simply false. I know of no tropical ecologist who would express confidence that we have an adequate understanding of what determines the distribution and abundance of organisms in the Neotropics. Obviously we know a lot. The hundreds of thousands of hours put in by the collective host of researchers, especially in the past two decades, have paid handsome dividends in the form of data, theory, and understanding. These studies, however insightful, are really nothing more than a collective foundation upon which to build. And a foundation is not a completed structure, only a beginning.

Much more basic research needs to be performed in the Neotropics. Basic research is not necessarily directed toward some immediate, pragmatic goal. It is research conducted only to gain knowledge. Often basic research, much of which is supported by public grants, is criticized as too esoteric. Why, for instance, would someone care that much about how far an oilbird flies in a given evening? Why support funding to mount radio transmitters on such creatures, only to chase them all around the rainforest in the dark of night listening for little beeps? But Roca's study (1994), briefly discussed on page 143, not only elucidated much about oilbird ecology, but also demonstrated that certain species of widely scattered rainforest trees may be strongly dependent on oilbirds for seed dispersal. The tree species richness of the forest is affected, indeed partly sustained, by this unique bird. The border between basic research and applied research intertwines; the two often meld together. More to the point, basic research is always ultimately pragmatic. The more we can understand about nature, the more we can make sound decisions on policy. It is impossible to know too much. It is only possible not to know enough.

There is so much still to be learned. Field studies in molecular genetics should help us to understand the evolutionary histories of various organisms and to formulate a better idea of how populations evolve in the Neotropics. Work in population genetics, using analysis of mitochondrial DNA as well as other techniques, should clarify distinctions among subspecies that will help sharpen species designations. To this end, there is a critical need for much additional work on the systematics of virtually all taxa in the Neotropics as well as for solid databases of inventories from various ecosystems. Such studies will ultimately have significant influence in conservation decisions. Currently there is an ambitious attempt to inventory all arthropods that occur at La Selva Biological Station in Costa Rica (Yoon 1995). Known by the acronym ALAS (Arthropods of La Selva), the project by mid-1995 had documented 6,151 species, represented by 56,881 specimens. The intention is to provide sufficient knowledge for making sound conservation decisions and preserving Costa Rica's biodiversity.

Even well-studied taxa are yielding the discovery of new species. Amazingly, a new species of primate, the satere marmoset (*Callithrix saterei*), was recently discovered in Amazonia between the Madeira and Tapajos rivers (Line 1996). Thus far its biology is poorly known and its range still undetermined. A new genus and species of ovenbird was recently described from the cocao-growing region of southeastern Bahia, Brazil (Pacheco et al. 1996). While it is unusual to find new bird species, it is extremely unusual to discover yet a new genus. The fact that the bird has been discovered attests to how much of a need there continues to be for basic systematic research. This small, insectivorous bird, *Acrobatornis fonsecai*, known as the pink-legged graveteiro, is restricted to cocao plantations that now cover much of the lowland Atlantic Forest. Unfortunately, because of a pathogen affecting cocao, owners of the plantations are selling out to lumberers, who are cutting for harvest the canopy trees that shade the cacao. This action, of course, threatens the future security of the graveteiro as well as many other species that occupy cocao plantations.

A close examination of Neotropical population studies shows that we have relatively little information about population structures and how life history characteristics influence vulnerability to extinction. Studies are now appearing that utilize Neotropical populations to test important evolutionary hypotheses. For example, the life-cycle stages of a Mexican population of *Cecropia obtusifolia* were studied for changes in frequency of enzyme polymorphisms, using electrophoretic techniques (Alvarez-Buylla and Garay 1994; Alvarez-Buylla et al. 1996). It was learned that the genetic structure of the cecropia populations changed significantly over the course of the life cycle, with a major loss of genetic differentiation from seedling stage to adult. Such studies bear heavily on our understanding of gap phase dynamics, the actions of natural selection, and the role of chance factors in affecting population genetic structure. Ultimately, such studies will be essential to understanding how best to conserve genetic variation in small populations subject to increasing amounts of fragmentation. As a second example, there is a concept, the metapopulation, that is proving useful in the analysis of some temperate populations. A metapopulation is a complex of local interacting populations, some of which are declining (sinks) and some of which are thriving (sources). Source populations disperse and bolster sink populations. Do metapopulations exist for Neotropical species, many of which, at least in Amazonia, are characterized by low populations and limited ranges? We know very little about the dispersal ecology of Neotropical species, and we have no firm idea if corridors will be successful in combating negative effects of forest fragmentation. We need much more study of the response of ecosystems to disturbance and the role played by natural disturbance in structuring Neotropical ecosystems. Understanding disturbance and recovery is of obvious potential use in restoration of degraded ecosystems. And finally (though this list is hardly exhaustive), recall that we still have not acquired much robust data on interspecific competition, diffuse competition, or the importance of predators and parasites as influences on species richness. Much about coevolution remains to be discovered. Yes, there remains so much to be learned.

Conservation Research

Research in conservation biology is obviously essential in securing a bright future for Neotropical ecosystems. The field of conservation biology is becoming increasingly sophisticated, informed by the principles of ecology, as they continue to be revealed by basic research. The journal *Conservation Biology* was first published in 1987 and has emerged as a major organ for communicating information in this youthful subdiscipline of ecology. Restoration ecology is still in its infancy, but it too should grow rapidly and thrive, as more research is performed.

Throughout the Neotropics, rapid assessment research continues to be greatly needed to help determine which tropical ecosystems most require preservation. And preservation is of extreme importance. To this end, monitoring studies, indeed many of them, are necessary to accumulate baseline data that are generally lacking for so many important Neotropical areas. Ideally, monitoring should be ongoing and should be done with the cooperation and support of local and national governments.

Conservation decisions need to be made, but how best to make them? What principles should apply in selecting areas in most need of protection? The SLOSS debate, cited above, was an attempt to apply information learned from fragmentation studies as well as island biogeography theory. But there are other approaches. Standard techniques of vegetation analysis have been employed in a study to ascertain how best to select conservation sites in Paraguay (Keel et al. 1993). This straightforward approach, which involved making a species inventory, considering the number of economically important plants and the number of rare, uncommon, and vulnerable species, made it possible to quantify the conservation worthiness of the various sites.

On a larger scale, a group of 100 biologists, physical scientists, and conservation planners combined their respective areas of expertise and evaluated Amazonia (nine countries) to produce a map that identified 104 priority areas for conservation (Prance 1990). These areas were ranked in priority, but, taken as a whole, all areas together occupied about 60% of the vast region.

Stotz et al. (1996) have developed a well-argued series of principles upon which to base conservation priorities. Though these researchers were discussing bird communities, it is clear that their principles apply broadly to virtually all taxa. They recommend focusing protection on assemblages of endemic species and habitat specialists. These species are obviously most vulnerable in most instances since they are already limited in population size. Second, focus on distinct communities rather than on sheer number of species. For example, although Amazonia is by far the most species-rich area, numerous other ecosystems (such as cerrado), less rich in species, are actually more threatened. Third, protect ecotones, tension zones, and gradients. These areas, where two or more ecosystems meet and overlap, support species from each of the various ecosystems as well as specialist species. As was learned in various studies of the Costa Rican cloud forest at Monteverde, many bird species migrate altitudinally with the seasons and thus require access to appropriate ecotones between ecosystems. Finally, preserve genetic and geographic diversity. This is an increasingly clear conclusion from the research now proliferating (though much more is needed) in systematics at the molecular level. Studies such as those dealing with cecropia, cited above, are models for what is needed for many other species.

Though it is arguable as to which regions and ecosystems and species are most threatened, Stotz et al. (1996), again focusing on birds, assert that the regions now under most threat are the Brazilian Atlantic Forest and the Northern Andes. These two regions have already experienced rather extreme loss of habitat. Extinction of endemic species seems imminent in these regions. Next at risk are the Greater Antilles, central South America, and Central Andes. The third group, at less risk, is considered to be the Lesser Antilles, Choco Lowlands (extreme northwestern South America west of the Andes), Gulf-Caribbean Slope, and Madrean Highlands (western and southern Mexico). The least current risk is Chiriqui-Darien Highlands (Panama), Southern Amazonia, Northern South America, Northern Amazonia, and the Pacific Arid Slope of Central America. But even these latter regions are considered vulnerable. Stotz et al. (1996) introduce the concept of "cornerstone habitats," identified as "regional habitats that possess significant conservation value, as measured by both their local and global role in harboring unique centers of diversity." As an example, the authors discuss arid lowland scrub, a habitat with 167 bird species of which 41 are restricted to that habitat and 32 (of the 41) are endemic to a single region. They then identify three "globally significant cornerstones," Central South America, Equatorial Pacific, and Baja-Sonora, which together harbor 20 of the 32 endemics, while no other region has more than 3. The concept of cornerstone habitats seems workable for taxa other than birds, though if merely applied to birds, many nonavian species would presumably also benefit.

Increasingly, ecologists are adopting the position that their research ought to be firmly directed toward meeting conservation goals, especially those studies that inventory and monitor, with the ultimate objective of protection of biodiversity. And conservation goals, however noble, are best served only when supported by rigorous science. Pearman (1995) and Pearman et al. (1995a), working with amphibians in a species-rich area near Misahualli along the Upper Napo River in Ecuador, have demonstrated that carefully designed, robust, and quantitative sampling protocols can be of immense importance in the establishment of ongoing programs to conserve biodiversity. Pearman argues that much of conservation research is weakened by its lack of predictive potential, the data being of limited use in planning sound conservation and management strategies. It is, for instance, potentially risky to rely on but a single taxon to ascertain broad priorities for conservation decisions. A study by Pearman et al. (1995b) showed only a weak correlation among biodiversity patterns of birds, butterflies, and amphibians from twenty-three study sites near Misahualli. Obviously, reliance on but a single taxon as the indicator in making decisions as to which of these areas to conserve would not necessarily be the right choice for preserving biodiversity among other taxa. Pearman et al. urge careful, quantitative sampling and monitoring of multiple groups in order to best determine management programs for the conservation of biodiversity.

Conservation science is built firmly on a foundation of basic ecology. But included among the agenda of most conservation biologists is not merely the study of biodiversity, but its preservation. Thus, these scientists are not neutral in reaction to events that threaten species and habitats. Scientists are normally portrayed as objective, detached researchers, but in reality that is rarely, if ever, true. Increasingly, ecologists are acknowledging that their values in favor of the preservation of biodiversity make them advocates as well as researchers. Some would argue that such advocacy puts ecologists on a "slippery slope," where their research goals and (presumably) their data and data analyses can easily be compromised by their personal philosophic values and advocacy for those values. This concern is neither a new one nor one about which there is widespread consensus. Many conservationists believe that history demonstrates with utter clarity how human populations have repeatedly abused the environment (Gore 1992). Once such a realization is accepted, it follows that ethics should focus increasingly on ecological concerns. But do species have rights? Do ecosystems have rights? Those questions aside, some scientists worry that it simply is not possible to do good science and at the same time be a committed advocate for issues that involve one's scientific work. Others, myself included, believe that advocacy for conservation and a commitment to the value of biodiversity in no way prevent one from performing science with honesty and skill. Quite the contrary, a brief perusal of those scientists who are, unarguably, the most respected leaders in research in Neotropical ecology will quickly reveal that most, if not all, are outspoken, articulate advocates for the preservation of biodiversity. For an insightful and thought-provoking treatment of the debate about conservation biology, values, and advocacy, see the special section in *Conservation Biology* 10:904–920 (June 1996).

Education   *Figure 87*

It is impossible to overstate how important education is in considering the future of Neotropical conservation. Janzen (1988b) has articulated a view in which he equates the value of wild ecosystems with such societal needs as "libraries, universities, museums, symphony halls, and newspapers." I like this analogy, in fact I treasure it, but I also realize that cultural resources such as museums and symphony halls are best supported and most appreciated by an *educated* public. The belief that rainforests, dry forests, cerrados, mangrove forests, and their numerous inhabitants have intrinsic value and should be preserved and protected *as a matter of ethics* has not gained widespread acceptance in Latin America (and it is unclear how much sympathy it would attract in North America). Sympathy for this view is tempered throughout much of Latin America by the pragmatics of poverty and overpopulation evident in most Neotropical countries. It is also, I fear, compromised by a historical lack of any kind of significant conservation ethic among the citizens. As Fearnside (1989) has accurately stated, arguments for rainforest preservation that are "explicitly noneconomic and nonutilitarian" are regarded pejoratively by most Brazilians as "poetry." I hope this view will eventually be changed by education. It must be if there is to be any realistic hope for future preservation of the full biodiversity of Neotropical ecosystems.

Indeed, there are numerous pragmatic and economic arguments that support rainforest preservation, and there are many economically sound and creative alternatives to deforestation (Anderson 1990; Redford and Padoch 1992). But an educated, enlightened citizen must also embrace the view that pragmatics is not everything. One can live without ever hearing Mozart, or viewing an original Monet, or reading Shakespeare or Darwin. But one's life is, I would argue, significantly (perhaps immeasurably) enhanced by the enjoyment gained from listening to Beethoven's Triple Concerto or seeing a good performance of "Swan Lake." The rainforest, the cerrado, the cloud forest, like other natural ecosystems, are each, indeed, analagous to libraries of fine art, fine music, fine literature. We can value these unique ecosystems for exactly that reason.

Just as there is an emotional component to human nature that can be awakened to respond to music and art, there may well be one that responds to nature, what E. O. Wilson (1984, 1992) has termed "biophilia." But such an appreciation for the intrinsic value of multitudes of creatures other than ourselves, is, I suspect, mostly an acquired and not an inherited taste. And it is acquired through education. It's fairly easy to be emotionally moved by the sight of rainforest from atop a canopy walkway in the early morning mist. It's a little harder to derive such an emotional rush when walking among the short, twisted trees of cerrado. For most of us, it may be even more difficult to conjure up an exultant feeling upon seeing one's first bushmaster, but believe me, it is possible. You learn about natural history, you do natural history, and both emotional attachment and intellectual satisfaction follow.

How can education best succeed? Given that much, if not most, of the research on Neotropical ecosystems has been done and continues to be done by scientists from the temperate zone, it becomes our moral responsibility (as well as in our pragmatic interests as ecologists, teachers, ecotourists, and conservationists) to disseminate the results of our studies not only to North Americans but also to Latin Americans. It is imperative that we make it clear to Latin American citizens that we visit the Neotropics, and their country in particular, to enjoy its natural history. We must convey to Latin Americans that there is real value in preservation—a value that can be measured in pride as well as in money. Teaching students in North America from elementary school through college about the ecology of tropical rainforests and other tropical ecosystems needs to be a continuous and fundamental objective of our work. Students from Latin America should be offered opportunities to study in North American institutions of higher learning so that they may return to their countries and educate others, as well as perform cutting-edge research. Note how many citations in this book have been authored by North Americans, but note also that, particularly in recent years, the number of Hispanic authors (especially in jointly authored papers) is increasing. Such collaborations are very encouraging. Many North American tropical ecologists offer graduate- or undergraduate-level college courses on Neotropical ecology, and many of these courses include a field segment in a Neotropical location. The logistics of such an endeavor, and I speak with experience here, are often anything but easy. But the effort is worth it. As I said in the introduction, there is nothing like being there, even if for a relatively short time. Absolutely nothing.

I have watched, with no little satisfaction, how, over the past two decades, the small Central American country of Belize (where I began my career as a tropical ecologist and for which I maintain a particular fondness) has become increasingly conscious of and sympathetic to conservation of its forests and their many inhabitants. Successful efforts have been made in Belize not only to preserve habitat but to educate the citizenry from elementary school on up. And, just as in the United States, there is much more that could be and needs to be done. When I netted and banded birds in Blue Creek Village, I tried to explain to the curious Kekchi-Mayan children and adults why such seemingly esoteric work was of interest and value not only to me, but to them. It wasn't easy, and I'm not confident that I met with much success. But I knew I had to try. It is critical to share knowledge. Other nations, and Costa Rica immediately comes to mind, have made similar successful efforts as Belize has to embrace conservation goals, to take national pride in their ecological resources, and to educate citizens about such matters. Some indications are that sympathy for conservation goals is currently growing stronger throughout Latin America.

Field-based opportunities for learning are proliferating. As mentioned earlier, there are numerous ecotour companies, many with superbly knowledgeable leaders, which offer high-quality trips throughout the Neotropics. Colleges are increasingly offering courses in tropical ecology that have at least the option of a field-based component. By way of example, I want to discuss briefly three organizations that offer different and, in my opinion, successful approaches to education. These are the Organization for Tropical Studies, Earthwatch, and the Children's Environmental Trust Foundation, International.

The Organization for Tropical Studies (Box 90630, Durham, NC 27708–0630), known popularly by its initials, OTS, is now in its fourth decade. OTS is a consortium of over fifty universities, mostly North American, united by their mutual interest in the study of Neotropical ecology. OTS supports three highly active biological stations in Costa Rica (La Selva, Las Cruces, Palo Verde). Essentially because of OTS, Costa Rica has emerged as a mecca for Neotropical research. Over 2,000 graduate students and professional scientists have taken more than 120 graduate-level courses at various OTS field sites. Many of these individuals have conducted basic research resulting in numerous publications.

Long-term studies are also being conducted at each of the biological stations. The professional journal *Biotropica* published a special issue to honor OTS' twenty-fifth anniversary, which featured a Symposium on Resource Availability and Tropical Ecosystems (*Biotropica* 24[2B], June 1992). OTS continues to offer courses for graduate students, emphasizing not only basic ecological research but also environmental policy making and environmental education. Undergraduates also can now benefit by enrolling in a newly initiated, semester-long program sponsored by OTS, taught entirely in Costa Rica, involving experience at each of the three biological stations. Courses are offered in tropical ecology, research techniques, and Latin American culture. It is abundantly clear that were it not for OTS, we would have much less understanding of Neotropical ecology.

The Center for Field Research and its affiliate Earthwatch (680 Mount Auburn Street, Box 403, Watertown, MA 02272) have been sponsoring research teams for a quarter of a century, many of them (my own included) in the Neotropics as well as other tropical and nontropical areas. Earthwatch estimates that it has sent around 50,000 volunteers into the field to assist principal investigators in their various research projects. The Earthwatch model is that laypersons pay to support field-based research. In return, they actually go into the field and participate in the very project that their money is supporting. Earthwatch programs are typically two weeks in duration, though a researcher may have several successive Earthwatch teams and thus maintain the project for many additional weeks. Volunteers learn the objective of the research, are trained in field and lab techniques as well as data analysis, and also learn much about the country and region in which they are working. Earthwatch acts as an effective catalyst between the public and scientists, bringing them together in a mutualism that has proven highly successful. Earthwatch volunteers demonstrate how science need not operate in a vacuum but, quite the contrary, is readily accessible to any interested person who is willing to learn. Earthwatch principal investigators have all submitted projects that undergo stringent peer review for the quality of the science and significance of the research. But Earthwatch PIs are also expected to take seriously their roles as teachers, as they have the responsibility to educate their volunteers as to the worth of the work they are being asked to perform. Earthwatch demonstrates that education and research can form a splendid partnership.

Children's Environmental Trust Foundation, International, known as CET (572 Alice Street, Zeeland, MI 49464), is a relatively recent organization founded in 1993 by three middle school teachers whose goal was to take children into the rainforest and teach them about its ecology. In addition, the founders believed that it was essential to have the North American children meet and interact with their counterparts in Latin America. With these goals firmly in mind, CET began sponsoring Children's Rainforest Workshops, the first of which took place in Amazonian Peru in 1992. The program involved the use of already existing ecotour facilities such that the children could readily access and safely experience the rainforest as well as have some adventure. The children who were participants in the initial Children's Rainforest Workshops helped in the actual construction of the canopy walkway maintained by the Amazonian Center for Environmental Education and Research (page 42).

The CET program involves serious commitment by each participant and bears no resemblance to a casual tour. Each child is part of a class that studies together weekly for an entire year, along with their teacher, so that they have learned much about the rainforest and the region prior to their arrival. They are well prepared. Once in the field they are taught by experienced, skilled instructors who are authorities on such subjects as general rainforest ecology, birds, and fishes. The children are a joy to teach. Having participated in this program as an instructor, I could not help but imagine how significantly the cause of rainforest conservation would be advanced if more children were to have the benefit of such a program. At the impressionable ages of these children, imagine what superb, lasting memories they must be acquiring, and what values must be being shaped in their minds. My hope is that CET will prosper and that others like it will follow.

A Final Word

I hope this book has interested you in Neotropical ecology and has impressed you with the conservation concerns that are so pressing throughout the Neotropics. Such awareness has, I believe, already helped somewhat to temper the rapid destruction of tropical forests. I am cautiously optimistic. Nearly a quarter of a century ago, Gomez-Pompa et al. (1972) called the tropical rainforest a "nonrenewable resource" and urged that international efforts be taken to protect "this gigantic pool of germplasm by the establishment of biological gene pool reserves from the different tropical rain forest environments of the world." That idea seems much less radical today than perhaps it did in 1972. But I think it's safe to say that it seems no less valid.

Appendix
"And Hey, Let's Be Careful Out There"

SOME regard the tropics as an equatorial Dante's Inferno, the so-called Green Hell. The rainforest: abominably hot, oppressively humid, crawling with vile scorpions, immense, hairy, ugly tarantulas, lethal, concealed snakes, marauding, biting, stinging, and generally irritating insects, not to mention countless legions of external and internal parasites awaiting to devour you from without and within. Well, calm down, don't get excited. If you know what you're doing, you have little about which to worry. On the contrary, you'll enjoy yourself and be quite safe in the rainforest. But, just as you need to learn the safety rules of the road in order to enjoy driving an automobile, so you need to learn basic precautions when moving about in a Neotropical forest. Here they are.

Trail Precautions

First, don't go alone. Tropical forests are often rather remote from convenient assistance. Sounds, such as calls for help, don't travel very far in dense forest. It is a wise move to have traveling companions, a form of human insurance of help if needed. Second, know where your feet and hands are at all times. Don't carelessly step over logs (or sit on them) without first checking where your feet (or derrière) are going to land. Many understory palms as well as other plants such as terrestrial bromeliads have sharp spines radiating out horizontally from the bark or along the leaf margins, so don't just grab onto a plant without looking first. There are also many plants that harbor aggressive ants that will attack you if you disturb the plant. Stay on trails. I once recklessly chased a pair of chestnut woodpeckers (*Celeus elegans*) through a Trinidad rainforest. I vaulted over logs, crashed through undergrowth, and eventually got great looks at the birds. I soon realized, however, that I had taken somewhat of a chance. I was alone, well off the trail, and I had not paid too much attention to where I was going. Trinidad hosts a fair number of bushmasters (page 317), very lethal snakes. When this fact dawned on me, I was ever so careful making my way back to the trail.

Take a rain poncho because sudden heavy downpours are routine, and remember that mildew is very common in high humidity environments, so make an effort to keep things from becoming "permanently wet" during your trip. For footwear, tall boots are really not necessary, but good-quality hiking shoes that will tolerate being wet and often very muddy will make your walking more comfortable. Some people like to wear lightweight rubber boots. If you have camera and/or binoculars, remember to include plastic bags so your equipment can be sealed and remain dry during rain showers.

And don't get lost. It would be bad.

Snakes

Snakes (also see chapter 13) are a major worry to most tropical visitors, often consuming more conversation time and creating more anxiety than is merited by the actual risk. Most tropical snakes are nonpoisonous, and most are not easy to find unless you are really looking and know just where and how to search. In reality, seeing a dangerous snake is, in itself, a rare event. Being struck by one is considerably rarer. Colwell (1985b) reports that 450,000 total person-hours of field work occurred at the Organization for Tropical Studies field sites in Costa Rican rainforest without one single poisonous snakebite. Of course, he mentioned this figure in relation to discussing his own experience after spoiling that record by being bitten by a fer-de-lance (a considerable misfortune, but it is heartening to know that he survived to write about the experience).

Snakes, including venomous species, are present throughout the Neotropics and, if you spend enough time, sooner or later your path and that of a venomous serpent will cross. Some poisonous snakes, like the fer-de-lance and bushmaster, are dwellers of the forest floor. Others, like the eyelash palm-pitviper, climb into low bushes and are threats to careless bushwackers. Coral snakes are potentially very lethal but generally remain hidden under rocks and logs. So be careful when you're moving rocks and logs. Poisonous snakes are usually well camouflaged and will strike only if threatened or accidentally stepped upon. Keep your eyes open and be conscious that there may be snakes. Never attempt to pick up a snake if you have the slightest doubt as to its identity (and remember that many venomous and nonvenomous species look alike). Baby poisonous snakes are, of course, small but potentially lethal—they contain identical venom as the adults, just less of it (and it doesn't take much to do real harm). Remember, even nonpoisonous snakes can bite severely, and such bites are often subject to bacterial infection.

If the worst happens, if you or someone in your party is bitten by a venomous snake, the first thing to do is try and be calm. I know that's hard to do, but you must try. It is essential that you determine if the snake has, in fact, injected venom. Sometimes they don't envenomate. Sometimes you think it's a poisonous species but it isn't. If not, fine, luck was belatedly with you. If envenomation has occurred, the condition will soon become obvious (pain and swelling, possible difficulty in breathing) and is serious no matter where the bite has occurred anatomically, and you must get the patient as quickly as possible to the nearest competent medical facility. If possible, keep the site of the wound lower than the heart and head. Sucking the poison from the wound is not effective and may be dangerous (if there are sores or cuts in the mouth). First aid for venomous snake bites is problematic at best. Some people advocate the use of electrical cattle stunners applied directly to the wound as a first aid measure. These have been shown not to be effective. For a more in-depth discussion of poisonous snakebite, see Hardy (1994), plus discussions by Watt and by Hardy in Campbell and Lamar (1989).

The largest Neotropical snakes are constrictors, species such as the boa constrictor and, the giant of them all, the anaconda (page 208). Constrictors are not poisonous but are potentially dangerous. They bite readily, and a bite may introduce infection and should be promptly cleaned with disinfectant. Because constrictors are long, thick, and thus strong, if you should attempt to pick one up, it may very well coil around you and begin constricting. You might be unpleasantly surprised at just how strong these serpents can be. Large constrictors are potentially lethal as they can actually cause you to suffocate. The easiest way to avoid this mishap is not to disturb constrictors. Most people are smart enough not to bother trying to pick up a 3–4.5-cm (10–15-ft)-long (sometimes more) anaconda, but a 1.5–1.8 m (5–6-ft)-long boa constrictor is more tempting. Resist temptation. Let them be.

Mosquitos and Other Biting Insects

Insects are usually not the nuisance they are reputed to be in the Neotropics. Anyone who has ever braved northern New England blackflies and mosquitos in spring has in all likelihood experienced worse insect irritation than will normally be encountered in the tropics. Nonetheless, mosquitos can be abundant, they do bite, and some carry malaria, yellow fever, or other potential maladies. However, insect repellent (particularly that containing strongly concentrated deet) is usually sufficient for protection. Mosquito netting or a tight tent is necessary for camping. In areas where mosquito-borne diseases occur, it is obviously wise to take protective innoculations in advance of the trip (malaria-preventative drugs must be taken before, during, and after the trip for a period of time—see below). Mosquitos are generally most abundant during the rainy season and usually much less so during dry season, though sudden rains can bring forth hordes in a short time. Be prepared.

There are also simulid flies related to blackflies that bite and leave little red blood spots, but these are generally not much more than a minor nuisance. Stingless sweat bees can be maddening when seemingly billions each want to land on you to lick your perspiration, but they don't wound you, nor do they carry vermin.

A few Neotropical cockroaches and beetles are nearly as big as mice and just about as dangerous. In other words, "no problema." Enjoy them.

Botflies

Once, upon returning from Belize, I noticed that a small insect bite on my forearm, presumably from a simulid fly, was enlarging rather than healing. Periodically the wound would feel like tiny hot needles were turning within. I was harboring a larval botfly (*Dermatobia hominis*) that had grown from an egg transported by a bloodsucking fly that had bitten me in Belize. The botfly maggot had hatched on my skin, burrowed inside, and was now using me as a source of shelter and sustenance. As it enlarges, a botfly maggot creates an obvious skin lesion, a condition called "specific myiasis" (Markell and Voge 1971). If left unattended, within forty to fifty days the larval fly will emerge to pupate. Discomfort increases as the larva grows because the insect turns, and its body is covered with sharp spines. I chose to remove the little maggot by simply covering the lesion with petroleum jelly, preventing the larva from breathing (a tiny abdominal breathing tube remains in contact with the air, though the rest of the insect is burrowed—the jelly blocked the breathing tube). I squeezed the dead larva from my arm. I was also told that botfly larvae can be removed by taping a piece of meat, preferably bacon fat, over the wound. The larva usually migrates from the human to the meat. This technique has actually been described in the *Journal of the American Medical Association*, where it was reported that within three hours after application of bacon fat, botfly larvae had moved far enough out of the wound into the meat that they could be removed with tweezers (Brewer et al. 1993). Amazonian Indians have an innovative way to kill botfly larvae. They employ the green cashew nut, which grows in Amazonian forests and contains a highly toxic oil (which is why cashews must be well roasted before eating them). The nut is cut in half and rubbed on the skin at the site of the larva. The oil terminates the little maggot, which is then pulled from the skin (Plotkin 1993). Botflies are not uncommon throughout the Neotropics, and insect wounds that increase in size are suspect.

Ants

Alfred Russel Wallace (1895) said it well: "Ants are found everywhere." Indeed, ants are omnipresent in the rainforest. Many are aggressive, both bite and sting, and come in large numbers. Never stand among an ant swarm crossing the trail, for they will climb onto your body and bite and/or sting you. Be warned that many ants are arboreal (including the nasty bullet ant) and may literally drop from the trees and attack. Ants also move around, especially army ants. On one occasion a traveling companion awoke to find himself and his sleeping bag pretty well covered by army ants, and on another occasion I had to remove several hundred army ants from someone's shoes after the small battalion of ants selected the footware as a suitable bivouac following a night's raiding. There is one particular ant of which to beware: the giant tropical ant (*Paraponera clavata*; page 328). This formidable, inch-long black ant, sometimes called a "bullet ant," is usually a solitary forager, and it packs a mighty sting. Don't toy with it.

Urticating Caterpillars

The caterpillars of many tropical butterfly and moth species are covered with sharp hairs, called urticating hairs, that cause itching, burning, and welts if they prick the skin, a reaction similar to that caused by stinging nettle. Do not touch any hairy caterpillars you may encounter. The fact that these fuzzy beasts sit on leaves and tree trunks is yet another reason for making sure you know where your hands are at all times. My friend Ted Davis fell down a muddy slope in Tingo Maria, Peru, and felt something odd in his pocket after the fall. Reaching in, he retrieved a large urticating caterpillar. His hand bothered him for weeks afterward.

Spiders, Scorpions, and Centipedes

Spiders, scorpions, and centipedes are common throughout the Neotropics. The biggest spiders are the wolf spiders and tarantulas. Both are large, hairy things, the sight of which tends to scare people. Neither is a serious danger. All spiders are poisonous, but the likelihood of being bitten is small, and, even if bitten, the likelihood of the bite doing any real harm is remote. Wolf spiders (family Lycosidae) have long legs and smallish bodies. Tarantulas (suborder Mygalomorpha, family Theraphosidae) are thick-bodied and hairy. They walk deliberately, almost sedately, highly dignified for a spider, using their first pair of legs rather like antennae, to explore their surroundings. One Amazonian species, popularly called the South American goliath birdeater, or simply the bird-eating spider (*Theraphosa blondi*), has an 18-cm (7-in) leg span. Amazon explorer Henry Walter Bates (1892) described seeing this huge

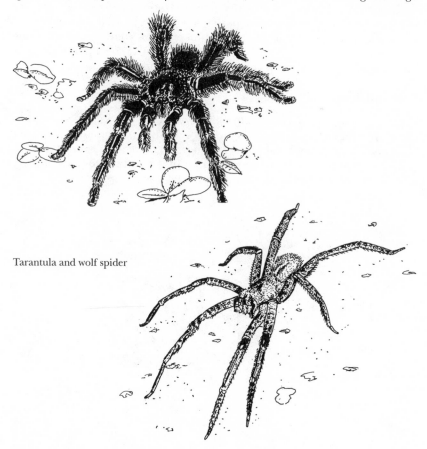

Tarantula and wolf spider

tarantula actually subduing a small bird. Remember, however, that you are much larger than a small bird. Most tarantula species bite only when highly irritated. A threatened animal will rear up on its back legs waving its front legs at its tormentor. Don't pick one up if it does this. But if you should, by some remote chance, be bitten by a tarantula, know that there are no documented fatalities from tarantula bites (Conniff 1996). Tarantulas have urticating hairs on their abdomens, an additional protection against predators (so don't pet them either). Both wolf spiders and tarantulas are generally nocturnal. Tarantulas do not spin webs but make burrows, often alongside trails, where they sit and wait for prey.

Scorpions sting rather than bite, and some, particularly the smallest, are very toxic, so you need to be careful with scorpions. Most are no more irritating than the spiders, however. Though a scorpion uses its two large, lobsterlike front claws to capture prey, it defends itself (and subdues prey) by stinging. The stinger is at the tip of the last segment on its flexible abdomen. Scorpions often hide in clothing and boots, so it's best to inspect these items carefully before dressing. At night I habitually stuff my shoes with socks to discourage scorpion invasion.

Centipedes are wormlike in shape but are arthropods and move swiftly on jointed legs. All are predators that possess a pair of formidable mandibles with which the larger species can give a nasty and poisonous bite. One species (*Scolopendra gigantea*) can reach lengths of 27 cm (10.5 in). Should you come upon this animal, do not pick it up (I doubt you'll be tempted).

Bees and Wasps

Bees and wasps are abundant throughout tropical areas, and if you are allergic to the stings of such creatures, you are well advised to bring an adrenalin kit, a precaution that is really essential. Hospital emergency rooms often take hours to reach from field sites, well beyond the onset time for anaphalactic shock. To make matters worse, many tropical bees and wasps tend to be rather

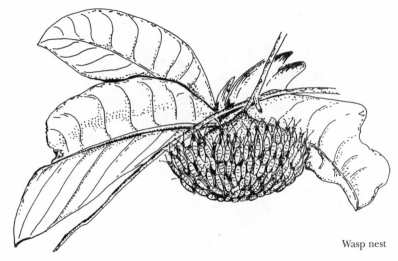

Wasp nest

aggressive (remember, so-called killer bees originated in South America as hybrids between African and South American bees, and they are both very aggressive and common in many areas). Species of moths and other insects derive protection from mimicking bees and wasps, and certain colonial birds locate their nests near wasp nests, again to derive protection by being close to the wasps (page 266). These adaptations are the evolutionary result of the simple fact that bees and wasps are, indeed, aggressive. Remember that.

Chiggers and Ticks

Chiggers are small mites abundant in brushy grassy areas but not as common inside rainforests. If you walk through fields, pastures, or grass, you have an excellent chance of acquiring chiggers. As larvae, they climb aboard a passerby and attach by inserting their mouthparts into the skin (they do not burrow into your body). The host tissue reacts with the chigger's saliva and itching results. Chigger bites are most common around areas where elastic from undergarments presses into the skin. Thus, the itching can be embarrassing as well as irritating. They really itch when present in numbers, and scratching can cause infection. You are advised to wear clothing that is tightly tucked in at the boots and waist to reduce chigger access. Many people habitually tuck pantlegs into socks or boots to minimize exposure to chiggers. Bites should be treated with alcohol. Powdered sulfur (which can be obtained from most pharmacies) dusted into boots and pantlegs is the best preventative.

Ticks are arachnids, closely related to spiders and mites. They are ectoparasites, attaching to passing animals to feed on the host's blood. They attach by burrowing into skin with their mouthparts and must be dislodged with care so as to prevent part of the animal from remaining in the wound. Nail polish remover is helpful in dislodging ticks. Tick wounds should be treated with disinfectant to prevent infection.

Infection

Open wounds, even very small ones, can easily become infected in the Neotropics. Bacteria thrive in the hot, humid climate. Be sure to cleanse any open wound thoroughly with disinfectant. In the field I carry Band-Aids plus a tube of topical antibiotic and treat any wound as soon as it happens. Powerful antibiotic preparations are usually available over the counter in most Latin American countries, but great care should be taken when using an antibiotic without a physician's prescription. Generally, it is best not to do so. And do make sure your tetanus vaccination is up to date before leaving for the Neotropics.

Dogs

Do not fraternize with dogs. Tropical dogs are usually malnourished and mangy, and often generously supplied with festering wounds of varying origins. They are as a general rule loaded with parasites and may easily pass eggs to humans if the human pets the dog or is licked by it. Tropical dogs may even be rabid, as rabies is not nearly so uncommon in parts of the tropics as it is in the United States. Cats tend to be more wary than dogs but are nonetheless common throughout the Neotropics and are subject to essentially the same potential problems as dogs. Don't pet the kitties either.

Food Precautions

"Don't drink the water" is a common warning throughout the global tropics. In general, take it seriously. While certain places do have treated and thus potable water, such is often not the case. Rely on bottled mineral water, Coca-Cola, and beer. These are generally safe. Of course you can drink all the coffee or tea you want as long as the water has been boiled first. And don't use ice.

Regarding food, the tropics is one of the few places where the really "healthy" foods are the fried foods. Fried fish and chicken and french fries (*papas fritos*) are safe to eat. All meats should be well done, and hamburger is to be avoided if the sanitation looks in the least questionable. It is not recommended that you consume any raw vegetables, raw or undercooked meat, or raw seafood, and salads are often best avoided. Well-cooked vegetables (such as rice and beans) are safe. Fruits are safe if you can peel them before eating them. Otherwise, the same cautions that prevail with vegetables apply. Dairy products are also potentially problematic. They may be unpasteurized and thus contaminated with bacteria. Breads, cookies, and crackers are safe.

Use your good judgment and common sense. There are many places in Latin America where you can safely dine on rare steak and a tasty salad, with a dessert of ice cream. But there are many places where doing so would be risky.

Montezuma's Revenge, or Turista

Minor diarrhea, often termed Montezuma's revenge or turista, is caused by a "change in intestinal flora," in most cases attributable to bacteria. The condition should last for no more than three to six days. If you lose fluids to diarrhea (and thus are suffering from infrequent urination, dry mouth, headache), remember that bananas and Coca-Cola, both normally readily available throughout the Neotropics, are each excellent sources of potassium, an essential electrolyte sometimes depleted with fluid loss. There are also oral rehydration kits available in packet form. Many tropical travelers rely on such over-the-counter treatments as Pepto-Bismol, Kaopectate, Imodium, or Lomotil when diarrhea ensues. Consult your physician and plan which remedies to include when you travel.

Truly Serious Diseases

Though generally uncommon (for well-prepared travelers), there are certain afflictions associated with the tropics that any visitor must try to avoid. Upon return from the tropics it is wise to report any unusual symptoms promptly. Any intestinal or other odd symptoms, particularly those that may show up soon after the trip, should be checked by a physician familiar with tropical diseases. It is important to state clearly both when and where you were in the tropics. The National Centers for Disease Control monitors tropical diseases geographically, and can be useful in supplying information on risks associated with various tropical areas.

Vaccines are available for typhoid fever, yellow fever, cholera, measles, mumps, rubella, polio, both hepatitis A and hepatitis B, and tetanus. Antimalarial drugs are also readily available. You should consult your physician about which of these you are advised to take.

Malaria is a mosquito-borne disease (genus *Anopheles*) but is caused by a tiny parasitic protozoan (*Plasmodium* spp.) that infects red blood cells. After reproducing, plasmodia emerge in synchrony, rupturing red blood cells and producing symptoms of high fever and intermittent severe chills. Malaria remains relatively widespread, and, unfortunately, drug-resistant strains are becoming more common. Usually, however, most malaria can be prevented by taking prophylactic treatment prior to, during, and following the trip. Some areas have resistant strains of plasmodia, so make sure your physician knows where you are going so that you receive the proper antimalarial drug. To learn where drug-resistant strains are, check with the Centers for Disease Control.

Typhoid fever is a serious bacterial disease caused by *Salmonella typhosa*. It is closely associated, as are other *Salmonella*-caused illnesses, with poor sanitation. Typhoid preventative (either injections or pills) should be taken prior to the trip.

Hepatitis A is a viral disease of the liver also associated with poor sanitation and thus is not uncommon throughout much of the tropics. Hepatitis B is not associated with poor sanitation but rather with contamination from body fluids or hypodermic needles. The chance of contracting hepatitis A is reduced by taking a gamma globulin injection just prior to departure. Separate vaccines are now available to reduce the chances of contracting hepatitis A and hepatitis B.

Cholera, caused by the bacterium *Vibrio comma*, probably originated in India but has spread widely throughout the world and can periodically flare into epidemic proportions in places in Latin America. It is highly contagious, transmitted in contaminated food or water. Severe cholera induces uncontrollable diarrhea and vomiting with extreme loss of fluids. Cholera patients must be constantly rehydrated with saline solution or the disease can be fatal. Cholera is not a problem throughout most of the Neotropics, but you should inquire if it is in the region you are planning to visit. If so, you should protect yourself by taking cholera vaccine prior to the trip.

Yellow fever is a potentially fatal viral disease transmitted by the mosquito *Aedes aegypli*. It was a major scourge during construction of the Panama Canal. Today yellow fever is less widespread, and inoculation is therefore needed only for certain countries. Check with the Centers for Disease Control to learn if your destination harbors yellow fever and if admission to the country requires proof of yellow fever innoculation.

Internal Parasites

Parasites, such as amoebas, trypanosomes, tapeworms, flukes, and roundworms, are common in areas of poor sanitation and can be very serious medical problems if not treated (Markell and Voge 1971).

A common and potentially serious amoebic infection is caused by *Entamoeba histolytica*. Called amoebic dysentery, the tiny protozoans can cause severe ulcerations of the intestines and can spread to other organ systems. Symptoms of intestinal distress such as abdominal pain and bloody stools should be checked by a physician. Amoebic dysentery is treated with drugs.

Trypanosomes are protozoan flagellates that infect the bloodstream, causing serious illness. The best-known trypanosome is *Trypanosoma rhodesiense*, responsible for the infamous "sleeping sickness" of Africa. In Latin America a somewhat similar affliction, Chagas' disease, is caused by *T. cruzi*. The protozoan is vectored by many insect species, but *Panstrongylus megistus*, the reduviid bug, is considered to be the most important. Chagas' disease can be fatal, though some people recover completely and others enter remission but continue to harbor the parasite. Symptoms include high fever, heart irregularities, and digestive difficulties. Chagas' disease is treated with drugs.

Related to trypanosomes are protozoans collectively called leishmanias. The most common leishmanias in the Neotropics cause severe and rapidly spreading skin lesions (somewhat similar to leprosy), most of which begin as sores. Unusual skin conditions, sores, rashes, etc. should be inspected by a physician. Leishmanias are vectored by various species of biting sand flies. Leishmaniasis is treated with drugs, especially antimony, but can be difficult to cure.

Flukes are parasitic flatworms, most of which occur in the Old World tropics. Some, however, are found in the Neotropics, among them the blood flukes. One species, *Schistosoma mansoni*, was accidentally established in the Neotropics during the period of the African slave trade. *S. mansoni* causes severe intestinal distress, bloody stools, and liver degeneration. Various drugs are used to kill the parasites. Some schistosome species cause swimmer's itch. Larval schistosomes in fresh water burrow into human skin, causing local irritation.

Tapeworms are also flatworms. Most occur only in the digestive system, but one species, the hydatid worm (*Echinococcus granulosus*), forms cysts in the liver and other organs. Tapeworms are acquired by eating undercooked meats or practicing poor sanitary habits. The hydatid worm, because it invades organs such as the liver, heart, and lung, can be fatal. It is normally found in canines. It alone is reason enough not to touch stray dogs. Hydatid cysts must be removed surgically. Tapeworm infections confined to the intestine are treated with drugs.

Roundworms are common in most tropical areas. Some species infect the intestinal system, others attack the blood and other tissues. Among the most common intestinal roundworms are the hookworms, the most common of which is *Necator americanus*. Hookworms invade the body by burrowing through skin, a good reason for not going barefoot in the tropics. They eventually find their way to the intestinal system, multiply, and produce symptoms such as colic, nausea, and abdominal pain. Severe infections cause bloody stools and overall lethargy due to blood loss. Drugs are used to eliminate the worms.

Another common roundworm, *Trichinella spiralis*, causes trichinosis, a painful and occasionally fatal muscle affliction. Ingestion of undercooked pork is the usual way in which *Trichinella* invades the human body. Drugs are used for treatment.

One roundworm group, collectively called filarial worms, causes several of the most serious diseases of the tropics, including elephantiasis and loa loa. Fortunately for the Neotropical traveler, most of these are generally confined to the Old World tropics. Some species do occur in Latin America, however, and can cause serious pathology of the subcutaneous tissues, sometimes leading to disfigurement. Larval worms can cause inflammation of the facial area, shoulders, and elsewhere, often quite painful. Drug therapy is required.

Be Aware, Be Prepared

Don't be scared. Tropical disease and parasite infection can be avoided by being aware, taking all prudent precautions while in the tropics, and planning ahead. Know where your trip is taking you and what maladies might await you. Does drug-resistant malaria occur there? Does yellow fever occur there? Check with the National Centers for Disease Control or a physician familiar with tropical medicine and seek professional advice concerning appropriate innoculations and preventative medicine. See your physician promptly if, after returning, you notice any unexplained symptoms. Proper preparation and diligence not only will protect you but will give you peace of mind and self-confidence, adding to the enjoyment of your journey.

Some Advice about Altitude Sickness

Many Andean visitors are affected by "high-altitude sickness," a condition caused by the lack of oxygen at high elevations. Susceptability to high-altitude sickness is variable from one individual to another. Some are never afflicted, some suffer only minor symptoms, and others are severely affected and must be removed to a lower elevation or the condition can actually become life threatening. Symptoms include headache, which can be unrelenting, nausea, and often sleeplessness. In severe cases there can be swelling of brain tissue and fluid in the lungs. Shortness of breath is normal at high elevations and is not a symptom of impending high-altitude sickness. High-altitude sickness is best avoided by being sure that you climb slowly, take time to regain your strengh if you feel exhausted, and make an effort not to overexert. It is very important to remain well hydrated and to ingest lots of carbohydrate. It is equally important that you avoid alcohol and tobacco. Medications, mostly sulfa drugs, are available that help reduce susceptibility, but they should be taken only after consultation with a physician. In some areas a weak coca tea is widely consumed to help eleviate symptoms of high-altitude sickness. Worked for me.

Basic Common Sense Regarding Local Politics and Other Realities

It is a very good idea to familiarize yourself with the political realities of the country or countries that you plan to visit. If you have doubts about the political stability or safety of a region, you can call a hot-line number at the U.S. Department of State that will give you a recorded message of all travel advisories (also available on the Internet). Note, however, that political situations can be volatile, arising quickly. You might want to familiarize yourself with the location of the American consulate in areas where you intend to be.

Many Latin American countries have police and/or army checkpoints scattered throughout. It is not at all uncommon to be stopped and asked to produce your travel documents and identification. Never be without your passport. When questioned, make sure you are polite and respectful. If you do not understand Spanish or Portuguese, apologize and emphasize with your body language and facial expressions that you wish to cooperate. Avoid any temptations to buy or use illicit drugs. The penalties for possession are extremely harsh, and you are not protected by the constitutional rights granted you in the United States. It would be your worst nightmare.

Conservation Ethics

For moral as well as legal reasons, please NEVER do any of the following things.

1. Buy or otherwise obtain historical cultural artifacts that belong to the heritage of the country in which you are a guest. I am obviously not talking about blowguns and carved toucans for sale at places like airport gift shops.

2. Buy or otherwise obtain any product made from or taken from a species of animal, especially an endangered species. It is against the law in the United States to import any product from an endangered species. In general, any animal souvenirs (pressed butterflies, insects, snake skin, dried caiman heads, bird feathers), even if not from endangered species, should be avoided. Buying them merely encourages further exploitation of the creatures.

Acronyms

ACEER	Amazon Center for Environmental Education and Research
ALAS	Arthropods of La Selva
BCI	Barro Colorado Island
BDFFP	Biological Dynamics of Forest Fragments Project
CET	Children's Environmental Trust Foundation, International
CITES	Convention on International Trade in Endangered Species of Wild Flora And Fauna
dbh	diameter breast height
FAO	Food and Agricultural Organization
GPP	gross primary productivity
IMAZON	Amazon Institute of Man and the Environment
INPA	National Institute for Research in Amazonia
LAI	leaf area index
mtDNA	mitochondrial DNA
NPP	net primary productivity
NPV	net present value
OTS	Organization for Tropical Studies
RAP	Rapid Assessment Program
TEM	Terrestrial Ecosystem Model
VAM	vesicular arbuscular mycorrhizae

Glossary

Adaptive radiation. The relatively recent evolution of many species from a common ancestor, with each species occupying a unique ecological niche.

Air plant. See epiphyte.

Allelochemic. A type of chemical synthesized by plants that confers some protection against herbivores or predators. Common examples are terpenoids and phenolic. See defense compound.

Allopatric. Term describing two populations (or species) that are geographically separated and thus cannot potentially interbreed.

Altiplano. High-elevation flat plain between eastern and western Andean ridges.

Apogeotropic roots. Roots that grow upward from the soil on the trunk of another tree.

Aposematic coloration. Conspicuous, usually bright animal coloration that typically signals that the animal is aggressive or toxic. Also called warning coloration.

Araucaria. Type of high forest dominated by Parana Pine (*Araucaria angustifolia*) found above 1,200 meters in southeastern Brazil.

Arboreal. Inhabitant of canopy and subcanopy, living in trees and shrubs.

Arrested litter. Litter that becomes trapped in epiphytes and the crowns of trees such as understory palms, and thus does not fall to the forest floor.

Arum. A group of plants to which philodendrons belong. Florescence organized in an arrangement where a leafy petal called a spathe surrounds a central spike of flowers called a spadix.

Automimicry. When a palatable organism gains protection from potential predators by proximity to other organisms of the same species that are unpalatable.

Batesian mimicry. A situation in which a palatable animal species comes to resemble an unpalatable species, thus gaining some protection from predation.

Biodiversity. The biological richness of an area or geographic region, usually defined as the number of species present.

Biogeochemical cycling. The continuous and cyclic movement of nutrient atoms such as calcium, nitrogen, phosphorus, and magnesium between the living and the nonliving components of the ecosystem.

Biomass. The total weight of living material in an ecosystem.

Blackwater river. A clear river that carries little or no sediment but is typically stained dark by phenolics, tannins, and other plant compounds. Also see igapo.

Broadleaf evergreen forest. A kind of forest where the trees are evergreen but with broad leaves, not needlelike leaves. Tropical moist forests are examples.

Bromeliad. A type of epiphyte characterized by a basal cluster of spikelike leaves and an elongated flower stalk.

Buttress. A tree root that extends out from the trunk as a phlangelike structure.

Caatinga. A Brazilian term for highly seasonal (with prolonged dry season) deciduous forest dominated by spiny trees and shrubs. Found extensively in eastern Brazil.

Campos. A Brazilian term (meaning fields) for any of several types of low and mid-elevational grassland that are usually seasonally wet.

Canopy. The uppermost layer of foliage in a forest.

Cauliflory. The characteristic of having flowers (and thus fruits) grow directly from the main trunk of a tree.

Centers of endemism. Areas with a proportionally large number of endemic species.

Cerrado. A Brazilian term (meaning closed) for a dense, dry woodland of small-stature trees and shrubs amidst savanna.

Chaco. An extensive flatland of dry shrubby woodlands, marshes, gallery forest and palm savanna found mostly in Brazil but also Paraguay and Bolivia.

Chiropterophilous. Bat-pollinated plants with characteristics adapted to attract bats and to aid bats in accessing nectar.

Clearwater river. A river with a low sediment load and moderate amounts of phenolics and related compounds. Intermediate between blackwater and whitewater conditions.

Cloud forest. A mountain forest that exists in perpetual mist, characterized by stunted trees with an abundance of epiphytic growth.

Coevolution. The evolutionary interaction of two or more species acting as selection pressures upon each other.

Colonizer plant species. Species with rapid dispersal capability, rapid growth rate, high reproductive output, and general physiological hardiness; effective at establishing in newly created gaps. Species typical of early ecological succession.

Commensal. An ecological relationship in which one species benefits from association with another but the other is neither harmed nor aided by the association.

Community. The total assemblage of plants, animals, and microbial organisms that interact within a given ecosystem.

Conspecific. Belonging to the same species (as "congeneric" refers to belonging to the same genera).

Cryptic coloration. Camouflaged appearance rendering the animal less visible.

Deciduous. The characteristic of dropping leaves during periods of stress due to dryness.

Decomposer food web. The diverse community of mostly microorganisms (fungi, bacteria, protozoans, and tiny animals) that rely on dead material as their principal energy source; also the minute animals that prey upon them.

Defense compound. A type of chemical synthesized by plants or, in some cases, by animals that confers some protection against herbivores or predators. Common examples are terpenoids and phenolics in plants, batrachotoxins in frogs.

Diffuse competition. The cumulative effect of slight competition by many species on a single species.

Drip tip. The sharply pointed tip of a typical tropical leaf, named for its tendency to drip rainwater.

Dry forest. A forest of short-stature trees that tend to drop leaves during dry season.

Ecological niche. The ecological role of a species in a community; includes consideration of physiological tolerances as well as interactions, both positive and negative, with other species.

Ecosystem. The total interacting living (biotic) and nonliving (abiotic) components of a given area.

Edaphic. Pertaining to the soil.

Emergent. A tree whose height well exceeds that of the average canopy tree, and thus whose crown is conspicuous above the canopy.

Endemic. Native to and found exclusively in one region or habitat type.

Endemic species. A species native to a specific region (and usually found only there).

Endemism. The percentage of endemic species that occur in a given habitat type.

Epiphyllus community. The small to minute flora and fauna that colonize tropical leaves and branches: lichens, mosses, liverworts, bacteria, fungi, protozoans, slime molds, others.

Epiphyte. A plant that lives on another plant but does not parasitize it. Examples are lichens, bromeliads, and orchids. Same as air plant.

Equilibrium plant species. Species characterized by moderate to poor dispersal abilities, and relatively slow growth, but generally shade tolerant and a good competitor. Species typical of late ecological succession.

Extrafloral nectaries. Nectar-rich bodies present on many tropical plants that are fed upon by ants and wasps but are not flowers.

Flash colors. Bright areas of coloration, normally concealed, on certain tree frogs, lepidopterans, and other animals that, when suddenly revealed, may act temporarily to confuse a potential predator.

Frequency dependent selection. A form or natural selection in which the evolutionary fitness of a trait varies with its abundance in the population.

Frugivore. An animal that eats primarily fruit.

Gallery forest. A generally lush forest that grows along a riverbank and floodplain.

Gap. An opening, large or small, natural or anthropogenic, in the canopy creating high light intensity.

Genetic drift. Gene frequency changes in a population due only to chance.

Herbivore. An animal that eats only plant material, such as a caterpillar or a fungus-garden ant.

Humification. The chemical conversion of dead material to humus, an amorphous, biochemically complex substance that represents the partial decomposition of plant material, mostly leaves.

Humus. The complex organic material resulting from the decomposition of forest leaf and branch litter.

Igapo. Seasonally flooded gallery forest along blackwater rivers such as the Rio Negro.

Jungle. A tangled, dense successional ecosystem consisting of many fast-growing, light-loving species.

Kin selection. When the Darwinian fitness of an organism includes genes shared in common with close relatives that cohabit the same immediate area.

Laterite. A kind of tropical soil high in aluminum and iron compounds, often reddish in color.

Laterization. The process whereby leaching caused by heavy rains plus high temperatures converts lateritic soils into hardened, bricklike material.

Latitudinal cline. Trend that increases or decreases with latitude. Example: alkaloid content in plant leaves tends to decrease with increasing latitude. Also see latitudinal diversity gradient.

Latitudinal diversity gradient. The strong tendency for various taxonomic groups to have their maximum species richness in the equatorial tropics, with species richness diminishing with increasing latitude (i.e., toward polar regions).

Leaching. The removal of nutrient atoms from soils brought about by the effect of rainwater washing through the soils, interacting with the clay component of the soil.

Leaf area index (LAI). The total area covered by leaves (photosynthetic surface) over a given area (usually a square meter) of ground.

Lek. An area in which several, and sometimes several dozen, males court passing females. May be concentrated or dispersed, covering a larger area where males are not in direct view of one another.

Liana. A type of woody vine that begins as a small shrub and entwines upward, entangling throughout the canopy.

Life zone. A recognizable band of vegetation within a certain altitudinal range along a mountain slope.

Llanos. A vast area of seasonal marshland in mostly in central and southern Venezuela but also in Colombia and northeastern Brazil.

Machete. A long-bladed knife used to cut tropical vegetation.

Mangrove. A group of unrelated tropical woody plants that are highly tolerant of immersion in salt water. They comprise a major ecosystem type along tropical coasts.

Milpa. Central American term (mostly Belizean) for slash-and-burn plot.

Mimicry. A situation in which an organism, through either appearance, behavior, or both, comes to resemble closely another, unrelated species that shares the same ecosystem. See Batesian mimicry, Mullerian mimicry, and mimicry complex.

Mimicry complex. A situation in which a group of several butterfly species, including some from different taxonomic families, converge in appearance.

Mixed flock. A foraging flock of birds comprised of several, and sometimes many, species.

Moist forest. A seasonal tropical forest receiving not less than 100 mm (approximately 4 in) of rainfall in any month for two out of three years, frost-free, with an average temperature of 24°C (75°F) or more. See rainforest.

Monolayered. A pattern in which leaves are arranged in a single layer in a shell around a tree. Characteristic of trees in shady areas.

Monopodial. A branching pattern in which lateral branches emerge from a tree bole such that the overall shape of the tree is lollipop-like.

Montane. Mountainous.

Mullerian mimicry. A situation in which two or more unpalatable species converge in appearance.

Multilayered. A pattern in which leaves are arranged in an overlapping manner. Characteristic of trees in well-lit areas.

Mutualism. A situation in which two or more species become evolutionarily interdependent in such a way that each benefits the other(s).

Mycorrhizae. A group of fungal species that are mutualists with trees. Fungi, which grow into the roots, aid in taking up minerals from the soil. Trees supply fungi with carbohydrates from photosynthesis.

Natural selection. The mechanism of evolutionary change first formally described by Charles Darwin and Alfred Russel Wallace. Argues that genetic characteristics best suited to a particular environment will be disproportionately passed to offspring because these characteristics confer greater survival value, and thus their possessors reproduce in greater numbers.

Nectarivore. An animal that eats primarily nectar, for example an adult butterfly or a hummingbird.

Neotropics. The term for the American or New World tropics.

Nitrogen fixation. The complex process by which microorganisms, often in association with root systems of certain vascular plants (such as legumes), acquire energy from the conversion of gaseous nitrogen to inorganic nitrate that is subsequently taken up by vascular plants.

Pampas. Temperate grasslands in the central and southern parts of South America.

Pantanal. An extensive area of seasonal grassland and marshland in southwestern Brazil and parts of Bolivia.

Paramo. Mountain shrublands found at higher elevations throughout the Andes.

Parapatric. When the ranges of two populations (or species) meet but do not extensively overlap.

Photosynthesis. The complex biochemical process by which green plants capture a small amount of the sun's light energy and incorporate it, with water and carbon dioxide, into energy-rich sugar compounds.

Population. A group of individuals all of which are the same species.

Prehensile. A type of tail found on certain Neotropical monkeys, anteaters, kinkajou, opossums, and some snakes that functions as a fingerlike limb. The sensitive and dextrous tail is capable of being curled around a branch, holding the animal securely.

Productivity. The total amount of photo-

synthesis that occurs in a glven ecosystem.

Prop root. A root that leaves the trunk or a
branch well above ground and helps
anchor the tree. Same as stilt root.

Puna. A high-elevation grassland found
throughout the Andes.

Pyrophyte. A general term for any species
or group of plants that is clearly
adapted to an environment in which
fire is frequent.

Rainforest. A very wet, essentially nonseasonal forest. See moist forest.

Refugia. Hypothezised shrunken areas of
rainforest that were scattered in Central and South America during the Ice
Age. So named because rainforest species found "refuge" in these rainforests, which were otherwise surrounded by savanna.

Savanna. An ecosystem that is primarily
grassland but with scattered trees and
shrubs.

Search image. A pattern recognized by
predators to aid in quickly identifying
potential prey.

Selection pressure. A characteristic of the
environment of an organism, either
abiotic or biotic, that influences the
probable survival of the organism.

Sexual selection. The process first described
by Charles Darwin in which females
mate preferentially with the most "attractive" males and/or in which males
compete among themselves for females. The result of female choice and
male/male competition is to select for
colorful and/or large males.

Slash-and burn-agriculture. A farming technique in which vegetation is cut and
then burned on small plots, thus opening the plots for planting crops as well
as fertilizing the soil.

Speciation. The process by which a species
evolves into two or more other species.

Species richness. The number of species
of a given taxon within a given ecosystem.

Stratification. The organization of trees in
a forest in different horizontal layers
such as canopy, subcanopy, shrub, and
herb layers.

Subspecies. A genetically distinct popula-

tion within a species that is not reproductively isolated from other populations within the species.

Succession. An ecological process in which
groups of fast growing species colonize
a disturbed area, eventually to be replaced by groups of slower-growing species that are good competitors and
that occupy the site indefinitely.

Superspecies. A group of species that are
morphologically similar and are considered to have evolved quite recently
from a common ancestor. Many examples among Amazonian and Andean
birds and mammals.

Surface roots. Roots that radiate out on,
rather than in, the soil. Most common
on nutrient poor soils.

Swidden agriculture. See slash-and-burn
agriculture.

Sympatric. When two populations' (or species') ranges extensively overlap.

Sympodial. A branching pattern in which
main branches radiate from the bole
of a tree much as the spokes of an
umbrella.

Taproot. A thick, deep central root.

Taxon. A group of organisms that are
members of the same evolutionary
group. For example, birds represent a
taxon, as do mammals, insects, and
flowering plants.

Taxonomy. The science of the classification of organisms.

Tepuis. A group of unique flat-topped,
geologically ancient table mountains
in southeastern Venezuela that remain
from the erosion of the Guianan
Shield. Most have endemic plant
species.

Tepuis. A group of geologically ancient
table mountains (flat-topped) in southeastern Venezuela. Not part of the
Andes.

Terra firme. An area of rainforest off the
floodplain and thus not subject to
flooding.

Transpiration. A metabolic process
whereby plants take up water and minerals from the soil, evaporating the
water via the leaves.

Urticating. Hairs that contain skin irri-
tants.

Vicariance. A situation in which populations are physically separated by some sort of natural barrier, such as a mountain range, unsuitable habitat, or river.

Varzea. A Brazilian term referring to riverine forests along whitewater rivers such as much of the Amazon and some of its tributaries.

Warning coloration. Obviousness of appearance associated with inpalatability or toxicity. Also called aposematic coloration.

Whitewater river. A sediment-rich river that is part of the Andes drainage pattern. See also varzea.

Zooxanthellae. Symbiotic photosynthetic dinoflagellate algae that live within corals and provide some photosynthate to the corals, while using the corals for shelter and as a nutrient source.

Neotropical Field Guides and Handbooks

Research and Tourist Facilities

Castner, J. L. 1990. *Rainforests: A guide to research and tourist facilities at selected tropical forest sites in Central and South America.* Gainesville, FL: Feline Press.

Birds

Alden, P. 1969. *Finding the birds in western Mexico.* Tucson: Univ. of Arizona Press. Unfortunately, now out of print, but still useful if you can find it.

Beavers, R. A. 1992. *The birds of Tikal: An annotated checklist for Tikal National Park and Peten, Guatemala.* College Station: Texas A&M Press.

Bond, J. 1993 (reissue). *A field guide to birds of the West Indies.* Boston: Houghton Mifflin Company.

Davis, L. I. 1972. *Field guide to the birds of Mexico and Central America.* Austin: Univ. of Texas.

Downer, A., and R. L. Sutton. 1990. *Birds of Jamaica, a photographic field guide.* Cambridge: Cambridge Univ. Press.

Dubs, B. 1992. *Birds of southwestern Brazil: Catalogue and guide to the birds of the Pantanal of Mato Grosso and its border areas.* Kusnacht, Switzerland: Bertrona-Verlag.

Dunning, J. S. 1989. *South American birds: A photographic aid to identification.* Newtown Square, PA: Harrowood Books.

Edwards, E. P. 1989. *A field guide to the Birds of Mexico, with Belize.* Sweet Briar, VA: E. P. Edwards.

ffrench, R. 1991. *A guide to the birds of Trinidad and Tobago.* Ithaca: Cornell Univ. Press.

Fjeldsa, J., N. Krabbe, and J. Fjeldsa. 1990. *Birds of the High Andes, a manual of the birds of the temperate zone of the Andes and Patagonia.* Svendborg, Denmark: Zoological Museum, University of Copenhagen, and Apollo Books.

Flisch, J. D. 1981. *Aves Brasileiras.* San Paulo: Dalgas-Ecoltec Ecologia Tecnica e Comercio Ltda.

Hilty, S. L., and W. L. Brown. 1986. *A guide to the birds of Colombia.* Princeton: Princeton Univ. Press.

Howell, S. N. G., and S. Webb. 1995. *A guide to the birds of Mexico and northern Central America.* Oxford: Oxford Univ. Press.

Isler, M. L., and P. R. Isler. 1987. *The tanagers: Natural history, distribution and identification.* Wash., DC: Smithsonian Inst. Press.

Koepcke, M. 1964. *The birds of the department of Lima, Peru.* Wynnewood, PA: Livingston.

Land, H. C. 1970. *Birds of Guatemala.* Wynnewood, PA: Livingston.

MeyerdeSchauensee, R., and W. H. Phelps, Jr. 1977. *A guide to the birds of Venezuela.* Princeton: Princeton Univ. Press.

Peterson, R. T., and E. L. Chaliff. 1973. *A field guide to Mexican birds.* Boston: Houghton Mifflin.

Raffaele, H. A. 1989. *A guide to the birds of Puerto Rico and the Virgin Islands.* Princeton: Princeton Univ. Press.

Ridgely, R. S., and J. A. Gwynne. 1989. *A guide to the birds of Panama with Costa Rica, Nicaragua, and Honduras, second edition.* Princeton: Princeton Univ. Press.

Ridgely, R. S., and G. Tudor. 1989. *The birds of South America, Vol. 1: The oscine passerines.* Austin: Univ. Texas Press.

———. 1994. *The birds of South America, Vol. 2: The suboscine passerines.* Austin: Univ. Texas Press.

More volumes by Ridgely and Tudor are in preparation until all species are treated. These volumes are essential for anyone interested in Neotropical ornithology.

Sick, H. 1993. *Birds in Brazil, a natural history.* Princeton: Princeton Univ. Press.

Smithe, F. D. 1966. *The birds of Tikal.* New York: Natural History Press.

Stiles, F. G., and A. F. Skutch. 1989. *A guide to the birds of Costa Rica.* Ithaca: Cornell Univ. Press.

Field guides to birds of Ecuador and Peru are in preparation as this volume goes to press.

Mammals

Eisenberg, J. F. 1989. *Mammals of the Neotropics: Vol. 1, The Northern Neotropics, Panama, Colombia, Venezuela, Guyana, Suriname, French Guiana.* Chicago: Univ. of Chicago Press.

Emmons, L. H. 1990. *Neotropical rainforest mammals: A field guide.* Chicago: Univ. of Chicago Press.

Redford, K. H., and J. F. Eisenberg. 1992. *Mammals of the Neotropics, the Southern Cone: Vol. 2, Chile, Argentina, Uruguay, Paraguay.* Chicago: Univ. of Chicago Press.

Reptiles and Amphibians

Campbell, J. A., and W. W. Lamar. 1989. *The venomous reptiles of Latin America.* Ithaca: Comstock Publishing Associates.

Lee, J. C. 1996. *Amphibians and reptiles of the Yucatan Peninsula.* Ithaca: Cornell Univ. Press.

Meyer, J. R., and C. F. Foster. 1996. *A guide to the frogs and toads of Belize.* New York: Krieger.

Rodriguez, L. O., and W. E. Duellman.

1994. *Guide to the frogs of the Iquitos Region, Amazonian Peru.* Lawrence: Nat. Hist. Mus., Univ. of Kansas.

Butterflies

D'Abrera, B. 1984. *Butterflies of South America.* Victoria, Australia: Hill House.

de la Maza Ramirez, R. 1991. *Mariposas Mexicanas.* Fondo de Cultura Economica.

DeVries, P. J. 1987. *The butterflies of Costa Rica and their natural history.* Princeton: Princeton Univ. Press.

Other Insects

Gagne, R. J. 1994. *The gall midges of the Neotropical region.* Ithaca: Cornell Univ. Press.

Hogue, C. L. 1993. *Latin American insects and entomology.* Berkeley: Univ. of California Press.

Penny, N. D., and J. R. Arias. 1982. *Insects of an Amazonian forest.* New York: Columbia Univ. Press.

Quintero, D., and A. Aiello, eds. 1992. *Insects of Panama and Mesoamerica: Selected studies.* Oxford: Oxford Univ. Press.

Plants

Croat, T. B. 1978. *Flora of Barro Colorado Island.* Palo Alto: Stanford Univ. Press.

Dressler, R. L. 1993. *Field guide to the orchids of Costa Rica and Panama.* Ithaca: Cornell Univ. Press.

Galloway, D. J. 1991. *Tropical lichens: Their systematics, conservation, and ecology.* Oxford: Oxford Univ. Press.

Gentry, A. H. 1993. *A field guide to the families and genera of woody plants of northwest South America.* Wash., DC: Conservation International.

Henderson, A., G. Galeano, and R. Bernal. 1995. *Field guide to the palms of the Americas.* Princeton: Princeton Univ. Press.

Lotschert, W. and G. Beese. 1981. *Collins guide to tropical plants.* London: Collins.

McQueen, J., and B. McQueen. 1993. *Orchids of Brazil.* Timber Press.

Schultes, R. E., and R. F. Raffauf. 1990. *The healing forest: Medicinal and toxic plants of Northwest Amazonia.* Portland, OR: Dioscorides Press.

Schultes, R. E., and A. Hofmann. 1992. *Plants of the gods: Their sacred, healing, and hallucinogenic powers.* Rochester, VT: Healing Arts Press.

Recommended Book-Length References

THE following list of books, updated and revised from that contained in the first edition is recommended to the reader who wishes to delve further into the literature about tropical biology. For convenience, I have divided it into three sections: historical, popular, and technical.

Historical References

Everyone should read these.

Bates, H. W. 1892. *The naturalist on the river Amazons.* London: John Murray. Classic account of Amazonian natural history, and quite wonderful. Available in libraries and recently reprinted in paperback (Penguin Nature Library).

Beebe, W. 1918. *Jungle peace.* London: Witherby.

——. 1921. *Edge of the jungle.* New York: Henry Holt. Both this and the previous volume contain short, delightful essays on tropical ecology. Classic.

Belt, T. [1874] 1985 reissue. *The naturalist in Nicaragua.* Chicago: Univ. of Chicago Press. One of the best of the classic exploratory accounts, focused entirely on Central America.

Chapman, F. M. 1938. *Life in an air castle.* New York: Appleton-Century. Highly readable with much information.

Darwin, C. R. [1906] 1959. *The voyage of the Beagle.* Reprint. London: J. M. Dent and Sons. One of the best classic accounts of travel throughout South America. Many reprinted editions are available. Must reading.

Wallace, A. R. 1895. *Natural selection and tropical nature.* London: Macmillan. This delightful book contains vivid descriptions, in glorious Victorian prose, of Wallace's experiences in Amazonia.

Waterton, C. [1825] 1983. *Wanderings in South America.* Reprint. London: Century Publishing. A very entertaining narrative by a rather eccentric but perceptive explorer.

Popular References

Ayensu, E. S., ed. 1980. *The life and mysteries of the jungle.* New York: Crescent Books. Lavish with color illustrations, large in format, this popular account (note title) is nonetheless full of interesting information.

Caufield, C. 1984. *In the rainforest.* Chicago: Univ. of Chicago Press. Discusses both tropical ecology and conservation issues for the general reader.

Collins, M., ed. 1990. *The last rainforests: A world conservation atlas.* New York: Oxford Univ. Press. This volume contains an abundance of maps and general information on the global tropics.

Denslow, J. S., and C. Padoch, eds.. 1988. *People of the tropical rain forest.* Berkeley: University of California Press. A multi-authored, well-illustrated volume that provides an insightful and highly readable introduction to anthropogenic issues as well as basic cultural anthropology in the global tropics.

Forsyth, A. 1990. *Portraits of the rainforest.* Altona, Manitoba: Camden House. Magnificent color photographs illustrate this coffee-table-sized book, but don't neglect Forsyth's highly reader-friendly narrative.

Forsyth, A., and K. Miyata. 1984. *Tropical nature.* New York: Charles Scribner's Sons. A charming and instructive series of essays on various facets of Neotropical ecology.

Goulding, M. 1990. *Amazon: The flooded forest.* London: Guild Publishing. An engaging and well-illustrated introduction to the various habits and natural history of the Amazon River, with an emphasis on floodplain ecology.

Goulding, M., N. J. H. Smith, and D. J. Mahar. 1996. *Floods of forturne: Ecology and economy along the Amazon.* New York: Columbia University Press. A broad and highly readable overview of the past history and current practices among the people whose livings are supported by the Amazon River and who affect the river. Strong emphasis on conservation issues.

Janzen, D. H. 1975. *Ecology of plants in the tropics.* London: Edward Arnold. Excellent though brief account of plant adaptations in the tropics. Emphasizes allelopathy, seed dispersal, pollination ecology.

Moffett, M. W. 1993. *The high frontier: Exploring the tropical rainforest canopy.* Cambridge: Harvard University Press. Discusses the newly emerging knowledge of canopy ecology, about which the author is one of the foremost authorities. Includes global tropics, not confined to Neotropics. Lavishly illustrated.

Myers, N. 1984. *The primary source: Tropical forests and our future.* New York: W. W. Norton. An overview of conservation issues involving the world's tropical areas. Excellent and concise presentations of the many complex issues, mostly revolving around the consequences of deforestation.

Perry, D. 1986. *Life above the jungle floor.* New York: Simon and Schuster. Layperson's account of Perry's unique research in the rainforest canopy.

Plotkin, M. J. 1993. *Tales of a shaman's apprentice: An ethnobotanist searches for new medicines in the Amazon rain forest.* New York: Viking (available in paperback from Penguin). Enjoyable and instructive narrative of Plotkin's initiation into the multiple uses of tropical plants by Amazonian peoples. Very highly recommended.

Smith, A. 1990. *Explorers of the Amazon.* Chicago: Univ. of Chicago Press. Not ecological but provides a fine synopsis of the lives of some of the most important historical figures in Amazonian exploration.

Snow, D. W. 1976. *The web of adaptation.* New York: Demeter Press Quadrangle. Superb brief discussions of sexual selection and frugivory among tropical bird species. Emphasizes manakins, cotingas, and oilbirds, all from the author's and his wife's work.

Wilson, E. O. 1992. *The diversity of life.* Cambridge: Belknap Press of Harvard Univ. Press. A widely acclaimed overview of evolutionary biology, ecology, and the importance of understanding biodiversity. Includes excellent discussions of tropical issues and examples.

Technical References

Anderson, A. B., ed. 1990. *Alternatives to deforestation: Steps toward sustainable use of the Amazon rain forest.* New York: Columbia Univ. Press. A multiauthored selection of papers, all of which focus on how to achieve sustainable use in Amazonia.

Buckley, P. A., M. S. Foster, E. S. Morton, R. S. Ridgely, and F. G. Buckley. 1985. *Neotropical ornithology.* Wash., DC: American Ornithologists' Union. This large volume contains sixty-three technical papers about tropical bird ecology.

Gentry, A. H., Ed. 1990. *Four Neotropical rainforests.* New Haven: Yale Univ. Press. Excellent techincal studies of the flora and fauna of rainforests in Panama, Costa Rica, Brazil, and Peru.

Hagan, J. M., III, and D. W. Johnston, eds. 1992. *Ecology and conservation of Neotropical migrant landbirds.* Wash., DC: Smithsonian Inst. Press. A large volume of technical and review papers from a major symposium. Provides a very robust overview of avian migrant ecology.

Janzen, D. H., ed. 1983. *Costa Rican natural history*. Chicago: Univ. of Chicago Press. An excellent general reference book obviously confined to Central America. It contains brief life histories of selected species from all major taxons plus very well written overviews of the ecology of climate, geography, agriculture, plants, mammals, reptiles and amphibians, and arthropods. In all, 174 contributors.

Leigh, E. G., Jr., A. S. Rand, and D. M. Windsor, eds. 1982. *The ecology of a tropical forest: Seasonal rhythms and long-term changes*. Wash., DC: Smithsonian Inst. This volume contains 32 papers that, if you read them all, will certainly convince you of the importance of seasonal variability in the tropics. The papers deal mostly with Barro Colorado Island in Panama.

Leith H., and M. J. A. Werger, eds. 1989. *Tropical rain forest ecosystems: Biogeographical and ecological studies (Ecosystems of the world, 14B)*. Amsterdam: Elsevier. A technical volume of 35 chapters, each written by experts in the various fields. Attempts a broad overview of global tropical rainforests. Valuable reference for comparing among Asian, African, and Neotropical rainforests.

MacArthur. R. H. 1972. *Geographical ecology: Patterns in the distribution of species*. New York: Harper & Row, reprinted by Princeton Univ. Press. A fine introduction to tropical ecology through a comparison with the temperate zone. Excellent discussion of species diversity. Somewhat mathematical.

McDade, L. A., K. S. Bawa, H. A. Hespenheide, and G. S. Hartshorn, eds. 1994. *La Selva: Ecology and natural history of a Neotropical rain forest*. Chicago: Univ. of Chicago Press. A multiauthored volume dealing with all aspects of the ecology of one of the most thoroughly studied Neotropical forests, written by those who, for the most part, were central figures in the research.

Prance, G. T., ed. 1982. *Biological diversification in the tropics*. New York: Columbia Univ. Press. This volume consists of 36 technical papers examining the rufugia theory for diversification in the tropics.

Prance, G. T., and T. E. Lovejoy. 1985. *Amazonia*. Oxford: Pergamon Press. An excellent collection of review articles written by various experts, many of which are cited in the text of this volume. Topics include geology and hydrology, botany and zoology, and applied human ecology in Amazonia.

Reagan, D. P., and R. B. Waide, eds. 1996. *The food web of a tropical rain forest*. Chicago: Univ. of Chicago Press. A highly laudable attempt to analyze all components of the food web of El Verde forest in Puerto Rico. Because Puerto Rico is an island, much of the analysis is not directly applicable to mainland rainforest. Nonetheless, the volume abounds in insightful information and serves as a model for the integration of diverse studies in understanding a complex ecosystem.

Redford, K., and C. Padoch, eds. 1992. *Conservation of Neotropical forests: Working from traditional resource use*. New York: Columbia Univ. Press. A multiauthored volume that brings together much information on how various groups of people utilize tropical forest.

Richards, P. W. 1952. *The tropical rain forest*. Cambridge: Cambridge Univ. Press. Dated now but still useful. Discusses soils and vegetation but no animal ecology. Written mostly for the professional botanist.

Robinson, J. G., and K. H. Redford, eds. 1991. *Neotropical wildlife use and conservation*. Chicago: Univ. of Chicago Press. Multiauthored volume that looks at individual case studies of effects of hunting, farming, ranching, etc. Pragmatic and realistic in focus.

Stotz, D. F., J. W. Fitzpatrick, T. A. Parker, III, and D. K. Moskovits. 1996. *Neotropical birds: Ecology and conservation*. Chicago: Univ. of Chicago Press. A superb volume that provides concise descriptions of all major Neotropical habitats as well as thoughtful discussion of priorities for conservation. Useful to any Neotropical ecologist, not just to ornithologists. Contains the ecological and distributional databases for all species of Neotropical birds largely compiled by the late Ted Parker.

Sutton, S. L., ed. 1983. *Tropical rain forest: Ecology and management.* Palo Alto: Blackwell Scientific. Discusses forest structure and function, and plant-animal interactions, through conservation issues.

Literature Cited

Abelson, P. H. 1975. Energy alternatives for Brazil. *Science* 189:417.

———. 1983. Rain forests of Amazonia. *Science* 221, editorial.

Abrahamson, D. 1985. Tamarins in the Amazon. *Science* 85:58–63.

Absy, M. L. 1985. The palynology of Amazonia: The history of the forests as revealed by the palynological record. In *Amazonia*, G. T. Prance and T. E. Lovejoy, eds. Oxford: Pergamon Press.

Akagi, H., O. Malm, F. J. P. Branches, Y. Kinjo, Y. Kashima, J. R. D. Guimaraes, R. B. Oliveira, K. Haraguchi, W. C. Pfeiffer, Y. Takizawa, and H. Kato. 1995. Human exposure to mercury due to goldmining in the Tapajos River Basin, Amazon, Brazil: Speciation of mercury in human hair, blood, and urine. In *Mercury as a Global Pollutant*, D. B. Porcella, J. W. Huckabee, and B. Wheatley, eds. Dordrecht: Kluwer.

Alderton, D. 1991. *Crocodiles & alligators of the world*. Oxford: Facts on File.

Allegretti, M. H. 1990. Extractive reserves: An alternative for reconciling development and environmental conservation in Amazonia. In *Alternatives to deforestation: Steps toward sustainable use of the Amazon rain forest*. A. B. Anderson, ed. New York: Columbia Univ. Press.

Alvarez-Buylla, E. R., and A. A. Garay. 1994. Population genetic structure of *Cepropia obtusifolia*, a tropical pioneer tree species. *Evolution* 48:437–453.

Alvarez-Buylla, E. R., A. Chaos, D. Pinero, and A. A. Garay. 1996. Demographic genetics of a pioneer tropical tree species: Patch dynamics, seed dispersal, and seed banks. *Evolution* 50:1155–1166.

Alvarez-Cordero, E., D. H. Ellis, and P. E. King. 1996. Long term satellite tracking of harpy eagles in Venezuela and Panama. Abstracts of the annual American Ornithologists' Union meeting, Boise State University, August 1996.

Anderson, A. B. ed. 1990. *Alternatives to deforestation: Steps toward sustainable use of the Amazon rain forest*. New York: Columbia Univ. Press.

Anderson, A. B., P. Magee, A. Gely, and M. A. G. Jardim. 1995. Forest management patterns in the floodplain of the Amazon estuary. *Cons. Bio.* 9:47–61.

Andersson, M. 1994. *Sexual selection*. Princeton: Princeton Univ. Press.

Andrade, G. I., and H. Rubio-Torgler. 1994. Sustainable use of the tropical rain forest: Evidence from the avifauna in a shifting-cultivation habitat mosaic in the Colombian Amazon. *Cons. Biol.* 8:545–554.

de Andrade, J. C., and J. P. P. Carauta 1982. The *Cecropia-Azteca* association: A case of mutualism? *Biotropica* 14:15.

Andrews, M. 1982. *Flight of the condor.* Boston: Little Brown.

Armesto, J. J., J. D. Mitchell, and C. Villagran. 1986. A comparison of spatial patterns of trees in some tropical and temperate forests *Biotropica* 18:1–11.

Askins, R. A. 1983. Foraging ecology of temperate-zone and tropical woodpeckers. *Ecology* 64:945–956.

———. 1995. Hostile landscapes and the decline of migratory songbirds. *Science* 267:1956–1957.

Askins, R. A., J. F. Lynch, and R. Greenberg. 1990. Population declines in migratory birds in eastern North America. In *Current Ornithology, Vol. 7,* D. M. Power, ed. New York: Plenum Press.

Augspurger, C. K. 1982. A cue for synchronous flowering. In *The ecology of a tropical forest,* E. G. Leigh, Jr., A. S. Rand, and D. M. Windsor, eds. Wash., DC: Smithsonian Inst. Press.

Austin, O. L. 1961. *Birds of the world.* New York: Golden Press.

Ayensu, E. S., ed. 1980. *The life and mysteries of the jungle.* New York: Crescent Books.

Ayers, J. M. 1986. Some aspects of social problems facing conservation in Brazil. *Trends in Ecol. Evol.* 1:48–49.

Bahn, P. G. 1992. Amazon rocks the cradle. *Nature* 355:588–589.

Baker, H. G. 1983. *Ceiba pentandra* (Ceyba, ceiba, Kapok tree). In *Costa Rican natural history,* D. H. Janzen, ed. Chicago: Univ. of Chicago Press.

Balick, M. J., 1985. Useful plants of Amazonia: A resource of global importance. In *Amazonia,* G. T. Prance and T. E. Lovejoy, eds. Oxford: Pergamon Press.

Balick, M. J., R. Arvigo, and L. Romero. 1994. The development of an ethnobiomedical forest reserve in Belize: Its role in the preservation of biological and cultural diversity. *Cons. Biol.* 8:316–317.

Balick, M. J., and P. A. Cox. 1996. *Plants, people, and culture: The science of ethnobotany.* New York: Scientific American Library.

Barbosa, A. C., A. A. Boischio, G. A. East, I. Ferrari, A. Goncalves, P. R. M. Silva, and T. M. E. da Cruz. 1995. Mercury contamination in the Brazilian Amazon: Environmental and occupational aspects. In *Mercury as a Global Pollutant,* D. B. Porcella, J. W. Huckabee, and B. Wheatley, eds. Dordrecht: Kluwer.

Barker, F. K., and S. M. Lanyon. 1996. A hypothesis of intrafamilial relationships in the toucans and New World barbet (Ramphastidae) based upon variation in cytochrome b. Abstracts of the American Ornithologists' Union annual meeting, Boise State University, August 1996.

Bates, H. W. 1862. Contributions of an insect fauna of the Amazon Valley. *Trans. Linn. Soc. London* 23:495–566.

———. 1863. *The naturalist on the river Amazons.* London: John Murray.

Bates, M. 1964. *The land and wildlife of South America.* New York: Time Inc.

Bawa, K. S., and L. A. McDade. 1994. Commentary. *In La Selva: Ecology and natural history of a Neotropical rain forest,* L. A. McDade, K. S. Bawa, H. A. Hespenheide, and G. S. Hartshorn, eds. Chicago: Univ. of Chicago Press.

Bazzaz, F. A., and S. T. A. Pickett. 1980. Physiological ecology of tropical succession: A comparative review. *Ann. Rev. Ecol. Syst.* 11: 287–310.

Beard, J. B. 1944. Climax vegetation in tropical America. *Ecology* 25:127–158.

Beard, J. S. 1953. The savanna vegetation of northern tropical America. *Ecol. Mono.* 23:149–215.

Beckerman, S. 1987. Swidden in Amazonia and the Amazon rim. In *Comparative farming systems,* B. L. Turner and S. B. Brush, eds. New York: Guilford Press.

Beehler, B. M., and M. S. Foster. 1988. Hotshots, hotspots, and female preference in the organization of lek mating sytems. *Amer. Nat.* 131:203–219.

Belwood, J. J., and G. K. Morris. 1987. Bat predation and its influence on calling behavior in Neotropical katydids. *Science* 238:64–67.

Benson, W. W. 1972. Natural selection for Mullerian mimicry in *Heliconius erato* in Costa Rica. *Science* 176:936–939.

———. 1985. Amazon ant plants. In *Amazonia,* G. T. Prance and T. E. Lovejoy, eds. Oxford: Pergamon Press.

Benson, W. W., K. S. Brown, and L. E. Gilbert. 1976. Coevolution of plants and herbivores: Passion flower butterflies. *Evolution* 29:659–680.

Bentley, B. L. 1976. Plants bearing extrafloral nectaries and the associated ant community: Interhabitat differences in the reduction of herbivore damage. *Ecology* 54:815–820.

———. 1977. Extrafloral nectaries and protection by pugnacious bodyguards *Ann. Rev. Ecol. Syst.* 8:407–427.

Bentley, B. L., and E. J. Carpenter. 1980. The effects of dessication and rehydration on nitrogen fixation by epiphylls in a tropical rainforest. *Microb. Ecol.* 6:109–113.

_____. 1984. Direct transfer of newly fixed nitrogen form free-living epiphyllus microorganisms to their host plant. *Oecologia* 63:52–56.

Bernays, E., and M. Graham. 1988. On the evolution of host specificity in phytophagous arthropods. *Ecology* 69:886–899.

Best, R. C., and V. M. F. da Silva. 1989. Amazon river dolphin, Boto *Inia geoffrensis* (de Blainville, 1817). In *Handbook of marine mammals, Vol. 4: River dolphins and the larger toothed whales*, Ridgway, S. H. and R. Harrison, eds. San Diego: Academic Press.

Bierregaard, R. O., Jr., T. E. Lovejoy, V. Kapos, A. Augusto de Santos, and R. W. Hutchings. 1992. The biological dynamics of tropical rainforest fragments. *Bioscience* 42:859–866.

Bishop, J. A., and L. M. Cook. 1975. Moths, melanism and clean air. *Sci. Amer.* 232:90–99.

Blake, J. G., F. G. Stiles, and B. A. Loiselle. 1990. Birds of La Selva Biological Station: habitat use, trophic composition, and migrants. In *Four Neotropical rainforests*, A. H. Gentry, ed. New Haven: Yale Univ. Press.

Bleiler, J. A., G. A. Rosenthal, and D. H. Janzen. 1988. Biochemical ecology of canavanine-eating seed predators. *Ecology* 69:427–433.

Blydenstein, J. 1967. Tropical savanna vegetation of the llanos of Colombia. *Ecology* 48:1–15.

Bonalume, R. 1991. Deforestation rate is falling. *Nature* 350:368.

Booth, W. 1989. Monitoring the fate of the forest from space. *Science* 243:1428–1429.

Boucher, D. H. 1983. Coffee (cafe). In *Costa Rican natural history*, D. H. Janzen, ed. Chicago: Univ. of Chicago Press.

_____. 1990. Growing back after hurricanes: Catastrophes may be critical to rain forest dynamics. *Bioscience* 40:163–166.

_____. 1991. Cocaine and the coca plant: Traditional and illegal uses. *Bioscience* 41:72–76.

Boulière, F. 1983. Animal species diversity in tropical forests. In *Tropical rain forest ecosystems: Structure and function*, F. B. Golley, ed. Amsterdam: Elsevier Scientific.

Boulière, F., and H. Hadley. 1970. The ecology of tropical savannas. *Ann. Rev. Ecol. Syst.* 1:125–152.

Brack, A. 1992. Nontimber forest products of the Peruvian Amazon. In *Sustainable harvest and marketing of rain forest products*, M. Plotkin and L. Famolare, eds. Wash., DC: Island Press.

Bradbury, J. W. 1981. The evolution of leks. In *Natural selection and social behavior: Research and new theory*, R. D. Alexander and D. W. Tinkle, eds. New York: Chicago Press.

Bradbury, J. W., and R. Gibson. 1983. Leks and mate choice. In *Mate choice*, P. Bateson, ed. Cambridge: Cambridge Univ. Press.

Brandon, C. 1983. *Noctilio leporinus* (Murcielago, pescador, fishing bulldog bat). In *Costa Rican natural history*, D. H. Janzen, ed. Chicago: Univ. of Chicago Press.

Brawn, J. D., S. K. Robinson, and W. D. Robinson. 1996. Research needs for conserving tropical versus temperate forest bird communities. Paper presented at the American Ornithologists Union annual meeting, Boise State University, August 1996.

Brewer, T. F., E. Gonzalez, and D. Felsenstein. 1993. Bacon therapy and furuncular myiasis. *Jour. Amer. Med. Assoc.* 270:2087–2088.

Bridges, E. L. 1949. *Uttermost part of the Earth: Indians of Tierra del Fuego*. New York: E. P. Dutton. [Reprinted by Dover Publications, NY].

Brokaw, N. V. L. 1982. Treefalls: Frequency, timing, and consequences. In *The ecology of a tropical rainforest*, E. G. Leigh, Jr., A. S. Rand, and D. M. Windsor, eds. Wash., DC: Smithsonian Inst. Press.

_____. 1985. Gap-phase regeneration in a tropical forest. *Ecology* 66: 682–687.

Brokaw, N. V. L. 1987. Gap-phase regeneration of three pioneer tree species in a tropical forest. *Journal of Ecology* 75:9–19.

Brooke, J. 1993. Gold miners and Indians: Brazil's frontier war. *New York Times,* September 7, 1993, A4.

Browder, J. O. 1992. The limits of extractivism. *Bioscience* 42:174–182.

Brower, A. V. Z. 1996. Parallel race formation and the evolution of mimicry in *Heliconius* butterflies: a phylogenetic hypothesis from mitochondrial DNA sequences. *Evolution* 50:195–221.

Brower, L. P. 1969. Ecological chemistry. *Sci Amer.* 220:22–29.

Brower, L. P., J. V. Z. Brower, and C. T. Collins. 1963. Experimental studies of mimicry: 7. Relative palatability of Mullerian mimicry among Neotropical butterflies of the subfamily Heliconiinae. *Zoologica* 48:65–84.

Brower, L. P., and J. V. Z. Brower. 1964. Birds, butterflies, and plant poisons: A study in ecological chemistry. *Zoologica* 49:137–159.

Brown, B. J., and J. L. Ewel. 1987. Herbivory in complex and simple tropical successional ecosystems. *Ecology* 68:108–116.

Brown, K. S., Jr., and G. G. Brown. 1992. Habitat alteration and species loss in Brazilian forests. In *Tropical deforestation and species extinction,* T. C. Whitmore and J. A. Sayer, eds. London: Chapman & Hall.

Brown, L., and L. L. Rockwood. 1986. On the dilemma of horns. *Nat. Hist.* 95: 54–62.

Brown, S., and A. E. Lugo. 1982. The storage and production of organic matter in tropical forests and their role in the global carbon cycle. *Biotropica* 14:161–187.

———. 1994. Rehabilitation of tropical lands: A key to sustaining development. *Restoration Ecology* 2:97–111.

Buckley, P. A., M. S. Foster, E. S. Morton, R. S. Ridgely, and F. G. Buckley, eds. 1985. *Neotropical ornithology.* Wash., DC: American Ornithologists' Union.

Burman, A. 1991. Saving Brazil's savannas. *New Scientist* 129:30–34.

Buschbacher, R., C. Uhl, and E. A. S. Serrao. 1988. Abandoned pastures in eastern Amazonia. II. Nutrient stocks in the soil and vegetation. *Jour. Ecol.* 76:682–699.

Bush, M. B., and P. A. Colinvaux. 1994. Tropical forest disturbance: paleoecological records from Darien, Panama. *Ecology* 75:1761–1768.

Cain, A. J., and P. M. Sheppard. 1954. Natural selection in *Cepaea. Genetics* 39:89–116.

Campbell, J. A., and W. W. Lamar. 1989. *The venomous reptiles of Latin America.* Ithaca: Comstock Publishing Associates.

Canby, T. Y. 1984. El Niño's ill wind. *Nat. Geog.* 162:143–183.

Capparella, A. P. 1985. Gene flow in a tropical forest bird: Effects of riverine barriers on the blue-crowned manakin (Pipra *coronata*). Abstracts of the 103d American Ornithologists' Union Meeting.

Carneiro, R. L. 1988. Indians of the Amazonian forest. In *People of the tropical rain forest,* J. S. Denslow and C. Padoch, eds. Berkeley: Univ. of Calif. Press.

Carroll, R. L. 1988. *Vertebrate paleontology and evolution.* New York: W. H. Freeman.

Castner, J. L. 1990. *Rainforests: A guide to research and tourist facilities at selected tropical forest sites in Central and South America.* Gainesville, FL: Feline Press.

Center, T. D., and C. D. Johnson. 1974. Coevolution of some seed beetles (Coleoptera: Bruchidae) and their hosts. *Ecology* 55:1096–1103.

Chagnon, N. 1992. *Yanomamo: The last days of Eden.* New York: Harcourt, Brace, Jovanovich.

Chazdon, R. L., and N. Fetcher. 1984. Photosynthetic light environments in a lowland tropical rain forest in Costa Rica. *J. Ecol.* 72:553–564.

Clark, D. A., 1994. Plant demography. In *La Selva: Ecology and natural history of a Neotropical rain forest,* L. A. McDade, K. S. Bawa, H. A. Hespenheide, and G. S. Hartshorn, eds. Chicago: Univ. of Chicago Press.

Clark, D. A. and D. B. Clark. 1987. Population ecology and microhabitat distribution of *Dipteryx panamensis*, a Neotropical rain forest emergent tree. *Biotropica* 19:236–244.

———. 1992. Life history diversity of canopy and emergent trees in a Neotropical rain forest. *Ecol. Mono.* 62:315–344.

Coimbra-Filho, A. F., and R. A. Mittermeier. 1981. *Ecology and Behavior of Neotropical Primates, Vol. 1.* Academia Brasileira de Ciencias.

Coley, P. D. 1982. Rates of herbivory on different tropical trees. In *The ecology of a tropical forest*, E. G. Leigh, Jr., A. S. Rand, and D. M. Windsor, eds. Wash., DC: Smithsonian Inst. Press.

———. 1983. Herbivory and defensive characteristics of tree species in a lowland tropical forest. *Ecol. Mono.* 53: 209–233.

———. 1984. Plasticity, costs, and antiherbivore effects of tannins in a Neotropical tree, *Cecropia peltata* (Moraceae). *Bull. Ecol. Soc. Amer.* 65:229.

Coley, P. D., J. P. Bryant, and F. S. Chapin III. 1985. Resource availability and plant antiherbivore defense. *Science* 230:895–899.

Colinvaux, P. A. 1989a. Ice-age Amazon revisited. *Nature* 340:188–189.

———. 1989b. The past and future Amazon. *Sci. Amer.* 259:102–108.

Colinvaux, P. A., P. E. De Oliveira, J. E. Moreno, M. C. Miller, and M. B. Brush. 1996. A long record from lowland Amazonia: Forest and cooling in glacial times. Science 274:85–88.

Collar, N. J., L. P. Gonzaga, N. Krabbe, A. Madrono Nietro, L. G. Naranjo, T. A. Parker III, and D. C. Wedge. 1992. *Threatened birds of the Americas: The ICBP/IUCN red data book.* Washington, DC: Smithsonian Inst. Press.

Collar, N. J., and A. T. Juniper. 1992. Dimensions and causes of the parrot conservation crisis. In *New World parrots in crisis: Solutions from conservation biology.* S. R. Beissinger and N. F. R. Snyder, eds. Wash., DC: Smithsonian Inst. Press.

Collins, M., ed. 1990. *The last rain forests.* New York: Oxford Univ. Press.

Colwell, R. K. 1973. Competition and co-existence in a simple tropical community. *Amer. Nat.* 107:737–760.

———. 1985a. Stowaways on the hummingbird express. *Nat. Hist.* 94:56–63.

———. 1985b. A bite to remember. *Nat. Hist.* 94:2–8.

Committee on Research Priorities in Tropical Biology. 1980. *Research priorities in tropical biology.* Wash., D.C.: National Academy of Science.

Condit, R., S. P. Hubbell, and R. B. Foster. 1992. Short-term dynamics of a Neotropical forest. *Bioscience* 42:822–828.

———. 1995. Mortality rates of 205 Neotropical tree and shrub species and the impact of a severe drought. *Ecol. Mono.* 65:419–439.

Connell, J. H. 1978. Diversity in tropical rain forests and coral reefs. *Science* 199:1302–1310.

Connell, J. H., and E. Orias. 1964. The ecological regulation of species diversity. *Amer. Nat.* 98:399–414.

Connell, J. H., and R. O. Slatyer. 1977. Mechanisms of succession in natural communities and their role in community stability and organization. *Amer Nat.* 111:1114–1119.

Conniff, R. 1996. Tarantulas: Earth tigers and bird spiders. *Nat. Geog.* 190:98–115.

Connor, E. F. 1986. The role of Pleistocene forest refugia in the evolution and biogeography of tropical biotas. *Trends in Ecol. Evol.* 1:165–168.

Cox, P. A., and M. J. Balick. 1994. The ethnobotanical approach to drug discovery. *Sci. Amer.* 270:82–87.

Cracraft, J. 1985. Historical biogeography and patterns of differentiation within the South America avifauna: Areas of endemism. *In Neotropical Ornithology*, P. A. Buckley, M. S. Foster, E. S. Morton, R. S. Ridgely, and F. G. Buckley, eds. Ornithol. Monogr. No. 36. Wash., DC: American Ornithologists' Union.

Cracraft, J., and R. O. Prum 1988. Patterns and processes of diversification: Speciation and historical congruence in some Neotropical birds. *Evolution* 42:603–620.

Croat, T. B. 1978. *Flora of Barro Colorado Island.* Palo Alto: Stanford Univ. Press.

Crump, M. L. 1983. *Dendrobates granuliferus* and *Dendrobates pumilio.* In *Costa Rican Natural History*, D. H. Janzen, ed. Chicago: Univ. Chicago Press.

D'Abrera, B. 1984. *Butterflies of South America.* Victoria, Australia: Hill House.

Dale, V. H., R. V. O'Neill, F. Southworth, and M. Pedlowski. 1994. Modeling effects of land management in the Brazilian Amazonian settlement of Rondonia. *Cons. Biol.* 8:196–206.

Dale, V. H., S. M. Pearson, H. L. Offerman, and R. V. O'Neill. 1994. Relating patterns of land-use change to faunal biodiversity in the Central Amazon. *Cons. Biol.* 8:1027–1036.

Daly, J. W., H. M. Garraffo, and T. F. Spande. 1993. Amphibian alkaloids. In *The Alkaloids, vol. 43*, G. A. Cordell, ed. San Diego: Academic Press.

Darwin, C. R. [1839] 1912 edition. *Journal of researches into the geology and natural history of the various countries visited bv H.M.S. Beagle . . . round the world under the command of Capt. Fitz Roy, R. N.* New York: Appleton.

———. 1842. *On the structure and distribution of coral reefs.* London: Scott.

———. 1859. *On the origin of species by means of natural selection of favored races in the struggle for life.* London: John Murray.

———. 1862. *On the various contrivances bv which British and foreign orchids are fertilised by insects, and on the good effects of intercrossing.* London: John Murray.

———. 1871. *The descent of man and selection in relation to sex.* London: John Murray.

———. [1906]. 1959. *The voyage of the Beagle.* Reprint. London: J. M. Dent and Sons.

Davis, E. W. 1983. Preparation ot the Haitian zombie poison. *Harvard Univ. Botanical Museum Leaflets* 29:139–149.

De Barcellos Falkenberg, D., and J. C. Voltilini. 1995. The montane cloud forest in southern Brazil. In *Tropical montane cloud forests: Ecological studies 110.* L. S. Hamilton, J. O. Juvik, and F. N. Scatena, eds. New York: Springer-Verlag.

DeGraaf, R. M., and J. H. Rappole. 1995. *Neotropical migratory birds: Natural history, distribution, and population change.* Ithaca: Cornell Univ. Press.

Delacour, J., and D. Amadon. 1973. *Curassows and related birds.* New York: American Museum of Natural History.

DeNault, L., and D. McFarlane. 1995. Reciprocal altruism between male vampire bats. *Animal Behavior* 49:855–856.

Denevan, W. M. 1976. The aboriginal population of Amazonia. In *The native population of the Americas in 1492*, W. M. Denevan, ed. Madison: Univ. of Wisc. Press.

Denslow, J. S. 1980. Gap partitioning among tropical rainforest trees. In *Tropical succession*, supplement to *Biotropica* 12:47–55.

Denslow, J. S., and G. S. Hartshorn. 1994. Tree-fall gap environments and forest dynamic processes. In *La Selva: Ecology and natural history of a Neotropical rain forest*, L. A. McDade, K. S. Bawa, H. A. Hespenheide, and G. S. Hartshorn, eds. Chicago: Univ. of Chicago Press.

Detwiler, R. D., and C. A. S. Hall. 1988. Tropical forests and the global carbon cycle *Science* 239:42–47.

DeVries, P. J. 1983. *Caligo memnon* (Buhito pardo, caligo, cream owl butterfly). In *Costa Rican natural history*, D. H. Janzen, ed. Chicago: Univ. of Chicago Press.

———. 1987. *The butterflies of Costa Rica and their natural history.* Princeton: Princeton Univ. Press.

———. 1990. Enhancement of symbiosis between butterfly caterpillars and ants by vibrational communication. *Science* 248:1104–1106.

———. 1992. Singing caterpillars, ants and symbiosis. *Sci. Amer.* 267:76–82.

———. 1994. Patterns of butterfly diversity and promising topics in natural history and ecology. In *La Selva: Ecology and natural history of a Neotropical rain forest*, L. A. McDade, K. S. Bawa, H. A. Hespenheide, and G. S. Hartshorn, eds. Chicago: Univ. of Chicago Press.

DeVries, P. J., and I. Baker. 1989. Butterfly exploitation of an ant-plant mutualism: Adding insult to herbivory. *Jour. New York Entomological Soc.* 97:332–340.

Dial, R., and J. Roughgarden. 1995. Experimental removal of insectivores from rain forest canopy: Direct and indirect effects. *Ecology* 76:1821–1834.

Diamond, J. M. 1976. Island biogeography and conservation: Strategy and limitations. *Science* 193:1027–1029.

Dietz, R. S., and J. C. Holden. 1972. The breakup of Pangaea. In *Continents adrift*, J. T. Wilson, ed. Readings from *Sci. Amer.* San Francisco: W. H. Freeman.

Dobzhansky, T. 1950. Evolution in the tropics. *Amer. Sci.* 38:209–221.

Doumenge, C., D. Gilmour, M. R. Perez, and J. Blockhus. 1995. Tropical montane cloud forests: Conservation status and management issues. In *Tropical montane cloud forests: Ecological studies 110.* L. S. Hamilton, J. O. Juvik, and F. N. Scatena, eds. New York: Springer-Verlag.

Dressler, R. L. 1968. Pollination by euglossine bees. *Evolution* 22:202–210.

———. 1981. *The orchids—Natural history and classification.* Cambridge: Harvard Univ. Press.

———. 1993. *Field guide to the orchids of Costa Rica and Panama.* Ithaca: Cornell Univ. Press.

Dubois, J. C. L. 1990. Secondary forests as a land-use resource in frontier zones of Amazonia. In *Alternatives to deforestation: Steps toward sustainable use of the Amazon rain forest.* A. B. Anderson, ed. New York: Columbia Univ. Press.

Duellman, W. E. 1992. Reproductive strategies of frogs. *Sci. Amer.* 267:80–87.

Duke, J. A. 1992. Tropical botanical extractives. In *Sustainable harvest and marketing of rain forest products,* M. Plotkin and L. Famolare, eds. Wash., DC: Island Press.

Dufour, D. L. 1990. Use of tropical rainforests by native Amazonians. *Bioscience* 40:652–659.

Dumbacher, J. P., B. M. Beehler, T. F. Spande, H. M. Garraffo, and J. W. Daly. 1992. Homobatrachotoxin in the Genus *Pitahui*: chemical defense in birds? *Science* 258:799–801.

Dyk, J. V. 1995. The Amazon. *Nat. Geo.* 187:2–39.

Ehrlich, P. R., and P. H. Raven. 1964. Butterflies and plants: A study in coevolution. *Evolution* 18:586–608.

———. 1967. Butterflies and plants. *Sci. Amer.* 216:104–113.

Ehrlich, P. R., and E. O. Wilson. 1991. Biodiversity studies: science and policy. *Science* 253:758–762.

Eisenberg, J. F. 1989. *Mammals of the Neotropics: Vol. 1, The Northern Neotropics, Panama, Colombia, Venezuela, Guyana, Suriname, French Guiana.* Chicago: Univ. of Chicago Press.

Eisner, T., and S. Nowicki. 1983. Spider web protection through visual advertisement: Role of the stabilimentum. *Science* 219:185–186.

Ellison, A. M., and E. L. Farnsworth. 1993. Seedling survivorship, growth, and response to disturbance in Belizean mangal. *Amer. Journ. Botany* 80:1137–1145.

Emlen, S. J., and L. W. Oring. 1977. Ecology, sexual selection, and the evolution of mating systems. *Science* 197:215–223.

Emmons, L. H. 1984. Geographic variation in densities and diversities of non-flying mammals in Amazonia. *Biotropica* 16:210–222.

———. 1987. Comparative feeding ecology of felids in a Neotropical rainforest. *Behav. Ecol. Sociobiol.* 20:271–283.

———. 1990. *Neotropical rainforest mammals: A field guide.* Univ. of Chicago Press.

Erwin, T. L. 1982. Tropical forests: Their richness in Coleoptera and other arthropod species. *Coleopterists' Bulletin* 36:74–75.

———. 1983. Beetles and other insects of tropical forest canopies at Manaus, Brazil, sampled by insecticidal fogging. In *Tropical rain forest: Ecology and management.* S. L. Sutton, T. C. Whitmore, and A. C. Chadwick, eds. London: Blackwell.

———. 1988. The tropical forest canopy: The heart of biotic diversity. In *Biodiversity*, E. O. Wilson, ed. Wash., DC: National Academy Press.

Estrada, A., R. Coates-Estrada, and C. Vaz-quez-Yanes. 1984. Observations of fruiting and dispersers of *Cecropia obtusifolia* at Los Tuxtlas, Mexico. *Biotropica* 16: 315–318.

Ewel, J. 1980. Tropical succession: Manifold routes to maturity. In *Tropical succession*, supplement to *Biotropica* 12:2–7.

———. 1983. Succession. In *Tropical rain forest ecosystems: Structure and function*, edited by F. B. Golley. Amsterdam: Elsevier Scientific.

Ewel, J., C. Berish, B. Brown, N. Price, and J. Raich. 1981. Slash and burn impacts on a Costa Rican wet forest site. *Ecology* 62:816–829.

Ewel, J., S. Gliessman, M. Amador, F. Benedict, C. Berish, R. Bermudez, B. Brown, A. Martinez, R. Miranda, and N. Price. 1982. Leaf area, light transmission, roots and leaf damage in nine tropical plant communities. *Agro-Ecosystems* 7:305–326.

Faaborg, J. 1982. Avian population fluctuations during drought conditions in Puerto Rico. *Wilson Bull.* 94:20–30.

Farnsworth, N. R. 1988. Screening plants for new medicines. In *Biodiversity*, E. O. Wilson, ed. Wash., DC: National Academy Press.

Fearnside, P. M. 1986. *Human carrying capacity of the Brazilian rainforest.* New York: Columbia Univ. Press.

———. 1987. Deforestation and international economic development projects in Brazilian Amazonia. *Conserv. Biol.* 1:214–221.

———. 1989. Extractive reserves in Brazilian Amazonia. *Bioscience* 39:387–393.

Fedducia, A. 1973. *Evolutionary trends in the Neotropical ovenbirds and woodhewers.* Wash., DC: American Ornithologists' Union.

Feigl, B. J., J. Melillo, and C. C. Cerri. 1995. Changes in the origin and quality of soil organic matter after pasture introduction in Rondonia (Brazil). *Plant and Soil* 175:21–29.

Feinsinger, P. 1978. Ecological interactions between plants and hummingbirds in a successional tropical community. *Ecol. Mono.* 48:269–87.

———. 1983. Variable nectar secretion in a *Heliconia* species pollinated by hermit hummingbirds. *Biotropica* 15:48–52.

Feinsinger, P., and R. K. Colwell. 1978. Community organization among Neotropical nectar-feeding birds. *Amer. Zool.* 18:779–95.

Fenton, M. B. 1992. Wounds and the origin of blood-feeding in bats. *Biol. Jour. of the Linnean Soc.* 47:161.

Fetcher, N., S. F. Oberbauer, and R. L. Chazdon. 1994. Physiological ecology of plants. In *La Selva: Ecology and natural history of a Neotropical rain forest*, L. A. McDade, K. S. Bawa, H. A. Hespenheide, and G. S. Hartshorn, eds. Chicago: Univ. of Chicago Press.

Few, R. 1994. *The atlas of the wild places.* Wash., DC: Smithsonian Inst. Press.

Fisher, M. J., I. M. Rao, M. A. Ayarza, C. E. Lascano, J. I. Sanz, R. J. Thomas, and R. R. Vera. 1994. Carbon storage by introduced deep-rooted grasses in South American savannas. *Nature* 371:236–238.

Fittkau, E. J., and H. Klinge. 1973. On biomass and trophic structure of the central Amazonian rain forest ecosystem. *Biotropica* 5:1–14.

Fitzpatrick, J. W. 1980a. Foraging behavior of Neotropical tyrant flycatchers. *Condor* 82:43–57.

———. 1980b. Wintering of North American tyrant flycatchers in the Neotropics. In *Migrant birds in the Neotropics: Ecology, behavior, distribution, and conservation*, ed. A. Keast and E. S. Morton. Wash., DC: Smithsonian Inst. Press.

———. 1985. Foraging behavior and adaptive radiation in the Tyrannidae. In *Neotropical ornithology*, P. A. Buckley, M. S. Foster, E. S. Morton, R. S. Ridgely, and F. G. Buckley, eds. Wash., DC: American Ornithologists' Union.

Fjeldsa, J., and N. Krabbe. 1990. *Birds of the high Andes.* Svendborg, Denmark: Zoological Museum, Univ. of Copenhagen and Apollo Books.

Flannery, K. V., ed. 1982. *Maya subsistence.* New York: Academic Press.

Fleming, T. H. 1983. *Piper* (Candela, candelillos, piper). In *Costa Rican natural history*, D. H. Janzen, ed. Chicago: Univ. of Chicago Press.

———. 1985a. Coexistence of five sypatric *Piper* (Piperaceae) species in a tropical dry forest. *Ecology* 66:688–700.

———. 1985b. A day in the life of a *Piper*-eating bat. *Nat Hist.* 94:52–59.

Fleming, T. H., R. Breitwisch, and G. H. Whitesides. 1987. Patterns of tropical vertebrate frugivore diversity. *Ann. Rev. Ecol. Syst.* 18:91–109.

Fong, P., and D. Lirman. 1994. Damage and recovery on a coral reef following Hurricane Andrew. *Nat. Geog. Soc. Res.* 10:246–248.

Forman, R. T. T. 1975. Canopy lichens with blue-green algae: A nitrogen source in a Colombian rain forest. *Ecology* 56:1176–1184.

Forshaw, J. M. 1973. *Parrots of the world.* Melbourne: Lansdowne Press.

Forsyth, A. 1994. Ted Parker: In memoriam. *Conservation Biol.* 8:293–294.

Foster, M. S. 1977. Odd couples in manakins: A study of social organization and cooperative breeding in *Chiroxiphia linearis. Amer. Nat.* 111:845–853.

Foster, R. B. 1982a. The seasonal rhythm of fruitfall on Barro Colorado Island. In *The ecology of a tropical forest*, E. G. Leigh, Jr., A. S. Rand, and D. M. Windsor, eds. Wash., DC: Smithsonian Inst. Press.

———. 1982b. Famine on Barro Colorado Island. In *The ecology of a tropical forest*, E. G. Leigh, Jr., A. S. Rand, and D. M. Windsor, eds. Wash., DC: Smithsonian Inst. Press.

———. 1990a. The floristic composition of the Rio Manu floodplain forest. In *Four Neotropical rainforests*, A. H. Gentry, ed. New Haven: Yale Univ. Press.

———. 1990b. Long-term change in the successional forest community of the Rio Manu floodplain. In *Four Neotropical rainforests*, A. H. Gentry, ed. New Haven: Yale Univ. Press.

Foster, R. B., and S. P. Hubbell. 1990. The floristic composition of the Barro Colorado Island forest. In *Four Neotropical rainforests*, A. H. Gentry, ed. New Haven: Yale Univ. Press.

Foster, R. B., T. A. Parker III, A. H.

Gentry, L. H. Emmons, A. Chicchon, T. Schulenberg, L. Rodriguez, G. Lamas, H. Ortega, J. Icochea, W. Wust, M. Romo, J. A. Castillo, O. Phillips, C. Reynel, A. Kratter, P. K. Donahue, and L. J. Barkley. 1994. *The Tambopata-Candamo Reserved Zone of southeastern Peru: A biological assessment.* Wash., DC: Conservation International.

Fragoso, J. M. V. 1991. The effect of hunting on tapirs in Belize. In *Neotropical wildlife use and conservation*, J. G. Robinson and K. H. Redford, eds. Chicago: Univ. of Chicago Press.

Furst, P. T., and M. D. Coe 1977. Ritual enemas. *Nat. Hist.* 86:88–91.

Futuyma, D. J. 1983. Evolutionary interaction among herbivorous insects and plants. In *Coevolution*, D. J. Futuyma and M. Slatkin, eds. Sunderland, MA: Sinauer.

Futuyma, D. J., and M. Slatkin, eds. 1983. *Coevolution.* Sunderland, MA: Sinauer.

Garrett, L. 1994. *The coming plague: Newly emerging diseases in a world out of balance.* New York: Farrar, Straus and Giroux.

Garwood, N. C. 1982. Seasonal rhythm of seed germination in a semi-deciduous tropical forest. In *The ecology of a tropical forest*, E. G. Leigh, Jr., A. S. Rand, and D. M. Windsor, eds. Washington, D.C.: Smithsonian Inst. Press.

Gentry, A. H. 1982. Neotropical floristic diversity: Phytogeographical connections between Central and South America, Pleistocene climatic fluctuations, or an accident of Andean orogeny? *Annals of the Missouri Botanical Garden* 69:557–593.

———. 1986a. Species richness and floristic composition of Choco region plant communities. *Caldasia* 15:71–91.

———. 1986b. Endemism in tropical versus temperate plant communities. In *Conservation biology: The science of scarcity and diversity*, M. E. Soule, ed. Sunderland, MA: Sinauer Associates.

———. 1988. Tree species richness of upper Amazon forests. *Proc. Nat. Acad. Sci. U.S.A.* 85:156–159.

———, ed. 1990a. *Four Neotropical rainforests.* New Haven: Yale Univ. Press.

Gentry, A. H. 1990b. Floristic similarities and differences between Southern Central America and Upper and Central Amazonia. In *Four Neotropical rainforests*, A. H. Gentry, ed. New Haven: Yale Univ. Press.

———. 1991. The distribution and evolution of climbing plants. In *The biology of vines*, F. E. Putz and H. A. Mooney, eds. Cambridge: Cambridge Univ. Press.

———. 1992. New nontimber forest products from Western South America. In *Sustainable harvest and marketing of rain forest products*, M. Plotkin and L. Famolare, eds. Wash., DC: Island Press.

———. 1993. *A field guide to the families and genera of woody plants of northwest South America*. Wash., DC: Conservation International.

Gentry, A. H., and J. L. Lopez-Parodi. 1980. Deforestation and increased flooding of the upper Amazon. *Science* 210:1354–1356.

Gentry, A. H., and J. Terborgh. 1990. Composition and dynamics of the Cocha Cashu "mature" floodplain forest. In *Four Neotropical rainforests*, A. H. Gentry, ed. New Haven: Yale Univ. Press.

George, U. 1989. Tepuis — Venezuela's islands in time. *Nat Geo.* 175:526–562.

Gheerbrant, A. 1992. *The Amazon: Past, present, and future*. New York: H. N. Abrams.

Ghiselin, M. T. 1985. A movable feaster. *Nat. Hist.* 94:54–61.

Gibbons, A. 1990. New view of early Amazonia. *Science* 248:1488–1490.

Gilbert, L. E. 1971. Butterfly-plant coevolution: Has *Passiflora adenopoda* won the selectional race with Heliconiine butterflies? *Science* 172:585–586.

———. 1975 Ecological consequences of a coevolved mutualism between butterflies and plants. In *Coevolution of animals and plants*, L. E. Gilbert and P. H. Raven, eds. Austin: Univ. of Texas Press.

———. 1982. The coevolution of a butterfly and a vine. *Sci. Amer.* 247:110–121.

———. 1983. Coevolution and mimicry. In *Coevolution*, D. J. Futuyma and M. Slatkin, eds. Sunderland, MA: Sinauer.

Gill, F. B. 1995. *Ornithology, 2d ed.* New York: W. H. Freeman and Co.

Gladfelter, W. B., J. C. Ogden, and E. H. Gladfelter. 1980. Similarity and diversity among coral reef fish communities: Virgin Islands vs. Marshall Islands. *Ecology* 61:1156–1168.

Glander, K. E. 1977. Poison in a monkey's Garden of Eden. *Nat. Hist.* 86:35–41.

Glynn, P. W. 1988. El Niño-Southern Oscillation 1982–1983: Nearshore population, community, and ecosystem responses. In *Ann. Rev. Ecol. Syst. 1988*. Palo Alto: Annual Reviews.

Goldizen, A. W. 1988. Tamarin and marmoset mating systems: Unusual flexibility. *Trends in Ecol. Evol.* 3:36–40.

Golley, F. B. 1983. Nutrient cycling and nutrient conservation. In *Tropical rain forest ecosystems: Structure and function*, F. B. Golley, ed. Amsterdam: Elsevier Scientific.

Golley, F. B., H. T. Odum, and R. F. Wilson. 1962. The structure and metabolism of a Puerto Rican red mangrove forest in May. *Ecology* 43:9–19.

Golley, F. B., J. T. McGinnis, R. G. Clements, G. I. Child, and H. J. Duever. 1969. The structure of tropical forests in Panama and Colombia. *BioScience* 19:693–696.

———. 1975. *Mineral cycling in a tropical moist forest ecosystem*. Athens: Univ. of Georgia Press.

Gomez-Pompa, A., C. Vazquez-Yanes, and S. Guevara. 1972. The tropical rain forest: A nonrenewable resource. *Science* 177:762–765.

Gomez-Pompa, A., H. L. Morales, E. J. Avilla, and J. J. Avilla. 1982. Experiences in traditional hyraulic agriculture. In *Maya subsistence*, K. V. Flannery, ed. New York: Academic Press.

Gomez-Pompa, A., and A. Kaus. 1990. Traditional management of tropical forests in Mexico. In *Alternatives to deforestation: Steps toward sustainable use of the Amazonian rain forest*, A. B. Anderson, ed. New York: Columbia Univ. Press.

Gore, A. 1992. *Earth in the balance: Ecology*

and the human spirit. Boston: Houghton Mifflin Co.

Gottlieb, O. R. 1985. The chemical uses and chemical geography of Amazon plants. In *Amazonia,* G. T. Prance and T. E. Lovejoy, eds. Oxford: Pergamon Press.

Gotwald, W. H., Jr. 1982. Army ants. In *Social insects, vol. 4,* H. R. Hermann, ed. New York: Academic Press.

———. 1995. *Army ants: The biology of social predation.* Ithaca: Cornell Univ. Press.

Gould, S. J., and N. Eldredge. 1977. Punctuated equilibria: The tempo and mode of evolution reconsidered. *Paleobiology* 3:115–151.

Goulding, M. 1980. *The fishes and the forest: Explorations in Amazonian natural history.* Berkeley: Univ. of Calif. Press.

———. 1985. Forest fishes of the Amazon. In *Amazonia,* G. T. Prance and T. E. Lovejoy, eds. Oxford: Pergamon Press.

———. 1990. *The flooded forest.* London: Guild Publishing.

———. 1993. Flooded forests of the Amazon. *Sci. Amer.* 266:114–120.

Goulding, M., N. J. H. Smith, and D. J. Mahar. 1996. *Floods of fortune: Ecology and economy along the Amazon.* New York: Columbia Univ. Press.

Graham, N. E., and W. B. White. 1988. The El Niño cycle: A natural oscillator of the Pacific Ocean-atmosphere system. *Science* 240:1293–1301.

Grajal, A., S. D. Strahl, R. Parra, M. G. Dominguez, and A. Neher. 1989. Foregut fermentation in the Hoatzin, a Neotropical leaf-eating bird. *Science* 245:1236–1238.

Graves, G. R., J. P. O'Neill, and T. A. Parker III. 1983. *Grallaricula ochraceifrons,* a new species of antpitta from northern Peru. *Wilson Bull.* 95:1–6.

Greenberg, R. 1987a. Development of dead leaf foraging in a tropical migrant warbler. *Ecology* 68:130–141.

———. 1987b. Seasonal foraging specialization in the Worm-eating Warbler. *Condor* 89:158–168.

Greene, H. W., and R. W. McDiarmid. 1981. Coral snake mimicry: Does it occur? *Science* 213: 1207–1211.

Groom, M. J., R. D. Podolsky, and C. A.

Munn. 1991. Tourism as a sustained use of wildlife: A case study of Madre de Dios, southeastern Peru. In *Neotropical wildlife use and conservation,* J. G. Robinson and K. H. Redford, eds. Chicago: Univ. of Chicago Press.

Grubb, P. J. 1977a. Control of forest growth and distribution on wet tropical mountains. *Ann. Rev. Ecol. Syst.* 8:83–107.

———. 1977b. The maintenance of species richness in plant communities: The importance of the regeneration niche. *Biol. Rev.* 52:107–145.

Gutierrez, P. C. 1994. Mitochondrial-DNA polymorphism in the oilbird (*Steatornis caripensis,* Steatornithidae) in Venezuela. *Auk* 111:573–578.

Hacon, S., P. Artaxo, F. Gerab, M. A. Yamasoe, R. C. Campos, L. F. Conti, and L. D. de Lacerda. 1995. Atmospheric mercury and trace elements in the region of Alta Floresta in the Amazon Basin. In *Mercury as a global pollutant,* D. B. Porcella, J. W. Huckabee, and B. Wheatley, eds. Dordrecht: Kluwer.

Haffer, J. 1969. Speciation in Amazonian forest birds. *Science* 165:131–137.

———. 1974. *Avian speciation in tropical South America.* Publication No. 14. Cambridge, MA: Nuttall Ornithology Club.

———. 1985. Avian zoogeography of the Neotropical lowlands. In *Neotropical ornithology,* P. A. Buckley, M. S. Foster, E. S. Morton, R. S. Ridgely, and F. G. Buckley, eds. Wash., DC: American Ornithologists' Union.

———. 1990. Geoscientific aspects of allopatric speciation. In *Vertebrates in the tropics,* G. Peters and R. Hutterer, eds. Bonn: Museum Alexander Koenig.

———. 1991. Mosaic distribution patterns of Neotropical forest birds and underlying cyclic disturbance processes. In *The mosaic-cycle concept of ecosystems, Ecological studies vol. 85,* H. Remmert, ed. New York: Springer Verlag.

———. 1993. Time's cycle and time's arrow in the history of Amazonia. *Biogeographica* 69:15–45.

Haffer, J., and J. W. Fitzpatrick. 1985. Geographic variation in some Amazonian forest birds. In *Neotropical ornithology*, P. A. Buckley, M. S. Foster, E. S. Morton, R. S. Ridgely, and F. G. Buckley, eds. Wash., DC: American Ornithologists' Union.

Hagan, J. M., III, and D. W. Johnston. 1992. *Ecology and conservation of Neotropical migrant landbirds*. Wash., DC: Smithsonian Inst. Press.

Halle, F., R. A. A. Oldman, and P. B. Tomlinson. 1978. *Tropical trees and forests: An architectural analysis*. Berlin: Springer-Verlag.

Hamilton, L. S., J. O. Juvik, and F. N. Scatena, eds. 1995. *Tropical montane cloud forests: Ecological studies 110*. New York: Springer-Verlag.

Hamilton, W. D. 1979. Wingless and fighting males in fig wasps and other insects. In *Reproduction, competition and selection in insects*, M. S. Blum and N. A. Blum, eds. New York: Academic Press.

Hammel, B. 1990. The distribution of diversity among families, genera, and habit types in the La Selva flora. In *Four Neotropical rainforests*, A. H. Gentry, ed. New Haven: Yale Univ. Press.

Hammond, N. 1982. *Ancient Maya civilization*. New Brunswick: Rutgers Univ. Press.

Hansen, M. 1983a. Chocolate (Cacao). In *Costa Rican natural history*, D. H. Janzen, ed. Chicago: Univ. of Chicago Press.

———. 1983b. Yuca (Yuca, cassava). In *Costa Rican natural history*, D. H. Janzen, ed. Chicago: Univ. of Chicago Press.

Harborne, J. B. 1982. *Introduction to ecological biochemistry*. New York: Academic Press.

Hardy, D. L., Sr. 1994. *Bothrops asper* (Viperidae) snakebite and field researchers in Middle America. *Biotropica* 26: 198–207.

Hart, R. D. 1980. A natural ecosystem analog approach to the design of a successional crop system for tropical forest environments. In *Tropical succession*, supplement to *Biotropica* 12:73–82.

Hartshorn, G. S. 1978. Tree falls and tropical forest dynamics. In *Tropical trees as living systems*, P. B. Tomlinson and M. H. Zimmerman, eds. London: Cambridge Univ. Press.

———. 1980. Neotropical forest dynamics. In *Tropical succession*, supplement to *Biotropica* 12:23–30.

———. 1983a. Plants. In *Costa Rican natural history*, D. H. Janzen, ed. Chicago: Univ. of Chicago Press.

———. 1983b. *Lecythis costaricensis* (Jacaro, Olla de Mono, Monkey Pot). In *Costa Rican natural history*, D. H. Janzen, ed. Chicago: Univ. of Chicago Press.

———. 1983c. *Manilkara zapota* (Nispero, chicle tree). In *Costa Rican natural history*, D. H. Janzen, ed. Chicago: Univ. of Chicago Press.

———. 1990. Natural forest management by the Yanesha Forestry Cooperative in Peruvian Amazonia. In *Alternatives to deforestation: Steps toward sustainable use of the Amazon rain forest*. A. B. Anderson, ed. New York: Columbia Univ. Press.

———. 1992. Forest loss and future options in Central America. In *Ecology and conservation of Neotropical migrant landbirds*. J. M. Hagan, III, and D. W. Johnston, eds. Wash., DC: Smithsonian Inst. Press.

Hartshorn, G. S., and B. E. Hammel. 1994. Vegetation types and floristic patterns. In *La Selva: Ecology and natural history of a Neotropical rain forest*, L. A. McDade, K. S. Bawa, H. A. Hespenheide, and G. S. Hartshorn, eds. Chicago: Univ. of Chicago Press.

Heisler, I. L. 1983. *Nyssodesmus python* (Milpes, large forest-floor millipede). In *Costa Rican natural history*, D. H. Janzen, ed. Chicago: Univ. of Chicago Press.

Heithaus, E. R., P. A. Opler, and H. B. Baker. 1974. Bat activity and pollination of *Bauhinia pauletia*: Plant-pollinator coevolution. *Ecology* 55:412–419.

Heithaus, E. R., T. H. Fleming, and P. A. Opler. 1975. Foraging patterns and resource utiliization in seven species of bats in a seasonal tropical forest. *Ecology* 56:841–854.

Henderson, A., G. Galeano, and R. Bernal. 1995. *Field guide to the palms of the*

Americas. Princeton: Princeton Univ. Press.

Herrera, R. 1985. Nutrient cycling in Amazonian forests. In *Amazonia,* G. T. Prance and T. E. Lovejoy, eds. Oxford: Pergamon Press.

Heyer, W. R., and L. R. Maxson. 1982. Distributions, relationships, and zoogeography of lowland frogs—The *Leptodactylus* complex in South America, with special reference to Amazonia. In *Biological diversification in the tropics,* G. T. Prance, ed. New York: Columbia Univ. Press.

Hill, K., and A. M. Hurtado. 1989. Hunter-gatherers of the New World. *Amer. Sci.* 77:436–443.

Hill R. S. 1992. *Nothofagus:* evolution from a southern perspective. *Trends in Ecol. Evol.* 7:190–194.

Hilty, S. L., and W. L. Brown. 1986. *A guide to the birds of Colombia.* Princeton: Princeton Univ. Press.

Hoglund, J. and R. V. Alatalo. 1995. *Leks.* Princeton: Princeton Univ. Press.

Hogue, C. L. 1993. *Latin American insects and entomology.* Berkeley: Univ. of California Press.

Holdridge, L. R. 1967. *Life zone ecology.* San José, Costa Rica: Tropical Science Center.

Holldobler, B., and E. O. Wilson. 1990. *The Ants.* Cambridge: Belknap Press of Harvard Univ.

Holloway, M. 1993. Sustaining the Amazon. *Sci. Amer.* 269:90–99.

Holthuijzen, A. M. A., and J. H. A. Boerboom. 1982. The *Cecropia* seedbank in the Surinam lowland rain forest. *Biotropica* 14:62–67.

Horn, H. S. 1971. *The adaptive geometry of trees.* Princeton: Princeton Univ. Press.

Horn, J. M., K. C. Spencer, and J. T. Smiley. 1984. The chemistry of extrafloral nectar of *Passiflora* and related species. *Bull. Ecol. Soc. Amer.* 65:265.

Horn, S. P. 1993. Postglacial vegetation and fire history in the Chirripo Paramo of Costa Rica. *Quaternary Res.* 40:107–116.

Horn, S. P., and R. L. Sanford, Jr. 1992. Holocene fires in Costa Rica. *Biotropica* 24:354–361.

Houghton, R. A. and G. M. Woodwell. 1989. Global climatic change. *Sci. Amer.* 260:36–44.

Howard, R., and A. Moore. 1980. *A complete checklist of the birds of the world.* Oxford: Oxford Univ. Press.

Howden, H. F. 1983. *Megasoma elephas* (Cornizuelo, rhinoceros beetle). In *Costa Rican natural history,* D. H. Janzen, ed. Chicago: Univ. of Chicago Press.

Howe, H. F. 1977. Bird activity and seed dispersal of a tropical wet forest tree. *Ecology* 58:539–550.

———. 1982. Fruit production and animal activity in two tropical trees. In *The ecology of a tropical forest,* E. G. Leigh, Jr., A. S. Rand, and D. M. Windsor, eds. Wash., DC: Smithsonian Inst. Press.

Howe, H. F., and G. F. Estabrook. 1977. On intraspecific competition for avian dispersers in tropical trees. *Amer. Nat.* 111:817–832.

Howell, D. J. 1976. Plant-loving bats, bat-loving plants. *Nat. Hist.* 85:52–59.

Howell, T. R. 1971. An ecological study of the birds of the lowland pine savanna and adjacent rain forest in northeastern Nicaragua. *Living Bird* 10:185–242.

Hrdy, S. B. 1981. *The woman that never evolved.* Cambridge: Harvard Univ. Press.

Hubbell, S. P. 1979. Tree dispersion, abundance, and diversity in a tropical dry forest. *Science* 203:1299–1309.

———. 1980. Seed predation and the coexistence of tree species in tropical forests. *Oikos* 35:214–229.

Hubbell, S. P., D. F. Wiemer, and A. Adejare. 1983. An antifungal terpenoid defends a Neotropical tree *(Hymenaea)* against attack by fungus-growing ants *(Atta). Oecologia* 60:321–327.

Hubbell, S. P., J. J. Howard, and D. F. Wiemer. 1984. Chemical leaf repellency to an attine ant: Seasonal distribution among potential host plant species. *Ecology* 65:1067–1076.

Hubbell, S. P., and R. B. Foster. 1986a. Canopy gaps and the dynamics of a Neotropical forest. In *Plant ecology,*

M. J. Crawley, ed. Oxford, England: Blackwell Scientific.

Hubbell, S. P., and R. B. Foster. 1986b. Commonness and rarity in a Neotropical forest: Implications for tropical tree conservation. In *Conservation biology: The science of scarcity and diversity*, M. E. Soule, ed. Sunderland, MA: Sinauer Associates.

———. 1986c. Biology, chance, and history and the structure of tropical rain forest tree communities. In *Community ecology*, J. Diamond and T. J. Case, eds. New York: Harper & Row.

———. 1990. Structure, dynamics, and equilibrium status of old-growth forest on Barro Colorado Island. In *Four Neotropical rain forests*, A. H. Gentry, ed. New Haven: Yale Univ. Press.

———. 1992. Short-term dynamics of a Neotropical forest: Why ecological research matters to tropical conservation and management. *Oikos* 63:48–61.

Huber, O. 1982. Significance of savanna vegetation in the Amazon territory of Venezuela. In *Biological diversification in the tropics*, G. T. Prance, ed. New York: Columbia Univ. Press.

———. 1987. Neotropical savannas: Their flora and vegetation. *Trends in Ecol. Evol.* 2:67–71.

Hughes, T. P. 1994. Catastrophes, phase shifts, and large scale degradation of a Caribbean coral reef. *Science* 265:1547–1551.

Hurlbert, K. J. 1994. A tribute to Alwyn H. Gentry. *Cons. Biol.* 8:291–292.

Idyll, C. P. 1973. The anchovy crisis. *Sci. Amer* 228:22–29.

Iltis, H. H., J. F. Doebley, R. Guzman, and B. Pazy. 1979. *Zea diploperennis* (Gramineae): A new teosinte from Mexico. *Science* 203:186–188.

Inchausti, P. 1995. Competition between perennial grasses in a Neotropical savanna: The effects of fire and of hydric-nutritional stress. *Jour. Ecol.* 83:231–243.

Intergovernmental Panel on Climate Change (IPCC) 1990. *Climate change: The IPCC scientific assessment*, J. T. Houghton, G. J. Jenkins, and J. J.

Ephraums, eds. Cambridge: Cambridge Univ. Press.

Irion, G. 1976. Die Entwicklung des zentral-und oberamazonischen Tieflands im Spat-Pleistozan und im Holozan (The development of the lowlands of central and uppen Amazonia in the Late Pleistocene and Holocene). *Amazonia* 6:67–79.

———. 1978. Soil infertility in the Amazonian rain forest. *Naturwissenschaften* 65:515–519.

———. 1989. Quaternary geological history of the Amazon Lowlands. In *Tropical forests*, L. B. Holm-Nielson, I. Neilson, and H. Basley, eds. London: Academic Press.

Isler, M. L., and P. R. Isler. 1987. *The tanagers: Natural history, distribution, and identification*. Wash., DC: Smithsonian Inst. Press.

Jacobs, M. 1988. *The tropical rain forest: a first encounter.* Berlin: Springer-Verlag.

Janos, D. P. 1980. Mycorrhizae influence tropical succession. *Biotropica* 12 (suppl.):56–64.

———. 1983. Tropical mycorrhizae, nutrient cycles, and plant growth. In *Tropical rain forest: ecology and management*, S. L. Sutton, T. C. Whitmore, and A. C. Chadwick, eds. Oxford: Blackwell Scientific.

Janos, D. P., C. T. Sahley, and L. H. Emmons. 1995. Rodent dispersal of vesicular-arbuscular mycorrhizal fungi in Amazonian Peru. *Ecology* 76:1852–1858.

Janzen, D. H. 1966. Coevolution of mutualism between ants and acacias in Central America. *Evolution* 20:249–275.

———. 1967. Synchronization of sexual reproduction of trees within the dry season in Central America. *Evolution* 21:620–637.

———. 1969a. Allelopathy by myrmecophytes: The ant *Azteca* as an allelopathic agent of *Cecropia*. *Ecology* 50:147–153.

———. 1969b. Seed-eaters versus seed size, number, toxicity, and dispersal. *Evolution* 23:1–27.

———. 1970. Herbivores and the number

of tree species in a tropical forest *Amer. Nat.* 104:501–528.

———. 1971a. Seed predation by animals. *Ann. Rev. Ecol. Syst.* 2:465–492.

———. 1971b. Euglossine bees as long-distance pollinators of tropical plants. *Science* 171:203–206.

———. 1972. The uncertain future of the tropics. *Nat Hist.* 81:80–89.

———. 1973. Tropical agroecosystems. *Science* 182:1212–1220.

———. 1974. Tropical blackwater rivers, animals, and mast fruiting in the Dipterocarpaceae. *Biotropica* 6:69–103.

———. 1975. *Ecology of plants in the tropics.* London: Edward Arnold.

———. 1976. Why are there so many species of insects? *Proc. XV Int. Cong. Ent.:* 84–94.

———. 1979. How to be a fig. *Ann. Rev. Ecol. Syst.* 10:13–52.

———. 1980a. Two potential coral snake mimics in a tropical deciduous forest. *Biotropica* 12:77–78.

———. 1980b. When is it coevolution? *Evolution* 34:611–612.

———. 1980c. Specificity of seed-attacking beetles in a Costa Rican deciduous forest. *Jour. Ecol.* 68:929–952.

———. ed. 1983a. *Costa Rican natural history.* Chicago: Univ. of Chicago Press.

———. 1983b. *Mimosa pigra* (Zarza, dormilona). In *Costa Rican natural history,* D. H. Janzen, ed. Chicago: Univ. of Chicago Press.

———. 1983c. *Brotogeris jugularis* (Perico, orange-chinned parakeet). In *Costa Rican natural history,* D. H. Janzen, ed. Chicago: Univ. of Chicago Press.

———. 1983d. *Tapirus bairdii* (Danto, danta, Baird's tapir). In *Costa Rican natural history,* D. H. Janzen, ed.. Chicago: Univ. of Chicago Press.

———. 1985. Plant defences against animals in the Amazonian rainforest. In *Amazonia,* G. T. Prance and T. E. Lovejoy, eds. Oxford: Pergamon Press.

———. 1988a. Tropical dry forests: The most endangered major tropical ecosystems. In *Biodiversity,* E. O. Wilson, ed. Wash., DC: National Academy Press.

———. 1988b. Tropical ecological and biocultural restoration. *Science* 239:243–244.

Janzen, D. H., and C. R. Carroll. 1983. *Paraponera clavata* (Bala, giant tropical ant). In *Costa Rican natural history,* D. H. Janzen, ed. Chicago: Univ. of Chicago Press.

Janzen. D. H., and C. C. Hogue. 1983. *Fulgora latenaria* (Machaca, peanuthead bug, lantern fly). In *Costa Rican natural history,* D. H. Janzen, ed. Chicago: Univ. of Chicago Press.

Janzen, D. H., and D. E. Wilson. 1983. Mammals. In *Costa Rican natural history,* D. H. Janzen, ed. Chicago: Univ. of Chicago Press.

Johnsgard, P. A. 1994. *Arena birds: sexual selection and behavior.* Wash., DC: Smithsonian Inst. Press.

Jordan, C. F. 1982. Amazon rain forests. *Amer. Sci.* 70:394–401.

———. 1985a. *Nutrient cycling in tropical forest ecosystems.* New York: J. Wiley.

———. 1985b. Soils of the Amazon rainforest. In *Amazonia,* G. T. Prance and T. E. Lovejoy, eds. Oxford: Pergamon Press.

———. 1986. Local effects of tropical deforestation. In *Conservation biology: the science of scarcity and diversity,* M. E. Soule, ed. Sunderland, MA: Sinauer.

Jordan, C. F., F. Golley, J. D. Hall, and J. Hall. 1979. Nutrient scavenging of rainfall by the canopy of an Amazonian rain forest. *Biotropica* 12:61–66.

Jordan, C. F., and R. Herrera. 1981. Tropical rain forests: Are nutrients really critical? *Amer. Nat.* 117:167–180.

Jordan, C. F., and J. R. Kline. 1972. Mineral cycling: Some basic concepts and their application in a tropical rain forest. *Ann. Rev. Ecol. Syst.* 3:33–50.

Jordano, P. 1983. Fig-seed predation and dispersal by birds. *Biotropica* 15:38–41.

Joyce, C. 1992. Western medicine men return to the field. *Bioscience* 42:399–403.

Junk, W. J., 1970. Investigations on the ecology and production-biology of the "floating meadows" (*Paspalo-Echinochloetum*) on the middle Amazon. *Amazonia* 2:449–495.

Junk, W. J. and K. Furch. 1985. The physical and chemical properties of Amazonian waters and their relationships with the biota. In *Amazonia*, G. T. Prance and T. E. Lovejoy, eds. Oxford: Pergamon Press.

Kalliola, R., J. Salo, M. Puhakka, and M. Rajasilta. 1991. New site formation and colonizing vegetation in primary succession on the western Amazon floodplains. *Jour. Ecol.* 79:877–901.

Kaplan, E. H. 1982. *A field guide to coral reefs.* Boston: Houghton Mifflin.

Karr, J. R. 1975. Production, energy pathways and community diversity in forest birds. In *Tropical ecological systems: Trends in terrestrial and aquatic research*, F. B. Golley and E. Medina, eds. New York: Springer-Verlag.

———. 1976. Seasonality, resource availability, and community diversity in tropical bird communities. *Amer. Nat.* 110:973–994.

Karr, J. R., D. W. Schemske, and N.V.L. Brokaw. 1982. Temporal variation in the understory bird community of a tropical forest. In *The ecology of a tropical forest*, E. G. Leigh, Jr., A. S. Rand, and D. M. Windsor, eds. Wash., DC: Smithsonian Institution Press.

Karr, J. R., and K. E. Freemark. 1983. Habitat selection and environmental gradients: Dynamics in the "stable" tropics. *Ecology* 64:1481–1494.

Kauffman, J. B., D. L. Cummings, and D. E. Ward. 1994. Relationships of fire, biomass, and nutrient dynamics along a vegetation gradient in the Brazilian cerrado. *Jour. Ecol.* 82:519–531.

Keast, A., and E. S. Morton, eds. 1980. *Migrant birds in the Neotropics: Ecology, behavior, distribution, and conservation.* Wash., DC: Smithsonian Inst. Press.

Keel, S., A. H. Gentry, and L. Spinzi. 1993. Using vegetation analysis to facilitate the selection of conservation sites in Eastern Paraguay. *Biol. Cons.* 7:66–75.

Keeler, K. H. 1980. Distribution of plants with extrafloral nectaries in temperate communities. *Amer. Midl. Nat.* 104: 274–280.

Kettlewell, B. 1973. *The evolution of melanisms: The study of a recurring necessity with special industrial melanism in the Lepidoptera.* Oxford: Oxford Univ. Press.

Kiltie, R. A. 1982. Bite force as a basis for niche differentiation between rain forest peccaries *(Tayassu tajacu* and *T. pecari). Biotropica* 14:188–195.

Kinzey, W. G. 1982. Distribution of primates and forest refuges. In *Biological diversification in the tropics*, G. T. Prance, ed. New York: Columbia Univ. Press.

Klein, B. C. 1989. Effects of forest fragmentation on dung and carrion beetle communities in Central Amazonia. *Ecology* 70:1715–1725.

Klinge, H., W. A. Rodrigues, E. Brunig, and E. J. Fittkau. 1975. Biomass and structure in a central Amazonian rain forest. In *Tropical ecological systems: Trends in terrestrial and aquatic research*, F. B. Golley and E. Medina, eds. New York: Springer-Verlag.

Knight, D. H. 1975. A phytosociological analysis of species rich tropical forest on Barro Colorado Island, Panama. *Ecol. Mono.* 45:259–284.

Kobe, R. K. 1995. Tropical-temperate comparisons of forest ecosystems productivity, based on tree growth and death. Abstracts of the 1995 Meeting of the Ecological Society of America (part 3).

Koford, C. B. 1957. The vicuna and the pina. *Ecol. Mono.* 27:153–219.

Kormondy, E. J. 1996. *Concepts of ecology.* 4th ed. Upper Saddle River, NJ: Prentice Hall.

Krantz, G. S. 1970. Human activities and megafaunal extinctions. *Amer. Sci.* 58:164–70.

Kricher, J. C. 1995. Black-and-white warbler. In *The Birds of North America, No. 158*, A. Poole and F. Gills, eds. Philadelphia: Academy of Natural Sciences, and Wash., DC: The American Ornithologists' Union.

Kricher, J. C., and W. E. Davis, Jr. 1986. Returns and winter site fidelity of North American migrants banded in Belize, Central America. *J. Field Ornithol.* 57:48–52.

———. 1987. No place like home. *Living Bird Quarterly* 6:24–27.

————. 1992. Patterns of avian species richness in disturbed and undisturbed habitats in Belize. In *Ecology and conservation of Neotropical migrant landbirds*, J. M. Hagan, III, and D. W. Johnston, eds. Wash., DC: Smithsonian Inst. Press.

Krieger, R. I., L. P. Feeny, and C. F. Wilkinson. 1971. Detoxication enzymes in the guts of caterpillars: An evolutionary answer to plant defenses? *Science* 172:579–581.

Kubitzki, K. 1985. The dispersal of forest plants. In *Amazonia*, G. T. Prance and T. E. Lovejoy, eds. Oxford: Pergamon Press.

Kunz, T. H. 1982. Roosting ecology of bats. In *Ecology of bats*, T. H. Kunz, ed. New York: Plenum Publishing Corp.

Kunz, T. H., P. V. August, and C. D. Burnett. 1983. Harem social organization in cave roosting *Artibeus jamaicensis* (Chiroptera: Phyllostomidae). *Biotropica* 15:133–138.

Kushlan, J. A., G. Morales, and P. C. Frohring. 1985. Foraging niche relations of wading birds in tropical wet savannas. In *Neotropical ornithology*, P. A. Buckley, M. S. Foster, E. S. Morton, R. S. Ridgely, and F. G. Buckley, eds. Wash., DC: American Ornithologists' Union.

Lack, D. 1947. *Darwin's finches*. Cambridge: Cambridge Univ. Press.

————. 1966. *Population studies of birds*. Oxford: Clarendon Press.

LaFay, H. 1975. The Maya: Children of time. *Nat. Geog.* 148:728–767.

Lal, R. 1990. Tropical soils: Distribution, properties and management. In *Tropical resources: Ecology and development*, J. I. Furtado, W. B. Morgan, J. R. Pfafflin, and K. Ruddle, eds. London: Harwood Assoc.

Land, H. 1970. *Birds of Guatemala*. Wynnewood, PA: Livingston.

Lande, R., and G. F. Barrowclough. 1987. Effective population size, genetic variation, and their use in population management. In *Viable populations for conservation*, M. E. Soule, ed. Cambridge: Cambridge Univ. Press.

Lavelle, P., E. Blanchart, A. Martin, S. Martin, A. Spain, F. Toutain, I. Barois, and R. Schaefer. 1993. A hierarchical model for decomposition in terrestrial ecosystems: Application to soils of the humid tropics. *Biotropica* 25:130–150.

Lean, J. and D. A. Warrilow. 1989. Simulation of the regional climatic impact of Amazon deforestation. *Nature* 342:411–413.

Leck, C. F. 1969. Observations of birds exploiting a Central American fruit tree. *Wilson Bull.* 81:264–269.

————. 1972. Seasonal changes in feeding pressures of fruit and nectar-eating birds in Panama. *Condor* 74:54–60.

————. 1987. Habitat selection in migrant birds: Seductive fruits. *Trends in Ecol. Evol.* 2(2):33.

Lee, D. W., J. B. Lowry, and B. C. Stone. 1979. Abaxial anthocyanin layer in leaves of tropical rain forest plants: Enhancer of light capture in deep shade. *Biotropica* 11:70–77.

Leigh, E. G., Jr. 1975. Structure and climate in tropical rain forest. *Ann. Rev. Ecol. Syst.* 6:67–86.

Leigh, E. G., Jr., A. S. Rand, and D. M. Windsor, eds. 1982. *The ecology of a tropical forest*. Wash., DC: Smithsonian Inst. Press.

Leigh, E. G., Jr., and D. M. Windsor. 1982. Forest production and regulation of primary consumers on Barro Colorado Island. In *The ecology of a tropical forest*, E. G. Leigh, Jr., A. S. Rand, and D. M. Windsor, eds. Wash., DC: Smithsonian Institution Press.

Leo, M. 1995. The importance of tropical montane cloud forest for preserving vertebrate endemism in Peru: The Rio Abiseo National Park as a case study. In *Tropical montane cloud forests: Ecological studies 110*. L. S. Hamilton, J. O. Juvik, and F. N. Scatena, eds. New York: Springer-Verlag.

Levey, D. J. 1985. Two ways to be a fruit-eating bird: Mashers versus gulpers. Abstracts of the 103d American Ornithologists' Union Meeting.

Levey, D. J., T. C. Moermond, and J. S. Denslow. 1984. Fruit choice in Neotropical birds: The effect of distance between fruits on preference patterns. *Ecology* 65:844–850.

Levey, D. J., and F. G. Stiles. 1994. Birds: Ecology, behavior, and taxonomic affinities. In *La Selva: Ecology and natural history of a Neotropical rain forest,* L. A. McDade, K. S. Bawa, H. A. Hespenheide, and G. S. Hartshorn, eds. Chicago: Univ. of Chicago Press.

Levey, D. J., T. C. Moermond, and S. S. Denslow. 1994. Frugivory: An overview. In *La Selva: Ecology and natural history of a Neotropical rain forest,* L. A. McDade, K. S. Bawa, H. A. Hespenheide, and G. S. Hartshorn, eds. Chicago: Univ. of Chicago Press.

Levin, D. A. 1971. Plant phenolics: An ecological perspective. *Amer. Nat.* 105:157–181.

———. 1976. Alkaloid-bearing plants: An ecogeographic perspective. *Amer. Nat.* 110: 261–284.

Levings, S. C., and D. M. Windsor. 1982. Seasonal and annual variation in litter arthropod populations. In *The ecology of a tropical forest,* E. G. Leigh, Jr., A. S. Rand, and D. M. Windsor, eds. Wash., DC: Smithsonian Inst. Press.

Lieberman, S. S., and C. F. Dock. 1982. Analysis of the leaf litter arthropod fauna of a lowland tropical evergreen forest (La Selva, Costa Rica). *Rev. Biol. Trop.* 30:27–34.

Lill, A. 1974. The evolution of clutch size and male "chauvinism" in the white-bearded manakin. *Living Bird* 13:211–231.

Limbaugh, C. 1961. Cleaning symbiosis. *Sci. Amer.* 205:42–49.

Line, L. 1996. New primate species discovered in Amazon rainforest. *New York Times,* June 19.

Lodge, D. J. 1996. Microorganisms. In *The food web of a tropical rain forest.* D. P. Reagan and R. B. Waide, eds. Chicago: Univ. of Chicago Press.

Loiselle, B. A., and J. G. Blake. 1992. Population variation in a tropical bird community. *Bioscience* 42:829–837.

Long, A. J. 1995. The importance of tropical montane cloud forests for endemic and threatened birds. In *Tropical montane cloud forests: Ecological studies 110.* L. S. Hamilton, J. O. Juvik, and F. N.

Scatena, eds. New York: Springer-Verlag.

Long, A., and P. S. Martin. 1974. Death of American ground sloths. *Science* 186:638–640.

Longman, K. A., and J. Jenik. 1974. *Tropical forest and its environment.* London: Longman.

Lopez-Parodi, J. 1980. Deforestation and increased flooding of the upper Amazon. *Science* 210: 1354–1355.

Lotschert, W., and G. Beese. 1981. *Collins guide to tropical plants.* London: Collins.

Lovejoy, T. E. 1974. Bird diversity and abundance in Amazon forest communities. *Living Bird* 13:127–192.

Lovejoy, T. E. 1995. Remarks delivered at Annual Meeting of the Ecological Society of America, Snowbird, Utah.

Lovejoy, T. E., R. O. Bierregaard, Jr., A. B. Rylands, J. R. Malcolm, C. E. Quintela, L. H. Harper, K. S. Brown, Jr., A. H. Powell, G. V. N. Powell, H. O. R. Schubart, and M. B. Hays. 1986. Edge and other effects of isolation on Amazon forest fragments. In *Conservation biology: The science of scarcity and diversity,* M. E. Soule, ed. Sunderland, MA: Sinauer.

Lovejoy, T. E., and R. O. Bierregaard, Jr. 1990. Central Amazonian forests and the minimum critical size of ecosystems project. In *Four Neotropical rainforests,* A. H. Gentry, ed. New Haven: Yale Univ. Press.

Lowe-McConnell, R. H. 1987. *Ecological studies in tropical fish communities.* Cambridge: Cambridge Univ. Press.

Lubin, Y. D. 1983. *Nasutitermes* (Comejan, hormiga blanca, nausute termite, arboreal termite). In *Costa Rican natural history,* D. H. Janzen, ed. Chicago: Univ of Chicago Press.

Lubin, Y. D., and G. G. Montgomery. 1981. Defenses of *Nasutitermes* termites (Isoptera, Termitidae) against *Tamandua* anteaters (Edentata, Myrmecophagidae). *Biotropica* 13:66–76.

Lucas, Y., F. J. Luizao, A. Chauvel, J. Rouiller, and D. Nahon. 1993. The relation between biological activity of the rain forest and mineral composition of soils. *Science* 260:521–523.

Lugo, A. E. 1988. Estimating reductions in the diversity of tropical forest species. In *Biodiversity*, E. O. Wilson, ed. Wash., DC: National Academy Press.

———. 1994. Preservation of primary forests in the Luquillo Mountains, Puerto Rico. *Cons. Biol.* 8:1122–1131.

Lugo, A. E., and S. C. Snedaker. 1974. The ecology of mangroves. *Ann. Rev. Ecol. Syst.* 5:39–64.

Lundberg, J. G., W. M. Lewis, and J. F. Saunders, Jr. 1987. A major food web component in the Orinoco River channel: Evidence from planktivorous electric fishes. *Science* 237:81–83.

MacArthur, R. H. 1965. Patterns of species diversity. *Biol. Rev.* 40:510–533.

———. 1972. *Geographical ecology: Patterns in the distribution of species*. Princeton: Princeton Univ. Press.

MacArthur, R. H., and J. MacArthur. 1961. On bird species diversity. *Ecology* 42:594–598.

MacArthur, R. H., and E. O. Wilson. 1963. An equilibrium theory of insular zoogeography. *Evolution* 17:373–387.

———. 1967. *The theory of island biogeography*. Princeton: Princeton Univ. Press.

McDade, L. A., K. S. Bawa, H. A. Hespenheide, and G. S. Hartshorn, eds. 1994. *La Selva: Ecology and natural history of a Neotropical rain forest*. Chicago: Univ. of Chicago Press.

McLarney, W. O. 1988. Guanacaste: The dawn of a park. *Nature Conservancy magazine* 38:11–15.

Magurran, A. E. 1988. *Ecological diversity and its measurement*. Princeton: Princeton University Press.

Mares, M. A. 1986. Conservation in South America: Problems, consequences, and solutions. *Science* 233:734–739.

Markell, E. K., and M. Voge. 1971. *Medical parasitology*. 3d ed. Philadelphia: Saunders.

Marquis, R. J. 1984. Leaf herbivores decrease fitness of a tropical plant. *Science* 226:537–539.

Marshall, L. G. 1988. Land mammals and the great American interchange. *Amer. Sci.* 76:380–388.

———. 1994. The terror birds of South America. *Sci. Amer.* 270:90–95.

Marshall, L. G., S. D. Webb, J. J. Sepkowski, and D. M. Raup. 1982. Mammalian evolution and the great American interchange. *Science* 215:1351–1357.

Martin, M. M. 1970. The biochemical basis of the fungus-attine ant symbiosis. *Science* 169:16–20.

Martin, M. M., and J. S. Martin. 1984. Surfactants: Their role in preventing the precipitation of proteins by tannins in insect guts. *Oecologia* 61:342–345.

Maxson, L. R., and W. R. Heyer. 1982. Leptodactylid frogs and the Brasilian Shield: An old and continuing adaptive relationship. *Biotropica* 14:10–14.

Maxson, L. R., and C. W. Myers. 1985. Albumin evolution intropical poison frogs (Dendrobatidae): A preliminary report. *Biotropica* 17:50–56.

May, R. M. 1988. How many species are there on Earth? *Science* 241:1441–1449.

———. 1992. How many species inhabit the Earth? *Sci. Amer.* 267:42–48.

Mayr, E. 1963. *Animal species and evolution.* Cambridge: Belknap Press, Harvard Univ.

Meade, R. H., and L. Koehnken. 1991. Distribution of the river dophin, tonina *Inia geoffrensis* in the Orinoco River Basin of Venezuela and Colombia. *Interciencia* 16:300–312.

Meade, R. H., J. M. Rayol, S. C. Da Conceicao, and J. R. G. Natividade. 1991. Backwater effects in the Amazon River Basin of Brazil. *Environ. Geol. Water Sci.* 18:105–114.

Medina, J. T. 1988. *The discovery of the Amazon.* New York: Dover.

Meggers, B. J. 1985. Aboriginal adaptations to Amazonia. In *Amazonia*, G. T. Prance and T. E. Lovejoy, eds. Oxford: Pergamon Press.

———. 1988. The prehistory of Amazonia. In *People of the tropical rain forest*, J. S. Denslow and C. Padoch, eds. Berkeley: Univ. Calif. Press.

Meyer-Abich, Adolf. 1969. *Alexander von Humholdt.* Bonn: Inter Nationes.

Meyer, J. L. 1990. A blackwater perspective on riverine ecosystem. *Bioscience* 40:643–651.

Meyer de Schauensee, R. 1966. *The species of birds of South America.* Wynnewood,

PA: Livingston, for the Academy of Natural Sciences, Philadelphia.

Miller, K., and L. Tangley. 1991. *Trees of life*. Boston: Beacon Press.

Milton, K. 1979. Factors influencing leaf choice by howler monkeys: A test of some hypotheses of food selection by generalist herbivores. *Amer. Nat.* 114:362–378.

———. 1981. Food choice and digestive strategies of two sympatric primate species. *Amer. Nat.* 117:496–505.

———. 1982. Dietary quality and demographic regulation in a howler monkey population. In *The ecology of a tropical forest*, E. G. Leigh, Jr., A. S. Rand, and D. M. Windsor, eds. Wash., DC: Smithsonian Inst. Press.

Mittermeier, R. A. 1978. South American river turtles: Saving their future. *Oryx* 14:222–230.

———. 1991. Hunting and its effect on wild primate populations in Suriname. In *Neotropical wildlife use and conservation*, J. G. Robinson and K. H. Redford, eds. Chicago: Univ. of Chicago Press.

Moermond, T. C., and J. S. Denslow. 1985. Neotropical avian frugivores: Patterns of behavior, morphology, and nutrition, with consequences for fruit selection. In *Neotropical Ornithology*, P. A. Buckley, M. S. Foster, E. S. Morton, R. S. Ridgely, and F. G. Buckley, eds. Wash., DC: American Ornithologists' Union.

Moffett, M. W. 1993. *The high frontier: Exploring the tropical rainforest canopy*. Cambridge: Harvard University Press.

———. 1995a. Poison-dart frogs. *Nat. Geog.* 187:98–111.

———. 1995b. Leafcutter ants. *Nat Geog.* 188:98–111.

Molion, L. C. B. 1991. Amazonia: burning and global climate impacts. In *Global biomass burning: Atmospheric, climatic and biospheric implications*. J. S. Lemine, ed. Cambridge: MIT Press.

Monroe, B. L., Jr., and C. G. Sibley. 1993. *A world checklist of birds*. New Haven: Yale Univ. Press.

Montgomery, G. G., ed. 1985. *The evolution and ecology of armadillos, sloths, and ver-*

milinguas. Wash., DC: Smithsonian Inst. Press.

Montgomery, G. G., and M. E. Sunquist. 1975. Impact of sloths on Neotropical forest energy flow and nutrient cycling. In *Ecological studies 11, Tropical ecological systems*, F. B. Golley and E. Medina, eds. Heidelberg: Springer-Verlag.

Moran, E. F., E. Brondizio, P. Mausel, and Y. Wu. 1994. Integrating Amazonian vegetation, land-use, and satellite data. *Bioscience* 44:329–338.

Morrison, T. 1974. *Land above the clouds*. London: Andre Deutsch.

———. 1976. *The Andes*. Amsterdam: Time-Life International.

Morton, E. S. 1973. On the evolutionary advantages and disadvantages of fruit eating in tropical birds. *Amer. Nat.* 107:8–22.

Moynihan, M. 1962. The organization and probable evolution of some mixed species flocks of Neotropical birds. *Smithsonian Miscellaneous Collection* 143:1–140.

———. 1976. *The new world primates: Adaptive radiation and the evolution of social behavior, languages, and intelligence*. Princeton: Princeton Univ. Press.

Muller, F. 1879. *Ituna* and *Thyridis:* A remarkable case of mimicry in butterflies. *Proc. Ent. Soc. London 1879:* 20–29.

Muller-Karger, F. E., C. R. McClain, and P. L. Richardson. 1988. The dispersal of the Amazon's water. *Nature* 333:56–59.

Munn, C. A. 1984. Birds of different feather also flock together. *Nat. Hist.* 11:34–42.

———. 1985. Permanent canopy and understory flocks in Amazonia: Species composition ancd population density. In *Neotropical ornithology*, P. A. Buckley, M. S. Foster, E. S Morton, R. S. Ridgely, and F. G. Buckley, eds. Wash., DC: American Ornithologists' Union.

———. 1992. Macaw biology and ecotourism, or "when a bird in the bush is worth two in the hand." In *New world parrots in crisis: Solutions from conservation biology*, S. R. Beissinger and

N. F. R. Snyder, eds. Wash., DC: Smithsonian Inst. Press.

Munn, C. A., and J. W. Terborgh. 1979. Multi-species territoriality in Neotropical foraging flocks. *Condor* 81:338–347.

Murray, K. G. 1988. Avian seed dispersal of three Neotropical gap-dependent plants. *Ecol. Mono.* 58:271–298.

Myers, C. W. and J. W. Daly. 1983. Dart-poison frogs. *Sci. Amer.* 248:120–133.

Myers, N., for the Committee on Research Priorities in Tropical Biology of the National Research Council. 1980. *Conversion of tropical moist forests.* Wash., DC: National Academy of Science.

———. 1986. Tropical deforestation and a mega-extinction spasm. In *Conservation biology: The science of scarcity and diversity,* M. E. Soule, ed. Sunderland, MA: Sinauer.

———. 1988. Tropical forests and their species. In *Biodiversity,* E.O. Wilson ed. Wash., DC: National Academy Press.

———. 1991. Tropical forests: present status and future outlook. *Climatic Change* 19:3–32.

Nadkarni, N. M. 1981. Canopy roots: convergent evolution in rainforest nutrient cycles. *Science* 214:1023–1024.

Nathanson, J. A. 1984. Caffeine and related methylxanthines: Possible naturally occurring pesticides. *Science* 226:184–186.

Nations, J. D. 1988. The Lacandon Maya. In *People of the tropical rain forest,* J. S. Denslow and C. Padoch, eds. Berkeley: Univ. of Calif. Press.

Neill, C., M. C. Piccolo, P. A. Steudler, J. M. Melillo, B. J. Feigl, and C. C. Cerri. 1995a. Nitrogen dynamics in soils of forests and active pastures in the western Brazilian Amazon Basin. *Soil Biol. Biochem.* 27:1167–1175.

Neill, C., M. C. Piccolo, C. C. Cerri, P. A. Steudler, J. M. Melillo, and M. Brito. 1995b. Net nitrogen mineralization and net nitrification rates in soils following deforestation for pasture across the southwestern Brazilian Amazon Basin landscape. *Oecologia,* in press.

Neill, C., B. Fry, J. M. Melillo, P. A. Steudler, J. F. L. Moraes, and C. C. Cerri. 1996. Forest- and pasture-derived carbon contributions to carbon stocks and microbial respiration of tropical pasture soils. *Oecologia* 107:113–119.

Nelson, B. W., C. A. C. Ferreira, M. F. de Silva, and M. L. Kawasaki. 1990. Endemism centres, refugia and botanical collection density in Brazilian Amazonia. *Nature* 345:714–716.

Nepstad, D., C. Uhl, and E. A. Serrao. 1990. Surmounting barriers to forest regeneration in abandoned, highly degraded pastures: A case study from Paragominas, Para, Brazil. In *Alternatives to deforestation: Steps toward sustainable use of the Amazon rain forest.* A. B. Anderson, ed. New York: Columbia Univ. Press.

Nicholaides, J. J., III, D. E. Bandy, P. A. Sanchez, J. R. Benites, J. H. Villachica, A. J. Coutu, and C. S. Valverde. 1985. Agricultural alternatives for the Amazon Basin. *Bioscience* 35:279–285.

Nijhout, H. 1994. Developmental perspectives on evolution of butterfly mimicry. *Bioscience* 44:148–157.

Nordin, C. F., Jr. and R. H. Meade. 1982. Deforestation and increased flooding in the upper Amazon. *Science* 215:426–427.

———. 1985. The Amazon and the Orinoco. *McGraw-Hill Yearbook of Science and Technology 1986,* 385–390. New York: McGraw-Hill.

Nores, M. 1992. Bird speciation in subtropical South America in relation to forest expansion and retraction. *Auk* 109:346–357.

Norris, R. 1988. Data for diversity. *Nature Conservancy Magazine* 38:4–10.

Norton, B. G. 1987. *Why preserve natural variety?* Princeton: Princeton Univ. Press.

Ojasti, J. 1991. Human exploitation of capybara. In *Neotropical wildlife use and conservation,* J. G. Robinson and K. H. Redford, eds. Chicago: Univ. of Chicago Press.

Oliveira, P. S., and H. F. Leitao-Filho. 1987. Extrafloral nectaries: Their taxonomic distribution and abundance in the woody flora of cerrado vegetation in southeast Brazil. *Biotropica* 19:140–148.

O'Neill, J. P, and G. R. Graves. 1977. A new genus and species of owl (Aves: Strigidae) from Peru. *Auk* 94:409–416.

O'Neill, T. 1993. New sensors eye the rain forest. *Nat. Geog.* 184:118–130.

Onuf, C. P., J. M. Teal, and I. Valiela. 1977. Interactions of nutrients, plant growth, and herbivory in a mangrove ecosystem. *Ecology* 58:514–526.

Opler, P. A. 1981. Polymorphic mimicry by a neuropteran. *Biotropica* 13:165–176.

Opler, P. A., H. G. Baker, and G. W. Frankie. 1980. Plant reproductive characteristics during secondary succession in Neotropical lowland forest ecosystems. In *Tropical succession*, supplement to *Biotropica* 12:40–46.

Oppenheimer, J. R. 1982. *Cebus capucinus:* home range, population dynamics, and interspecific relationships. In *The ecology of a tropical forest*, E. G. Leigh, Jr., A. S. Rand, and D. M. Windsor, eds. Wash., DC: Smithsonian Inst. Press.

Oster, G., and S. Oster. 1985. The great breadfruit scheme. *Nat. Hist.* 94:35–41.

Otte, D 1980. Beetles adorned with horns. *Nat Hist.* 89:34–41.

Pacheco, J. F., B. M. Whitney, and L. Gonzago. 1996. A new genus and species of furnarid (Aves: Furnariidae) from the cocoa-growing region of southeastern Brazil. *Wilson Bull.* 108:397–433.

Padoch, C. 1988. People of the floodplain and forest. In *People of the tropical rainforest*, J. S. Denslow and C. Padoch, eds. Berkeley: Univ. of California Press.

Papageorgis, C. 1975. Mimicry in Neotropical butterflies. *Amer. Sci.* 63:522–532.

Parker, G. G. 1994. Soil fertility, nutrient acquisition, and nutrient cycling. In *La Selva: Ecology and natural history of a Neotropical rain forest*, L. A. McDade, K. S. Bawa, H. A. Hespenheide, and G. S. Hartshorn, eds. Chicago: Univ. of Chicago Press.

Parker, T. A., III, and J. P. O'Neill. 1985. A new species and new subspecies of *Thryothorus* wren from Peru. In *Neotropical Ornithology*, P. A. Buckley, M. S. Foster, E. S. Morton, R. S. Ridgely, and F. G. Buckley, eds. Wash., DC: American Ornithologists' Union.

Parker, T. A., III, D. F. Stotz, and J. W. Fitzpatrick. 1996. Ecological and distributional databases. In *Neotropical birds: Ecology and conservation*, by D. F. Stotz, J. W. Fitzpatrick, T. A. Parker III, and D. K. Moskovits. Chicago: Univ. of Chicago Press.

Pearman, P. B. 1995. An agenda for conservation research and its application, with a case-study from Amazonian Ecuador. *Envir. Cons.* 22:39–43.

Pearman, P. B., A. M. Velasco, and A. Lopez. 1995a. Tropical amphibian monitoring: A comparison of methods for detecting inter-site variation in species' composition. *Herpetologica* 51:325–337.

Pearman, P. B., M. Guerrero G., T. D. Sisk, and D. D. Murphy. 1995b. Correlation patterns among groups proposed as biological indicators. Abstracts of the 1995 Meeting of the Ecological Society of America (Part 3).

Pearson, D. L. 1977. A pantropical comparison of bird community structure on six lowland forest sites. *Condor* 79:232–244.

———. 1982. Historical factors and bird species richness. In *Biological diversification in the tropics*, G. T. Prance, ed. New York: Columbia Univ. Press.

Pearson, D. L., and J. A. Derr. 1986. Seasonal patterns in lowland forest floor arthropod abundance in southeastern Peru. *Biotropica* 18:244–256.

Pederson, H. B., and H. Balslev. 1992. The economic botany of Ecuadorean palms. In *Sustainable harvest and marketing of rain forest products*, M. Plotkin and L. Famolare, eds. Wash., DC: Island Press.

Penny, N. D., and J. R. Arias. 1982. *Insects of an Amazonian forest.* New York: Columbia Univ. Press.

Peres, C. A., 1994. Indigenous reserves and nature conservation in Amazonian forests. *Cons. Biol.* 8:586–588.

Peres, C. A. and J. Terborgh. 1995. Amazonian nature reserves: An analysis of the defensibility of existing conservation units and design criteria for the future. *Cons. Biol.* 9:34–46.

Perfecto, I., R. A. Rice, R. Greenberg, and M. E. Van der Voort. 1996. Shade coffee: A disappearing refuge for biodiversity. Bioscience 46:598–608.

Perrins, C. M., and A. L. A. Middleton, eds. 1985. *The encyclopedia of birds.* New York: Facts on File.

Perry, D. R. 1978. Factors influencing arboreal epiphytic phytosociology in Central America. *Biotropica* 10:235–237.

———. 1984. The canopy of the tropical rain forest. *Sci. Amer.* 251:138–147.

Peters, C. M., A. Gentry, and R. Mendelsohn. 1989. Valuation of an Amazonian rainforest. *Nature* 339:655–656.

Peterson, R. T., and E. L. Chaliff. 1973. *A field guide to Mexican birds.* Boston: Houghton Mifflin.

Phillips, O. 1991. The ethnobotany and economic botany of tropical vines. In *The biology of vines,* F. E. Putz and H. A. Mooney, eds. Cambridge: Cambridge Univ. Press.

Phillips, O., A. H. Gentry, C. Reynel, P. Wilkin, and C. Galvez-Durand B. 1994. Quantitative ethnobotany and Amazonian conservation. *Cons. Biol.* 8:225–248.

Pianka, E. R. 1966. Latitudinal gradients in species diversity: A review of concepts. *Amer. Nat.* 100:33–45.

Picado, C. 1913. Les bromeliacees epiphytes considerees comme milieu biologique. *Bull. Sci. France Belgique ser.* 7 47:216–360.

Pickett, S. T. A., and P. S. White, eds. 1985. *The ecology of natural disturbance and patch dynamics.* New York: Academic Press.

Pimm, S. L., and R. A. Askins. 1995. Forest losses predict bird extinctions in eastern North America. *Proc. Natl. Acad. Sci.* 92:9343–9347.

Pires, J. M., and G. T. Prance. 1985. The vegetation types of the Brazilian Amazon. In *Amazonia,* G. T. Prance and T. E. Lovejoy, eds. Oxford: Pergamon Press.

Plotkin, M. J. 1988. The outlook for new agricultural and industrial products from the tropics. In *Biodiversity,* E. O. Wilson, ed. Wash., D.C.: National Academy Press.

———. 1993. *Tales of a shaman's apprentice: An ethnobotanist searches for new medicines in the Amazon rain forest.* New York: Viking.

Plotkin, M. and L. Famolare, eds. 1992. *Sustainable harvest and marketing of rain forest products.* Wash., DC: Island Press.

Porter, J. W. 1972. Patterns of species diversity in Caribbean reef corals. *Ecology* 53:745–748.

———. 1976. Autotrophy, heterotrophy, and resource partitioning in Caribbean reef-building corals. *Amer. Nat.* 110:731–42.

Posey, D. A. 1982. Keepers of the forest. *Garden* 6:18–24.

Post, W. M., T. Peng, W. R. Emanuel, A. W. King, V. H. Dale, and D. L. DeAngelis. 1990. The global carbon cycle. *Amer. Sci.* 78:310–326.

Possehl, S. 1996. Buzz, honk, click: Bats reveal presence to new sonar device. *New York Times,* June 11.

Pough, F. H. 1973. Lizard energetics and diet. *Ecology* 54:837–844.

Poulin, R. 1995. Phylogeny, ecology, and the richness of parasite communities in vertebrates. *Ecology* 65:283–302.

Pounds, J. A., and M. L. Crump. 1994. Amphibian declines and climate disturbance: The case of the golden toad and the harlequin frog. *Cons. Biol.* 8:72–85.

Prance, G. T., ed. 1982a. *Biological diversification in the tropics.* New York: Columbia Univ. Press.

———. 1982b. Forest refuges: Evidence from woody angiosperms. In *Biological diversification in the tropics,* G. T. Prance, ed. New York: Columbia Univ. Press.

———. 1985a. The changing forests. In *Amazonia,* G. T. Prance and T. E. Lovejoy, eds. Oxford: Pergamon Press.

———. 1985b. The pollination of Amazonian plants. In *Amazonia,* G. T. Prance and T. E. Lovejoy, eds. Oxford: Pergamon Press.

———. 1990a. The floristic composition of the forests of Central Amazonian

Brazil. In *Four Neotropical rainforests,* A. H. Gentry, ed. New Haven: Yale Univ. Press.

Prance, G. T. 1990b. Consensus for conservation. *Nature* 345:384.

Prance, G. T., W. A. Rodrigues, and M. F. da Silva. 1976. Inventario florestal de um hectare de mata de terra firme, km 30 da Estrada Manaus-Itacoatiara. *Acta Amazonica* 6:9–35.

Prestwich, G. D., and B. L. Bentley. 1981. Nitrogen fixation by intact colonies of the termite *Nasutitermes corniger. Oecologia* 49:249–251.

Prestwich, G. D., B. L. Bentley, and E. J. Carpenter. 1980. Nitrogen sources for Neotropical nausite termites: Fixation and selective foraging. *Oecologia* 46: 397–401.

Prum, R. O. 1994. Phylogenetic analysis of the evolution of alternative social behavior in the manakins (Aves: Pipridae). *Evolution* 48:1657–1675.

Prum, R. O. and W. E. Lanyon. 1989. Monophyly and phylogeny of the *Schiffornis* group (Tyrannoidea). *Condor* 91:444–461.

Putz, F. E. 1984. The natural history of lianas on Barro Colorado Island, Panama. *Ecology* 65:1713–1724.

Putz, F. E., and N. M. Holbrook. 1988. Tropical rain forest images. In *People of the tropical rain forest,* J. S. Denslow and C. Padoch, eds. Berkeley: Univ. of Calif. Press.

Putz, F. E., and H. A. Mooney, eds. 1991. *The biology of vines.* Cambridge: Cambridge Univ. Press.

Quinn, A. F., and A. Hastings. 1987. Extinction in subdivided habitats. *Cons. Biol.* 1:198–208.

Raich, J. W., E. B. Rastetter, J. M. Melillo, D. W. Kicklighter, P. A. Steudler, B. J. Peterson, A. L. Grace, B. Moore III, and C. J. Vorosmarty. 1991. Potential net primary productivity in South America: Application of a global model. *Ecol. Appl.* 1:399–429.

Raikow, R. J. 1994. A phylogeny of the woodcreepers (Dendrocolaptinae). *Auk* 111:104–114.

Rankin-De-Merona, J. M., R. W. Hutchings, and T. E. Lovejoy. 1990. Tree

mortality and recruitment over a five-year period in undisturbed upland rainforest in the Central Amazon. In *Four Neotropical rainforests,* A. H. Gentry, ed. New Haven: Yale Univ. Press.

Rappole, J. H. 1995. *The ecology of migrant birds: A Neotropical perspective.* Wash., DC: Smithsonian Inst. Press.

Rappole, J. H., and D. H. Warner. 1980. Ecological aspects of avian migrant behavior in Veracruz, Mexico. In *Migrant birds in the Neotropics,* A. Keast and E. S. Morton, eds. Wash., DC: Smithsonian Inst. Press.

Rappole, J. H., E. S. Morton, T. E. Lovejoy, III, and J. L. Ruos. 1983. *Nearctic avian migrants in the Neotropics.* Wash., DC: U.S. Department of the Interior, Fish and Wildlife Service.

Rappole, J. H., M. A. Ramos, and K. Winkler. 1989. Wintering Wood Thrush mortality in southern Veracruz. *Auk* 106:402–410.

Rathcke, B. J., and R. W. Poole. 1975. Coevolutionary race continues: Butterfly larval adaptation to plant trichomes. *Science* 187:175–176.

Raven, P. H. 1988. Our diminishing tropical forests. In *Biodiversity,* E. O. Wilson, ed. Wash., DC: National Academy Press.

Raven, P. H., and D. Axelrod. 1975. History of the flora and fauna of Latin America. *Amer. Sci.* 63:420–429.

Ray, T. S., and C. C. Andrews. 1980. Antbutterflies: Butterflies that follow army ants to feed on antbird droppings. *Science* 210:1147–1148.

Reagan, D. P., and R. B. Waide, eds. 1996. *The food web of a tropical rain forest.* Chicago: Univ. of Chicago Press.

Redford, K. H., 1992. The empty forest. *Bioscience* 42:412–422.

Redford, K. H., and J. G. Robinson. 1987. The game of choice: Patterns of Indian and colonial hunting in the neotropics. *Am. Anthropol.* 89:650–667.

Redford, K. H. and J. F. Eisenberg. 1992. *Mammals of the Neotropics, the Southern Cone: Volume 2, Chile, Argentina, Uruguay, Paraguay.* Chicago: Univ. of Chicago Press.

Reice, S. R. 1994. Nonequilibrium determinants of biological community structure. *Amer. Sci.* 82:424–435.

Remsen, J. V., Jr. 1984. High incidence of "leapfrog" pattern of geographic variation in Andean birds: Implications for the speciation process. *Science* 224:171–172.

———. 1990. *Community ecology of Neotropical kingfishers.* Univ. Calif. Pub. 124:1–116.

Remsen, J. V., Jr., and T. A. Parker III. 1983. Contribution of river created habitats to bird species richness in Amazonia. *Biotropica* 15:223–231.

———. 1985. Bamboo specialists among Neotropical birds. Abstracts of the 103d American Ornithologists' Union Meeting.

Repetto, R. 1990. Deforestation in the tropics. *Sci. Amer.* 262:36–42.

Rettenmeyer, C. W. 1983. *Eciton burchilli* and other army ants (Hormiga arriera, army ants). In *Costa Rican natural history,* D. H. Janzen, ed. Chicago: Univ. of Chicago Press.

Richards, P. W. 1952. *The tropical rain forest.* Cambridge: Cambridge Univ. Press.

———. 1973. The tropical rain forest. *Sci. Amer.* 229:58–67.

Richey, J. E., C. Nombre, and C. Deser. 1989. Amazon River discharge and climate variability: 1903 to 1985. *Science* 246:101–103.

Ricklefs, R. E. 1969a. The nesting cycle of songbirds in tropical and temperate regions. *Living Bird* 8:165–175.

———. 1969b. Natural selection and the development of mortality rates in young birds. *Nature* 223:422–425.

———. 1970. Clutch size in birds: Outcome of opposing predator and prey adaptations. *Science* 168:599–600.

———. 1977. Environmental heterogeneity and plant species diversity: A hypothesis. *Amer. Nat.* 111:376–381.

Ridgely, R. S., and J. A. Gwynne. 1989. A *guide to the birds of Panama with Costa Rica, Nicaragua, and Honduras,* second ed. Princeton: Princeton Univ. Press.

Ridgely, R. S., and G. Tudor. 1989. *The Birds of South America, Vol 1.* Austin: Univ. of Texas Press.

———. 1994. *The Birds of South America, Vol 2.* Austin: Univ. of Texas Press.

Robbins, C. S., J. W. Fitzpatrick, and P. B. Hamel. 1992. A warbler in trouble: *Dendroica cerulea.* In *Ecology and conservation of Neotropical migrant landbirds,* J. M. Hagan III and D. W. Johnston, eds. Wash., DC: Smithsonian Inst. Press.

Robbins, M. B., G. H. Rosenberg, and F. S. Molina. 1994. A new species of cotinga (Cotingidae: *Doliornis*) from the Ecuadorian Andes, with comments on plumage sequences in *Doliornis* and *Ampelion. Auk* 111:1–7.

Robinson, J. G., and K. H. Redford, eds. 1991. *Neotropical wildlife use and conservation.* Chicago: Univ. of Chicago Press.

Robinson S. K. 1985a. Coloniality in the yellow-rumped cacique as a defense against nest predators. *Auk* 102:506–519.

———. 1985b. The yellow-rumped cacique and its associated nest pirates. In *Neotropical ornithology,* P. A. Buckley, M. S. Foster, L. S. Morton, R. S. Ridgely, and F. G. Buckley, eds. Wash., DC: American Ornithologists' Union.

———. 1986. Social security for birds. *Nat. Hist.* 95(3):39–47.

Robinson, S. K., and J. Terborgh. 1990. Bird communities of the Cocha Cashu Biological Station in Amazonian Peru. In *Four Neotropical rainforests,* A. H. Gentry, ed. New Haven: Yale Univ. Press.

Roca, R. L. 1994. Oilbirds of Venezuela: Ecology and conservation. Cambridge, MA: Publications of the Nuttall Ornithological Club no. 24 (R. A. Paynter, Jr., ed.).

Rockwood, L. L. 1976. Plant selection and foraging patterns in two species of leaf-cutting ants (*Atta*). *Ecology* 57:48–61.

Rodriguez, G. 1987. Structure and production in Neotropical mangroves *Trends in Ecol. Evol.*: 2:264–267.

Roosevelt, A. C. 1989. Lost civilizations of the lower Amazon. *Nat. Hist.* 98:75–83.

Roosevelt, A. C., R. A. Housley, M. Imazio da Silveira, S. Maranca, and R. Johnson. 1991. Eighth millennium pottery from a prehistoric shell midden in the Brazilian Amazon. *Science* 254:1621–1624.

Rosenberg, G. H. 1985. Birds specialized on Amazon river islands. Abstracts of the 103d American Ornithologists' Union Meeting.

Ross, C. A., ed. 1989. *Crocodiles and alligators.* Oxford: Facts on File.

Ruddle, K. 1970. The hunting technology of the Maraca indians. *Antropologica* 25:21–63.

Rutzler, K., and I. C. Feller. 1987. Mangrove swamp communities. *Oceanus* 30:16–24.

———. 1996. Caribbean mangrove swamps. *Sci. Amer.* 274:94–99.

Ryan, C. A. 1979. Proteinase inhibitors. In *Herbivores: Their interaction with secondary plant metabolites,* G. A. Rosenthal and D. H. Janzen, eds. New York: Academic Press.

St. John, T. V. 1985. Mycorrhizae. In *Amazonia,* G. T. Prance and T. E. Lovejoy, eds. Oxford: Pergamon Press.

Salafsky, N., B. L. Dugelby, and J. W. Terborgh. 1993. Can extractive reserves save the rain forest? An ecological and socioeconomic comparison of non-timber forest product extraction systems in Peten, Guatemala, and West Kalimantan, Indonesia. *Cons. Biol.* 7:39–52.

Salati, E. 1985. The climatology and hydrology of Amazonia. In *Amazonia,* G. T. Prance and T. E. Lovejoy, eds. Oxford: Pergamon Press.

Salati, E., and P. B. Vose. 1984. Amazon Basin: A system in equilibrium. *Science* 225:129–138.

Salick, J., R. Herrera, and C. F. Jordan. 1983 Termitaria: Nutrient patchiness in nutrient-deficient rain forests. *Biotropica* 15:1–7.

Sanford, R. L., Jr. 1987. Apogeotropic roots in an Amazon rain forest. *Science* 235:1062–1064.

Sanford, R. L., Jr., J. Saldarriaga, K. Clark, C. Uhl, and R. Herrera. 1985. Amazonian rain-forest fires. *Science* 227:53–55.

Sanford, R. L., Jr., P. Paaby, J. C. Luvall, and E. Phillips. 1994. Climate, geomorphology, and aquatic systems. In *La Selva: Ecology and natural history of a Neotropical rain forest,* L. A. McDade, K. S. Bawa, H. A. Hespenheide, and G. S.

Hartshorn, eds. Chicago: Univ. of Chicago Press.

Sarmiento, G. 1983. The savannas of Tropical America. In *Tropical savannas,* F. Bourlière, ed. New York: Elsevier.

Sarmiento, G., and M. Monasterio. 1975. A critical consideration of the environmental conditions associated with the occurrence of savanna ecosystems in tropical America. In *Tropical ecological systems: Trends in terrestrial and aquatic research,* F. B. Golley and E. Medina, eds. New York: Springer-Verlag.

Schoener, T. W. 1971. Large-billed insectivorous birds: A precipitous diversity gradient. *Condor* 73:154–161.

Schulenberg, T. S., and M. D. Williams. 1982 A new species of antpitta (*Grallaria*) from northern Peru. *Wilson Bull.* 94:105–113.

Schultes, R. E. 1992. Ethnobotany and technology in the northwest Amazon: a partnership. In *Sustainable harvest and marketing of rain forest products,* M. Plotkin and L. Famolare, eds. Wash., DC: Island Press.

Schultes, R. E., and A. Hoffmann. 1992. *Plants of the gods: Their sacred, healing, and hallucinogenic powers.* Rochester, VT: Healing Arts Press.

Schultes, R. E., and R. F. Raffauf. 1990. *The healing forest: Medicinal and toxic plants of the northwest Amazonia.* Portland, OR: Dioscorides Press.

Schwartz, P., and D. W. Snow. 1978. Display and related behavior of the wire-tailed manakin. *Living Bird* 17:51–78.

Scott, D. A., J. Proctor, and J. Thompson. 1992. Ecological studies on a lowland evergreen rain forest on Maraca Island, Roraima, Brazil. II: Litter and nutrient cycling. *Jour. Ecol.* 80:705–717.

Scott, N. J., and S. Limerick. 1983. Reptiles and amphibians. In *Costa Rican natural history,* D. H. Janzen, ed. Chicago: Univ. of Chicago Press.

Scott, P. E., and R. F. Martin. 1983. Reproduction of the turquoisebrowed motmot at archaeological ruins in Yucatan. *Biotropica* 15:8–14.

Seigler, D., and P. W. Price. 1976. Secondary compounds in plants: Primary functions. *Amer Sci* 110:101–105.

Sherry, T. W. 1984. Comparative dietary ecology of sympatric, insectivorous Neotropical flycatchers. *Ecol. Mono.* 54:313–338.

Short, L. L. 1982. *Woodpeckers of the world.* Greenville, DE: Delaware Museum of Natural History.

Shoumatoff, A. 1978. *The rivers Amazon.* San Francisco: Sierra Club Books.

Shoumatoff, A. 1986. *In southern light: Trekking through Zaire and the Amazon.* New York: Simon and Schuster.

Shukla, J., C. Nobre, and P. Sellers. 1990. Amazonian deforestation and climate change. *Science* 247:1322–1325.

Sibley, C. G., and J. E. Ahlquist. 1983. Phylogeny and classification of birds based on the data of DNA-DNA hybridization. *Current ornithology*, R. F. Johnston, ed. New York: Plenum Press.

———. 1990. *Phylogeny and classification of birds, a study in molecular evolution.* New Haven: Yale Univ. Press.

Sibley, C. G., and B. L. Monroe, Jr. 1990. *Distribution and taxonomy of birds of the world.* New Haven: Yale Univ. Press.

Sick, H. 1967. Courtship behavior in manakins (Pipridae): A review. *Living Bird* 6:5–22.

———. 1993. *Birds in Brazil, a natural history.* Princeton: Princeton Univ. Press.

Siemans, A. H. 1982. Prehispanic agricultural use of the wetlands of northern Belize In *Maya subsistence*, F. V. Fleming, ed. New York: Academic Press.

Silman, M. R. 1996. Regeneration from seed in a Neotropical rain forest. Ph.D. dissertation, Dept. of Zoology, Duke Univ.

Silva, J. F., J. Raventos, H. Caswell, and M. C. Trevisan. 1991. Population responses to fire in a tropical savanna grass, *Andropogon semiberbis*: A matrix model approach. *Jour. Ecol.* 79:345–356.

Simberloff, D. S. 1984. Mass extinction and the destruction of moist tropical forests. *Zh. Obshch. Biol.* 45:767–778.

Simberloff, D. S., and L. G. Abele. 1976. Island biogeography theory and conservation practice. *Science* 191:285–286.

Simpson, B. B., and J. Haffer. 1978. Speciation patterns in the Amazonian forest biota. *Ann. Rev. Ecol. Syst.* 9:497–518.

Simpson, G. G. 1980. *Splendid isolation: The curious history of South American mammals.* New Haven: Yale Univ. Press.

Sioli, H., ed. 1984. *The Amazon: Limnology and landscape ecology of a mighty tropical river and its basin.* Dordrecht: W. Junk Publishers.

Skole, D. L. and C. Tucker. 1993. Tropical deforestation and habitat fragmentation in the Amazon: satellite data from 1978 to 1988. *Science* 260:1905–1910.

Skole, D. L., W. H. Chomentowski, W. A. Salas, and A. D. Nobre. 1994. Physical and human dimensions of deforestation in Amazonia. *Bioscience* 44:314–322.

Skutch, A. F. 1954. *Life histories of Central American birds.* Pacific Coast Avifauna No. 31. Berkeley: Cooper Ornithological Society.

———. 1960. *Life histories of Central American birds II.* Pacific Coast Avifauna No. 34. Berkeley: Cooper Ornithological Society.

———. 1967. *Life histories of Central American highland birds.* Publication No. 7. Cambridge: Nuttall Ornithology Club.

———. 1969. *Life histories of Central American birds III.* Pacific Coast Avifauna No. 35. Berkeley: Cooper Ornithological Society.

———. 1972. *Studies of tropical American birds.* Publication No. 10. Cambridge: Nuttall Ornithology Club.

———. 1973. The *life of the hummingbird.* New York: Crown.

———. 1981. *New studies of tropical American birds.* Publication No. 19. Cambridge: Nuttall Ornithology Club.

———. 1983. *Birds of tropical America.* Austin: Univ of Texas Press.

———. 1985. *The life of the woodpecker.* Ithaca: Cornell Univ. Press.

———. 1989. *The life of the tanager.* Ithaca: Cornell Univ. Press.

———. 1991. *The life of the pigeon.* Ithaca: Cornell Univ. Press.

———. 1996. *Antbirds and ovenbirds.* Austin: Univ. of Texas Press.

Smil, V. 1979. Energy flows in the developing world. *Am. Sci.* 67:522–531.

Smiley, J. T. 1985. *Heliconius* caterpillar mortality during establishment on plants with and without attending ants. *Ecology* 66:845–849.

Smith, A. 1990. *Explorers of the Amazon.* Chicago: Univ. of Chicago Press.

Smith, N. J. H. 1981. Colonization lessons from a tropical forest. *Science* 214:755–760.

———. 1981. *Man, fishes, and the Amazon.* New York: Columbia Univ. Press.

Smith, S. M. 1975. Innate recognition of coral snake pattern by a possible avian predator. *Science* 187:759–760.

———. 1977. Coral snake pattern rejection and stimulus generalisation by naive great kiskadees (Aves: Tyrannidae). *Nature* 265:535–536.

Smith, T. J. 1987. Seed predation in relation to tree dominance and distribution in mangrove forests. *Ecology* 68:266–273.

Smythe, N., W. E. Glanz, and E. G. Leigh, Jr. 1982. Population regulation in some terrestrial frugivores. In *The ecology of a tropical forest*, E. G. Leigh, Jr., A. S. Rand, and D. M. Windsor, eds. Wash., DC: Smithsonian Inst. Press.

Snow, B. K., and D. W. Snow. 1979. The ochre-bellied flycatcher and the evolution of lek behavior. *Condor* 81:286–292.

Snow, D. W. 1961 The natural history of the oilbird, *Steatornis caripensis*, in Trinidad, W. I. Part 1. General behavior and breeding habits. *Zoologica* 46:27–48.

———. 1962a. A field study of the black-and-white manakin, *Manacus manacus*, Trinidad. *Zoologica* 47:65–104.

———. 1962b. The natural history of the oilbird, *Steatornis caripensis*, in Trinidad, W. I. Part 2. Population, breeding ecology, food. *Zoologica* 47:199–221.

———. 1966 A possible selective factor in the evolution of fruiting seasons in tropical forest. *Oikos* 15:274–281.

———. 1971. Observations on the purple-throated fruit-crow in Guyana. *Living Bird* 10:5–18.

———. 1976. *The web of adaptation.* Ithaca: Cornell Univ. Press.

———. 1982. *The cotingas: Bellbirds, umbrel-labirds, and other species.* Ithaca: Cornell Univ. Press.

Solbrig, O. T., and M. D. Young. 1992. Toward a sustainable and equitable future for savannas. *Environment* 34:6–15, 32–35.

Sollins, P., F. Sancho M., R. Mata Ch., and R. L. Sanford, Jr. 1994. Soils and soil processes. In *La Selva: Ecology and natural history of a Neotropical rain forest*, L. A. McDade, K. S. Bawa, H. A. Hespenheide, and G. S. Hartshorn, eds. Chicago: Univ. of Chicago Press.

Soule, M. E. 1980. Thresholds for survival: Maintaining fitness and evolutionary potential. In *Conservation biology: An evolutionary-ecological perspective*, M. E. Soule and B. A. Wilcox, eds. Sunderland, MA: Sinauer Associates.

Sowls, L. K. 1984. *The peccaries.* Tucson: Univ. of Arizona Press.

Spencer, K. C 1984. Chemical correlates of coevolution: The *Passiflora/Heliconius* interaction. *Bull. Ecol. Soc. Amer.* 65:231.

Stiles, E. W. 1980. Patterns of fruit presentation and seed dispersal in bird-disseminated woody plants in the eastern deciduous forest. *Amer. Nat.* 116:670–688.

———. 1984. Fruit for all seasons. *Nat. Hist* 93:42–53.

Stiles, F. G. 1975. Ecology, flowering phenology and hummingbird pollination of some Costa Rican *Heliconia* species. *Ecology:* 56:285–301.

———. 1977. Coadapted competitors: The flowering seasons of humming-bird-pollinated plants in a tropical forest. *Science* 198:1177–1178.

———. 1983. *Heliconia latispatha* (Plata-nillo, wild plantain). In *Costa Rican natural history*, D. H. Janzen, ed. Chicago: Univ. of Chicago Press.

Stoddart, D. R. 1969. Post-hurricane changes in the British Honduras reefs and cays. *Atoll Research Bulletin* 131:1–25.

Storer, R. W. 1969. What is a tanager? *Living Bird* 8:127–136.

Stotz, D. F., J. W. Fitzpatrick, T. A. Parker, III, and D. K. Moskovits. 1996. *Neotropi-*

cal birds: *Ecology and conservation.* Chicago: Univ. of Chicago Press.

Stradling, D. J. 1978. The influence of size on foraging in the ant *Atta cephalotes,* and the effect of some plant defense mechanisms. *Jour. Animal Ecol.* 47:173–188.

Strahl, S. D. 1985. Correlates of reproductive success in communal Hoatzins (*Opisthocomus hoazin*). Abstracts of the 103d American Ornithologists' Union Meeting.

Strahl, S. D. and A. Schmitz. 1990. Hoatzins: cooperative breeding in a folivorous Neotropical bird. In *Cooperative breeding in birds,* P. B. Stacey and W. D. Koenig, eds. Cambridge: Cambridge Univ. Press.

Stuiver, M. 1978. Atmospheric carbon dioxide and carbon reservoir changes. *Science* 199:253–258.

Tanner, J. E., T. P. Hughes, and J. H. Connell. 1994. Species coexistence, keystone species, and succession: A sensitivity analysis. *Ecology* 75:2204–2219.

Terborgh, J. 1974. Preservation of natural diversity: the problem of extinction prone species. *Bioscience* 24:715–722.

———. 1986. Keystone plant resources in the tropical forest. In *Conservation biology: the science of scarcity and diversity,* M. E. Soule, ed. Sunderland, MA: Sinauer.

———. 1989. *Where have all the birds gone?* Princeton: Princeton Univ. Press.

———. 1992. Maintenance of diversity in tropical forests. *Biotropica* 24(2b): 283–292.

Terborgh, J., and J. S. Weske. 1975. The role of competition in the distribution of Andean birds. *Ecology* 56:562–576.

Terborgh, J., and A. W. Goldizen. 1985. On the mating system of the cooperatively breeding saddle-backed tamarin (*Saguinus fuscicollis*). *Behav. Ecol. Sociobiol.* 16:293–299.

Terborgh, J., S. K. Robinson, T. A. Parker, III, C. A. Munn, and N. Pierpont. 1990. Structure and organization of an Amazonian forest bird community. *Ecol. Monog.* 60:213–238.

Terborgh, J., R. B. Foster, and P. V. Nunez

1996. Tropical tree communities: A test of the nonequilibrium hypothesis. *Ecology* 77:561–567.

Thompson, J., J. Proctor, V. Viana, W. Milliken, J. A. Ratter, and D. A. Scott. 1992. Ecological studies on a lowland evergreen rain forest on Maraca Island, Roraima, Brazil. I: Physical environment, forest structure, and leaf chemistry. *Jour. Ecol.* 80:689–703.

Thomsen, J. B. and A. Brautigam. 1991. Sustainable use of Neotropical parrots. In *Neotropical wildlife use and conservation.* J. G. Robinson and K. H. Redford, eds. Chicago: Univ. of Chicago Press.

Timson, J. 1993. How vampire bats acquired a taste for blood. *New Scientist* 137:18.

Toledo, V. M. 1977. Pollination of some rainforest plants by nonhovering birds in Veracruz, Mexico. *Biotropica* 9:262–267.

Tomlinson, P. B. 1995. *The botany of mangroves.* Cambridge: Cambridge Univ. Press.

Trail, P. W. 1985a. Courtship disruption modifies mate choice in a lek-breeding bird. *Science* 227:778–779.

———. 1985b. A lek's icon: The courtship display of a Guianan cock-of-the-rock. *Amer. Birds* 39:235–240.

Tramer, E. J. 1974. On latitudinal gradients in avian diversity. *Condor* 76:123–130.

Traylor, M. A., Jr., and J. W. Fitzpatrick. 1982. A survey of the tyrant flycatchers. *Living Bird* 19:7–50.

Turner, B. L., II, and P. D. Harrison. 1981. Prehistoric raised-field agriculture in the Maya lowlands. *Science* 213: 339–405.

———, eds. 1983. *Pulltrouser swamp: Ancient Maya habitat, agriculture, and settlement in northern Belize.* Austin: Univ. of Texas Press.

Turner, D. C. 1975. *The vampire bat: A field study in behavior and ecology.* Baltimore: Johns Hopkins Univ. Press.

Turner, J. R. G. 1971. Studies of Mullerian mimicry and its evolution in burnet moths and heliconid butterfiies. In *Ecological genetics and evolution,* R. Creed, ed. Oxford, England: Blackwell Scientific.

Turner, J. R. G. 1975. A tale of two butterflies. *Nat. Hist.* 84:29–37.

———. 1981. Adaptation and evolution in *Heliconius*: A defense of neo-Darwinism. *Ann. Rev. Ecol. Syst.* 12:99–121.

———. 1988. The evolution of mimicry: A solution to the problem of punctuated equilibrium. *Amer. Nat.* 131, Supplement:S42-S66.

Tuttle, M. D., and M . J . Ryan. 1981. Bat predation and the evolution of frog vocalizations in the neotropics. *Science* 214:677–678.

Uhl, C. 1987. Factors controlling succession following slash and burn agriculture in Amazonia. *Ecology* 75:377–407.

———. 1988. Restoration of degraded lands in the Amazon Basin. *In Biodiversity*, E. O. Wilson, ed. Wash., DC: National Academy Press.

Uhl, C., and C. F. Jordan. 1984. Succession and nutrient dynamics following forest cutting and burning in Amazonia. *Ecology* 65:1476–1490.

Uhl, C., K. Clark, N. Dezzeo, and P. Maquirino. 1988a. Vegetation dynamics in Amazonian treefall gaps. *Ecology* 69:751–763.

Uhl, C., J. B. Kauffman, and D. L. Cummings. 1988b. Fire in the Venezuelan Amazon 2: Environmental conditions necessary for forest fires in the evergreen rainforest of Venezuela. *Oikos* 53:176–184.

Uhl, C., R. Buschbacher, and E. A. S. Serrao. 1988c. Abandoned pastures in eastern Amazonia. I. Patterns of plant succession. *Jour. Ecol.* 76:663–681.

Uhl, C., and I. C. G. Vieira. 1989. Ecological impacts of selective logging in the Brazilian Amazon: A case study from the Paragominas region in the state of Para. *Biotropica* 21:98–106.

Uhl, C., D. Nepstad, R. Buschbacher, K. Clark, B. Kauffman, and S. Subler. 1990. Studies of ecosystem response to natural and anthropogenic disturbances provide guidelines for designing sustainable land-use systems in Amazonia. In *Alternatives to deforestation: Steps toward sustainable use of the Amazon rain forest*, A. B. Anderson, ed. New York: Columbia Univ. Press.

Utley, J. F., and K. Burt-Utley. 1983. Bromeliads (Pina silvestre, pinuelas, chiras, wild pineapple). In *Costa Rican natural history*, D. H. Janzen, ed. Chicago: Univ. of Chicago Press.

Vandermeer, J. 1983a. African oil palm (Palma de aceite). In *Costa Rican natural history*, D. H. Janzen, ed. Chicago: Univ. of Chicago Press.

———. 1983b. Banana (Platano, banano). In *Costa Rican natural history*, D. H. Janzen, ed.. Chicago: Univ. of Chicago Press.

———. 1983c. Coconut (Coco). In *Costa Rican natural history*, D. H. Janzen, ed. Chicago: Univ. of Chicago Press.

Van Devender, R. W. 1983. *Basiliscus basiliscus* (Chisbala, garrobo, basilisk, Jesus Christ lizard). In *Costa Rican natural history*, D. H. Janzen, ed. Chicago: Univ. of Chicago Press.

Van Valen, L. 1973. A new evolutionary law. *Evol. Theory* 1:1–30.

Velga, M. M., J. A. Meech, and N. Onate. 1994. Mercury pollution from deforestation. *Nature* 368:816–817.

Vickers, W. T. 1988. Game depletion hypothesis of Amazonian adaptation: Data from a native community. *Science* 239:1521–1522.

———. 1991. Hunting yields and game composition over ten years in an Amazon Indian territory. In *Wildlife use and conservation*, J. G. Robinson and K. H. Redford, eds. Chicago: Univ. of Chicago Press.

Vitousek, P. M., 1984. Litterfall, nutrient cycling, and nutrient limitation in tropical forests. *Ecology* 65:285–298.

———. 1994. Beyond global warming: Ecology and global change. *Ecology* 75:1861–1876.

Vitousek, P. M. and R. L. Sanford. 1986. Nutrient cycling in moist tropical forest. *Ann. Rev. Ecol. Syst.* 17:137–167.

Vitousek, P. M., P. R. Ehrlich, A. E. Ehrlich, and P. A. Matson. 1986. Human appropriation of the products of photosynthesis. *Bioscience* 36:368–373.

Vogt, K. A., C. C. Grier, and D. J. Vogt. 1986. Production, turnover, and nutrient dynamics of above- and below-ground detritus of world forests. In *Adv. Ecol. Res.* 15, A. Macfadyen and E. D. Ford, eds. London: Academic Press.

Wahle, C. M. 1980. Detection, pursuit, and overgrowth of tropical gorgonians by milleporid hydrocoarals: Perseus and Medusa revisited. *Science* 209:689–691.

Wallace, A. R. 1853. *Palm trees of the Amazon and their uses.* London: J. Van Voorst.

Wallace, A. R. 1876. *The geographical distribution of animals. With a study of the relations of living and extinct faunas as elucidating the past changes in the earth's surface.* New York: Harper & Bros.

———. 1895. *Natural selection and tropical nature.* London: Macmillan.

Walsh, G. E. 1974. Mangroves: A review. In *Ecology of Halophytes,* R. J. Reimold and W. H. Queens, eds. New York: Academic Press.

Walter, H. 1971. *Ecology of tropical and subtropical vegetation.* New York: Van Nostrand Reinhold.

———. 1973. *Vegetation of the earth in relation to climate and the ecophysiological conditions.* London: English Universities Press.

Walterm, K. S. 1983. Orchidaceae (orquideas, orchids). In *Costa Rican natural history,* D. H. Janzen, ed. Chicago: Univ. of Chicago Press.

Waterton, C. [1825]. 1983. *Wanderings in South America.* Reprint. London: Century Publishing.

Webb, S. D. 1978. A history of savanna vertebrates in the new world. Part II: South America and the great interchange. *Ann. Rev. Ecol. Syst.* 9:393–426.

Weber, N. A. 1972. The attines: The fungus-culturing ants. *Amer. Sci.* 60:448–456.

Werner, D. I. 1991. The rational use of green iguanas. In *Neotropical wildlife use and conservation,* J. G. Robinson and K. H. Redford, eds. Chicago: Univ. of Chicago Press.

Wetterer, J. K. 1994. Nourishment and evolution in fungus-growing ants and their fungi. In *Nourishment and Evolu-*

tion in Insect Societies, J. H. Hunt and C. A. Nalepa, eds. Boulder: Westview Press.

Wheelwright, N. T. 1983. Fruits and the ecology of resplendent quetzals. *Auk* 100:286–301.

———. 1985. Fruit size, gape width, and the diets of fruit-eating birds. *Ecology* 66:808–818.

———. 1988. Fruit-eating birds and bird-dispersed plants in the tropics and temperate zone. *Trends in Ecol. Evol.* 3:270–274.

Wheelwright, N. T., and G. H. Orians. 1982. Seed dispersal by animals: Constraints with pollen dispersal, problems of terminology, and constraints on co-evolution. *Amer. Nat.* 119:402–413.

Wheelwright, N. T., W. A. Haber, K. G. Murray, and C. Guindon. 1984. Tropical fruit-eating birds and their food plants: A survey of a Costa Rican lower Montane forest. *Biotropica* 16:173–192.

Whittaker, R. H. 1975. *Communities and ecosystems.* 2d ed. New York: Macmillan.

Whittaker, R. H., and P. P. Feeny. 1971. Allelochemics: Chemical interactions between species. *Science* 171:757–770.

Wiebes, J. T. 1979. Co-evolution of figs and their insect pollinators. *Ann. Rev. Ecol. Syst.* 10:1–12.

Wilk, R. R. 1991. *Household ecology: Economic change and domestic life among the Kekchi Maya in Belize.* Tucson: Univ. of Arizona Press.

Wilkinson, G. S. 1990. Food sharing in vampire bats. *Sci. Amer.* 263:76–82.

Williams-Linera, G. 1983. Biomass and nutrient content in two successional stages of tropical wet forest in Uxpanapa, Mexico. *Biotropica* 15:275–284.

Willis, E. O. 1966. The role of migrant birds at swarms of army ants *Living Bird* 5:187–232.

———. 1967. The behavior of the bicolored antbirds. *Univ. of California Publ. Zool.* 79:1–127.

———. 1974. Populations and local extinctions of birds on Barro Colorado Island, Panama. *Ecol. Monog.* 44:153–169.

Willis, E. O., and Y. Oniki. 1978. Birds and army ants. *Ann. Rev. Ecol. Syst.* 9:243–263.

Wilson, D. E. 1989. Bats. In *Tropical rain forest ecosystems: Biogeographical and ecological studies (Ecosystems of the world, 14B)*, W. Leith and M. J. A. Werger, eds. Amsterdam: Elsevier.

Wilson, E. O. 1971. *The insect societies.* Cambridge: Belknap Press of Harvard Univ.

———. 1978. *On human nature.* Cambridge: Harvard Univ. Press.

———. 1984. *Biophilia.* Cambridge: Harvard Univ. Press.

———. 1987. The arboreal ant fauna of Peruvian Amazon forests: A first assessment. *Biotropica* 19:245–251.

———, ed. 1988. *Biodiversity.* Wash., DC: National Academy Press.

———. 1988. The current state of biological diversity. In *Biodiversity*, E. O. Wilson, ed. Wash., DC: National Academy Press.

Wilson, E. O. 1991. Rain forest canopy: The high frontier. *Nat. Geog.* 180:78–107.

———. 1992. *The diversity of life.* New York: W.W. Norton.

Winkler, K., J. H. Rappole, and M. A. Ramos. 1990. Population dynamics of the Wood Thrush in southern Veracruz, Mexico. *Condor* 92:444–460.

Witman, J. D. 1992. Physical disturbance and community structure of exposed and protected reefs: A case study from St. John, U.S. Virgin Islands. *Amer. Zoologist* 32:641–654.

Wolf, L. L. 1975. "Prostitution" behavior in a tropical hummingbird. *Condor* 77:140–144.

Woodley, J. D., E. A. Chornesky, P. A. Clifford, J. B. C. Jackson, L. S. Kaufman, N. Knowlton, J. C. Lang, M. P. Pearson, J. W. Porter, M. C. Rooney, K. W. Rylaarsdam, V. J. Tunnicliffe, C. M. Wahle, J. L. Wulff, A. S. G. Curtis, M. D. Dallmeyer, B. P. Jupp, M. A. R. Koehl, J. Neigel, and E. M. Sides. 1981. Hurricane Allen's impact on Jamaican coral reefs. *Science* 214:749–761.

Woodwell, G. H. 1978. The carbon dioxide question. *Sci. Amer.* 238:34–43.

Woodwell, G. H., J. E. Hobbie, R. A. Houghton, J. M. Melillo, B. Moore, B. J. Peterson, and G. R. Shaver. 1983. Global deforestation: Contribution to atmospheric carbon dioxide. *Science* 222:1081–1086.

World Conservation Monitoring Centre. 1992. *Global biodiversity: Status of the Earth's living resources*, B. Groombridge, ed. London: Chapman and Hall.

Worthington, A. 1982. Population sizes and breeding rhythms of two species of manakins in relation to food supply. In *The ecology of a tropical forest*, E. G. Leigh, Jr., A. S. Rand, and D. M. Windsor, eds. Wash., DC: Smithsonian Inst. Press.

Wunderle, J. M., Jr. 1982. The timing of the breeding season in the Bananaquit (*Coereba flaveola*) on the island of Grenada, W.I. *Biotropica* 14:124–131.

Yoon, C. K. 1995. Monumental inventory of insects in a Costa Rican forest is under way. *New York Times*: July 11, 1995, C4.

Young, A. M. 1971. Wing coloration and reflectance in *Morpho* butterflies as related to reproductive behavior and escape from avian predators. *Oecologia* 7:209–222.

Zahl, P. A. 1975. Hidden worlds in the heart of a plant. *Nat. Geog.* 147:388–397.

Zeil, W. 1979. *The Andes: A geological review.* Berlin: Gerbruder Borntraeger.

Zimmerman, P. R., J. P. Greenberg, S. O. Wandiga, and P. J. Crutzen. 1982. Termites: A potentially large source of atmospheric methane, carbon dioxide, and molecular hydrogen. *Science* 218:563–565.

Zink, R. M. and M. C. McKitrick. 1994. The debate over species concepts and its implications for ornithology. *Auk* 111:701–719.

Zucker, W. V. 1983. Tannins. Does structure determine function? An ecological perspective. *Amet Nat.* 121:355–365.

Index

abaxial anthocyanin, 59
acacia (*Acacia* spp.), 71, 131–32, 232;
  bull's horn, 149
Ache tribe, 173–75
acouchy (*Myroprocta* spp.), 304
Acre, Brazil, 336, 338
adaptation, 77–79, 126, 142
adaptive radiation, 98–106, 203, 279, 323
agave, 232
agouti, 27, 94, 135, 177, 303–4, 367;
  Dasyprocta fuliginosa, 38; *Dasyprocta
  punctata*, 11
agriculture: crops, 176–77, 179, 183–88,
  341–42; large-scale, 66, 335, 336;
  modernization of, 346; nonindigenous
  farmers and, 181–82; prehistoric, 182–
  83; slash-and-burn (swidden), 66, 175–
  79, 235, 361, 362; successional crop system, 179–80
aguajale, 197
air plants. *See* epiphytes
alfisols, 52
algae, 245, 247–49
alkaloids, 145, 150–52, 163, 167, 168
allelochemics. *See* defense compounds
alligator, American (*Alligator mississippiensis*), 209
allspice (*Pimenta dioica*), 359
alpaca (*L. G. pacos*), 224
Alpha Helix, 164
Alta Floresta, Brazil, 335, 343, 344
altiplano, 218, 224–25
altitude sickness, 387
Amacuro Delta, 190
Amadon, D., 254
Amazon Basin: climate in, 4, 355; deforestation in, 335–38, 340–68; fish in, 201;
  monkeys in, 297; nature reserves in,
  367–68; in Pleistocene, 115; precipitation in, 5, 8, 14; rainforest productivity
  in, 45; regeneration patterns in, 66–68;
  rivers of, 191; soils in, 52, 53, 65; species
  richness in, 88, 120–25
Amazon Center for Environmental Educa-

tion and Research (ACEER), 42–43,
  375–76
Amazon Institute of Man and the Environment (IMAZON), 345
Amazon River, 17, 49, 57, 109, 167, 190–
  93, 356
Amazons, 198–200
amino acids, 27, 147; L-canaline, 155;
  L-canavanine, 147, 155; proline, 130,
  156–57; tyrosine, 130
amoebic dysentery, 386
amphibians, 89, 322–24
anaconda, 229, 312, 318, 319, 379; golden
  brown (*Eunectes murinus*), 208; turtle
  (*Chelus fimbriata*), 210–11; side-necked
  (Pelomedusidae), 210; Stupendemys,
  210
anchoveta (*Engraulis ringens*), 15
Andagoya, Colombia, 8
Andean Development Corporation, 336
Andes Mountains, 12, 13, 218–25, 371;
  and geographic isolation, 197; soils in,
  52, 57; source of Amazon, 190; tree species richness in, 120–25
Angel Falls, 190, 225–26
Angel, Jimmy, 225
ant followers, 265, 282–85, 348. *See also*
  antbird
ant-plants (*myrmecophytes*), 130–32
ant: Acromyrmex, 133; army (*Eciton
  burchelli; Labidus praedator*), 39–40, 95,
  282–83, 285, 326–28, 332, 380; and ant-
  eating plants, 130–32; Atta, 133, 134;
  Azteca, 131; biting (*Azteca* spp.), 72–73;
  fungus garden (*Atta cephalotes*), 133–35,
  147; giant tropical or bullet (*Paraponera*
  spp.), 328, 380; group predation by,
  327; leafcutter, 38, 42, 133–35, 146, 148,
  154; *Pseudomyrmex ferruginea*, 131–32,
  149
antbird, 110, 223, 284–85; bare-crowned,
  294; bicolored (*Gymnopithys leucaspis*),
  285; blackish (*Cercomacra nigrescens*),
  109, 110; dusky (*C. tyrannina*), 109, 110;

antbird (*cont.*)
ground (Formicaridae), 252, 279; Myrmotherula, 92; ocellated (*Phaenostictus mcleannani*), 285; spotted (*Hylophylax naevioides*), 285; typical (Thamnophilidae), 252, 279; 284, 286; white-plumed (*Pithys albifrons*), 39–40, 285

anteater, 112, 236; giant (*Myrmecophaga tridactyla*) , 229, 230, 236; 307–8; northern tamandua (*Tamandua mexicana*), 308; southern or collared tamandua (*T. tetradactyla*), 307

antpitta, 222, 284; giant (*Grallaria gigantea*), 223

antshrike, 110, 265; barred (*Thamnophilus doliatus*), 284; flycatching (*Thamnomanes*, 285); great (*Taraba major*), 284

antthrush, 284; black-faced (*Formicarius analis*), 285, 286; rufous-capped (*Folmicarius colma*), 39

anurans, 322–24

aposematic coloration. *See* warning coloration

aracari (*Pteroglossus*), 117, 260; collared (*P. torquatus*), 281

Arecife, 233

Arima Valley, Trinidad, 92, 274, 340

armadillo (*Dasypus novemicinctus*), 112, 114, 309–10

arrested litter, 55

arrow poison, 166

arthropods, 89, 94

arum, ornamental (*Monstera deliciosa*), 30

Askins, Robert A., 96

assassin bug, 36

Atacama-Sechura Desert, 14, 219

Atlantic Coastal Forest, 338–39, 350, 352, 368, 371

Austin, O. L., 252

automimicry, 160

bacteria, 49, 383, 384

Bahia, Brazil, 369

balche, 168

Balick, M. J., 164, 166

bamboo, 96–97

banana (*Musa acuminata; M. balbisiana*), 184, 342

Banks, Joseph, 185

barbet, 260

bare-eye, black-spotted (*Phlegopsis nigromanulata*), 39

Barra del Colorado-Tortuguero, 339

barracuda (*Sphyraena barracuda*), 246

Barro Colorado Island (BCI), Panama: birds in, 293; canopy trees in, 149; forest demographics in, 59, 62–63; extinction on, 353; fruit trees on, 138; gaps in, 10, 58, 59; leaf area index in, 46; leaf-cutting ants on, 133; monkeys on, 297, 301; seasons in, 9, 11–12; species richness on, 35, 94

bat (Microchiroptera): adaptive radiation by, 98–102; Carollia, 70; and cockroaches, 329; common vampire (*Desmodus rotundus*), 100–1; competition avoidance by, 92; courtship and mating, 102; echolocation by, 99; Ectophylla alba, 102; false vampire (*Vampyrum spectrum*), 99; fringe-lipped (*Trachops cirrhosus*), 86; fruit-eating, 100; greater fishing bulldog (*Noctilio leporinus*), 100; as kapok pollinator, 73; and mammalian diversity, 87; nectarivorious, 100; pollination by, 27, 127, 129–30; predation by, 86; Puerto Rican (*Artibeus jamaicensis*), 102; as resource for predators, 95–96; as seed consumer, 27; sloths on, 308; *Thyroptera tricolor*, 102; vampire (*Diaemus youngii*), 100

Bates, Henry Walter, 9, 54, 159, 210, 251, 252, 381–82

batrachotoxins, 81–82, 170

Bauhinia pauletia, 129–30

Beagle, H.M.S., 219

beans (*Phaseolus vulgaris*), 184

bear, spectacled (*Tremarctos ornatus*), 222–23

Beard, J. S., 228

bee, 32, 127, 128, 266, 382–83; euglossine, 128, 348

beech, southern (*Nothofagus*), 112

Beehler, B. M., 278

Beese, G., 28

beetle (*Cyclocephala* spp.), 128, 36–37, 94, 379; bruchid (*Merobruchus columbinus*), 154–55, 297; harlequin (*Acrocinus longimanus*), 329; rhinoceros (*Megasoma elephas*), 329–30

Belém, Brazil, 8, 195, 336

Belize Association of Traditional Healers, 165

Belize City, Belize, 8

Belize, 8, 19, 99, 122, 137, 339, 365; agriculture in, 183; and conservation, 374; savanna in, 229, 234; spiny palm in, 148

bellbird (*Procnias*), 272–74, 277; bare-throated (*P. nudicollis*), 273; bearded (*P. averano*), 137; three-wattled (*P. tricarunculata*), 273; white (*P. alba*), 273

Belt, Thomas, 115

Beltian bodies, 1341

Benson, Woodruff, 158

Bierregaard, R. O., 348

biodiversity, 362, 372; and diffuse competition, 85; between habitat (beta), 88–89;

hyper, 87; loss of, 350–54; of mammals, 97; regional (gamma), 89; of trees, 67; within habitat (alpha), 88, 90. *See also* species richness
biogeochemical cycling, 49
Biological Dynamics of Forest Fragments Project, 347, 362
Biotropica, 375
birds: bill size and shape, 214, 92, 103–4; as carnivores, 113; cave dwelling by, 142; foraging flocks of, 251, 265, 284; fruit and nectar feeders, 139–40, 260–70, 277, 279; insect-arthropod feeders, 279–85; large ground dwellers, 253–56; and leks, 278–79; long-distance migrants, 291–94, 347, 353; plumage color, 271, 256–60; predatory, 286–91; along rivers and streams, 211–17; and seasonal changes, 10; and selection pressures, 277–78; sexual selection among, 270–77; size gradation of, 216; species richness of, 35–36, 95, 109–10, 124; 251–52; suboscines and oscines, 279; territoriality of, 293; voices of, 272–73. *See also* sexual selection
bittern, pinnated, 230
Bladen Reserve, 339
Bligh, William, 185
Blue Creek Village, Belize, 138, 178–79, 364
Bolivia, 18, 167, 360. *See also* Pantanal
bonefish (*Albula vulpes*), 246
booby (*Sula* spp.), 20; brown (*Sula leucogaster*), 242
borhyaenoids, 112, 114
botfly (*Dermatoba hominis*) , 301, 379–80
boto. *See* dolphin
bottle brush tree (*Callistemom lanceolatus*), 26
Brack, A., 358
Brazil, 132, 337–38; campos cerrados of, 231; cloud forests in, 220, 221; deforestation in, 3, 357; headhunters in, 171; indigenous population of, 172; savanna in, 236; species richness in, 34. *See also* Manaus; Pantanal
Brazil National Institute for Research in Amazonia, 347
Brazil nut tree (*Bertholletia excelsa*), 25, 27, 128, 187, 194
Brazilian Shield, 53–54
Brazilian Society of Zoology, 352
breadfruit tree (*Artocarpus altilis*), 185
breadnut tree (*Brosmium alicastrum*), 66, 183
Brokaw, N. V. L., 59–60, 71
bromeliads (Bromeliaceae), 31–32, 33, 42; *Puya raimondii*, 224

Browder, J. O., 360
Brower, Lincoln, 158, 160
Brown, G. G., 352
Brown, K. S., Jr., 352
Brown, Keith, 118
Brown, S., 360
Buckley, P. A., 252
bunting, indigo (*Passerina cyanea*), 294
butterfly, 36: antbutterfly, 332; Batesian mimicry by, 159–60; black witch (*Ascalapha odorata*), 331; brush-footed (Nymphalidae), 155; heliconid, 118, 130, 153, 155–59; *Heliconius charitonius*, 156; *H. erato*, 156, 158, 160, 161; *H. melpomene*, 160, 161; *Mechanitis isthmia*, 157; Melinaea, 153–54; monarch (*Danaus plexippus*), 155, 158; morpho (*Morpho didius*), 38, 79, 331–32; myrmecophilous caterpillars, 132; North American viceroy (*Basilarchia archippus*), 159–60; owl (*Caligo memnon*), 331; painted lady (*Vanessa virginiensis*), 155; pipevine swallowtail (*Battus philenor*), 155; *Pseudosphinx tetrio*, 153; swallowtail (*Eurytides*), 332; *Thisbe irenea*, 132; toxic compounds in, 83; *Zaretis itys*, 80
Byrsonima crassifolia, 228

caatinga, 232
cacao (*Theobroma cacao*), 26, 128, 185–86, 342, 369
cacique, 265–66; yellow-rumped (*Cacicus cela*), 266
cactus (*Lophophora* spp.), 168, 232
Caesalpinia brasiliensis, 71
Caesearia sylvestris, 228
caffeine, 145
caiman; black (*Melanosuchus niger*), 210; broad-snouted (*Caiman latirostris*), 210; common (*C. crocodilus*), 209–10; Cuvier's dwarf (*Paleosuchus palperbrosus*), 210; Schneider's dwarf (*P. trigonatus*), 210; spectacled, 229
Calama, Chile, 14
calcium oxalate, 148
Calliandra surinamensis, 71
Campbell, J. A., 314, 315–17
campos cerrados, 132, 231, 238, 371
Canaima Falls, Venezuela, 56
canals (*chinampas*), 183
Candela. See piper
Candellillos. See piper
cannonball tree (*Couroupita guanianensis*), 26
canopy walkway, 42, 375
capuchin, white-throated (faced), 11
capybara (*Hydrochaeris hydrochaeris*), 112, 114, 207–8, 229, 231

caracara: black (*Daptrius ater*), 194; black and white mountain (*Phalcobaenus albogularis*), 224; crested (*Polyborus plancus*), 237; red-throated (*D. americanus*), 194; yellow-headed (*Milvago chimachima*), 194, 289
carbon, 363
carbon dioxide, 354–55
carbon fixation, 97
cardiac glycosides, 147, 149, 158
Caribbean National Forest, 340
carne asada tree (*Andira inermis*), 152
Carvajal, Gaspar de, 199, 200
caryophylene epoxide, 147
Casearia corymbosa, 138
cassava. *See* manioc
cat, 384
catbird, gray (*Dumetella carolinensis*), 292, 293
caterpillar, 380–81
catfish , 230; red-tailed, 18; *Vandellia cirrhosa* (*candiru*), 201, 205
cattle ranching, 339, 340–41, 348, 363
cauliflory, 26, 129
cavy, patagonian, 238
cecropia (*Cecropia* spp.), 40, 47, 66, 71–73, 194, 292–93, 308; *C. insignis*, 59, 71; *C. obtusifolia*, 72, 369; *C. peltata*, 146; defense compounds of, 151; extrafloral nectaries of, 131
cedar tree (*Cedrela odorata*), 175
Cedrelia, 66
Cenozoic Era, 112, 119
Center for Field Research, 375
Center, T. D., 154
centipede (*Scolopendra gigantea*), 382
Central America, 339
chachalaca, plain (*Ortalis vetula*), 254
Chaco, 18, 228
Chaga's disease, 386
Chalif, E. L., 257
Chan Chich Lodge, Belize, 255, 365
characins, 205; *Semaprochilodus*, 205
chat, yellow-breasted (*Icteria virens*), 292
chicle tree (*Manilkara zapota*), 25, 148, 359–60
chigger, 383
Children's Environmental Trust Foundation, International (CET), 375–76
Children's Rainforest Workshops, 375
Chile, 14
chinchilla (*Chinchilla* spp.), 224, 303
chiropterophily, 129–30
Choco, 371
Choco Indians, 35, 81, 323–24
cholera, 385
cicada, 37
Cinchona, 164

cinnamon (*Ocotea quixos*), 199
Ciudad Bolivar, Venezuela, 189
Clark, Deborah, 33
climate, 4–16, 62, 114, 124, 233, 354–57
Clusia, 221
coatimundi (*Nasua* spp.), 310; *Nasua ilarica*, 11, 136; South American coati (*N. nasua*) , 38, 311; white-nosed coati (*N. narica*), 311
Coca-Cola, 167, 384
cocaine, 166–67
Cocha Cashu Biological Station, Peru, 35, 52, 60, 123, 353, 364
cock-of-the-rock: Andean (*Rupicola peruviana*), 223; Guianan (*R. Rupicola*), 223, 269, 270–72, 277
cockroach, 94, 379; giant (*Blaberus giganteus*), 328–29
Cockscomb Basin Jaguar Reserve, 339
Cocoa Plum Cay, Belize, 244
Coe, M. D., 168
coevolution, 127–35, 249, 261
coffee (*Coffea arabica*), 186, 346–47
Colinvaux, P. A., 118, 119
Colombia, 81, 117, 167, 220, 337, 351; bird species richness in, 124; deforestation in, 3; Indians of, 177; precipitation in, 8
Colwell, R. K., 263, 378
Combacopsis quinatum, 267
competition, 85, 280; interspecific, 90–93, 370
conch, queen (*Strombus gigas*), 246
condor, Andean (*Vultur gryphus*), 224
Connell, J. H., 121, 250
Conservation Biology, 370, 372
conservation ethics, 388
Conservation International, xi
convergence, 161
Cook, Captain, 185
cooperative behavior, 277
coot, giant (*Fulica gigantea*), 225
coral, 247; boulder (*Montastrea annularis*), 246, 248; brain (*Diploria*), 248, 249; elkhorn (*Acropora palmata*), 246, 248; finger (*Porites*), 248; gorgonian (*Gorgonia* spp.), 249; sheet (*Agaricia*), 248; starlet (*Siderastrea*), 248; stinging (*Millepora*), 248, 249
coral reef, 20, 122, 247–50
coral tree (*Erythrina* spp.), 26
cormorant (*Phalacrocorax olivaceus*), 193
cornerstone habitats, 371
corydoras, bronze (*Corydoras geneus*), 201
Costa Rica, 93, 132, 135, 136, 138, 220, 339; anurans in, 322; bats in, 99; chemical prospecting in, 164; and conservation, 374; fire in, 176; fruit eating by

birds in, 139–40; monkeys in, 152, 297; orchids in, 32; seasonal changes in, 10; soils in, 52, 57; species richness in, 36, 124. *See also* La Selva Biological Station

cotinga, 95, 135, 222, 268–69, 279; Amazonian umbrellabird (*Cephalopterus ornatus*), 194, 269; hestnut-bellied (*Doliornis remseni*), 252; pompadour (*Xipholena punicea*), 42; spangled (*Cotinga cayana*), 43. See also bellbird; cock-of-the rock; fruit crow; piha

Cox, P. A., 164

crab, 31; land hermit (*Coenobita clypeatus*), 242; mangrove fiddler (*Macropipus puber*), 242; mangrove tree (*Aratus pisonii*), 242

Cracraft, J., 119

Croat, T. B., 28

crocodile: American (*C. acutus*), 209; Cuban (*C. rhombifer*), 209; Morelet's (*C. moreletii*), 209; Orinoco (*C. intermedius*), 209

crossbill, red (*Loxia curvirostra*), 237

cryptic coloration (crypsis), 79–81, 126, 162, 249, 277, 280, 330

ctenosaur (*Ctenosaura*), 321

Cuba, 261, 339–40

Cueva del Guacharo, Venezuela, 143

curare (*Curarea toxicofera*), 165–66, 170

curassow, 27, 95, 109; great, 255; helmeted (*Pauxi pauxi*), 25; *Mitu salvini*, 175; razor-billed (*M. mitu*), 107, 125

Curatella americana, 228

Cusco-Amazonico Lodge, 364

cyanogenic glycosides, 146–47, 149, 151, 156, 162, 184

cyanohydrins, 156

D'Abrera, B., 331

Darien, Panama, 46, 69, 339, 371

Darwin's finches, 98, 109

Darwin, Charles, 238, 270; on adaptive radiation, 98; on beetles, 329; on coevolution, 127; on earthquakes, 12; and evolution, 75–77; on rainforest, 16, 21–22; on reefs, 248; and species diversity, 87

Davis, William E., Jr., 137, 293, 380–81

decomposer food web, 47

decomposition, 47–49, 51, 52, 54, 55

deer: marsh, 230; pampas (*Ozotoceras bezoarticus*), 237–38; red brocket, 230

defense compounds (allelochemics), 56, 249, 334; alkaloids, 145, 150, 151, 152, 163, 167; amino acids, 147, 155, 156; calcium oxalate, 148; cardiac glycosides, 147, 149, 158; cyanogenic glycosides, 146–47, 149, 151, 156, 162, 184; hydrogen cyanide, 334; and insects, 153–59;

and monkeys, 151–53; phenolics, 146, 151; saponins, 146; as selection pressure, 153; tannins, 146; terpenoids, 147. *See also* mimicry

deforestation, 3, 5, 334; in Amazonia, 335–38, 340–68; in Atlantic Coastal Forest, 338–39; in Caribbean islands, 339–40; and cattle ranching, 340–41; in Central America, 339; economics of, 346–47; and ecotourism, 363–66; effects of, 347–60; and future priorities, 368–76; and human resettlement, 341–43; and logging, 345–46; and mercury pollution, 343–45; and mining, 343–44; and national parks, 366–68; and reclamation, 360–63; and termite abundance, 326

Delacour, J., 254

Denslow, Julie, 59

DeVries, P. J., 132, 158, 330

dipper, white-capped (*Cinclus leucocephalus*), 225

Dipteryx panamensis, 60, 61

disease, 342, 384–85

disturbance, 17, 63–69, 121, 249, 370

diversity gradients, 86–98

Dobzhansky, T., 87

Doedicurus, 113

dog, 383–84

dolphin: pink river or boto (*Inia geoffrensis*), 205–7; gray or tucuxi (*Sotalia fluviatilis*)

Dominican Republic, 340

dove (Columbiformes), 27, 255; ruddy quail, 256

Doyle, Sir Arthur Conan, 226

Dressler, R. L., 32

drought, 62–63, 233

dry season. *See* seasons

duck, torrent (*Merganetta armata*), 225

eagle: harpy (*Harpia harpyia*), 42, 253, 286, 289–90, 353, 367; crowned (*Harpyhaliaetus coronatus*), 336; hawk (*Spizaetus ornatus*), 289; white hawk (*S. melanoleucus*), 289

Earthwatch, 375

echolocation, 99, 141, 142

ecological niche, 91, 120–21

ecological succession, 68–69, 151, 179, 338

ecosystems, 17–20, 246

ecotourism, 268, 339, 363–66

Ecuador, 3, 36, 164, 174, 186, 220, 337, 351, 352, 371

edge effects, 349

education, 372–76

eel, electric (*Electrophorus electris*), 201

egret (Ardeidae), 213–14, 229, 236; white-necked (*Ardea cocoi*), 214

Ehrlich, P. R., 353–54
Eisenberg, J. H., 295
Eisner, Thomas, 164
El Salvador, 339
elfinwood, 221
Emmons, L. H., 295, 296
"Empty Forest, The" (Redford), 366–67
endangered species, 297–300, 302, 310, 352
endemism, 118, 338–39, 350, 351–52, 364, 371
Entamoeba histolytica, 386
entisols, 52
epiphytes (air plants), 29, 30–32, 42, 51, 95, 132
Equatorial Counter Current, 7
equilibrium species, 65–66
Erwin, Terry, 36, 37
Erythroxylum coca, var. ipadu, 166; *E. novogranatense*, 166
ESA theory, 98
Espeletias, 224
Estrada, A., 72
ethnobotany, 162–68
ethnozoology, 162–68
euphonia bird, 32
Euphorbia leucocephala, 148
evapotranspiration, 5
Everglades, 236–37
evolution: and adaptive radiation patterns, 98–106; convergent, 115; parallel, 161; process of, 75–83; and selection pressures, 83–86; in South America, 111–20; and speciation, 106–11; and species richness, 86–98, 120–25. *See also* coevolution
"Evolution in the Tropics" (Dobzhansky), 87
Ewel, J., 179
exploitationist hypothesis, 131
Explorer's Inn Reserve, Peru, 36, 364, 365
extinction, 7–8, 62–63, 93, 113–14, 268, 281, 350–53, 371
extractivism, 357–60
extrafloral nectaries, 72, 130–32, 156–57

falcon: bat (*Falco rufigularis*), 286, 288; forest (*Micrastur*), 286, 289; laughing (*Herpetotheres cachinnans*), 288–89; orange-breasted (*Falco deiroleucus*), 69
Fearnside, P. M., 359
Feinsinger, P., 263
female preference model, 278
ferns, 33
ferralsols, 52
Ficus (*Ficus* spp.), 30, 66, 128–29, 140; *Ficus ovalis*, 267
fig wasp (Agaonidae), 128–29
fire, 176, 233, 234

fish, 27, 198, 200–205
Fitzpatrick, J. W., 102, 105, 118
FitzRoy, Captain, 219
flagelliflory, 129
flamboyant tree (*Delonix regia*), 26, 71
flamingo: Andean (*Phoenicoparrus andinus*), 225; Chilean (*P. ruber*), 225; James's (*P. jamesi*), 225
flash colors, 324
flatworm, 31
floating meadows, 196
flood cycle, 192–93
flowers, 26–27
fluke, 386 (*Shistosoma mansoni*), 386
flycatcher, tyrant, 111, 140, 279, 280–81; bill shapes of, 103–4; boat-billed (*Megarhynchus pitangua*), 103; diets of, 105–6; eastern kingbird (*Tyrannus tyrannus*), 105–6; flatbilled, 92; fork-tailed (*Muscivora tyrannus*), 237; gray-capped (*Myiozetetes grandadensis*), 104; great kiskadee (*Pitangus sulphuratus*), 82–83, 103, 105; least (*Empidonax minimimus*), 291; lesser kiskadee (*Pitangus lictor*), 103–4; northern bentbill (*Oncostoma cinereigulare*), 103; northern royal (*Onychorhynchus mexicanus*), 103, 105; ochre-bellied (*Pipromorpha oleaginea*), 106, 137; piratic (*Legatus leucophaius*), 266; rusty-margined (*Myiozetetes cayanensis*), 104; social (*M. similis*), 103, 104; sulfur-rumped (*Myiobius sulphureipygius*), 137; sulphury (*Tyrannopsis sulphurea*), 197; vermilion (*Pyrocephalus rubinus*), 237; white-ringed (*Conopias parva*), 104; white-throated spadebill (*Platyrinchus mystaceus*), 103, 105
foliar nectaries, 131
food precautions, 384
food web, 246
forest: cloud, 15, 220–23, 263, 351; dry, 18, 20, 232, 351; elfin, 221–23; equilibrium, 66; floodplain (varzea), 17–18, 169–70, 172, 192, 194, 197–98; fragmentation of, 347–50; gallery (igapo), 17, 18, 54, 56; moist, 17, 351; monsoon, 8; white soil, 150–51. *See also* jungle
Foster, M. S., 277, 278
Foster, R. B., 12, 34–35, 58, 66, 122
fox: crab-eating, 230; gray, 236
frangipani (*Plumeria* spp.), 26, 153
frequency-dependent selection, 93
frigatebird (man-o-war, *Fregata magnificens*), 20; magnificent, 241–42
frog: *Chiasmocleis shudikarensis*, 349; dart-poison (*Dendrobates pumilio*), 81–82, 323–24; Dendrobatidae, 81; gaudy-leaf (*Agalychinis callidryas*), 323; glass (*Cen-*

trolenella),323; harlequin (*Atelopus varius*), 8; *Phyllobates terribilis*, 81; *Phyllomedusa*, 349; and predation, 86; smoky (*Leptodactylus pentadactylus*), 118, 324; species richness of, 36; toxic compounds in, 81–82, 170, 324; tree, 31, 43, 227, 323
frugivores, 135, 267, 268, 277, 279, 353
fruit crow, 269; bare-necked (*Gymnoderus foetidus*), 194; crimson (*Haematoderus militaris*), 42; purple-throated (*Querula purpurata*), 136
fruit, 9, 11–12, 135–43, 184–85, 277
Fundacao SOS Mata Atlantica, 339
Fundcao Nacional do Indio (FUNAI), 172
fungi, 48–49, 50–51, 342. *See also* mycorrhizae
Furst, P. T., 168

Galapagos Archipelago, 77, 98
gallery forests. *See* forests
Gallon Jug, Belize, 99
gap-phase dynamics, 124
gaps, 10, 33, 57–60, 62; cecropia growth in, 72; colonizing species in, 149; creation of, 61; described, 40; and jungle, 17; kapok tree in, 73; as resource for plants and animals, 95; succession in, 64–65; vegetation in, 29
genes, 76–79, 106, 108–9, 159, 161, 277, 369
genetic drift, 108–9
Gentry, A. H., xi, 28, 35, 88, 120, 351, 356, 358
Glander, K. E., 152, 301
Glossophaga soricina, 130
glyptodonts, 112, 113, 114
gnateater, 279
gnatwren, long-billed, 168
gold mining, 172, 173, 343
Gomez-Pompa, A., 376
gomphotheres, 27, 112, 114
Gondwanaland, 111–12, 226
Goodyear, Charles, 188
goose, Andean (*Chloephaga melanoptera*), 225
Gottlieb, O. R., 164
Gotwald, W. H., Jr., 327
Gran Sabana, 18, 225
grasses: *Andropogon gayanus*, 235; *Brachiaria humidicola*, 235; *Echonochloa polystachya*, 196; manatee (*Syringodium filiforme*), 245; *Paspalum repens*, 196; turtle (*Thalassia testudinum*), 245
grassland, 18, 20, 231–32. *See also* pampas; paramo; puna; savanna
graveteiro, pink-legged (Acrobatornis fonsecai), 369

grazing rates, 10
Great American Interchange, 113–15
Greater Antilles, 220, 258, 291, 339–40, 368, 371
grebe: Atitlan (*Podilymbus gigas*), 225; pied-billed (*P. podiceps*), 225
greenhouse effect, 354–55
Grenada, 10, 340
gross primary productivity (GPP), 45
grunt, French (*Haemulon flavolineatum*), 245
guacharo. See oilbird
guan, 27, 95; black-throated piping (*Pipile jacutinga*), 352; horned (*Oreophasis derbianus*), 255; piping (*Pipile pipile*), 255; Spix's, 254
Guanacaste, Costa Rica, 129–30, 133, 154, 158, 360–61
guanaco (*Lama guanacoe*), 224
Guatemala, 96, 220, 339
Guianan Shield, 52, 53–54, 189, 190, 225
guinea pig (*Cavia aperea*), 303
Gulf of Paria, 190
Gulf-Caribbean Slope, 371
gull, black-headed Andean (*Larus serranus*), 225
Gynoxys, 221

Haffer, J., 118, 119, 235
Haiti, 340
hallucinogens, 164, 167–68
Hamilton, William, 129
Hammond, N., 183
Hart, Robert, 179
hatchetfish, silver (*Gasteropelecus levis*), 201
hawk: black-collared (*Busarellus nigricollis*), 194, 231, 288; common black (*Beautiogallus anthracinus*), 287; crane (*Geranospiza caerulescens*), 287; great black (*Beautiogallus urubitinga*), 287; roadside (*Buteo magnirostris*), 288; savanna (*Heterospizas meridionalis*), 237, 287; Swainson's (*Buteo swainsoni*), 291; white (*Leucopternis albicollis*), 287, 288; zone-tailed (*Buteo albonotatus*), 287
headstander, pearl (*Chilodus punctatus*), 201
Healing Forest, The (Schultes and Raffauf), 165
heliconia (*Heliconia* spp.), 26, 33, 40, 69–70, 102, 112, 194, 263
Henderson, A., 28
hepatitis, 385
heron (Ardeidae), 229, 236; agami, or chestnut-bellied (*Agamia agami*), 214; black-crowned night-heron (*Nycticorax*

heron (*cont.*)
  nycticorax), 214–15; boat-billed (*Coch
  learius cochlearius*), 214–15; capped
  (*Pilherodias pileatus*), 214; striated (*Butorides striatus*), 214; tiger (*Tigrisoma* spp.),
  214
Heston, Charlton, 326–27
hoatzin (*Opisthocomus hoatzin*), 120, 196,
  211–13
Honduras, 234, 318, 339
Hooker, Joseph, 77
hookworm (*Necator americanus*), 386–87
hornbill (Bucerotidae), 260
hornworm, tobacco (*Manduca sexta*), 145
hotshot model, 278–79
hotspot model, 278–79
Hubbell, S. P., 34–35, 58, 66, 72, 122
human settlement, 169–88, 341–43
Humboldt, Alexander von, 4, 15, 16
humification, 48
hummingbird, 32, 95, 127, 223; bearded
  helmetcrest (*Oxypogon guerinii*), 261;
  bee (*Calypte helenae*), 261; booted racket-
  tail (*Ocreatus underwoodii*), 261–62;
  flowerpiercers, 264; giant (*Patagonia
  gigas*), 224, 261; green-tailed trainbearer (*Lesbia nuna*), 224; long-tailed
  hermit (*Phaethornithinae superciliosus*),
  40, 70, 111, 262; Peruvian sheartail
  (*Thaumastura coa*), 224; purple-collared
  woodstar (*Myrtis fanny*), 224; purple-
  throated carib (*Eulampis jugularis*),
  263; ruby topaz (*Chrysolampis mosquitus*), 262; sword-billed (*Ensifera ensifera*),
  261
humus, 48, 49, 235
hunting and gathering, 171–75
hunting, 365, 366–67
hurricanes, 244
Huxley, Thomas Henry, 76
hylaea. *See* rainforest

ibis, 217, 229, 231, 236; buff-necked
  (*Theristicus caudatus*), 237; sharp-tailed
  (*Cercibis oxycerca*), 237
Ice Age, 115–20
Icteridae, 265–66
igapo. *See* forest
iguana (*Iguana iguana*), 38, 39, 320
Imeri Refugium, 119
Imitaca Forest, Venezuela, 290
inceptisols, 52
infection, 383
insects, 379–80; and plants, 153–59; as
  seed consumers, 27; species richness of,
  36–37
intermediate disturbance hypothesis, 121,
  122

International Zoological Expedition, 364
interspecific competition hypothesis, 90–
  93, 95
Intertropical Convergence (ITC), 5–6, 7
intoxicants, 167–68
invertebrates, 325–33
Iquitos, Peru, 351, 356, 358; Amazon River
  at, 190, 191; climate in, 4–6; deforestation near, 336; and ecotourism, 364;
  markets, 170; precipitation in, 8; rainforest described, 22; rainy season in,
  192; river islands at, 195; rubber production in, 188
island biogeography, 347–48
Isle of Pines, 261
Isler, M. L., 264
Isler, P. R., 264
isolation: allopatric, 117; geographic
  (vicariance), 107–8, 111, 117, 119; reproductive, 106

jacamar, 217; rufous-tailed (*Galbula ruficauda*), 158
jacana: northern (*Jacana spinosa*), 215;
  wattled (*J. jacana*), 215
Jacaranda, 26
jaguar (*Panthera onca*), 39, 80, 94, 229,
  312, 353, 367
jaguarundi (*Felis yagouroundi*), 236, 313
Jamaica, 340
Janzen, D. H., 72–73, 95, 131–32, 144,
  151, 154, 267, 328, 360, 372
Javari tribe, 172
jay, blue (*Cyanocitta cristata*), 158
Johnson, C. D., 154
Jordan, Carl F., 54, 55
jump dance, 274, 276
jungle, 17, 63, 64–66
Jurassic period, 111–12

Kaplan, E. H., 247
kapok tree (*Ceiba pentandra*), 26–28, 34,
  66, 73–74, 194
katydid, 79, 94
Kekchi Indians, 178–79, 364
keystone species, 353
Kiltie, R. A., 306
kin selection, 159, 277, 303
kingbird, tropical (*Tyrannus melancholicus*), 197
kingfisher, 89–90; Amazon (*Chloroceryle
  amazona*), 38, 216; green (*C. americana*), 216; green-and-rufous (*C. inda*),
  216; pygmy (*C. aenea*), 216; ringed
  (*Megaceryle torquata*), 90, 215–16
kinkajou (*Potos flavus*), 11–12, 38, 78, 310
Kinzey, W. G., 118
kite: double-toothed (*Harpagus bidentatus*),

96; hook-billed (*Chondrohierax uncina-tus*), 287; pearl (*Gampsonyx swainsonii*), 286, 287; snail (*Rostrhamus sociabilis*), 287; swallow-tailed (*Elanoides forficatus*), 287; white-tailed (*Elanus leucurus*), 287

Knight, D. H., 34, 66

L-DOPA, 147

La Selva Biological Station, Costa Rica, 6, 151, 369; arrested litter in, 55; distur-bance frequency at, 33; forest demo-graphics in, 60–61; fire in, 176; fruit trees in, 138; gaps in, 58; heliconia in, 70; seasons in, 9; species richness in, 34, 35–36; soil types, 52

Lacandon Mayans, 180

Lake Atitlan, 225

Lake Pata, 119

Lake Titicaca, 218, 219, 225

Lamar, W. W., 314, 315–17

Landsat Thematic Mapper, 337, 338

lantern fly (*Fulgora laternaria*), 330

Las Claritas, Venezuela, 343

latasols, 52

laterization, 53

latex, 148

latitudinal cline, 150

leaching, 49–50

leaf area index, 46

leaf litter, 46, 51, 61, 74, 136, 246

leaves, 26; defense methods of, 149; drip tips, 49; dry clusters, 95; 65; trichomes on, 148

Leck, C. F., 138

legumes (Leguminosae), 35, 70–71, 134, 147

leishmanias, 386

leks, 262, 269, 272, 274–75, 278–79

lepidopterans (Riodinidae; Lycaenidae), 79, 132, 330–32

Lesser Antilles, 220, 340, 371

Leuhea seemannii, 36

Levey, D. J., 139

Levin, D. A., 150

lianas, 29–30, 58, 163, 166

life zones, 15–16

light, 61, 64

Lill, Alan, 277

Limerick, S., 322

litopterns, 113, 114

lizard: anolis (*Anolis; Norops*), 94, 320; basi-lisk (*Basiliscus spp.*), 320–21; caiman (*Dracaena guianensis*), 322; common tegu (*Tupinambis teguixin*), 322; gecko (Gekkonidae), 322; northern tegu (*T. nigropunctatus*), 322; red tegu (*T. rufes-cens*), 322; tropidarus, 321

llama (*Lama guanicoe glama*), 224

Llanos, 18, 190, 208, 209, 213, 228–30, 233, 236, 308

logging, 172, 345–46, 367

Lost World, The (Doyle), 226

Lotschert, W., 28

Lovejoy, T. E., 348

Lugo, A., 360

Luquillo Mountains, 340

Lyell, Charles, 77

macaw, 267–68; blue-and-yellow (*Ara araraura*), 22, 268, 365; chestnut-fronted (*A. severa*), 267; clay licks, 365–66; glaucous (*A. glaucus*), x; great green (*A. ambigua*), 135, 267; hyacinth (*Ano-dorhynchus hyacinthinus*) , x, 231, 268; military (*A. militaris*), 267; red-and-green (*A. chloroptera*), 267, 365; scarlet (*A. macao*), 38, 267, 365

Macrauchenia, 113

Madeira River, 17

madera negra tree (*Gliricidia sepium*), 152

Madres de Dios, Peru, 364

magnolia, 150

magpie jay (*Calocitta formosa*), 158

mahogany trees (*Swietenia macrophylla; S. humilis*), 28

maize (*Zea mays*), 183

malaria, 342, 385

mammals, 87; anteaters, 307–8; aquatic, 205–7; armadillos, 309–10; bats, 95–96, 98–102, 129–30; felines, 312–13; marsu-pials, 112, 313–14; monkeys, 295–305; peccaries, 305–6; raccoons, 310–11; ro-dents, 303–5; sloths, 308–9; weasels, 311–12

man-o'-war bird. *See* frigatebird

Man-o'-War Cay, Belize, 241–44

manakin, 10, 80, 135, 139, 269–70, 274–75, 279; bearded, 109; blue-backed (*Chi-roxiphia*), 275–76; blue-crowned (*Pipra coronata*), 109; golden-headed (*P. erythorocephala*), 137, 275; swallow-tailed, 275–76; white-bearded (*Manacus manacus*), 137, 274–75; wire-tailed (*P. filicauda*), 276

manatee: Amazonian (*Trichechus inun-guis*), 206–7; West Indian (*T. manatus*), 106, 245

Manaus, Brazil, 56, 57, 336; canopy tower, 42; capuchins near, 299; climate in, 4–6; and ecotourism, 363–64; market in, 181; precipitation in, 8; rainy season in, 192; rubber production in, 188; species rich-ness in, 35; tree mortality rate in, 61

mango (*Magifera indica*), 185

mangrove, 29, 146, 239–44; black (*Avicen-nia germinans*), 241, 243, 244; button-

mangrove (*cont.*)
wood (*Conocarpus erecta*), 241, 243; red
(*Rhizophora mangle*), 240, 243, 244, 246;
white (*Laguncularia racemosa*), 241, 243
manioc (*Manihot esculenta*), , 147, 162,
170, 176–77, 181, 184, 342
mantispid (*Climaciella brunnea*), 160
Manu Lodge, 364, 365
Manu National Park, 52, 105, 364
mara (*Dolichotis patagonum*), 238
Maraca Island, Brazil, 55
Marajo Island, Brazil, 169
margay (*Felis wiedii*), 80, 91, 312–13
Martin, M. M., 134
matas (tree groves), 229, 233
Mato Grosso, Brazil, 172
Mayans, 178–80, 182–83, 186
mayapple (*Podophyllum peltatum*), 163
Meade, R. H., 356
meadowlark, eastern (*Sturnella magna*),
237
Meggers, B. J., 181
Mendes, Chico, 336, 358
Mendoza, Chile, 219
Merck Pharmaceutical Company, 164
mercury pollution, 343–45
Merriam, C. Hart, 15–16
methylmercury, 343, 344
methylxanthine, 145
Mexico, 371
Miconia (*Melasomataceae*), 33; *M. argentea*,
59, 92–93
Middleton, A. L. A., 252
Milankovitch cycles, 119
milk tree (*Brosimum alicastrum*), 69; *B.
utile*, 27
milkweed, 155, 158, 160
Miller, Bruce, 99
Miller, Carolyn, 99
millipede (*Nyssodesmus python*), 332–33
milpas, 178
Milton, Katherine, 153
mimicry, 79, 157, 287, 317, 318, 328, 331;
Batesian, 159–60; complexes, 161–62;
evolution of, 160–61; Mullerian, 160. *See
also* cryptic coloration
mimosa (*Mimosa pigra, M. pudica*), 70–71
mineral cycling, 52–55, 310
minerals, 176, 234
Minimum Critical Size of Ecosystems Project, 347, 348
mining, 343–44. *See also* gold mining
mites (*Rhinoseius*), 265
mitochondrial DNA, 142, 161, 369
Monasterio, G., 232, 235
monkey pot tree (*Lecythis costaricensis*), 27
monkey, 112; bare-eared squirrel (*Saimiri
ustus*), 297; bearded saki (*Chiropotes

spp.*), 298; black spider (*Ateles paniscus*),
298; brown-headed or Colombian black
spider (*A. fusciceps*), 299; brown howler
(*A. caraya*), 300; capuchin (*Cebus*), 296–
97; Central American spider (*A. geoffroyi*), 298; Central American squirrel
(*S. oerstedii*), 297; common squirrel (*S.
sciureus*), 297; common woolly (*L.
lagothricha*), 300; dusky titi (*Callicebus
moloch*), 297; and epiphytes, 31; Geoffrey's tamarin (*S. geoffroyi*), 302;
Goeldi's (*Callimico goeldii*), 296; golden
lion tamarin (*Leontopithecus rosalia*),
302; Guianan saki (*Pithecia pithecia*),
298; howler (*Alouatta* spp.) , 37, 84, 152,
300–301; marmoset (*Callithrix* spp.;
Cebuellu spp.), 301–2; mating patterns
of, 302; Mexican black howler (*A.
pigra*), 300; midas tamarin (*S. midas*),
302; night or douroucouli (*Aotes* spp.),
297–98; and plants, 151–53; prehensile
tail of, 78; pygmy marmoset (*Cebuella
pygmaea*), 198, 296, 301–2; red-handed
howler (*A. belzebul*), 300; red howler
(*A. seniculus*), 300; saddleback tamarin
(*S. fusciocollis*), 302; saki (*Pithecia
monachus*), 43; as seed consumer, 27;
satere marmoset (*Callithrix saterei*), 369;
spider (*Ateles* spp.), 153, 298–99;
tamarin (*Saguinus* spp.), 296, 301–3;
tufted-ear marmoset (*C. jacchus*), 301;
uakari (*Cacajao* spp.), 298; white-bellied
spider (*A. belzebuth*), 298–99; white-
faced (*Cebus capucinus*), 12; white-
throated capuchin (*C. capuchinus*), 297;
woolly (*Lagothrix* spp.), 175, 300; wooly
spider or muriqui (*A. arachnoides*), 299;
yellow-tailed woolly (*L. favicauda*), 300
Monroe, B. L., 252
Monteverde cloud forest, 138, 371
Montezuma's revenge, 384
Mora excelsa, 34
Morabelli Creek, Guyana, 24
morning glory tree (*Ipomoena arborescens*),
26
Morton, Eugene, 136
mosquito, 31, 379, 385
moth (*Biston bitularia*), 81
motmot, 258; blue-crowned (*Momotus
momota*), 258; rufous (*Baryphthengus
ruficapillus*), 40; tody (*Hylomanes momotula*), 258; turquoise-browed (*Eumomota
superciliosa*), 82–83, 85
Mt. Aconcagua, 219
Mt. Cotopaxi, 218
Mt. Darwin, 219
Mt. Mercedario, 219
Mt. Roraima, 225

Mt. Tupungato, 218
mountains, 12, 13, 14–16. *See also* Andes
    Mountains
mouse (*Mus musculus*), 114, 303; spiny
    pocket (Heteromys; Liomys), 305
Muller, Fritz, 160
Mullerian bodies, 131
Mundurucu tribe, 171–72
Munn, C. A., 265, 365–66
Musaceae, 112
mushroom, 168
mutualism, 126–27
mycorrhizae, 31, 32, 49, 50–51, 177. *See
    also* fungi
Myers, N., 350, 352–53

Naked Jungle, The, 326–27
National Centers for Disease Control, 384–
    85, 387
National Council of Rubber Tappers, 358
National Indian Foundation, 172
National Museum of Natural History, 347
national parks, 349, 366–68
natural selection, 75–77, 84
Naturalist on the River Amazons, The
    (Bates), 9
nectarivorous species, 353
needlefish (*Strongylura notata*), 246
Nepstad, D., 362
nest predation, 85–86, 277
net primary productivity (NPP), 45, 46,
    249
Nicaragua, 220, 234, 237, 339
Nicholaides, J. J., 361
nightjar: whip-poor-will (*Caprimulgus vo-
    ciferous*), 141
Nino, El, 7–8, 12, 15, 62
nitrogen cycling, 363
nitrogen fixation, 55–56
nomadism, 174–75. *See also* agriculture,
    slash-and-burn
nonequilibrium hypothesis, 123
Nordin, C. F., Jr., 356
Norfolk Island pine (*Araucaria excelsa*), 26
notoungulates, 112, 113, 114
nunbird (*Monasa nigrifrons*), 39, 280–81
nutmeg (Myristicaceae), 167
nutrient cycling, 47–49, 55

ocelot (*Felis pardalis*), 38–39, 80, 91, 94,
    312–13
oilbird (*Steatornis caripensis*), 140–43, 353,
    368
olingo (*Bassaricyon gabbii*), 310, 311
oncilla (*Felis tigrina*), 313
Oniki, Y., 285
Opler, Paul, 160
opossum, 78, 112, 114; common (*Didelphis*

marsupialis), 314; Virginia (*D. virgini-
    ana*), 313; water (*Chironectes minimus*),
    314; woolly (*Caluromys*), 314
orchids (Orchidaceae), 32, 42, 51, 128,
    226; *Cattleya*, 32; Vanilla, 32
Ordaz, Diego de, 190
Orellana, Francisco de, 170, 199–200
Organization for Tropical Studies, 374–
    75, 378
Origin of Species (Darwin), 75–77, 87, 127
Orinoco River, 17, 167, 189–90, 192, 206
oriole, 140, 265–66; Baltimore (*Icterus gal-
    bula*), 292; moriche (*I. chrysocephalus*),
    197; orchard (*I. spurius*), 292; troupial
    (*I. icterus*), 266
oropendola, 265–66; russet-backed (*Psaro-
    colius angustifrons*), 266
oscar (*Astronotus ocellatus*), 201
otter, giant (*Pteronura brasiliensis*), 207,
    230; southern river (*Lutra longicaudis*),
    207, 230
ovenbird, 197, 279, 282, 283–84, 293;
    thornbird (*Pacellodomus*), 283
owl, barn (*Tyto alba*), 234; black-and-white
    (*Ciccaba nigrolineata*), 96, 290; mottled
    (*C. virgata*), 290; pygmy (*Glaucidium*),
    291; spectacled (*Pulsatrix perspicillata*),
    290
oxbows, 195–96
oxisols, 52, 233

paca (*Agouti paca*), 12, 94, 136, 304, 367;
    nocturnal (*Cuniculis paca*), 11
Pacific Arid Slope, 371
Padre Island, Peru, 195
palm, 22, 28, 182; African oil (*Elaeis
    guineensis*), 187; *Aphandra natalia*, 186;
    Astrocaryum chambira, 186, 198; *Chamae-
    dorea* spp., 359–60; coconut (*Cocos nu-
    cifera*), 27, 187; morich (*Moriche
    flexuosa*), 170, 186, 194, 197, 228; muru-
    muru or tucum (*Astrocaryum aculeatum*),
    186; warree (*Acrocomia vinifera*), 148–
    49
palm-swift, fork-tailed (*Reinarda squamata*),
    197
palmcreeper, point-tailed (*Berrlepschia
    rikeri*), 197
pampas, 231–32, 237–38
Panama, 86, 132, 220; birds in, 103, 273;
    land bridge, 113; liana growth in, 30;
    monkeys in, 297; seasonal changes in,
    10. *See also* Barro Colorado Island
Pantanal, 18, 208, 209, 228, 230–31, 238,
    308
Papageorgis, Christine, 161–62
papaya (*Carica papaya*), 185
Para, Brazil, 67–68, 343, 345

Paraguay, 18, 173–75
parakeet (*Brotogeris jugularis*), 267
paramo, 15, 176, 223–34
parasites, 94–95, 126, 342, 370, 383, 385–87
Parker, Theodore A., III, xi, 252
parrot, 27, 37, 95, 135, 140, 266–68; amazon (*Amazonia*), 267; parrotlet (*Forpus*), 267; *Pionopsitta*, 119; at risk of extinction, 268
parrotfish, 246
passionflower (*Passiflora* spp.), 30, 147, 153, 155–57, 351; *P. adenopoda*, 157
pastures, 67–68
Pearman, P. B., 371–72
peccary, 27, 38, 136, 349; collared (*Tayassu tajacu*), 11, 305; white-lipped (*T. pecari*), 305–6
pelican, brown (*Pelicanus occidentalis*), 20, 241
Pentaclethra macroloba, 34, 59
penuliflory, 129
peppers (*Capsicum* spp.), 184
peripatus, 333
periwinkle, mangrove (*Littorina angulifera*), 242
Perrins, C. M., 252
Perry, D. R., 31
Peru, 337
Peru, 66, 88, 92, 167, 337, 365–66. *See also* Iquitos
Peten, Guatemala, 359
Peterson, R. T., 257
peyote, 168
phenolics, 146
Phillips, O., 182
philodendron, 30, 148
Phorusrhacoids, 113
phosphorus, 54
photosynthesis, 44–45, 47, 65, 247
Phyllostornus discolor, 130
Pickett, S. T. A., 121
pigeon (Columbiformes), 27, 255
piha, 277; screaming (*Lipaugus vociferans*), 38, 269, 272
pine, Caribbean (*Pinus caribaea*), 20, 228
pineapple (*Ananas comosus*), 185
piper (Piperaceae), 33; (*Piper* spp.), 70; *P. arieianum*, 151
piranha, 230; red (*Sarrasalmus natterei*), 201, 205
pitcher plant, 226
Pithecelobium saman, 154
pitohui (*Pitohui dichrous, P. kirkocephalus, P. ferrugineus*), 82
pitviper. *See* snake
Piwualli, 164
Pizarro, Gonzalo, 199

Planalto, 53
plant ecology, 44
plant identification, 28–29
plate tectonics, 12, 14, 111–12
platyrrhines. *See* monkeys
Pleistocene faunal exchange, 313
Pleistocene, 115, 118, 119, 142, 235
Plotkin, Mark, 163, 164, 165
Podocarpus, 221
politics, 387–88
pollination, 9, 27, 32, 92–93, 127–29
Polylepis trees, 221–22
polyphenols, 151
porcupine, 78, 112–13, 114; Brazilian (*Coendou prehensilis*), 304; North American (*Erethizon dosatum*), 303
potato (*Solanum tuberosum*), 184
potoo (*Nyctibius* spp.), 194; common (*Nyctibius griseus*), 80
Pough, F. H., 321
poui, pink (*Tabebuia pentaphylla*), 26
Poulin, R., 94–95
precipitation, 5, 6, 8, 49–50; recycling, 14
predation hypothesis, 93–95, 105
predators, 85–86, 93–94, 142, 370
productivity, 44–47, 97–98
productivity-resources hypothesis, 95–98
Prosopis caldenia, 232
protectionist hypothesis, 131
protein, 155
Prum, R. O., 119, 279
prussic acid, 184
pseudoscorpion, 329
Pseudosphinx tetrio, 83
Psychotria (Rubiaceae), 33
Puerto Maldonado, 365
Puerto Rico, 10, 57, 94, 150, 222
puffbird: white-necked (*Notharctus macrorhynchos*), 280; white-whiskered or brown (*Malacoptila panamensis*), 280–81
Pulltrouser Swamp, Belize, 183
puma (mountain lion; cougar; *Puma concolor*), 94, 313
puna, 15, 224–25
pyrophytes, 234

quinine, 164
Quito, Ecuador, 199, 218

raccoon, crab-eating (*Procyonidae cancrivorus*), 230, 310
rain shadow effect, 14–15
rainforest: animals in, 37–41; canopy, 41–43; demographics of, 60–61; disturbance and ecological succession in, 63–69; drought in, 62–63; as ecosystem, 16–17; gaps in, 33; identifying plants in,

28–29; leaching in, 49–50; mycorrhizae in, 50–51; mineral cycling in, 52–55; nitrogen fixation in, 55–56; nutrient cycling in, 47–49, 55; productivity of, 44–47; recycling in, 51; river types, 56–57; similarity to coral reefs, 249–50; soil types, 52–55; species richness in, 33–37; structural complexity of, 22–23; successional plants in, 69–74; typical trees in, 23–28; vines and epiphytes in, 29–32. *See also* gaps; La Selva; Tikal

rainy season. *See* seasons

Rapid Assessment Programs, xi

Rappole, John, 293

rat: rice (*Oryzomys* spp.); spiny (*Proechimys* spp.), 51, 304; tree (*Echimys* spp.), 304

ratites, 112

Raven, P. H., 350

ray, spotted eage (*Aetobatus narinari*), 241

reciprocal altruism, 101

reclamation, 360–63

recycling, 51

Redford, K. H., 295, 366–67

redstart, American (*Setophaga ruticilla*), 292

reduviid bug (*Panstrongylus megistus*), 386

refuges (*refugia*), 115–20

regeneration pathways, 362

Reice, S. R., 121

reptiles, 89, 314–22

Reserva Ducke, Brazil, 35

resguardos (reserves), 172

resin, 164

restoration ecology, 360–63, 370–72

Rhabdolichops zareti, 205

rhea, 112, 336; common (*Rhea americana*), 238; Darwin's (*Pterocnemia pennata*), 238

rice (*Oryza sativa*), 183–84, 342

Ridgely, R. S., 252, 270

Rio Apure, 189

Rio Arauca, 189

Rio Carhuasanta, 190

Rio Jurua, 191

Rio Jutai, 191

Rio Madeira, 191, 198

Rio Manu, 35, 66, 365–66

Rio Maranon, 190, 337

Rio Meta, 189

Rio Napo, 42, 109, 164, 191, 199–200, 337, 371

Rio Negro, 17, 56, 57, 67, 77, 189, 191, 199

Rio Palenque Biological Station, 351, 352

Rio Paraguay, 230

Rio Platano Biosphere Reserve, 339

Rio Purus, 191

Rio Solimoes, 17, 190

Rio Tapajos, 18, 34, 191

Rio Teles Pires, 335

Rio Tigre, 190

Rio Tocantins, 18, 195

Rio Trombetas, 191

Rio Ucayali, 190, 337

Rio Xingu, 18, 191

Rio Yarapo, 201

riverine ecosystem, 17–18

rivers: blackwater, 17–18, 56–57, 150–51, 191, 198; clearwater, 18; habitat of, 193–98; natural history of, 200–17; whitewater, 17, 56–57, 191, 198. *See also specific rivers*

Roca, R. L., 143, 368

rodents, 238

Rondonia, Brazil, 336, 338, 345, 362, 363

roots: apogeotropic, 55; buttressed, 54–55, 197; epiphyte mat, 31, 54; prop, 24–25, 240; stilted, 197

Roraima, Brazil, 55, 172, 226, 338, 345

rotenone, 163–64, 201

roundworm (*Trichinella spiralis*), 386–87

rubber tree (*Hevea brasiliensis*), 148, 187–88

rubber, 359

Rural Workers' Union, 358

safety, 378–87

salamander, 31, 322

Salar de Coipasa, 219

Salar de Uyuni, 219

sally-gleaning, 102–3

salt flats, 219

Samanea saman, 71

San Carlos de Rio Negro, Venezuela, 46, 60–61

sandbars, 194–95

Sanford, R. L., 52

Santarem, Brazil, 169

saponins, 146

sargeant major, black-striped (*Abudefduf saxatilis*), 245

Sarmiento, G., 232, 235

satellite imagery, 337

savanna, 18, 64, 114, 172, 228–38

schoolmaster (*lutjanus apodus*), 245

Schultes, Richard Evans, 163, 164, 165

Schwartz, P., 274

scorpion, 382

Scott, N. J., 322

screamer, horned (*Anhima cornuta*), 200, 213

sea cucumber (*Isostichopus badionotus*), 246

sea star (*Oreaster reticulatus*), 246

sea urchin, longspined (*Diadema antillarum*), 241

seagrass, 20, 245–56

search images, 94;

seasons, 6, 8–12, 16–17, 58; dry, 84, 189–90; growing, 46–47; rainy, 189–90, 192; and specialization, 97

seed bank, 66

seed shadow hypothesis, 122

seedeater, paramo (*Catamenia homochroa*), 227

seeds, 64, 65; colonization patterns and survivorship of, 123; dispersal of, 92, 136, 143, 138–39, 198, 265, 269, 297, 362, 368; and forest regeneration, 66–67; predation of, 94, 136, 154, 244, 267

selection pressures, 83–86, 117, 153, 277, 280; abiotic, 83–84; biotic, 84–86

Selva Lacandona, Mexico, 180

seriema, 113, 231

Serra Pelada gold mine, 343

sexual selection, 249, 272, 273, 276–78, 301, 329

Shaman Pharmaceuticals, 164

shaman, 163–64, 167, 170

shrubs, 33, 92–93

Sick, H., 252, 274

sickle cells, 78, 79

silicon, 149

silk-cotton tree. *See* kapok tree

Silman, Miles, 123–24

Simpson, George Gaylord, 112

Siona-Secoya Indian community, 174

skunk cabbage (*Symplocarpus foetidus*), 148

skunk: hog-nosed (*Conepatus*), 312; hooded (*Mephitis macroura*), 312; spotted (*Spilogale putorius*), 312

Skutch, A. F., 252, 257, 281–82, 264

SLOSS debate, 349–50, 370

sloth, 12, 126–27; giant ground (*Megatherium*), 112, 113; ground, 27; maned (*Bradypus torquatus*), 352; three-toed (*Bradypus variegatus*), 40, 79, 80, 308–9, 310; tree, 112; two-toed (*Choloepus*), 309

snail, 31

snake: anaconda (*Eunectes murinus*), 208, 229, 312, 318, 319, 379; boa constrictor (*Boa constrictor*), 79, 319, 379; bushmaster (*Lachesis muta*), 317, 378; chunk-headed (*Imantodes cenchoa*), 320; constrictors, 318–19; copperhead (*Agkistrodon contortrix*), 315; coral (*Micrurus; Micruoides*) , 82–83, 318; emerald green boa (*Boa canina*), 319; eyelash palmpitviper (*Bothriopsis schlegelii*), 316; fer-de-lance (*Bothrops asper*), 315–16, 378; forest-pitvipers (*Bothriopsis*), 316; indigo (*Drymarchon corais*), 319; jumping pit-

viper (*P. nummifer*), 316; king, 83; *Lampropeltis triangulum hondurensis*, 318; montane or hognosed pitviper (*Porthidium*), 316; neotropical rattlesnake (*Crotalus durissus*), 316–17; palmpitvipers (*Bothriechis*), 316; pitvipers, 315–17; rainbow boa (*Epicrates cenchris*), 319; speckled forest-pitviper (*B. taeniata*), 316; two-striped forest-pitviper (*B. bilineata*), 316; vine (*Oxybelis*), 319; water moccasin (*A. piscivorous*), 315

snakebite, 315–17, 378–79

Snow, Barbara, 137, 272

Snow, David, 136, 137, 141, 252, 273, 274, 277

sociality, 141, 142

soils, 48, 51, 52–55, 176, 233–34

sounds, 37–38

Southern Oscillation, 7–8

soybeans, 347

sparrow, chipping (*Spizella passerina*), 237

specialization, 94, 97, 153, 279–80

speciation, 14, 89–90, 106–11, 119, 350

species richness: of birds, 95; in blackwater rivers, 198; of coral reef, 247–48; and deforestation, 348; and evolution, 86–98, 120–25; and fire, 234; and foliage height, 95; and insect feeding, 279–80; in rainforest, 33–37; and savannas, 235, 236; of successional areas, 66; of trees, 62–63, 368. *See also* biodiversity

specific myiasis, 380

spider: bird-eating (*Theraphosa blondi*), 381–82; orb-weaving, 78–79, 94; tarantula, 381; wolf, 381

spoonbill, 236

squash (*Cucurbita* spp.), 184

squid, reef (*Sepioteuthis sepioidea*), 246

squirrel: Amazon dwarf (*Microsciurus flaviventer*), 43, 305; Amazonian red-tailed (*S. granatensis*), 304; Central American variegated (*S. variegatoides*), 305; tree (*Sciurus* spp.), 304–5

stabilimenta, 78

stability-time hypothesis, 89–90

Stann Creek Valley, Belize, 58

Stiles, F. G., 70

stinking toe tree (*Hymenaea courbaril*), 27, 134, 147

stork, 229, 236–37; jabirou (*Jabiru mycteria*), 213, 214, 237; maguari (*Euxenura maguari*), 213, 237; wood (*Mycteria americana*), 213, 237

Stotz, D. F., 222, 253, 361, 370–71

strangler figs, 30

subduction, 14

succession rate, 361

successional patterns, 67–68
sugar cane (*Saccharum officinarum*), 184
sunbittern (*Eurypyga helias*), 213
sundew, 226
Surinam, 71
swallow: rough-winged (*Stelgidopteryx ruficollis*), 194; white-banded (*Atticora fasciata*), 194; white-winged (*Tachycineta albiventer*), 194
swamps, 197
Symposium on Resource Availability and Tropical Ecosystems, 375

table mountains (*tepuis*), 190, 225–27, 352
tamandua (*Tamandua mexicana*), 10
tamarin, golden lion (*Leontopithecus chrysomelas*), 352; saddleback (*Saguinus fuscicollis*), 41
Tambo Lodge, 364
Tambopata Reserve, Peru, 198
tanager, 43, 135, 139, 140, 223; ant (*Habia*), 265; bananaquit (*Coereba flaveola*), 10, 265; blue-gray (*Thraupis episcopus*), 264; bush, 223; euphonia (*Euphonia*), 264–65; golden, 139; grass-green (*Chlorornis riefferii*), 223; hepatic (*Piranga flava*), 237; honeycreeper, 265; mountain (*Hemispingus superciliaris*), 111; paradise (*Tangara chilensis*), 264; scarlet-billed mountain (*Aniognathus igniventris*), 223; summer (*P. rubra*), 292; white-winged shrike (*Lanio versicolor*), 265
tannins, 146
tapaculo, 279
tapeworm (*Echinoccus granulosus*), 386
tapir (Tapiridae), 306–7; Baird's (*Tapirus bairdii*), 11, 27, 107–8, 306; Brazilian (*T. terrestris*), 108, 307; mountain (*T. pinchaque*), 108, 307
Tarapoto Island, Peru, 195
tayra, 236
teal, puna (*Anas versicolor*), 225
temperature, 6, 49, 64, 115. *See also* climate
tepuis. See table mountains
Terborgh, J., 94, 349, 353
termite (*Nasutitermes* sp.), 10, 53, 281, 325–26
tern: large-billed (*Phaetusa simplex*), 193; yellow-billed (*Sterna superciliaris*), 193
terpenoids, 147
terra firme, 18, 172, 347
Terra Nova Rain Forest Reserve, Belize, 165
Terrestrial Ecosystem Model (TEM), 45
terror birds, 113, 114

tetra, 230; cardinal (*Cheirodon axelrodi*), 201; neon (*Hyphessobrycon innesi*), 201
Tetragastris panamensis, 138
Thoatherium, 113
thornwood, 232
tick, 383
Tierra del Fuego, 219
tiger, saber-toothed, 115
Tikal, Guatemala, 27, 68–69, 183, 255; World Heritage Site, 339
Timarca Island, Peru, 195
timbo (*Lonchocarpus utilis*), 201
tinamou, 27, 253–54; great (*Tinamus major*), 38, 253
tityra, masked (*Tityra semifasciata*), 138
toad, giant marine (*Bufo marinus*), 324; golden (*B. periglenes*), 8
Tobago, 340
tomato (*Lycopersican esculentum*), 184
Tomlinson, P. B., 239
topography, 109–11
toucan, 27, 135, 139, 258–60; chestnut-mandibled (*Ramphastos swainsonii*), 260; Cuvier's (*R. cuvieri*), 140, 259; keel-billed (*R. sulfuratus*), 259, 260; mountain, 223; Selenidera, 119; yellow-ridged (*R. culminatus*), 259–60
toucanet (*Aularchorhyinchus*), 139, 260
toxicity, 27
Toxodon, 113
trade winds, 5, 7
trail precautions, 377–78
Trail, Pepper, 271–72
Transamazon highway system, 172, 336, 338, 341, 342, 345
transpiration, 49
Traylor, M. A., 102, 105
trees: boles of, 22, 23; buttresses of, 24–25; canopy, 122; demographics of, 60–61, 123–24; emergent, 61; flowers of, 9, 26–27; fruits and seeds of, 27–28, 43; leaves of, 26; pioneer species of, 10; roots of, 25; species richness of, 97, 120–25, 197; stature of, 23–24; toxic, 152–53; trunks and crowns, 25–26
Trema micrantha, 59
tribal warfare, 171–72
Trichilia cuneata, 138
Trinidad, 34, 92, 106, 220, 273, 274–75, 340
trogon (Trogoniformes), 27, 139, 257–58; collared (*Trogon collaris*), 42; resplendent quetzal (*Pharomachrus mocinno*), 139, 223, 257–58; violaceous (*T. violaceous*), 257; white-tailed (*T. viridis*), 38
tropicbird, white-tailed, 20
trumpeter (*Psophia crepitans*), 175, 255–56

trypanosome (*Trypanosoma rhodesiense*), 386
tucotucos (*Ctenoys* spp.), 238
tucuxi. *See* dolphin
Tudor, G., 252
Tukanoan Indians, 177
turista, 384
turkey: ocellated (*Agriocharis ocellata*), 255; wild (*Meleagris gallopavo*), 255
Turner, John R. G., 160
turtle, 245; anaconda (*Chelus fimbriata*), 210–11; arran (*Podocnemis expansa*), 210–11; hawksbill (*Eretmochelys imbricata*), 246
Twin Cays, Belize, 240
typhoid fever, 385
tyrannulet, white-throated (*Mecocerculus leucophrys*), 227

Uhl, C., 66, 362–63
ultisols, 52, 233
UNESCO Man and the Biosphere Reserve, 364
United Nations Conference on Environment and Development, 354
upwelling, 15

Venezuela, 18, 225. *See also* Llanos
Veracruz, Mexico, 65
vesicular-arbuscular mycorrhizae (VAM), 51–52
vicariance. *See* isolation
Vickers, W. T., 175
vicuna (*Vicogna vicugna*), 224
vines, 29–30, 95
Virola surinamensis, 138, 164, 167
virus, 85
viscacha, mountain (*Lagidium peruanum*), 224
Vitousek, P. M., 52
voice, 300–301
voodoo, 324
Voyage of the Beagle, The, 238
vulture, 237; black (*Coragyps atratus*), 286; greater yellow-headed (*Cathartes melambrotus*), 286; king (*Sarcoamphus papa*), 286; lesser yellow-headed (*Cathartes burrovianus*), 286; turkey (*Cathartes aura*), 286

walking stick, 80
Wallace, Alfred Russel, 16, 28, 34, 77, 200, 295
Wanderings in South America (Waterton), 79
warbler, 227; Bachman's (*Vermivora bachmanii*), 293; black-and-white (*Mniotilta varia*), 292; blackburnian (*Dendroica fusca*), 223; cerulean (*D. cerulea*), 292;

chestnut-sided (*Dendroica pensylvanica*), 294; elfin-woods (*D. angelae*), 222; hooded (*Wilsonia citrina*), 293; Kentucky (*Oporornis formosus*), 292, 293; Kirtland's (*D. kirtlandii*), 293; mangrove (*D. erithacorides*), 243; Tennessee (*Vermivora peregrina*), 292; worm-eating (*Helmintheros vermivorus*), 72, 292; yellow (*D. petechia*), 243
warning (aposematic) coloration, 81–83, 126, 158–59, 162, 249, 324
wasp, 266, 382–83
waterlily, Victoria (*Victoria amazonica*), 18, 128
Waterton, Charles, 79, 165–66
weasel: grison (*Galictis vittata; G. cuja*), 312; tayra (*Eira barbara*), 311
weevil, 27
Wheaton College, ix
Wheelwright, N. T., 139
White, P. S., 121
"Why Are There So Many Species of Insects?" (Janzen), 95
Wickham, Henry, 188
Willis, E. O., 285, 293
willow (*Salix*), 194–95
Wilson, Edward O., 36, 98, 350, 353–54, 373
Wolf, L. L., 263
wolf, maned (*Chrysocyon brachyurus*), 238
wood thrush (*Hylocichla mustelina*), 292, 293
wood-rail, gray-necked (*Aramides cajanea*), 194
woodcreeper, 110–11, 279; barred (*Dendrocolaptes certhia*), 40; long-billed (*Nasica longirostris*), 282; scythebill (*Campylorhamphus*), 282; tawny-winged (*Dendrocincla anabatina*), 282; wedge-billed (*Glyphorynchus spirurus*), 282, 283
woodlands, 20
woodpecker, 281–83; black-cheeked (*Melanerpes pucherani*), 96; chestnut (*Celeus elegans*), 377; cream-colored (*Celeus flavus*), 281; golden-olive (*Piculus rubiginosus*), 96; imperial (*Campephilus imperialis*), 281; ivory-billed (*Campephilus*), 281; pale-billed (*Campephilus guatemalensis*), 281; piculet (*Picumnus*), 281; yellow-tufted (*Melanerpes cruemtatis*), 281
World Resources Institute, 336, 337
World Wildlife Fund, 347
Wourali vine, 166
wren, 222

xate, 359–60
Xingu tribe, 172

Yacumama Lodge, 201, 364
Yahgan (Yamana) Indians, 219
Yanomamo Indians, 163, 167, 172–73, 344
Yanomamo, Peru, 120
yellow fever, 385

yellowthroat, northern (*Geothlypis trichas*), 292
Yucatan, Mexico, 85
Yurimaguas, Peru, 361–62

zooplankton, 205

ABOUT THE AUTHOR

John Kricher is Professor of Biology at Wheaton College, where he is the first recipient of the Bojan Hamlin Jennings Chair in the Natural Sciences and Mathematics. Among the many books he has written are *A Field Guide to Ecology of Eastern Forests* and *A Field Guide to Ecology of Western Forests.*